Universitext

Universitext is a series of textbooks that presents material from a wide variety of mathematical disciplines at master's level and beyond. The books, often well class-tested by their author, may have an informal, personal even experimental approach to their subject matter. Some of the most successful and established books in the series have evolved through several editions, always following the evolution of teaching curricula, into very polished texts.

Thus as research topics trickle down into graduate-level teaching, first textbooks written for new, cutting-edge courses may make their way into *Universitext*.

More information about this series at http://www.springer.com/series/223

Stephen J. Gustafson · Israel Michael Sigal

Mathematical Concepts of Quantum Mechanics

3rd edition

Stephen J. Gustafson
Department of Mathematics
University of British Columbia
Vancouver, Canada

Israel Michael Sigal
Department of Mathematics
University of Toronto
Toronto, Canada

ISSN 0172-5939   ISSN 2191-6675   (electronic)
Universitext
ISBN 978-3-030-59561-6   ISBN 978-3-030-59562-3   (eBook)
https://doi.org/10.1007/978-3-030-59562-3

Mathematics Subject Classification: 81S22, 81Q10, 81Q15, 81S40, 81S30, 81Q20, 81U05, 81U10, 81T17, 35Q40, 35Q70, 35P25, 35P15, 47A40, 47A11, 47A40, 47A75

This Springer imprint is published by the registered company Springer Nature Switzerland AG
The registered company address is: Gewerbestrasse 11, 6330 Cham, Switzerland

Preface

Preface to the third edition

In this edition we have expanded the elementary part of the book to make it self-contained, and have added new intermediate level material related to recent developments. This new material fits nicely into the general structure of Quantum Mechanics, as well as to our book.

In particular, on the intermediate level, we added sections on the time-dependent Born-Oppenheimer approximation, adiabatic theory, geometrical phases, Aharonov-Bohm effect and density functional theory.

We also added a sections on quantum open systems, expanded some others, and organized the relevant sections into a separate chapter under this name. This chapter develops some fundamental concepts lying at the heart of quantum information theory, presently perhaps the fastest growing area of physics. It also comes closest to issues of foundations, which are not considered in this book. However, we refer to an excellent recent paper [116] for an in-depth discussion of the issues involved, and many references.

As in the previous editions, we tried to stay at the most elementary mathematical level possible, and did not pursue generalizations and mathematical questions arising naturally in the subject. The latter is done in the excellent books of F. Strocchi ([285]), L. Takhtajan ([287]), G. Teschl ([288]), J. Dimock ([79]), B. C. Hall, ([153]) and G. Dell'Antonio ([76]).

Consequently, prerequisites for this edition are the same as for the previous ones: introductory real and complex analysis and elementary differential equations. This book could be used for senior level undergraduate, as well as graduate, courses in both mathematics and physics departments.

The new material makes the book more flexible as a source in designing courses guided by different interests and needs. This edition could be used for introductory, intermediate and advanced courses; some of the sections would serve for introductions to geometrical methods in Quantum Mechanics, to quantum information theory and to quantum electrodynamics/field theory (Sections 7.5-7.7, 7.9, 12.4; Chapters 18-19 and 20-24, respectively).

Acknowledgment: The authors are grateful to R. Frank, M. Lemm, B. Nachtergaele and S. Teufel for reading parts of the new material and making many pertinent remarks.

Vancouver/Toronto,   *Stephen Gustafson*

May 2020   *Israel Michael Sigal*

Preface to the second edition

One of the main goals motivating this new edition was to enhance the elementary material. To this end, in addition to some rewriting and reorganization, several new sections have been added (covering, for example, spin, and conservation laws), resulting in a fairly complete coverage of elementary topics.

A second main goal was to address the key physical issues of stability of atoms and molecules, and mean-field approximations of large particle systems. This is reflected in new chapters covering the existence of atoms and molecules, mean-field theory, and second quantization.

Our final goal was to update the advanced material with a view toward reflecting current developments, and this led to a complete revision and reorganization of the material on the theory of radiation (non-relativistic quantum electrodynamics), as well as the addition of a new chapter.

In this edition we have also added a number of proofs, which were omitted in the previous editions. As a result, this book could be used for senior level undergraduate, as well as graduate, courses in both mathematics and physics departments.

Prerequisites for this book are introductory real analysis (notions of vector space, scalar product, norm, convergence, Fourier transform) and complex analysis, the theory of Lebesgue integration, and elementary differential equations. These topics are typically covered by the third year in mathematics departments. The first and third topics are also familiar to physics undergraduates. However, even in dealing with mathematics students we have found it useful, if not necessary, to review these notions, as needed for the course. Hence, to make the book relatively self-contained, we briefly cover these subjects, with the exception of Lebesgue integration. Those unfamiliar with the latter can think about Lebesgue integrals as if they were Riemann integrals. This said, the pace of the book is not a leisurely one and requires, at least for beginners, some amount of work.

Though, as in the previous two issues of the book, we tried to increase the complexity of the material gradually, we were not always successful, and first in Chapter 13, and then in Chapter 20, and especially in Chapter 21, there is a leap in the level of sophistication required from the reader. One may say the book proceeds at three levels. The first one, covering Chapters 1 -11, is elementary; the second one, covering Chapters 13 - 18, is intermediate; and the last one, covering Chapters 20 - 24, advanced.

During the last few years since the enlarged second printing of this book, there have appeared four books on Quantum Mechanics directed at mathematicians:

F. Strocchi, *An Introduction to the Mathematical Structure of Quantum Mechanics: a Short Course for Mathematicians*. World Scientific, 2005.

L. Takhtajan, *Quantum Mechanics for Mathematicians*. AMS, 2008.

L.D. Faddeev, O.A. Yakubovskii, *Lectures on Quantum Mechanics for Mathematics Students. With an appendix by Leon Takhtajan*. AMS, 2009.

J. Dimock, *Quantum Mechanics and Quantum Field Theory*. Cambridge Univ. Press, 2011.

These elegant and valuable texts have considerably different aims and rather limited overlap with the present book. In fact, they complement it nicely.

Acknowledgment: The authors are grateful to I. Anapolitanos, Th. Chen, J. Faupin, Z. Gang, G.-M. Graf, M. Griesemer, L. Jonsson, M. Merkli, M. Mück, Yu. Ovchinnikov, A. Soffer, F. Ting, T. Tzaneteas, and especially J. Fröhlich, W. Hunziker and V. Buslaev for useful discussions, and to J. Feldman, G.-M. Graf, I. Herbst, L. Jonsson, E. Lieb, B. Simon and F. Ting for reading parts of the manuscript and making useful remarks.

Vancouver/Toronto,   *Stephen Gustafson*
May 2011   *Israel Michael Sigal*

Preface to the enlarged second printing

For the second printing, we corrected a few misprints and inaccuracies; for some help with this, we are indebted to B. Nachtergaele. We have also added a small amount of new material. In particular, Chapter 11, on perturbation theory via the Feshbach method, is new, as are the short sub-sections 14.1 and 14.2 concerning the Hartree approximation and Bose-Einstein condensation. We also note a change in terminology, from "point" and "continuous" spectrum, to the mathematically more standard "discrete" and "essential" spectrum, starting in Chapter 6.

Vancouver/Toronto,   *Stephen Gustafson*
July 2005   *Israel Michael Sigal*

From the preface to the first edition

The first fifteen chapters of these lectures (omitting four to six chapters each year) cover a one term course taken by a mixed group of senior undergraduate and junior graduate students specializing either in mathematics or physics. Typically, the mathematics students have some background in advanced analysis, while the physics students have had introductory quantum mechanics. To satisfy such a disparate audience, we decided to select material which is interesting from the viewpoint of modern theoretical physics, and which illustrates an interplay of ideas from various fields of mathematics such as operator theory, probability, differential equations, and differential geometry. Given our time constraint, we have often pursued mathematical content at the expense of rigor. However, wherever we have sacrificed the latter, we have tried to explain whether the result is an established fact, or, mathematically speaking, a conjecture, and in the former case, how a given argument can be made rigorous. The present book retains these features.

Vancouver/Toronto,   *Stephen Gustafson*
Sept. 2002   *Israel Michael Sigal*

Contents

1

Physical Background

The main ingredients of any physical theory are the state space, the dynamical law giving the evolution of states, and the description of the results of measurements. The mathematical description of these structures is extracted from experiments. We describe here the foundational experiments and ideas of quantum mechanics.

The starting point of quantum mechanics was Planck's idea that electro-magnetic radiation is emitted and absorbed in discrete amounts – quanta. Einstein ventured further by suggesting that the electro-magnetic radiation itself consists of quanta, or particles, which were then named photons. These were the first quantum particles and the first glimpse of wave-particle duality. Then came Bohr's model of an atom, with electrons moving on fixed orbits and jumping from orbit to orbit without going through intermediate states. This culminated first in Heisenberg, and then in Schrödinger quantum mechanics, with the next stage incorporating quantum electro-magnetic radiation accomplished by Jordan, Pauli, Heisenberg, Born, Dirac and Fermi.

To complete this thumbnail sketch we mention two dramatic experiments. The first one was conducted by E. Rutherford in 1911, and it established the planetary model of an atom with practically all its weight concentrated in a tiny nucleus ($10^{-13} - 10^{-12}$ cm) at the center and with electrons orbiting around it. The electrons are attracted to the nucleus and repelled by each other via the Coulomb forces. The size of an atom, i.e. the size of electron orbits, is about 10^{-8} cm. The problem is that in classical physics this model is unstable.

The second experiment is the scattering of electrons off a crystal conducted by Davisson and Germer (1927), G.P. Thomson (1928) and Rupp (1928), after the advent of quantum mechanics. This experiment is similar to Young's 1805 experiment confirming the wave nature of light. It can be abstracted as the double-slit experiment described below. It displays an interference pattern for electrons, similar to that of waves.

In this introductory chapter, we present a very brief overview of the basic structure of quantum mechanics, and touch on the physical motivation for

© Springer-Verlag GmbH Germany, part of Springer Nature 2020
S. J. Gustafson and I. M. Sigal, *Mathematical Concepts of Quantum Mechanics*, Universitext, https://doi.org/10.1007/978-3-030-59562-3_1

the theory. A detailed mathematical discussion of quantum mechanics is the focus of the subsequent chapters.

1.1 The Double-Slit Experiment

Suppose a stream of electrons is fired at a shield in which two narrow slits have been cut   (see Fig. 1.1.) On the other side of the shield is a detector screen.

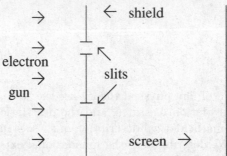

Fig. 1.1. Experimental set-up.

Each electron that passes through the shield hits the detector screen at some point, and these points of contact are recorded. Pictured in Fig. 1.2 and Fig. 1.3 are the intensity distributions observed on the screen when either of the slits is blocked.

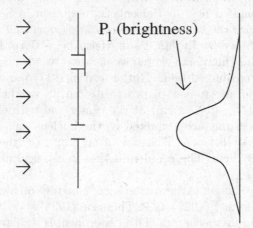

Fig.1.2. First slit blocked.

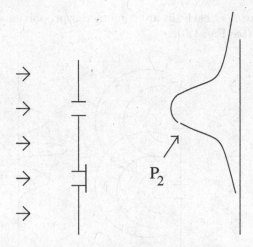

Fig.1.3. Second slit blocked.

When both slits are open, the observed intensity distribution is shown in Fig. 1.4.

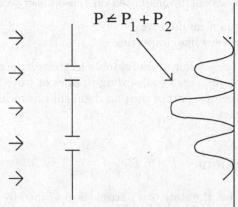

Fig.1.4. Both slits open.

Remarkably, this is not the sum of the previous two distributions; i.e., $P \neq P_1 + P_2$. We make some observations based on this experiment.

1. We cannot predict exactly where a given electron will hit the screen, we can only determine the distribution of locations.
2. The intensity pattern (called an *interference pattern*) we observe when both slits are open is similar to the pattern we see when a wave propagates through the slits: the intensity observed when waves E_1 and E_2 (the waves here are represented by complex numbers encoding the amplitude and

phase) originating at each slit are combined is proportional to $|E_1 + E_2|^2 \neq |E_1|^2 + |E_2|^2$ (see Fig. 1.5).

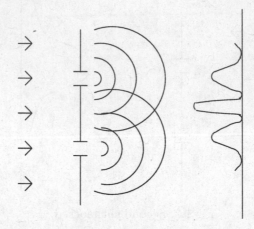

Fig.1.5. Wave interference.

We can draw some conclusions based on these observations.

1. Matter behaves in a random way.
2. Matter exhibits wave-like properties.

In other words, the behaviour of individual electrons is intrinsically random, and this randomness propagates according to laws of wave mechanics. These observations form a central part of the paradigm shift introduced by the theory of quantum mechanics.

1.2 Wave Functions

In quantum mechanics, the state of a particle is described by a complex-valued function of position and time, $\psi(x, t)$, $x \in \mathbb{R}^3$, $t \in \mathbb{R}$. This is called a *wave function* (or *state vector*). Here $\mathbb{R}^d$ denotes d-dimensional Euclidean space, $\mathbb{R} = \mathbb{R}^1$, and a vector $x \in \mathbb{R}^d$ can be written in coordinates as $x = (x_1, \ldots, x_d)$ with $x_j \in \mathbb{R}$.

In light of the above discussion, the wave function should have the following properties:

1. $|\psi(\cdot, t)|^2$ is the probability distribution for the particle's position. That is, the probability that a particle is in the region $\Omega \subset \mathbb{R}^3$ at time t is $\int_\Omega |\psi(x, t)|^2 dx$. Thus we require the normalization $\int_{\mathbb{R}^3} |\psi(x, t)|^2 dx = 1$.
2. ψ satisfies some sort of wave equation.

For example, in the double-slit experiment, if ψ_1 gives the state beyond the shield with the first slit closed, and ψ_2 gives the state beyond the shield with the second slit closed, then $\psi = \psi_1 + \psi_2$ describes the state with both slits open. The interference pattern observed in the latter case reflects the fact that $|\psi|^2 \neq |\psi_1|^2 + |\psi_2|^2$.

1.3 State Space

The space of all possible states of the particle at a given time is called the *state space*. For us, the state space of a particle will usually be the square-integrable functions:

$$L^2(\mathbb{R}^3) := \{\psi : \mathbb{R}^3 \to \mathbb{C} \mid \int_{\mathbb{R}^3} |\psi(x)|^2 dx < \infty\}$$

(we can impose the normalization condition as needed). This is a vector space, and has an inner-product given by

$$\langle \psi, \phi \rangle := \int_{\mathbb{R}^3} \bar{\psi}(x)\phi(x)dx.$$

In fact, it is a Hilbert space (see Section 25.2 for precise definitions and mathematical details).

1.4 The Schrödinger Equation

We now give a motivation for the equation which governs the evolution of a particle's wave function. This is the celebrated *Schrödinger equation*. An evolving state at time t is denoted by $\psi(x,t)$, with the notation $\psi(t)(x) \equiv \psi(x,t)$.

Our equation should satisfy certain physically sensible properties:

1. *Causality*: The state $\psi(t_0)$ at time $t = t_0$ should determine the state $\psi(t)$ for all later times $t > t_0$.
2. *Superposition principle*: If $\psi(t)$ and $\phi(t)$ are evolutions of states, then $\alpha\psi(t) + \beta\phi(t)$ (α, β constants) should also describe the evolution of a state.
3. *Correspondence principle*: In "everyday situations," quantum mechanics should be close to the classical mechanics we are used to.

The first requirement means that ψ should satisfy an equation which is first-order in time, namely

$$\frac{\partial}{\partial t}\psi = A\psi \tag{1.1}$$

for some operator A, acting on the state space. The second requirement implies that A must be a *linear* operator.

We use the third requirement – the correspondence principle – in order to find the correct form of A. Here we are guided by an analogy with the transition from wave optics to geometrical optics.

$$\text{Wave Optics} \quad \rightarrow \text{Geometrical Optics}$$

$$\updownarrow \qquad\qquad\qquad\qquad \updownarrow$$

$$\text{Quantum Mechanics} \rightarrow \text{Classical Mechanics}$$

In everyday experience we see light propagating along straight lines in accordance with the laws of geometrical optics, i.e., along the characteristics of the equation

$$\frac{\partial \phi}{\partial t} = \pm c |\nabla_x \phi| \qquad\qquad (c = \text{speed of light}), \qquad\qquad (1.2)$$

known as the *eikonal equation*. On the other hand we know that light, like electro-magnetic radiation in general, obeys Maxwell's equations which can be reduced to the wave equation (say, for the electric field in the complex representation)

$$\frac{\partial^2 u}{\partial t^2} = c^2 \Delta u, \qquad\qquad (1.3)$$

where $\Delta = \sum_{j=1}^{3} \partial_j^2$ is the Laplace operator, or the *Laplacian* (in spatial dimension three).

The eikonal equation appears as a high frequency limit of the wave equation when the wave length is much smaller than the typical size of objects. Namely we set $u = a e^{\frac{i\phi}{\lambda}}$, where a and ϕ are real and $O(1)$ and $\lambda > 0$ is the ratio of the typical wave length to the typical size of objects. The real function ϕ is called the eikonal. Substitute this into (1.3) to obtain

$$\ddot{a} + 2i\lambda^{-1}\dot{a}\dot{\phi} - \lambda^{-2}a\dot{\phi}^2 + i\lambda^{-1}a\ddot{\phi}$$

$$= c^2(\Delta a + 2i\lambda^{-1}\nabla a \cdot \nabla\phi - \lambda^{-2}a|\nabla\phi|^2 + \lambda^{-1}a\Delta\phi)$$

(where dots denote derivatives with respect to t). In the short wave approximation, $\lambda \ll 1$ (with derivatives of a and ϕ $O(1)$), we obtain

$$-a\dot{\phi}^2 = -c^2 a|\nabla\phi|^2,$$

which is equivalent to the eikonal equation (1.2).

An equation in classical mechanics analogous to the eikonal equation is the Hamilton-Jacobi equation

$$\frac{\partial S}{\partial t} = -h(x, \nabla S), \qquad\qquad (1.4)$$

where $h(x, k)$ is the classical Hamiltonian function, which for a particle of mass m moving in a *potential* V is given by $h(x, k) = \frac{1}{2m}|k|^2 + V(x)$, and $S(x, t)$ is the classical *action*. We would like to find an evolution equation which would lead to the Hamilton-Jacobi equation in the way the wave equation led to the eikonal one. We look for a solution to equation (1.1) in the form $\psi(x, t) = a(x, t)e^{iS(x,t)/\hbar}$, where $S(x, t)$ satisfies the Hamilton-Jacobi equation

(1.4) and $\hbar$ is a parameter with the dimensions of action, small compared to a typical classical action for the system in question. Assuming a is independent of $\hbar$, it is easy to show that, to leading order, ψ then satisfies the equation

$$i\hbar\frac{\partial}{\partial t}\psi(x,t) = -\frac{\hbar^2}{2m}\Delta_x\psi(x,t) + V(x)\psi(x,t). \tag{1.5}$$

This equation is of the desired form (1.1). In fact it is the correct equation, and is called the *Schrödinger equation*. The small constant $\hbar$ is *Planck's constant*; it is one of the fundamental constants in nature. For the record, its value is roughly

$$\hbar \approx 6.6255 \times 10^{-27} \text{ erg sec.}$$

The equation (1.5) can be written as

$$\boxed{i\hbar\frac{\partial}{\partial t}\psi = H\psi} \tag{1.6}$$

where the linear operator H, called a *Schrödinger operator*, is given by

$$\boxed{H\psi := -\frac{\hbar^2}{2m}\Delta\psi + V\psi.}$$

Example 1.1 Here are just a few examples of potentials.

1. Free motion : $V \equiv 0$.
2. A wall: $V \equiv 0$ on one side, $V \equiv \infty$ on the other (meaning $\psi \equiv 0$ here).
3. The double-slit experiment: $V \equiv \infty$ on the shield, and $V \equiv 0$ elsewhere.
4. The Coulomb potential : $V(x) = -\alpha/|x|$ (describes a hydrogen atom).
5. The harmonic oscillator : $V(x) = \frac{m\omega^2}{2}|x|^2$.

We will analyze some of these examples, and others, in subsequent chapters.

1.5 Classical Limit

Now we go in the opposite direction and pass from a quantum description to a classical one. Thus we assume that the ratio $\alpha := \hbar/(\text{classical action}) \to 0$ (the classical limit). Changing to dimensionless units, we arrive at the dimensionless Schrödinger equation

$$i\alpha\frac{\partial}{\partial t}\psi(x,t) = -\frac{\alpha^2}{2m}\Delta_x\psi(x,t) + V(x)\psi(x,t), \tag{1.7}$$

where (abusing the notation) m and x, t stand for the dimensionless parameter and variables. For the classical limit, we would like to set $\alpha = 0$ in (1.7), but we see that the limiting equation becomes trivial: $V(x)\psi(x,t) = 0$. So we have to do something more subtle.

Write ψ as $\psi = ae^{\frac{iS}{\alpha}}$ where a and S are real-valued functions. Plug this into the Schrödinger equation and take the real and imaginary parts of the result to obtain (assuming a^{-1} is not too singular)

$$\frac{\partial S}{\partial t} = -\frac{1}{2m}|\nabla S|^2 - V + \frac{\alpha^2}{2m}a^{-1}\Delta a, \qquad (1.8)$$

$$\frac{\partial a}{\partial t} = -\frac{1}{m}\nabla S \cdot \nabla a - \frac{1}{2m}a\Delta S. \qquad (1.9)$$

Problem 1.2 Derive these equations.

Now, we can set $\alpha = 0$ in (1.8) and (1.9) (classical limit). If we denote $S_* := S\big|_{\alpha=0}$ and $a_* := a\big|_{\alpha=0}$, then (1.8) gives

$$\frac{\partial S_*}{\partial t} + \frac{1}{2m}|\nabla S_*|^2 + V = 0, \qquad (1.10)$$

which is the Hamilton-Jacobi equation of Classical Mechanics. Hence S_* is the classical action. The equation (1.9) can be rewritten as

$$\frac{\partial a^2}{\partial t} = -\mathrm{div}(\frac{1}{m}(\nabla S)a^2). \qquad (1.11)$$

This is a classical transport equation. Thus at $\alpha = 0$, we arrive at two classical equations, (1.10) and (1.11).

Let a_* be a solution to (1.9) with $S = S_*$. We show in Section 2.5 that, for small α, the function

$$\psi^{\mathrm{sc}} = a_* e^{\frac{iS_*}{\alpha}}$$

('sc' stands for the semi-classical), constructed from purely classical objects gives, gives a good approximation to the exact wave function $\psi = ae^{\frac{iS}{\alpha}}$.

Let $\rho = a^2 = |\psi|^2$. By the interpretation of Quantum Mechanics, this is the probability density. By Classical Mechanics, $v = \frac{\nabla S}{m}$ is the classical velocity. Hence, we interpret $j := \rho v = a^2\frac{\nabla S}{m}$ as the probability current density. Now, equation (1.11) can be rewritten as

$$\frac{d\rho}{dt} = -\mathrm{div}j, \qquad (1.12)$$

This is the law of conservation of probability in differential form (for more details, see Subsection 3.5).

The equations (1.10) and (1.12) (at $\alpha = 0$) describe classical non-interacting particles of density $\rho(x,t) = |a_*|^2(x,t)\big|_{\alpha=0}$ and velocity $v(x,t) = \frac{\nabla S_*(x,t)}{m}$. Taking the gradient of (1.10) and using that $\nabla|\nabla S_*|^2 = 2(\nabla S_* \cdot \nabla)\nabla S_*$ and $\nabla S_* = mv$, we obtain an equation for the velocity v:

$$m\frac{dv}{dt} = -\nabla V, \qquad (1.13)$$

where $\frac{dv}{dt} = (\partial_t + v \cdot \nabla)v$ is the material derivative. This is the classical Newton's equation for particle flow in the potential $V(x)$.

To take into account quantum corrections, one has to retain the last term on the r.h.s. of (1.8) (or treat it by a perturbation theory). We do not pursue this here, but take it up in the semi-classical analysis later on.

Next we consider a stationary state $\psi(x,t) = e^{-\frac{iEt}{\alpha}}\phi(x)$, where $H\phi = E\phi$. Then $S = -Et + \chi$ and $a = |\phi|$ where χ is the argument of ϕ: $\phi = |\phi|e^{i\chi} = ae^{i\chi}$. Hence,

$$\frac{\partial S}{\partial t} = -E \quad \text{and} \quad \frac{\partial a}{\partial t} = 0.$$

These equations, together with Eqs (1.8) and (1.9), imply that

$$|\nabla S|^2 - 2m(E - V) = \frac{\alpha^2 \Delta a}{a} \tag{1.14}$$

and

$$\mathrm{div}(a^2 \nabla S) = 0. \tag{1.15}$$

In the regime $\alpha = \hbar/(\text{classical action}) \to 0$ (the classical limit) we obtain a stationary flow of particle fluid. Hence in the classical limit, $v = \frac{\nabla S_*}{m}$ is interpreted as velocity, and $k = \nabla S_*$ as momentum. Note that (1.14) implies that in the classical limit, $|k| = \sqrt{2m(E - V)}$ (in the classically allowed region $V(x) \le E$).

2

Dynamics

The purpose of this chapter is to investigate the existence and a key property – conservation of probability – of solutions of the Schrödinger equation for a particle of mass m in a potential V. The relevant background material on linear operators is reviewed in the Mathematical Supplement Chapter 25.

We recall that the Schrödinger equation,

$$i\hbar\frac{\partial\psi}{\partial t} = H\psi \tag{2.1}$$

where the linear operator $H = -\frac{\hbar^2}{2m}\Delta + V$ is the corresponding Schrödinger operator, determines the evolution of the particle state (the wave function), ψ. We supplement equation (2.1) with the initial condition

$$\psi|_{t=0} = \psi_0 \tag{2.2}$$

where $\psi_0 \in L^2(\mathbb{R}^3)$. The problem of solving (2.1)- (2.2) is called an *initial value problem* or a *Cauchy problem*.

Both the existence and the conservation of probability do not depend on the particular form of the operator H, but rather follow from a basic property – *self-adjointness*. This property is rather subtle, so for the moment we just mention that self-adjointness is a strengthening of a much simpler property – *symmetry*. A linear operator A acting on a Hilbert space $\mathcal{H}$ is *symmetric* if for any two vectors in the domain of A, $\psi, \phi \in D(A)$,

$$\langle A\psi, \phi\rangle = \langle \psi, A\phi\rangle.$$

2.1 Conservation of Probability

Since we interpret $|\psi(x,t)|^2$ at a given instant in time as a probability distribution, we should have

© Springer-Verlag GmbH Germany, part of Springer Nature 2020
S. J. Gustafson and I. M. Sigal, *Mathematical Concepts of Quantum Mechanics*, Universitext, https://doi.org/10.1007/978-3-030-59562-3_2

$$\int_{\mathbb{R}^3} |\psi(x,t)|^2 dx = \int_{\mathbb{R}^3} |\psi(x,0)|^2 dx = 1 \qquad (2.3)$$

at all times, t. If (2.3) holds, we say that *probability is conserved*.

Theorem 2.1 Solutions $\psi(t)$ of (2.1) with $\psi(t) \in D(H)$ conserve probability if and only if H is symmetric.

Proof. Suppose $\psi(t) \in D(H)$ solves the Cauchy problem (2.1)-(2.2). We compute

$$\frac{d}{dt}\langle \psi, \psi \rangle = \langle \dot\psi, \psi \rangle + \langle \psi, \dot\psi \rangle = \langle \frac{1}{i\hbar} H\psi, \psi \rangle + \langle \psi, \frac{1}{i\hbar} H\psi \rangle$$
$$= \frac{1}{i\hbar}[\langle \psi, H\psi \rangle - \langle H\psi, \psi \rangle]$$

(here, and often below, we use the notation $\dot\psi$ to denote $\partial\psi/\partial t$). If H is symmetric then this time derivative is zero, and hence probability is conserved. Conversely, if probability is conserved for all such solutions, then by a version of the *polarization identity*,

$$\langle \psi, \phi \rangle = \frac{1}{4}(\|\phi + \psi\|^2 - \|\phi - \psi\|^2 - i\|\phi + i\psi\|^2 + i\|\phi - i\psi\|^2), \qquad (2.4)$$

whose proof is left as an exercise below, we have $\frac{d}{dt}\langle \psi(t), \phi(t) \rangle = 0$, for any two solutions $\psi(t)$ and $\phi(t)$. This implies $\langle H\psi, \phi \rangle = \langle \psi, H\phi \rangle$ for all $\psi, \phi \in D(H)$ (since we may choose $\psi_0 = \psi$ and $\phi_0 = \phi$). This, in turn, implies H is a symmetric operator. The latter fact follows from $\qquad \qquad \square$

Problem 2.2 Prove (2.4).

Problem 2.3 Show that the following operators on $L^2(\mathbb{R}^3)$ (with their natural domains) are symmetric:

1. x_j (that is, multiplication by x_j);
2. $p_j := -i\hbar\partial_{x_j}$;
3. $H_0 := -\frac{\hbar^2}{2m}\Delta$;
4. the multiplication operator by $f(x)$ ($\psi(x) \to f(x)\psi(x)$), provided $f : \mathbb{R}^3 \to \mathbb{R}$ is bounded;
5. $f(p) := \mathcal{F}^{-1} f(k)\mathcal{F}$, where $f : \mathbb{R}^3 \to \mathbb{R}$ is bounded and $\mathcal{F}$ denotes Fourier transform;
6. integral operators $Kf(x) = \int K(x,y)f(y)\, dy$ with $K(x,y) = \overline{K(y,x)}$ and, say, $K \in L^2(\mathbb{R}^3 \times \mathbb{R}^3)$.

2.2 Self-adjointness

As was mentioned above the key property of the Schrödinger operator H which guarantees existence of dynamics is its self-adjointness. We define this notion here. More detail can be found in Section 25.5 of the mathematical supplement.

Definition 2.4 A linear operator A acting on a Hilbert space $\mathcal{H}$ is *self-adjoint* if A is symmetric and $\mathrm{Ran}(A \pm i1) = \mathcal{H}$.

Note that the condition $\mathrm{Ran}(A \pm i1) = \mathcal{H}$ is equivalent to the fact that the equations

$$(A \pm i)\psi = f \qquad (2.5)$$

have solutions for all $f \in \mathcal{H}$. The definition above differs from the one commonly used (see Section 25.5 of the Mathematical Supplement and e.g. [244]), but is equivalent to it. This definition isolates the property one really needs and avoids long proofs which are not relevant to us.

Example 2.5 The operators in Problem 2.3 are all self-adjoint.

Proof. We show this for $p = -i\hbar\partial_x$ on the space $L^2(\mathbb{R})$. This operator is symmetric, so we compute $\mathrm{Ran}(-i\hbar\partial_x + i)$. That is, we solve

$$(-i\hbar\partial_x + i)\psi = f,$$

which, using the Fourier transform (see Section 25.14), is equivalent to $(k + i)\hat{\psi}(k) = \hat{f}(k)$, and therefore

$$\psi(x) = (2\pi\hbar)^{-1/2} \int_{\mathbb{R}} e^{ikx/\hbar} \frac{\hat{f}(k)}{k+i} \, dk.$$

Now for any such $f \in L^2(\mathbb{R})$,

$$(1 + |k|^2)^{1/2}|\hat{\psi}(k)| = |\hat{f}(k)| \in L^2(\mathbb{R}),$$

so ψ lies in the Sobolev space of order one, $H^1(\mathbb{R}) = D(-i\hbar\partial_x)$, and therefore $\mathrm{Ran}(-i\hbar\partial_x + i1) = L^2$. Similarly $\mathrm{Ran}(-i\hbar\partial_x - i1) = L^2$. $\square$

Problem 2.6 Show that x_j, $f(x)$ and $f(p)$, for f real and bounded, and Δ are all self-adjoint on $L^2(\mathbb{R}^3)$ (with their natural domains).

In what follows we omit the identity operator 1 in expressions like $A - z1$.

The next result establishes the self-adjointness of Schrödinger operators.

Theorem 2.7 Assume that V is real and bounded. Then $H := -\frac{\hbar^2}{2m}\Delta + V(x)$, with $D(H) = D(\Delta)$, is self-adjoint on $L^2(\mathbb{R}^3)$.

Proof. It is easy to see (just as in Problem 2.3) that H is symmetric. To prove $\mathrm{Ran}(H \pm i) = \mathcal{H}$, we will use the following facts proved in Sections 25.4 and 25.5 of the mathematical supplement:

1. If an operator K is bounded and satisfies $\|K\| < 1$, then the operator $1 + K$ has a bounded inverse.
2. If A is symmetric and $\mathrm{Ran}(A - z) = \mathcal{H}$ for some z, with $\mathrm{Im}\, z > 0$, then it is true for every z with $\mathrm{Im}\, z > 0$. The same is true for $\mathrm{Im}\, z < 0$.

3. If A is self-adjoint, then $A - z$ is invertible for all z with $\text{Im} \, z \neq 0$, and

$$\|(A - z)^{-1}\| \leq \frac{1}{|\text{Im} \, z|}. \tag{2.6}$$

Since H is symmetric, it suffices to show that $\text{Ran}(H + i\lambda) = \mathcal{H}$, for some $\lambda \in \mathbb{R}$, $\pm\lambda > 0$, i.e. to show that the equation

$$(H + i\lambda)\psi = f \tag{2.7}$$

has a unique solution for every $f \in \mathcal{H}$ and some $\lambda \in \mathbb{R}$, $\pm\lambda > 0$. Write $H_0 = -\frac{\hbar^2}{2m}\Delta$. We know H_0 is self-adjoint, and so $H_0 + i\lambda$ is one-to-one and onto, and hence invertible. Applying $(H_0 + i\lambda)^{-1}$ to (2.7), we find

$$\psi + K(\lambda)\psi = g,$$

where $K(\lambda) = (H_0 + i\lambda)^{-1}V$ and $g = (H_0 + i\lambda)^{-1}f$. By (2.6), $\|K(\lambda)\| \leq \frac{1}{\lambda}\|V\|$. Thus, for $|\lambda| > \|V\|$, $\|K(\lambda)\| < 1$ and therefore $\mathbf{1} + K(\lambda)$ is invertible, according to the first statement above. Similar statements hold also for $K(\lambda)^T := V(H_0 + i\lambda)^{-1}$. Therefore

$$\psi = (\mathbf{1} + K(\lambda))^{-1}g.$$

Moreover, it is easy to see that

$$(H_0 + i\lambda)(\mathbf{1} + K(\lambda)) = (\mathbf{1} + K(\lambda)^T)(H_0 + i\lambda)$$

and therefore $\psi = (H_0 + i\lambda)^{-1}(\mathbf{1} + K(\lambda)^T)^{-1}f$ (*show this*). So $\psi \in D(H_0) = D(H)$. Hence $\text{Ran}(H + i\lambda) = \mathcal{H}$ and H is self-adjoint, by the second property above. $\square$

Unbounded potentials. The Coulomb potential $V(x) = \frac{\alpha}{|x|}$ is not bounded. We can extend the proof of Theorem 2.7 to show that Schrödinger operators with real potentials with Coulomb-type singularities are still self-adjoint. More precisely, we consider a general class of potentials V satisfying for all $\psi \in D(H_0)$

$$\|V\psi\| \leq a\|H_0\psi\| + b\|\psi\| \tag{2.8}$$

(H_0-bounded potentials) for some a and b with $a < 1$.

Problem 2.8 Show that $V(x) = \frac{\alpha}{|x|}$ satisfies (2.8) with $a > 0$ arbitrary and b depending on a. *Hint*: Write $V(x) = V_1(x) + V_2(x)$ where

$$V_1(x) = \begin{cases} V(x) & |x| \leq 1 \\ 0 & |x| > 1 \end{cases}, \qquad V_2(x) = \begin{cases} 0 & |x| \leq 1 \\ V(x) & |x| > 1. \end{cases}$$

Use that $\|V_1\psi\| \leq \sup|\psi|\|V_1\|$, that by the Fourier transform $\sup|\psi| \leq (\int(|k|^2 + c)^{-2}dk)^{-1/2}(\|\Delta\psi\| + c\|\psi\|)$, and the fact that $\int(|k|^2 + c)^{-2}dk \to 0$ as $c \to \infty$.

Theorem 2.9 Assume that H_0 is a self-adjoint operator and V is a symmetric operator satisfying (2.8) with $a < 1$. Then the operator $H := H_0 + V$ with $D(H) = D(H_0)$ is self-adjoint.

Proof. As in the proof of Theorem 2.7, it suffices to show that $\|V(H_0 - i\lambda)^{-1}\| < 1$, provided λ is sufficiently large. Indeed, (2.8) implies that

$$\|V(H_0 - i\lambda)^{-1}\phi\| \le a\|H_0(H_0 - i\lambda)^{-1}\phi\| + b\|(H_0 - i\lambda)^{-1}\phi\|. \qquad (2.9)$$

Now, since $\|H_0\phi\|^2 \le \|H_0\phi\|^2 + |\lambda|^2\|\phi\|^2 = \|(H_0 - i\lambda)\phi\|^2$ and $\|(H_0 - i\lambda)^{-1}\phi\| \le |\lambda|^{-1}\|\phi\|$, we have that

$$\|V(H_0 - i\lambda)^{-1}\phi\| \le a\|\phi\| + b|\lambda|^{-1}\|\phi\|. \qquad (2.10)$$

Since $a < 1$ we take λ such that $a + b|\lambda|^{-1} < 1$, which gives $\|V(H_0 - i\lambda)^{-1}\| < 1$. After this we continue as in the proof of Theorem 2.7. $\square$

Problem 2.10 Prove that the operator $H := -\frac{\hbar^2}{2m}\Delta - \frac{\alpha}{|x|}$ (the Schrödinger operator of the hydrogen atom with infinitely heavy nucleus) is self-adjoint.

Theorem 2.9 has the following easy and useful variant

Theorem 2.11 Assume that H_0 is a self-adjoint, positive operator and V is symmetric and satisfies $D(V) \supset D(H_0)$ and

$$\langle\psi, V\psi\rangle \le a\langle\psi, H_0\psi\rangle + b\|\psi\|, \qquad (2.11)$$

with $a < 1$. Then the operator $H := H_0 + V$ with $D(H) = D(H_0)$, is self-adjoint.

For a proof of this theorem see e.g. [245], Theorem X.17.

Now we present the following more difficult result, concerning Schrödinger operators whose potentials grow with x:

Theorem 2.12 Let $V(x)$ be a continuous function on $\mathbb{R}^3$ satisfying $V(x) \ge 0$, and $V(x) \to \infty$ as $|x| \to \infty$. Then $H = -\frac{\hbar^2}{2m}\Delta + V$ is self-adjoint on $L^2(\mathbb{R}^3)$

The proof of this theorem is fairly technical, and can be found in [162], for example.

Remark 2.13 Here and elsewhere, the precise meaning of the statement "the operator H is self-adjoint on $L^2(\mathbb{R}^d)$" is as follows: there is a domain $D(H)$, with $C_0^\infty(\mathbb{R}^d) \subset D(H) \subset L^2(\mathbb{R}^d)$, for which H is self-adjoint, and H (with domain $D(H)$) is the unique self-adjoint extension of $-\frac{\hbar^2}{2m}\Delta + V(x)$, which is originally defined on $C_0^\infty(\mathbb{R}^d)$. The exact form of $D(H)$ depends on V. If V is bounded or relatively bounded as above, then $D(H) = D(\Delta) = H^2(\mathbb{R}^d)$.

By definition, every self-adjoint operator is symmetric. However, not every symmetric operator is self-adjoint. Nor can every symmetric operator be extended uniquely to a larger domain on which it is self-adjoint. For example, the Schrödinger operator $A := -\Delta - c/|x|^2$ with $c > 1/4$ is symmetric on the domain $C_0^\infty(\mathbb{R}^3 \setminus \{0\})$ (the infinitely differentiable functions supported away from the origin), but does not have a unique self-adjoint extension (see [245]). It is usually much easier to show that a given operator is symmetric than to show that it is self-adjoint, since the latter question involves additional domain considerations.

2.3 Existence of Dynamics

We consider the Cauchy problem (2.1)- (2.2) for an abstract linear operator H on a Hilbert space $\mathcal{H}$. Here $\psi = \psi(t)$ is a differentiable path in $\mathcal{H}$.

Definition 2.14 We say the *dynamics exist* if for all $\psi_0 \in \mathcal{H}$ the Cauchy problem (2.1)- (2.2) has a unique solution which conserves probability.

The main result of this chapter is the following

Theorem 2.15 The dynamics exist if and only if H is self-adjoint.

We sketch here a proof only of the implication which is important for us, namely that *self-adjointness of H implies the existence of dynamics*, with details relegated to the mathematical supplement Section 25.6 (for a proof of the converse statement see [244]). We derive this implication from the following result:

Theorem 2.16 If H is a self-adjoint operator, then there is a unique family of bounded operators, $U(t) := e^{itH/\hbar}$, having the following properties for $t, s \in \mathbb{R}$:

$$i\hbar \frac{\partial}{\partial t} U(t) = H U(t) = U(t) H, \tag{2.12}$$

$$U(0) = \mathbf{1}, \tag{2.13}$$

$$U(t)U(s) = U(t+s), \tag{2.14}$$

$$\|U(t)\psi\| = \|\psi\|. \tag{2.15}$$

Theorem 2.16 implies the part of Theorem 2.15 of interest to us here. Indeed, the family $\psi(t) := U(t)\psi_0$ is the unique solution of the Cauchy problem (2.1)-(2.2), and also conserves probability. (The uniqueness follows from (2.3).)

The operator family $U(t) := e^{-itH/\hbar}$ is called the *propagator* or *evolution operator* for the equation (2.1). The properties recorded in the equations (2.14) and (2.15) are called the *group* and *isometry* properties. The operator $U(t) = e^{-iHt/\hbar}$ is furthermore invertible (since $U(t)U(-t) = \mathbf{1}$). Moreover, one can show that it preserves the inner product:

$$\langle U(t)\psi, U(t)\phi \rangle = \langle \psi, \phi \rangle,$$

for all $\psi, \phi \in L^2$, i.e. it is *unitary* (see Definition 25.29). Such a family is called a *one-parameter unitary group*.

Sketch of a proof of Theorem 2.16. We begin by discussing the exponential of a bounded operator. For a bounded operator, A, we can define the operator e^A through the familiar power series

$$e^A := \sum_{n=0}^{\infty} \frac{A^n}{n!}$$

which converges absolutely since

$$\sum_{n=0}^{\infty} \frac{\|A^n\|}{n!} \le \sum_{n=0}^{\infty} \frac{\|A\|^n}{n!} = e^{\|A\|} < \infty.$$

With this definition, for a bounded operator A, it is not difficult to prove (2.12) - (2.14) for $U(t) = e^{-itA/\hbar}$, and if A is also self-adjoint, (2.15).

Problem 2.17 For A bounded, prove (2.12) - (2.14) for $U(t) = e^{-itA/\hbar}$, and, if A is self-adjoint, also (2.15).

Now for an *unbounded* but *self-adjoint* operator A, we may define the bounded operator e^{iA} by approximating A by bounded operators. Since A is self-adjoint, the operators

$$A_\lambda := \frac{1}{2}\lambda^2[(A + i\lambda)^{-1} + (A - i\lambda)^{-1}] \qquad (2.16)$$

are well-defined and bounded for $\lambda > 0$. Using the bound (2.6), implied by the self-adjointness of A, we show that the operators A_λ approximate A in the sense that

$$A_\lambda \psi \to A\psi \quad \text{as} \quad \lambda \to \infty \quad \text{for } \psi \in D(A). \qquad (2.17)$$

Since A_λ is bounded, we can define the exponential e^{iA_λ} by power series as above. One then shows that the family $\{e^{iA_\lambda}, \lambda > 0\}$ is a Cauchy family, in the sense that

$$\left\| \left(e^{iA_{\lambda'}} - e^{iA_\lambda} \right) \psi \right\| \to 0 \qquad (2.18)$$

as $\lambda, \lambda' \to \infty$ for all $\psi \in D(A)$. This Cauchy property implies that for any $\psi \in D(A)$, the vectors $e^{iA_\lambda}\psi$ converge to some element of the Hilbert space as $\lambda \to \infty$. Thus we can define

$$e^{iA}\psi := \lim_{\lambda \to \infty} e^{iA_\lambda}\psi \qquad (2.19)$$

for $\psi \in D(A)$. It follows from (2.15) that $\|e^{iA}\psi\| \le \|\psi\|$ for all ψ in $D(A)$, which is dense in $\mathcal{H}$. Thus we can extend this definition of e^{iA} to all $\psi \in \mathcal{H}$. This defines the exponential e^{iA} for any self-adjoint operator A.

If H is self-adjoint, then so is $Ht/\hbar$ for every $t \in \mathbb{R}$. Hence the conclusions above apply to $Ht/\hbar$. This defines the propagator $U(t) = e^{-iHt/\hbar}$. Using Problem 2.17 and (2.19), we can prove that $U(t)$ has the properties (2.12) – (2.15). This implies Theorem 2.16. $\square$

The theorem above, together with the Fourier transform (see Section 25.14), also provides one method of defining functions of self-adjoint operators:

Definition 2.18 Let A be a self-adjoint operator, and $f(\lambda)$ be a function on $\mathbb{R}$ whose inverse Fourier transform, $\check{f}$, is integrable: $\int_{\mathbb{R}} |\check{f}(t)| dt < \infty$. Then the operator

$$f(A) := (2\pi\hbar)^{-1/2} \int_{\mathbb{R}} \check{f}(t) e^{-iAt/\hbar} dt \qquad (2.20)$$

is well-defined, is bounded, and is self-adjoint if f is real. It is a function of the operator A.

Example 2.19 Theorem 2.16 allows us to define exponentials of the self-adjoint operators on $L^2(\mathbb{R}^3)$ with which we are familiar: x_j, $p_j := -i\hbar\partial_{x_j}$, $H_0 := -\frac{\hbar^2}{2m}\Delta$, $f(x)$ and $f(p)$ (for f a real function).

Problem 2.20 (i) Determine how the operators e^{ix_j} and e^{ip_j} act on functions in $L^2(\mathbb{R}^3)$.

(ii) Show that the families $e^{ix_j a/\hbar}$, $a \in \mathbb{R}$, and $e^{ip_j b/\hbar}$, $b \in \mathbb{R}$, are one parameter groups of unitary operators on $L^2(\mathbb{R}^3)$.

Problem 2.21 For any ψ_0 in our Hilbert space, define the operator $e^{-\frac{iHt}{\hbar}}$: $\psi_0 \to \psi$, the solution to the Schrödinger equation,

$$i\hbar\frac{\partial\psi}{\partial t} = H\psi, \qquad (2.21)$$

with the initial condition ψ_0. Show that (a) $e^{-\frac{iHt}{\hbar}}$ preserves the inner product,

$$\langle e^{-\frac{iHt}{\hbar}}\psi, e^{-\frac{iHt}{\hbar}}\phi\rangle = \langle\psi, \phi\rangle,$$

(b) the operators $e^{-\frac{iHt}{\hbar}}$ and $e^{\frac{iHt}{\hbar}}$ are inverses of each other and (c) the adjoint of $e^{-\frac{iHt}{\hbar}}$ is given by $e^{\frac{iHt}{\hbar}}$, i.e.

$$\langle e^{-\frac{iHt}{\hbar}}\psi, \phi\rangle = \langle\psi, e^{\frac{iHt}{\hbar}}\phi\rangle.$$

Hint: For the first relation in (a), prove first that $e^{-\frac{iHt}{\hbar}}$ is isometry, preserves the norm, and then use the parallelogram relation between norms and inner products show that it preserves also the inner product. For the second relation in (a), consider the differential equation for

$$e^{\frac{iHt}{\hbar}} e^{-\frac{iHt}{\hbar}} \psi$$

and for the third relation, show that it follows from the first two. (For H bounded, one can also use the representation

$$e^{-\frac{iHt}{\hbar}} = \sum_0^\infty \frac{1}{n!}(-\frac{iHt}{\hbar})^n,$$

where the series $\sum_0^\infty \frac{1}{n!}(-\frac{iHt}{\hbar})^n$ is absolutely convergent.)

To summarize, if H is self-adjoint, then the operators $U(t) := e^{-iHt/\hbar}$ exist and are unitary for all $t \in \mathbb{R}$ (since $Ht/\hbar$ is self-adjoint). Moreover, the family $\psi(t) := U(t)\psi_0$ is the unique solution of the equation (2.1), with the initial condition $\psi(0) = \psi_0$, and it satisfies $\|\psi(t)\| = \|\psi_0\|$. Thus for the Schrödinger equation formulation of quantum mechanics to make sense, the Schrödinger operator H must be self-adjoint. As was shown in Theorem 2.9, Schrödinger operators $H := -\frac{\hbar^2}{2m}\Delta + V(x)$ with potentials $V(x)$ satisfying (2.8) are self-adjoint, and therefore generate unitary dynamics.

2.4 The Free Propagator

We conclude this chapter by finding the *free propagator* $U(t) = e^{iH_0t/\hbar}$, i.e. the propagator for Schrödinger's equation in the absence of a potential. Here $H_0 := -\frac{\hbar^2}{2m}\Delta$ acts on $L^2(\mathbb{R}^3)$. The tool for doing this is the Fourier transform, whose definition and properties are reviewed in Section 25.14.

Let $g(k) = e^{-\frac{a|k|^2}{2}}$ (a *Gaussian*), with $Re(a) \geq 0$. Then setting $p := -i\hbar\nabla$ and using Definition 25.78 and Problem 25.75 (part 1) from Section 25.14, we have

$$g(p)\psi(x) = (2\pi a\hbar^2)^{-3/2} \int e^{-\frac{|x-y|^2}{2a\hbar^2}} \psi(y)dy.$$

Since $-\hbar^2\Delta = |p|^2$, we can write this as

$$(e^{a\hbar^2\Delta/2}\psi)(x) = (2\pi a\hbar^2)^{-3/2} \int e^{-\frac{|x-y|^2}{2a\hbar^2}} \psi(y)dy. \qquad (2.22)$$

Taking $a = \frac{it}{m\hbar}$ here, we obtain an expression for the Schrödinger evolution operator $e^{-iH_0t/\hbar}$ for the Hamiltonian of a free particle, $H_0 = -\frac{\hbar^2}{2m}\Delta$:

$$(e^{-iH_0t/\hbar}\psi)(x) = \left(\frac{2\pi i\hbar t}{m}\right)^{-3/2} \int_{\mathbb{R}^3} e^{\frac{im|x-y|^2}{2\hbar t}} \psi(y)dy. \qquad (2.23)$$

One immediate consequence of this formula is the pointwise decay (in time) of solutions of the free Schrödinger equation with integrable initial data:

$$\left|e^{-iH_0t/\hbar}\psi(x)\right| \leq \left(\frac{2\pi\hbar t}{m}\right)^{-3/2} \int_{\mathbb{R}^3} |\psi(y)|dy. \qquad (2.24)$$

As another consequence, we make a connection between the free Schrödinger evolution, and the classical evolution of a free particle. Using the relation $|x-y|^2 = |x|^2 - 2x \cdot y + |y|^2$, we obtain

$$e^{-iH_0 t/\hbar}\psi(x) = \left(\frac{2\pi i\hbar t}{m}\right)^{-3/2} e^{i\frac{m|x|^2}{2\hbar t}} \int e^{-i\frac{mx}{t}\cdot y/\hbar}\left(e^{i\frac{m|y|^2}{2\hbar t}}\psi(y)\right)dy.$$

Denoting $\psi_t(y) := e^{i\frac{m|y|^2}{2\hbar t}}\psi(y)$, we have

$$e^{-iH_0 t/\hbar}\psi(x) = \left(\frac{it}{m}\right)^{-3/2} e^{i\frac{m|x|^2}{2\hbar t}}\widehat{\psi}_t(mx/t) \tag{2.25}$$

where (as usual) $\widehat{\psi}$ denotes the Fourier transform of ψ. One can show that if $\hat{\psi}(k)$ is localized near $k_0 \in \mathbb{R}^d$, then so is $\hat{\psi}_t(k)$ for large t, and therefore the right hand side of (2.25) is localized near the point

$$x_0 = v_0 t, \qquad \text{where} \qquad v_0 = k_0/m,$$

i.e., near the classical trajectory of the free particle with momentum k_0.

2.5 Semi-classical Approximation

In this section, we describe the semi-classical approximation of the quantum propagator. As in Section 1.5, we consider the dimensionless Schrödinger equation in dimensionless units,

$$i\alpha\frac{\partial}{\partial t}\psi(x,t) = -\frac{\alpha^2}{2m}\Delta_x\psi(x,t) + V(x)\psi(x,t), \tag{2.26}$$

where m and x,t are the dimensionless parameter and variables and α is the ratio $\alpha := \hbar/(\text{classical action}) \to 0$ (the classical limit).

As in Section 1.5, assuming an initial condition of the form $\psi_0 := a_0 e^{iS_0/\alpha}$, we write ψ as $\psi = ae^{iS/\alpha}$, where a and S are real-valued functions. Plug this into equation (2.26) and take the real and imaginary parts of the result to obtain (assuming a^{-1} is not too singular)

$$\frac{\partial S}{\partial t} = -\frac{1}{2m}|\nabla S|^2 - V + \frac{\alpha^2}{2m}a^{-1}\Delta a, \tag{2.27}$$

$$\frac{\partial a}{\partial t} = -\frac{1}{m}\nabla S\cdot\nabla a - \frac{1}{2m}a\Delta S. \tag{2.28}$$

For a discussion of the physical meaning of Eqs. (2.27)-(2.28), see Section 1.5.

Setting $\alpha = 0$ in (2.27) and denoting the solution of the resulting equation by S_* gives

$$\frac{\partial S_*}{\partial t} + \frac{1}{2m}|\nabla S_*|^2 + V = 0, \tag{2.29}$$

which is the Hamilton-Jacobi equation of Classical Mechanics. Hence S_* is the classical action. Now, we consider (2.28) with $S = S_*$ and denote by $a = a_*$ the solution of the resulting equation, i.e.

$$\frac{\partial a_*}{\partial t} = -\frac{1}{m}\nabla S_* \cdot \nabla a_* - \frac{1}{2m}a_*\Delta S_*, \tag{2.30}$$

Eq. (2.30) is a transport equation along the classical trajectories of the vector field $\nabla S \cdot \nabla$ (i.e. the solution a is obtained by transporting the initial conditions a_0 along the characteristics of $\nabla S \cdot \nabla$). Eqs. (2.29) and (2.30) are solved by the method of characteristics, with a_* given by an explicit expression (see [77, 92, 141, 250]).

With the purely classical objects S_* and a_* solving the equations (2.29) and (2.30), we associate the wave function

$$\psi^{\mathrm{sc}} := a_* e^{iS_*/\alpha},$$

gives the leading order approximation to ψ, called the *semi-classical* or *WKB* *(for Gregor Wentzel, Hendrik Anthony Kramers, and Léon Brillouin) approximation.* ('sc' for 'semi-classical'.) We show below that, as long as Eqs. (2.29) and (2.30) have solutions, the difference of the exact solution and semiclassical approximation satisfies the estimate

$$\|\psi_t - \psi_t^{\mathrm{sc}}\| \le Ct\alpha, \tag{2.31}$$

where we display the time-dependence as the subindex t. This is a remarkable estimate which shows that a function constructed from purely classical objects gives a good approximation to the quantum wave function. Since $\psi_t = U(t)\psi_0$, (2.31) gives a semi-classical estimate of the propagator $U(t)$.

We now prove estimate (2.31) justifying the semi-classical approximation sketched above. Instead of writing solutions to (2.26) as $\psi_t = ae^{iS/\alpha}$, with a and S real-valued functions (satisfying (2.27) and (2.28)), we look for them in the form

$$\psi_t = ae^{iS_*/\alpha},$$

where S_* solves equation (2.29), which does not involve a and α. Plugging this into (2.26), we find the equation for a:

$$\frac{\partial a}{\partial t} = La, \quad L \equiv L(t) := -\frac{1}{m}\nabla S_* \cdot \nabla - \frac{1}{2m}\Delta S_* + \frac{i\alpha}{2m}\Delta. \tag{2.32}$$

($L(t)$ depends on t through S_*.) Thus, as long as Eq (2.29) has a solution, Eq. (2.32) is equivalent to Eq. (2.26). We observe that, by (2.30), a_* solves (2.32) to the order $O(\alpha)$,

$$(\partial_t - L)a_* = -\frac{i\alpha}{2m}\Delta a_*, \tag{2.33}$$

and the semiclassical approximation $\psi_t^{\mathrm{sc}} := a_* e^{iS_*/\alpha}$ solves (2.26) to the order $O(\alpha^2)$:

$$\left(i\alpha\frac{\partial}{\partial t} - H_\alpha\right)\psi_t^{\mathrm{sc}} = -e^{\frac{iS_*}{\alpha}}\frac{\alpha^2}{2m}\Delta a_* \tag{2.34}$$

where $H_\alpha := -\frac{\alpha^2}{2m}\Delta_x + V(x)$.

To estimate the difference between the exact solution $\psi_t = ae^{iS_*/\alpha}$ and the semiclassical approximation $\psi_t^{sc} := a_* e^{iS_*/\alpha}$, we let $U(t) := e^{-iH_\alpha t/\alpha}$ be the propagator for (2.26). Writing $U(-t)\psi_t^{sc}$ as the integral of its derivative, using

$$i\alpha\partial_t U(-t)\psi_t^{sc} = U(-t)(i\alpha\partial_t - H_\alpha)\psi_t^{sc}$$

and (2.34) and applying $U(t)$ to the result, we obtain

$$\psi_t^{sc} - U(t)\psi_0 = \frac{1}{i\alpha}\int_0^t U(t-s)(e^{iS_*/\alpha}\frac{\alpha^2}{2m}\Delta a_*)ds. \qquad (2.35)$$

Taking the norm in (2.35) and using that $\|\int_0^t f(s)ds\| \leq \int_0^t \|f(s)\|ds$ and that $U(t)$ is unitary yields

$$\|U(t)\psi_0 - \psi_t^{sc}\| \leq \frac{1}{\alpha}\int_0^t \|\frac{\alpha^2}{2m}\Delta a_*\|ds. \qquad (2.36)$$

The last inequality, together with the relation $\psi_t = U(t)\psi_0$ and the fact that $\|\Delta a_*\| < \infty$, gives (2.31).

3

Observables

Observables are the quantities that can be experimentally measured in a given physical framework. In this chapter, we discuss the observables of quantum mechanics.

3.1 The Position and Momentum Operators

We recall that in quantum mechanics, the state of a particle at time t is described by a wave function $\psi(x,t)$. The probability distribution for the position, x, of the particle, is $|\psi(\cdot,t)|^2$. Thus the mean value of the position at time t is given by $\int x|\psi(x,t)|^2 dx$ (note that this is a vector in $\mathbb{R}^3$). If we define the coordinate multiplication operator

$$x_j : \psi(x) \mapsto x_j\psi(x)$$

then the mean value of the jth component of the coordinate x in the state ψ is $\langle \psi, x_j\psi \rangle$.

Recall that the evolution state $\psi(x,t)$ obeys the Schrödinger equation

$$i\hbar\frac{\partial\psi}{\partial t} = H\psi. \tag{3.1}$$

For $\psi(x,t)$ solving (3.1), we compute

$$\frac{d}{dt}\langle\psi, x_j\psi\rangle = \langle\dot\psi, x_j\psi\rangle + \langle\psi, x_j\dot\psi\rangle = \langle\frac{1}{i\hbar}H\psi, x_j\psi\rangle + \langle\psi, x_j\frac{1}{i\hbar}H\psi\rangle$$

$$= \langle\psi, \frac{i}{\hbar}Hx_j\psi\rangle - \langle\psi, x_j\frac{i}{\hbar}H\psi\rangle = \langle\psi, \frac{i}{\hbar}[H, x_j]\psi\rangle$$

where $[A,B] := AB - BA$ is the *commutator* of A and B. (More about the commutators later.) Since $H = -\frac{\hbar^2}{2m}\Delta + V$, and $\Delta(x\psi) = x\Delta\psi + 2\nabla\psi$, we find

© Springer-Verlag GmbH Germany, part of Springer Nature 2020
S. J. Gustafson and I. M. Sigal, *Mathematical Concepts of Quantum Mechanics*, Universitext, https://doi.org/10.1007/978-3-030-59562-3_3

$$\frac{i}{\hbar}[H, x_j] = -\frac{i\hbar}{m}\nabla_j,$$

leading to the equation

$$\frac{d}{dt}\langle\psi, x_j\psi\rangle = \frac{1}{m}\langle\psi, -i\hbar\nabla_j\psi\rangle.$$

As before, we denote the operator $-i\hbar\nabla_j$ by p_j. As well, we denote the mean value $\langle\psi, A\psi\rangle$ of an operator A in the state ψ by $\langle A\rangle_\psi$. Then the above becomes

$$\boxed{m\frac{d}{dt}\langle x_j\rangle_\psi = \langle p_j\rangle_\psi} \tag{3.2}$$

which is reminiscent of the definition of the classical momentum. We call the operator p the *momentum operator*. In fact, p_j is a self-adjoint operator on $L^2(\mathbb{R}^3)$. (As usual, the precise statement is that there is a domain on which p_j is self-adjoint. Here the domain is just $D(p_j) = \{\psi \in L^2(\mathbb{R}^d) \mid \frac{\partial}{\partial x_j}\psi \in L^2(\mathbb{R}^d).)$

Using the Fourier transform, we compute the mean value of the momentum operator

$$\langle\psi, p_j\psi\rangle = \langle\hat\psi, \widehat{p_j\psi}\rangle = \langle\hat\psi, k_j\hat\psi\rangle = \int_{\mathbb{R}^3} k_j|\hat\psi(k)|^2 dk.$$

This, and similar computations, show that $|\hat\psi(k)|^2$ is the probability distribution for the particle momentum.

3.2 General Observables

Definition 3.1 An *observable* is a self-adjoint operator on the state space $L^2(\mathbb{R}^3)$.

We have already met six observables: the position operators, x_1, x_2, x_3, and the momentum operators, p_1, p_2, p_3 (combined into vector-observables x and p). The Schrödinger operator,

$$H = -\frac{\hbar^2}{2m}\Delta + V,$$

is self-adjoint and therefore, by our definition, is an observable. But what is the physical quantity it stands for (or 'observes')? We find the answer below.

In general, we interpret $\langle A\rangle_\psi$ as the average of the observable A in the state ψ. The reader is invited to derive the following equation for the evolution of the mean value of an observable:

$$\frac{d}{dt}\langle A\rangle_\psi = \langle\frac{i}{\hbar}[H, A]\rangle_\psi. \tag{3.3}$$

Problem 3.2 Check that for any observable, A, and for any solution ψ of the Schrödinger equation, Eq (3.3) holds.

Eq (3.3) is analogous to the classical Hamilton equation, $\frac{d}{dt}a = \{h,a\}$, for a classical observable, a, i.e. a function on the phase space. Here $\{f,g\} = \sum_{j=1}^{3}(\frac{\partial f}{\partial k_j}\frac{\partial g}{\partial x_j} - \frac{\partial f}{\partial x_j}\frac{\partial g}{\partial k_j})$ is the Poisson bracket and $h = h(x,k)$ is the classical Hamiltonian with $h(x(t),k(t))$ giving the energy of a solution $(x(t),k(t))$ (see Section 4.1 and supplemental Section 4.7 for more details). For this reason, H is interpreted as the observable of energy and is called the *Hamiltonian operator*, or *quantum Hamiltonian*.

Now, we would like to apply (3.3) to the momentum operator, $p = -i\hbar\nabla$. Simple computations give $[\Delta, p] = 0$ and $[V, p] = i\hbar\nabla V$, so that $\frac{i}{\hbar}[H,p] = -\nabla V$ and hence

$$\frac{d}{dt}\langle p_j \rangle_\psi = \langle -\nabla_j V \rangle_\psi. \qquad (3.4)$$

This is a quantum mechanical mean-value version of Newton's equation of classical mechanics. Or, if we include Equation (3.2), we have a quantum analogue of the classical Hamilton equations. Note also that, since $-\hbar^2\Delta = |p|^2$ (here $|p|^2 = \sum_{j=1}^{3} p_j^2$), we have

$$H = -\frac{\hbar^2}{2m}\Delta + V = \frac{1}{2m}|p|^2 + V,$$

which is in an agreement with our interpretation of x and p as the observables of the co-ordinate and momentum, respectively, and H, as a Hamiltonian operator.

Next, we address the question: What is the probability, $\text{Prob}_\psi(A \in \Omega)$, that measured values of the physical observable represented by A in a state ψ land in an interval $\Omega \subset \mathbb{R}$? As in the probability theory, this is given by the expectation

$$\text{Prob}_\psi(A \in \Omega) = \langle \chi_\Omega(A) \rangle_\psi \qquad (3.5)$$

of the observable $\chi_\Omega(A)$, where $\chi_\Omega(\lambda)$ is the characteristic function of the set Ω (i.e. $\chi_\Omega(\lambda) = 1$, if $\lambda \in \Omega$ and $\chi_\Omega(\lambda) = 0$, if $\lambda \notin \Omega$) and the operator-function $\chi_\Omega(A)$ can be defined according to the formula (2.20) and a limiting procedure which we skip here. We call $\chi_\Omega(A)$ the characteristic function of the operator A. This definition can be justified using spectral decompositions of the type (25.53) of Section 25.11, but we will not go into this here.

Finally, by the analogy with above, we introduce the *angular momentum observable*, $L := (L_1, L_2, L_3)$, with $L_j = (x \times p)_j$. The operator-vector L can be written as

$$L = x \times p.$$

3.3 The Heisenberg Representation

The framework outlined up to this point is called the *Schrödinger representation* of quantum mechanics. Chronologically, quantum mechanics was first formulated in the Heisenberg representation, which we now describe. For an observable A, define

$$A(t) := e^{itH/\hbar} A e^{-itH/\hbar}.$$

Let $\psi(t)$ be the solution of Schrödinger's equation with initial condition ψ_0: $\psi(t) = e^{-itH/\hbar}\psi_0$. Since $e^{-itH/\hbar}$ is unitary, we have, by simple computations which are left as an exercise,

$$\langle A \rangle_{\psi(t)} = \langle A(t) \rangle_{\psi_0} \tag{3.6}$$

and

$$\frac{d}{dt} A(t) = \frac{i}{\hbar}[H, A(t)]. \tag{3.7}$$

Problem 3.3 Prove equations (3.6) and (3.7).

This last equation is called the *Heisenberg equation* for the time evolution of the observable A. In particular, taking x and p for A, we obtain the quantum analogue of the Hamilton equations of classical mechanics:

$$m\dot{x}(t) = p(t), \qquad \dot{p}(t) = -\nabla V(x(t)). \tag{3.8}$$

In the Heisenberg representation, then, the state is fixed (at ψ_0), and the observables evolve according to the Heisenberg equation. Of course, the Schrödinger and Heisenberg representations are completely equivalent (by a unitary transformation).

Remark 3.4 A note of caution: defining commutators is a subtle business. We have to assume the set $D(A) \cap D(H)$ is dense in $L^2(\mathbb{R}^3)$ and that $A : D(H) \cap D(A) \to D(H)$ and $H : D(H) \cap D(A) \to D(A)$. Checking the last two properties is cumbersome. To avoid this, we can choose to understand equations involving commutators in the sense of expectations or quadratic forms. For instance, we assume the set $D(A) \cap D(H)$ is dense in $L^2(\mathbb{R}^3)$ and interpret (3.7) as (3.3), for all $\psi \in D(H) \cap D(A)$, or as $\frac{d}{dt}\langle \psi, A\phi \rangle = \langle \psi, \frac{i}{\hbar}[H, A]\phi \rangle$, for all $\psi, \phi \in D(H) \cap D(A)$, by defining

$$\langle \psi, i[H, A]\phi \rangle := i \left(\langle H\psi, A\phi \rangle - \langle A\psi, H\phi \rangle \right),$$

for any two self-adjoint operators and for all $\psi, \phi \in D(A) \cap D(H)$.

3.4 Conservation Laws

We say that an observable A (or, more precisely, the physical quantity represented by this observable) is *conserved* if its average, $\langle A\rangle_{\psi(t)}$, is independent of t,

$$\langle A\rangle_{\psi(t)} = \langle A\rangle_{\psi(0)}, \qquad (3.9)$$

for every $\psi(t)$ which solves the Schrödinger equation

$$i\hbar\partial_t\psi = H\psi. \qquad (3.10)$$

Differentiating relation (3.9) w.r. to and recalling (3.6)-(3.7), we conclude that an observable A is conserved if and only if $\langle\psi,[A,H]\psi\rangle = 0$, $\forall\psi$, which in turn implies[1] that A commutes with the Schrödinger operator H, i. e.

$$[A,H] = 0.$$

We consider several examples. Since obviously $[H,H] = 0$, we have $\langle H\rangle_{\psi(t)}$ = constant for any solution to (3.10), which is the mean-value version of the conservation of energy, or $H(t) = H$.

As the second example, we take the obvious relation $[1,H] = 0$, which gives the conservation of the total probability: $\langle 1\rangle_{\psi(t)} = \|\psi(t)\|^2$ = constant for any solution to (3.10).

Furthermore, $\frac{i}{\hbar}[x_j,H] = p_j \neq 0$, which says that the particle position is never conserved, i.e. a quantum particle cannot be localized at a point.

The relation $\frac{i}{\hbar}[p_j,H] = -\partial_{x_j}V \neq 0$ implies that p_j is conserved if and only if V is independent of x_j. Taken for all j's, this gives the first Newton's law: a particle will be in the state of the motion with a constant velocity as long as no force is acting on it, i.e. $\nabla V(x) = 0$.

Problem 3.5 Prove that (a) the momentum is conserved if and only if the potential $V(x)$ is constant; and (b) the angular momentum is conserved if and only if the potential $V(x)$ is spherically symmetric, i.e. $V(Rx) = V(x)$ for any rotation R in $\mathbb{R}^3$ around the origin, i.e. $V(x) = W(|x|)$ for some function $W(r)$ on the interval $[0,\infty)$.

Most of the conservation laws come from symmetries of physical systems. First, an operator U on a Hilbert space $\mathcal{H}$ is called *unitary* if it preserves the inner product: $\langle U\psi, U\phi\rangle = \langle\psi,\phi\rangle$, for all $\psi,\phi \in \mathcal{H}$ (see Definition 25.29). We say that a unitary operator U is a symmetry of (3.10) if U maps $D(H)$ into itself and

---

[1] Showing this is actually not trivial. If B is a non-negative operator, then using square roots one can show that $\langle\psi,B\psi\rangle = 0$ implies $B = 0$. If B is self-adjoint but not necessary positive, then one can reduce the problem to proving the desired property for the spectral projections of B, which are non-negative operators.

ψ_t is a solution to (3.10) $\rightarrow U\psi_t$ is a solution to (3.10).

If U is a symmetry, then inverting U in the equation $i\hbar\partial_t U\psi_t = HU\psi_t$ gives $i\hbar\partial_t\psi_t = U^{-1}HU\psi_t$, which, together with the original Schrödinger equation, (3.10), implies $U^{-1}HU\psi = H\psi$ for any $\psi \in D(H)$. Hence, if U is a symmetry, then U commutes with H:

$$U^{-1}HU = H.$$

Most often symmetries are organized in continuous families which are, or could be reduced to, one-parameter groups U_s, $s \in \mathbb{R}$, of unitary operators. This means that each U_s is a unitary operator, $U_s\psi$ is continuous in s, for every ψ, and

$$U_0 = 1, \qquad U_sU_t = U_{s+t} \tag{3.11}$$

(cf. (2.13)-(2.14)). For a one-parameter group U_s, one defines the generator as the operator, A, defined by

$$A\psi := i\partial_s U_s\psi\big|_{s=0}, \tag{3.12}$$

for those ψ's for which the derivative on the r.h.s exist. (Note that the definition above differs from the standard one by the factor i and that U_s satisfies the equation $i\partial_s U_s = AU_s$.) Then (ignoring domain questions)

U_s is a symmetry of (3.10) $\rightarrow A$ commutes with H ($[H, A] = 0$).

Indeed, the fact that U_s is a symmetry implies that, after dropping i, $U_s^{-1}HU_s = H$. Differentiating the last equation with respect to s at $s = 0$ and using (3.12), we arrive, after dropping i, at

$$[H, A] = 0.$$

(For the definition of the commutators, see Remark 3.4.)

Examples of one-parameter groups, which could serve as symmetry groups, and their generators:

- Spatial translation: $U_s^{\text{transl}} : \psi(x) \rightarrow \psi(x + se_j)$, $s \in \mathbb{R}$, where $e_1 := (1, 0, 0), e_2 := (0, 1, 0), e_3 := (0, 0, 1)$, with generator $\frac{1}{\hbar}p_j = -i\nabla_{x_j}$.
- Spatial rotation: $U_s^{\text{rot}} : \psi(x) \rightarrow \psi((R_s^j)^{-1}x)$, where R_s^j is the counterclockwise rotation around the j-axis by the angle $s \in [0, 2\pi)$, with generator

$$\frac{1}{\hbar}L_j = (x \times (-i\nabla_x))_j.$$

- Gauge transformation: $U_s^{\text{gauge}} : \psi(x) \rightarrow e^{is}\psi(x)$, $s \in \mathbb{R}$, with generator $i1$.

Problem 3.6 Show that the families, $U_s^{\text{transl}}, U_s^{\text{rot}}, U_s^{\text{gauge}}$, defined above, are one-parameter groups of unitary operators, find their generators and write the differential equations for the groups.

Hence the *translational, rotational and gauge symmetries imply the conservation of momentum, angular momentum and probability*, respectively.

The fact that L_j's are the generators of rotations justifies the term the *angular momentum* operator for $L = (L_1, L_2, L_3) = x \times p$.

We summarize a part of the discussion above as

- Time translation invariance (V is independent of t) $\Rightarrow$ conservation of energy
- Space translation invariance (V is independent of x) $\Rightarrow$ conservation of momentum
- Space rotation invariance (V is rotation invariant, i.e. is a function of $|x|$) $\Rightarrow$ conservation of angular momentum
- Gauge invariance (invariance under the transformation $\psi \rightarrow e^{i\alpha}\psi$) $\Rightarrow$ conservation of probability.

Discussion. More generally, one can consider symmetries associated with groups. In the last three examples these are the groups of translations and rotations of $\mathbb{R}^3$, denoted T^3 or $\mathbb{R}^3$, and $O(3)$, respectively (forming together the group of rigid motions of $\mathbb{R}^3$), and the group $U(1)$ of complex numbers of unit modulus, serving as a gauge group. (For particles with internal degrees of freedom, specifically with a spin to be considered later on, the gauge group is the group $SU(m)$ of complex, unitary, $m \times m$ matrices, for an appropriate m.)

We represent these groups by unitary operators on the state space $L^2(\mathbb{R}^3)$: $U_y^{\text{transl}} : \psi(x) \rightarrow \psi(x+y)$, $y \in \mathbb{R}^3$,

$$U_R^{\text{rot}} : \psi(x) \rightarrow \psi(R^{-1}x), \ R \in O(3), \tag{3.13}$$

and $U_\alpha^{\text{gauge}} : \psi(x) \rightarrow e^{i\alpha}\psi(x)$, $e^{i\alpha} \in U(1)$, for the spatial translations and rotations and the gauge transformations. These operators satisfy

$$U_y^{\text{transl}}U_{y'}^{\text{transl}} = U_{y+y'}^{\text{transl}} \quad \text{and} \quad U_R^{\text{rot}}U_{R'}^{\text{rot}} = U_{RR'}^{\text{rot}},$$

and similarly for U_α^{gauge}. The operators U_y^{transl} and U_R^{rot} give unitary representations of the groups of translations and rotations of $\mathbb{R}^3$, respectively (the first one is commutative, or abelian, the second one is not). They can be written as products of one parameter groups, so that the analysis relevant for us can be reduced to the latter case.

Problem 3.7 Find the conditions under which the above groups are symmetry groups of the Schrödinger equation.

We will consider the group of rotations, $O(3)$, in more detail. Note that the rotations, R, are represented by orthogonal matrices (i.e. real matrices satisfying $R^T R = 1$) of determinant 1. The counter-clockwise rotations R_θ^ω around the axis along a unit vector ω by the angles $\theta \in [0, 2\pi)$ form a one-parameter group, and its generator is given by

$$\partial_\theta|_{\theta=0} R_\theta^\omega x = \omega \times x.$$

Problem 3.8 (a) Prove the above statement. Hint: Prove that $\partial_\theta|_{\theta=0} R_\theta^{e_j} x = e_j \times x$ and use that $\omega = \sum \omega_j e_j$. (b) Show that operators (matrices) A_j given by $A_j x := e_j \times x$ satisfy the commutation relations

$$[A_1, A_2] = A_3, \ [A_2, A_3] = A_1, \ [A_3, A_1] = A_2. \tag{3.14}$$

Now, $\partial_\theta|_{\theta=0} \psi(R_\theta^\omega x) = \nabla\psi(x) \cdot (\omega \times x) = \omega \cdot (x \times \nabla\psi(x))$. Hence

$$i\hbar\partial_\theta|_{\theta=0} U_{R_\theta^\omega}^{\text{rot}} = -i\hbar\omega \cdot (x \times \nabla) = \omega \cdot L, \tag{3.15}$$

which proves one of the statements in Problem 3.6.

Furthermore, it is easy to see for $H := -\frac{\hbar^2}{2m}\Delta + V(x)$ that

$$U_R^{\text{rot}\,-1} H U_R^{\text{rot}} = -\frac{\hbar^2}{2m}\Delta + V(Rx). \tag{3.16}$$

Problem 3.9 Prove the last relation.

Relations (3.15) and (3.16) imply the proof of statement (b) in Problem 3.5.

Unlike the components of the position and momentum operators, x_j and p_j, the components of the angular momentum one do not mutually commute:

$$\frac{i}{\hbar}[L_k, L_l] = \epsilon^{klm} L_m. \tag{3.17}$$

Here ϵ^{klm} is the Levi-Chivita symbol: $\epsilon^{123} = 1$ and ϵ^{klm} changes sign under the permutation of any two indices. This is related to the fact that U_R^{rot} is a representation of the group of rotations of $\mathbb{R}^3$ and, unlike the group of translations, the group of rotations is non-abelian.

Eq. (3.17) implies $\frac{1}{i\hbar}L_j$ are the generators of representation (3.13) of the group $SO(3)$. The vector space spanned by $\frac{1}{i\hbar}L_j$, equipped with the commutator relation $[\cdot, \cdot]$, is the representation on $L^2(\mathbb{R}^3)$ of the Lie algebra $so(3)$ of the group $SO(3)$.

3.5 Conserved Currents

With each conserved observable A and a state ψ, one can associate a density and a current (of A in ψ):

$$\rho_A(\psi) := \bar{\psi}A\psi, \qquad j_A(\psi) := \frac{i\hbar}{2m}\left((\nabla\bar{\psi})A\psi - \bar{\psi}\nabla A\psi\right).$$

These functions satisfy an equation, called (the differential form of) the conservation law, along any solution $\psi(t)$ of the Schrödinger equation:

Proposition 3.10 (Differential conservation laws) *Let A be a conserved observable, i.e. $D(H) \cap D(A)$ is dense in $L^2(\mathbb{R}^3)$ and $[H, A] = 0$. If $\psi(t)$ is a solution of the Schrödinger equation (3.1), then*

$$\partial_t \rho_A(\psi(t)) = -\text{div} j_A(\psi(t)). \tag{3.18}$$

Proof. We omit the arguments t and ψ and use the Leibniz rule and Schrödinger equation (3.10) to compute formally

$$\partial_t \rho_A = (\partial_t \bar{\psi}) A \psi + \bar{\psi} A \partial_t \psi = \overline{\frac{1}{i\hbar} H \psi} A \psi + \bar{\psi} \frac{1}{i\hbar} A H \psi.$$

Since A and H commute, and since V is real and therefore $\overline{\frac{1}{i\hbar} V \psi} A \psi + \bar{\psi} \frac{1}{i\hbar} V A \psi = 0$, this gives

$$\partial_t \rho_A = -\frac{i\hbar}{2m} \Delta \bar{\psi} A \psi + \frac{i\hbar}{2m} \bar{\psi} \Delta A \psi.$$

By the Leibnitz rule, $\Delta \bar{\psi} A \psi = \text{div}((\nabla \bar{\psi}) A \psi) - (\nabla \bar{\psi}) \cdot \nabla A \psi$ and $\bar{\psi} \Delta A \psi = \text{div}(\bar{\psi} \nabla A \psi) - (\nabla \bar{\psi}) \cdot \nabla A \psi$, which implies $\partial_t \rho_A = \frac{i\hbar}{2m} \text{div}(-(\nabla \bar{\psi}) A \psi + \bar{\psi} \nabla A \psi)$, giving (3.18). □

Integrating equation (3.18) in x and using the Gauss theorem, one obtains the global conservation law,

$$\partial_t \int \rho_A(\psi(t)) = 0.$$

We illustrate the general result on the 'trivial' observable $A = 1$ which gives the differential form of the conservation of probability. For $A = 1$, the associated density $\rho_1(\psi) \equiv \rho(\psi)$ and the current $j_1(\psi) \equiv j(\psi)$ are $\rho(\psi)(x) := |\psi(x)|^2$ and $j(\psi)(x) = \frac{\hbar}{m} \text{Im}(\bar{\psi}(x) \nabla \psi(x))$. Then (3.18) gives the differential form of the conservation of probability law:

$$\partial_t |\psi|^2 = -\text{div} j(\psi). \tag{3.19}$$

This result shows how the probability distribution changes under the Schrödinger equation, and provides the formula for the probability current:

$$j(\psi)(x) = \frac{\hbar}{m} \text{Im}(\bar{\psi}(x) \nabla \psi(x)).$$

4

Quantization

In this chapter, we discuss the procedure of passing from classical mechanics to quantum mechanics. This is called "quantization" of a classical theory.

4.1 Quantization

To describe a *quantization* of classical mechanics, we start with the Hamiltonian formulation of classical mechanics (see supplemental Section 4.7 for more details), where the basic objects are as follows:

1. The *phase space* (or state space): $\mathbb{R}^3_x \times \mathbb{R}^3_k$.
2. The *Hamiltonian*: a real function $h(x,k)$ on $\mathbb{R}^3_x \times \mathbb{R}^3_k$ (which gives the energy of the classical system).
3. *Classical observables*: (real) functions on $\mathbb{R}^3_x \times \mathbb{R}^3_k$.
4. *Poisson bracket*: a bilinear form mapping each pair of classical observables, f, g, to the observable (function)

$$\{f,g\} = \sum_{j=1}^{3} \left(\frac{\partial f}{\partial k_j} \frac{\partial g}{\partial x_j} - \frac{\partial f}{\partial x_j} \frac{\partial g}{\partial k_j} \right).$$

5. *Canonically conjugate variables*: co-ordinate functions, x_i, k_i, satisfying

$$\{x_i, x_j\} = \{k_i, k_j\} = 0; \qquad \{k_i, x_j\} = \delta_{ij}. \qquad (4.1)$$

6. *Classical dynamics*: Hamilton's equations,

$$\dot{x} = \{h, x\}, \qquad \dot{k} = \{h, k\}. \qquad (4.2)$$

The corresponding fundamental objects in quantum mechanics are the following:

© Springer-Verlag GmbH Germany, part of Springer Nature 2020
S. J. Gustafson and I. M. Sigal, *Mathematical Concepts of Quantum Mechanics*, Universitext, https://doi.org/10.1007/978-3-030-59562-3_4

1. The *state space*: $L^2(\mathbb{R}_x^3)$.
2. The quantum Hamiltonian: a *Schrödinger operator*, $H = h(x, p)$ acting on the state space $L^2(\mathbb{R}^3)$.
3. *Quantum observables*: (self-adjoint) operators on $L^2(\mathbb{R}_x^3)$.
4. *Commutator*: a bilinear form mapping each pair of operators acting on $L^2(\mathbb{R}_x^3)$ into the commutator, $\frac{i}{\hbar}[\cdot, \cdot]$.
5. *Canonically conjugate operators*: co-ordinate operators x_i, p_i satisfying

$$[x_i, x_j] = [p_i, p_j] = 0; \qquad \frac{i}{\hbar}[p_i, x_j] = \delta_{ij}. \tag{4.3}$$

6. The *dynamics* of the quantum system can be described by the Heisenberg equations

$$\dot{x} = \frac{i}{\hbar}[H, x], \qquad \dot{p} = \frac{i}{\hbar}[H, p].$$

The relations (4.3) are called the *canonical commutation relations*. To quantize classical mechanics we pass from the canonically conjugate variables, x_i, k_i, satisfying (4.1) to the canonically conjugate operators, x_i, p_i, $i = 1, 2, 3$, satisfying (4.3):

$$x_i, \ k_i \qquad \longrightarrow \qquad x_i, \ p_i. \tag{4.4}$$

Hence with classical observables $f(x, k)$, we associate quantum observables $f(x, p)$. This is a fairly simple procedure if $f(x, p)$ is a sum of a function of x and a function of p, but rather subtle otherwise. It is explained in the next section.

If the classical Hamiltonian function is $h(x, k) = |k|^2/2m + V(x)$, the corresponding quantum Hamiltonian is the Schrödinger operator

$$H = h(x, p) = \frac{|p|^2}{2m} + V(x) = -\frac{\hbar^2}{2m}\Delta + V(x).$$

Similarly, we pass from the classical angular momentum, $l_j = (x \times k)_j$, to the *angular momentum operators*, $L_j = (x \times p)_j$.

The following table provides a summary of the classical mechanical objects and their quantized counterparts:

Object	CM	QM
state space	$\mathbb{R}_x^3 \times \mathbb{R}_k^3$ and Poisson bracket	$L^2(\mathbb{R}_x^3)$ and commutator
evolution of state	path in phase space	path in $L^2(\mathbb{R}^3)$
observable	real function on state space	self-adjoint operator on state space
result of measuring observable	deterministic	probabilistic
object determining dynamics	Hamiltonian function	Hamiltonian (Schrödinger) operator
canonical coordinates	functions x and k	operators x (mult.) and p (differ.)

Quantization of classical systems does not lead to a complete description of quantum systems. As was noted in the previous chapter, quantum mechanical particles might have also internal degrees of freedom, such as spin, which have no classical counterparts and therefore cannot be obtained as a result of quantization of a classical system. To take these degrees of freedom into account one should modify ad hoc the quantization procedure above, or add new quantization postulates as is done in the relativistic theory.

4.2 Quantization and Correspondence Principle

The correspondence between classical observables and quantum observables,

$$f(x,k) \to f(x,p),$$

is a subtle one. It is easy to see that a classical observable $f(x)$ is mapped under quantization into the operator of multiplication by $f(x)$, and an observable $g(k)$, into the operator $g(p)$, defined for example using the Fourier transform and the three-parameter translation group, $e^{-ip \cdot x/\hbar}$:

$$g(p) := (2\pi\hbar)^{-3/2} \int \breve{g}(x) e^{-ip \cdot x/\hbar} dx$$

where $\breve{g}$ is the inverse Fourier transform of g (see Definition 2.18). However, the following simple example shows the ambiguity of this correspondence for more general functions of x and k. The function $x \cdot k = k \cdot x$ could, for example, be mapped into any of the following distinct operators:

$$x \cdot p, \quad p \cdot x, \quad \frac{1}{2}(x \cdot p + p \cdot x).$$

This ambiguity can be resolved by requiring that the quantum observables obtained by a quantization of real classical observables are self-adjoint (or at least symmetric) operators. This selects the symmetric term above, for example. The corresponding quantization is called the *Weyl quantization*. In this case the operator $A = a(x, p)$ associated with the classical observable $a(x, k)$ is given by

$$A = (2\pi\hbar)^{-3} \int\int \hat{a}(\xi, x') e^{i(\xi \cdot x - x' \cdot p)/\hbar} d\xi dx', \tag{4.5}$$

where $\hat{a}$ is the Fourier transform of a in x and the inverse Fourier transform in k and therefore

$$a(x, k) = (2\pi\hbar)^{-3} \int\int \ddot{a}(\xi, \eta) e^{i(\xi \cdot x - \eta \cdot k)/\hbar} d\xi d\eta. \tag{4.6}$$

Reversing the quantization procedure, one would like to show that classical mechanics arises from quantum mechanics in the limit as $\hbar/$(classical action) $\to 0$. Assume we have passed to physical units in which a typical classical action in our system is 1, so that $\hbar$ is now the ratio of the Planck constant to the classical action. First one would like to show that a product of quantum observables is given by a product of classical ones, e.g.

$$a(x, p)b(x, p) = (ab)(x, p) + O(\hbar), \tag{4.7}$$

and therefore the former can be identified with latter. Assuming that the classical observables a and b satisfy $\int(|\xi| + |\eta|)|\hat{a}(\xi, \eta)| d\xi d\eta < \infty$, and similarly for b (this condition is considerably stronger than needed), one can easily prove (4.7). Indeed, we use the Baker-Campbell-Hausdorff formula

$$e^A e^B = e^{A + B + \frac{1}{2}[A, B]}, \tag{4.8}$$

provided $[A, B]$ is a multiple of the identity, which can be verified by computing

$$\begin{aligned} \partial_s(e^{sA} e^{sB}) &= (A + e^{sA} B e^{-sA}) e^{sA} e^{sB} \\ &= (A + B + \int_0^s dr e^{rA} [A, B] e^{-rA}) e^{sA} e^{sB} \\ &= (A + B + s[A, B]) e^{sA} e^{sB}. \end{aligned}$$

Using (4.8), (4.5) and the relation $[\xi \cdot x - \eta \cdot p, \xi' \cdot x - \eta' \cdot p] = \hbar\omega$, where $\omega := \xi \cdot \eta' - \eta \cdot \xi'$, we compute for $A := a(x, p)$ and $B := b(x, p)$,

$$AB = (2\pi\hbar)^{-6} \int \cdots \int \hat{a}(\xi, \eta) \hat{b}(\xi', \eta') e^{i\Phi/\hbar} e^{i\omega/\hbar} d\xi d\eta d\xi' d\eta'. \tag{4.9}$$

where $\Phi := (\xi + \xi') \cdot x - (\eta + \eta') \cdot p$. Now, we expand $e^{i\omega} = 1 + O(|\omega|)$ and evaluate the contribution of the first term using property 5 of the Fourier transform given in Section 25.14. Together with the definition of the convolution ($f *$

$g)(x) := \int_{\mathbb{R}^d} f(y)g(x-y)dy$ (see (25.56) of Section 25.14), this gives, after changing the variables of integration as $\xi \to \xi - \xi'$, $\eta \to \eta - \eta'$,

$$(2\pi\hbar)^{-6} \int \cdots \int \hat{a}(\xi,\eta)\hat{b}(\xi',\eta')e^{i\Phi/\hbar}d\xi d\eta d\xi' d\eta'$$

$$= (2\pi\hbar)^{-6} \int\int (\hat{a} * \hat{b})(\xi,\eta)e^{i(\xi\cdot x - \eta\cdot p)/\hbar}d\xi d\eta = ab.$$

The remainder is simply estimated by taking the absolute value under the integral.

Under the stronger condition $\int(|\xi|^2+|\eta|^2)|\hat{a}(\xi,\eta)|d\xi d\eta < \infty$, one can prove the stronger statement,

$$a(x,p)b(x,p) = (ab - \frac{i}{2}\hbar\{a,b\})(x,p) + O(\hbar^2). \tag{4.10}$$

To prove this, we expand $e^{i\omega} = 1 + i\omega + \frac{1}{2}(i\omega)^2 + O(|\omega|^2)$ and use Properties 3 - 5 of the Fourier transform in Section 25.14, to evaluate the contribution of $i\omega$. This is done similarly to the first term above, once we write $\xi\hat{a}(\xi,\eta) = -i\hbar\widehat{\nabla_x a}(\xi,\eta)$, $\eta\hat{a}(\xi,\eta) = i\hbar\widehat{\nabla_k a}(\xi,\eta)$ and similarly for b. The remainder term here is treated similarly to the remainder above. Equation (4.10) implies that in the next order, the commutators give Poisson brackets:

$$i[a(x,p), b(x,p)] = \hbar\{a,b\}(x,p) + O(\hbar^2). \tag{4.11}$$

Equation (4.11) allows one to connect the quantum and classical evolutions. Indeed, let ϕ_t be the flow generated by the Hamilton equations (4.2), i.e. the map $\phi_t : (x_0,k_0) \to$ the solution of (4.2) with the initial conditions (x_0,k_0) (see supplemental Section 4.7), and let $\alpha_t^{cl}a = a \circ \phi_t$ and $\alpha_t(A) := e^{iHt/\hbar}Ae^{-iHt/\hbar}$ be the evolutions of classical and quantum observables (α_t^{cl} is called the Liouville dynamics, and α_t is nothing but the Heisenberg dynamics.) Denote the Weyl quantization map given in (4.5) by Q, so that $A = Q(a)$. One can show that for a certain class of classical observables a, we have

$$\alpha_t(Q(a)) = Q(\alpha_t^{cl}a) + O(\hbar), \tag{4.12}$$

for $t \leq C\frac{\hbar}{\sup|V|}$, as $\hbar \to 0$. Given (4.11), a proof of (4.12) is fairly simple. We give it here modulo one classical estimate. Using the Duhamel principle (i.e. writing $\alpha_{-t}(Q(\alpha_t^{cl}(a)))-Q(a)$ as the integral of derivative $\partial_s\alpha_{-s}(Q(\alpha_s^{cl}(a))) = \alpha_{-s}([H,Q(\alpha_s^{cl}(a))] - Q(\{h,\alpha_s^{cl}(a)\})))$, we obtain

$$\alpha_t(A) - Q(a \circ \phi_t) = \int_0^t ds\, \alpha_{t-s}(R(a \circ \phi_s)), \tag{4.13}$$

where $R(a) := \frac{i}{\hbar}[H,A] - Q(\{h,a\})$. Since $\|\alpha_t(A)\| = \|A\|$, this gives

$$\|\alpha_t(A) - Q(a \circ \alpha_t^{cl})\| \leq \int_0^t ds\|R(a \circ \alpha_s^{cl})\|. \tag{4.14}$$

Note that the full evolution α_t drops out of the estimate. Using estimate (4.11) for the remainder $R(a)$ and an appropriate estimate for the classical observable $\alpha_s^{cl} a = a \circ \phi_s$, we arrive at (4.12).

In mathematics, operators obtained by a certain quantization rule from functions $f(x, k)$ satisfying certain estimates are called *pseudodifferential operators*, while the functions themselves are called *symbols*. Differential operators with smooth coefficients, as well as certain integral and singular integral operators are examples of pseudodifferential operators. The relation (4.11) is one of the central statements coming from pseudodifferential calculus.

4.3 A Particle in an External Electro-magnetic Field

Here we apply the rules discussed in the previous sections to describe a quantum particle moving in an external electro-magnetic field. Of course, if the external field is purely electric, E, then it is a potential field, $E = -\nabla \Phi$, for some $\Phi : \mathbb{R}^3 \mapsto \mathbb{R}$, and it fits within the framework we have considered above with $V(x) = \Phi(x)$.

Suppose, then, that both electric and magnetic fields, E and B, are present. These are vector fields on $\mathbb{R}^3$, which could be time-dependent: $B, E : \mathbb{R}^{3+1} \mapsto \mathbb{R}^3$. We know from the theory of electro-magnetism (Maxwell's equations) that these fields can be expressed in terms of vector and scalar potentials $A : \mathbb{R}^3 \mapsto \mathbb{R}^3$, and $\Phi : \mathbb{R}^3 \mapsto \mathbb{R}$, via

$$E = -\nabla \Phi - \partial_t A, \qquad B = \mathrm{curl} A$$

(we are using units in which the speed of light, c, is equal to one; for more details see Subsection 4.7). For simplicity, in what follows we assume that the electric and magnetic fields E and B are time-independent.

It is shown in Subsection 4.7 that the classical Hamiltonian function for a particle of charge e subject to the fields E and B is,

$$h(x, k) = \frac{1}{2m} (k - eA(x))^2 + e\Phi(x).$$

According to our general quantization procedure, we replace the classical canonical variables x and k by the quantum canonical operators x and p. The resulting Schrödinger operator is

$$H(A, \Phi) = \frac{1}{2m} (p - eA)^2 + e\Phi, \tag{4.15}$$

acting on $L^2(\mathbb{R}^3)$. The self-adjointness of $H(A, \Phi)$ can be established by using *Kato's inequality* (see [73]).

We now consider the Schrödinger equation with this Hamiltonian:

$$i\hbar \partial_t \psi = H(A, \Phi)\psi. \tag{4.16}$$

It, like the Hamiltonian $H(A, \Phi)$, depends of the magnetic and electric potentials A and Φ, rather than on the magnetic and electric fields $B := \mathrm{curl}A$ and $E := \nabla\Phi$ (in the static case).

However, as we recall that in the theory of electro-magnetism, the vector potential A is not uniquely determined by the magnetic field B. In fact, if we add the gradient of any function χ to A (a *gauge transformation*), we obtain the same magnetic field B:

$$\mathrm{curl}(A + \nabla\chi) = \mathrm{curl}A = B.$$

Similar ambiguity exists for the electric potential Φ. This is distinct from the classical mechanics where A and Φ enter into the Newton's equations only through the magnetic and electric fields $B := \mathrm{curl}A$ and $E := \nabla\Phi$.

What saves the day is the *gauge invariance* of (4.16), which is its key feature: if ψ satisfies equation (4.16), then $\psi_\chi := e^{ie\chi/\hbar}\psi$ satisfies

$$i\hbar\partial_t\psi_\chi = H(A + \nabla\chi, \Phi + \partial_t\chi)\psi_\chi. \tag{4.17}$$

This property follows from the relation

$$H(A + \nabla\chi, \Phi) = e^{ie\chi/\hbar}H(A, \Phi)e^{-ie\chi/\hbar}. \tag{4.18}$$

Problem 4.1 Check that equation (4.18) holds.

Thus if A and $\tilde{A}$ differ by a gradient vector-field $\nabla\chi$, then the operators $H(A, \Phi)$ and $H(\tilde{A}, \Phi)$ are unitarily equivalent via the unitary map

$$\psi \mapsto e^{ie\chi/\hbar}\psi$$

on $L^2(\mathbb{R}^3)$. Thus the two Hamiltonians are physically equivalent. Of course, this is to be expected as A and $\tilde{A}$ correspond to the same magnetic field.

One can impose restrictions (called *gauge conditions*) on the vector potential A in order to remove some, or all, of the freedom involved in the choice of A. A common choice is $\mathrm{div}A = 0$, known as the *Coulomb gauge*. By an appropriate gauge transformation, the Coulomb gauge can always be achieved.

The gauge invariance of (4.16) seems to say that physically, the magnetic potential does not matter, only the magnetic field is important. Indeed this is so, but only in physical spaces which have, like $\mathbb{R}^d$, the special mathematical property of being *simply connected* in the sense that any closed path can be shrunk to a point; i.e. they have no holes. We return to this phenomenon in Section 7.6 on the Aharonov-Bohm effect.

4.4 Spin

Quantum mechanical particles may also have internal degrees of freedom, which have no classical counterpart. Mathematically, this mean that the wave functions $\psi(x)$ take values not in $\mathbb{C}$, but in a higher dimensional (complex) space V (with $\dim_{\mathbb{C}} V = n > 1$). The state space in this case is $L^2(\mathbb{R}^3; V)$. For V we take a finite-dimensional, complex inner-product space. (When we want to be specific, we take $V = \mathbb{C}^n$.) Then we have

$$L^2(\mathbb{R}^3; V) = \underbrace{L^2(\mathbb{R}^3; \mathbb{C})}_{\text{space of external degrees of freedom}} \otimes \underbrace{V}_{\text{space of internal degrees of freedom}} .$$

Here, on the r.h.s. we have the tensor product of two Hilbert spaces, which can be thought of as the space with a basis $\{\phi_i v_j\}$ given by products of basis elements of each factor, equipped with the corresponding inner product $\langle \phi v, \psi w \rangle = \langle \phi, \psi \rangle \langle v, w \rangle$. It can be identified with the space of square integrable functions with values in V ($x \to \phi(x) \in V$) on the l.h.s..

Now the Schrödinger equation is invariant under a larger group of gauge transformations, $U_g^{\text{gauge}} : \psi(x) \to g\psi(x)$, $g \in U(n)$, where $U(n)$ is a group of complex, unitary, $n \times n$ matrices.

Consider the simplest non-trivial case $n = 2$, and restrict ourselves to the subgroup $SU(2)$ of $U(2)$ with determinant one (the special unitary group). The Lie algebra $su(2)$ of the group $SU(2)$ is the space of trace zero, anti-hermitian ($A^* = -A$) matrices acting on $\mathbb{C}^2$, equipped with the commutator. We can choose a basis, A_1, A_2, A_3, in $su(2)$, satisfying the commutation relations

$$[A_1, A_2] = A_3, \ [A_2, A_3] = A_1, \ [A_3, A_1] = A_2, \qquad (4.19)$$

which have already come up in (3.14) while considering the Lie algebra $so(3)$ of the group of rotations $SO(3)$. In fact, the Lie algebras $su(2)$ and $so(3)$ are isomorphic, while one can find a Lie group homomorphism $SU(2) \to SO(3)$, with kernel $\{\pm 1\}$; i.e., $SU(2)/\{\pm 1\}$ and $SO(3)$ are Lie group isomorphic (see Remark 4.4 below).

If we define the hermitian matrices $S_j := -i\hbar A_j$, we obtain

$$[S_k, S_l] = i\hbar \epsilon^{klm} S_m, \qquad (4.20)$$

where, recall, ϵ^{klm} is the Levi-Chivita symbol: $\epsilon^{123} = 1$ and ϵ^{klm} changes sign under permutation of any two indices. These are exactly the same relations as for the angular momentum operators (3.17). Consequently, the observable $S = (S_1, S_2, S_3)$ is thought of as internal angular momentum, and is called the *spin*.

However, there is an important difference between the spin and angular momentum observables. We will show later on in Theorem 6.28 that the operator $S^2 := \sum_j S_j^2$ has eigenvalues $\lambda = \hbar^2 r(r+1)$, $r = 0, \frac{1}{2}, 1, \frac{3}{2}, 2, \ldots$, of multiplicity $2r + 1$. We see that, unlike the angular momentum, the spin can take half-integer values. This means that if V is an eigenspace of S^2 with a half-integer r, then the spin observables $S_j, j = 1, 2, 3$, acting on V cannot be represented by generators of rotations on $L^2(\mathbb{R}^3)$.

We say that a *particle has spin* r if and only if the internal spin space V is an eigenspace, V_r, of S^2 (in some representation) with eigenvalue $\lambda = \hbar^2 r(r+1)$. By Theorem 6.28, V_r has dimension $2r+1$ and $r \in \{0, \frac{1}{2}, 1, \frac{3}{2}, 2, \dots\}$.

As a basis in V_r, it is convenient to use a basis of the eigenvectors of the third component, S_3, of the spin operator. We write elements of V_r in this basis as $\psi(x, s)$, $s = -r, \dots, r$, where $\psi(x, s)$ satisfies $S_3\psi(x, s) = \hbar s\psi(x, s)$, or as vectors $\psi(x) = (\psi_{-r}(x), \dots, \psi_r(x))$, where each ψ_j belongs to the familiar one-particle space $L^2(\mathbb{R}^3) = L^2(\mathbb{R}^3; \mathbb{C})$, with the identification $\psi(x, s) \leftrightarrow \psi_s(x)$. (Usually such functions are written as columns, but for typographical simplicity we write them as rows.) This identifies V_r with the space $\mathbb{C}^{2r+1}$ and $L^2(\mathbb{R}^3; \mathbb{C}) \otimes V_r = L^2(\mathbb{R}^3; V_r)$, with $L^2(\mathbb{R}^3; \mathbb{C}) \otimes \mathbb{C}^{2r+1} = L^2(\mathbb{R}^3; \mathbb{C}^{2r+1})$.

For $r = \frac{1}{2}$, it is convenient to write S_j as $S_j = \frac{\hbar}{2}\sigma_j$, where σ_j are the Pauli matrices

$$\sigma_1 = \begin{pmatrix} 0 & 1 \\ 1 & 0 \end{pmatrix}, \sigma_2 = \begin{pmatrix} 0 & -i \\ i & 0 \end{pmatrix}, \sigma_3 = \begin{pmatrix} 1 & 0 \\ 0 & -1 \end{pmatrix}. \tag{4.21}$$

For $r = \frac{1}{2}$, the spin operators S_j act on $V_{\frac{1}{2}}$ as

$$S_1\psi(x, s) = \hbar|s|\psi(x, -s), \quad S_2\psi(x, s) = -i\hbar s\psi(x, -s), \quad S_3\psi(x, s) = \hbar s\psi(x, s).$$

It is an experimental fact that all particles belong to one of the following two groups: particles with integer spins, or *bosons*, and particles with half-integer spins, or *fermions*. (The particles we are dealing with – electrons, protons and neutrons – are fermions, with spin $\frac{1}{2}$, while photons, which we will deal with later, are bosons, with spin 1. Nuclei, though treated as point particles, are composite objects whose spin could be either integer or half-integer.)

Problem 4.2 1) Find the generators, r_1, r_2, r_3, of the rotations, $R_1(\varphi)$, $R_2(\varphi)$, $R_3(\varphi)$, around the x_1-, x_2-, x_3-axes. 2) Show that these generators satisfy commutation relations (4.19).

As an example, consider the rotations $R_3(\varphi)$ around the x_3-axis, given by

$$R_3(\varphi) = \begin{pmatrix} \cos\varphi & \sin\varphi & 0 \\ -\sin\varphi & \cos\varphi & 0 \\ 0 & 0 & 1 \end{pmatrix}, \tag{4.22}$$

The generators of these rotations are given by $r_i := \partial_\varphi R_i(\varphi)\big|_{\varphi=0}$. For $R_3(\varphi)$, we have

$$r_3 = \begin{pmatrix} 0 & 1 & 0 \\ -1 & 0 & 0 \\ 0 & 0 & 1 \end{pmatrix}. \tag{4.23}$$

As we mentioned above, the algebra $su(2)$ is isomorphic to the algebra of the rotation group $so(3)$. We can define this isomorphism, $\phi : so(3) \rightarrow su(2)$, starting with $\phi(r_i) = S_i$ and then extending to the entire $so(3)$ by linearity.

Problem 4.3 Show that the map $\phi : so(3) \rightarrow su(2)$ defined above is an algebra isomorphism.

Remark 4.4 Though the algebras $so(3)$ and $su(2)$ are isomorphic, the groups $SU(2)$ and $SO(3)$ are not: there is a Lie group homomorphism, $\phi : SU(2) \rightarrow SO(3)$, s.t. $\phi^{-1}(1) = \{\pm 1\}$. Indeed, define the map $h : \mathbb{R}^3 \rightarrow \{$ traceless hermitian 2×2 matrices$\}$ by $h_x := \sum_j x_j \sigma_j$ and define $\phi : u \rightarrow R$, where $R = \phi(u)$ solves the equation

$$h_{Rx} = u h_x u^*,$$

for all $x \in \mathbb{R}^3$. Using the equations $h_{Rx} = u h_x u^*$ and $\det h_x = -|x|^2$, we compute

$$|Rx|^2 = -\det h_{Rx} = -\det h_x = |x|^2,$$

giving $|Rx| = |x|$, i.e. R preserves the Euclidean norm $|x|$. Hence R is a rotation of the space $\mathbb{R}^3$ and therefore is an element of the orthogonal group $O(3)$. One can show that $\phi(u)$ has determinant 1, i.e. $\phi(u) \in SO(3)$, and therefore $\phi : SU(2) \rightarrow SO(3)$. To show that ϕ is a group homomorphism, we use the equation $h_{Rx} = u h_x u^*$ to compute

$$(uu')h_x(uu')^* = uu'h_x(u')^*u^* = uh_{R'x}u^* = h_{RR'x}.$$

Finally, since the transformation $T_u h = u h u^*$ satisfies $T_u h = u h u^* = T_{-u} h$, we see that ϕ maps u and $-u$ into the same element of $SO(3)$.

The spin interacts with an external magnetic field. The energy of this interaction has a form similar to that of an orbiting classical charge. For a charge moving in a circular orbit, the classical energy of interaction with a magnetic field B is $-\mu \cdot B(x)$, $\mu := g \frac{e}{2m} l$, where e and m are the charge and mass of the particle and l is its (classical) angular momentum. In quantum mechanics, this interaction (in the case of spin $r = \frac{1}{2}$) is

$$-\mu \cdot B(x), \ \mu := g \frac{e}{2m} S,$$

where g is called the gyromagnetic ratio. Based on classical mechanics, one expects $g = 1$, but it turns out that $g = 2$ (plus small corrections, due to creation and annihilation of photons out of and into the vacuum, if the electromagnetic field is quantized).

4.5 Many-particle Systems

Now we consider a physical system consisting of n particles which interact pairwise via the potentials $V_{ij}(x_i - x_j)$, where x_j is the position of the j-th particle. Examples of such systems include atoms or molecules – i.e., systems consisting of electrons and nuclei interacting via Coulomb forces.

In classical mechanics such a system is described by the particle coordinates, x_j and momenta k_j, $j = 1, \ldots, n$, so that the classical state of the system is given by the pair (x, k) where $x = (x_1, \ldots, x_n)$ and $k = (k_1, \ldots, k_n)$ and the state space, also called the *phase-space*, of the system is $\mathbb{R}_x^{3n} \times \mathbb{R}_k^{3n}$, or a subset thereof. The dynamics of this system is given by the classical Hamiltonian function

$$h(x, k) = \sum_{j=1}^{n} \frac{1}{2m_j} |k_j|^2 + V(x)$$

where m_j is the mass of the j-th particle and V is the total potential of the system, and the standard Poisson brackets

$$\{f, g\} = \sum_{j=1}^{n} \left(\frac{\partial f}{\partial k_j} \cdot \frac{\partial g}{\partial x_j} - \frac{\partial f}{\partial x_j} \cdot \frac{\partial g}{\partial k_j} \right)$$

where $\frac{\partial f}{\partial x_j}$ and $\frac{\partial f}{\partial k_j}$ are the gradients in x_j and k_j, respectively. Since in our case the particles interact only with each other and by two-body potentials $V_{ij}(x_i - x_j)$, V is given by

$$V(x) = \frac{1}{2} \sum_{i \neq j} V_{ij}(x_i - x_j). \tag{4.24}$$

Quantizing this system in exactly the same way as the one-particle one, we associate with particle coordinates x_j, and momenta k_j, the quantum coordinates x_j, and momenta $p_j := -i\hbar \nabla_{x_j}$, which are operators. And so the classical Hamiltonian $h(x, k)$ leads to the Schrödinger operator $H_n := h(x, p)$, $p = (p_1, \ldots, p_n)$, i.e.

$$H_n = \sum_{j=1}^{n} \frac{1}{2m_j} |p_j|^2 + V(x), \tag{4.25}$$

acting on $L^2(\mathbb{R}^{3n})$. This is the Schrödinger operator, or quantum Hamiltonian, of the $n-$particle system.

Example 4.5 Consider a molecule with N electrons of mass m and charge $-e$, and M nuclei of masses m_j and charges $Z_j e$, $j = 1, \ldots, M$. In this case, the Schrödinger operator, H_{mol}, is

$$H_{mol} = \sum_{1}^{N} \frac{1}{2m} |p_j|^2 + \sum_{1}^{M} \frac{1}{2m_j} |q_j|^2 + V(x, y) \tag{4.26}$$

acting on $L^2(\mathbb{R}^{3(N+M)})$. Here $x = (x_1, \ldots, x_N)$ are the electron coordinates, $y = (y_1, \ldots y_M)$ are the nucleus coordinates, $p_j = -i\hbar \nabla_{x_j}$ is the momentum of the j-th electron, $q_j = -i\hbar \nabla_{y_j}$ is the momentum of the j-th nucleus, and

$$V(x,y) = \frac{1}{2}\sum_{i\neq j}\frac{e^2}{|x_i - x_j|} - \sum_{i,j}\frac{e^2 Z_j}{|x_i - y_j|} + \frac{1}{2}\sum_{i\neq j}\frac{e^2 Z_i Z_j}{|y_i - y_j|} \qquad (4.27)$$

is the sum of Coulomb interaction potentials between the electrons (the first term on the r.h.s.), between the electrons and the nuclei (the second term), and between the nuclei (the third term). For a neutral molecule, we have

$$\sum_{j=1}^{M} Z_j = N.$$

If $M = 1$, the resulting system is called an atom, or Z-atom ($Z = Z_1$).

4.6 Identical Particles

Quantum many-particle systems display a remarkable new feature. Unlike in classical physics, identical particles (i.e., particles with the same masses, charges and spins, or, more generally, which interact in the same way) in quantum physics are indistinguishable. Assume we have n identical particles (of spin 0). Classically, states of such a system are given by $(x_1, k_1, \ldots, x_n, k_n)$, with the state space, also called the *phase-space*, being $\prod_1^n(\mathbb{R}_x^3 \times \mathbb{R}_k^3) = \mathbb{R}_x^{3n} \times \mathbb{R}_k^{3n}$. Naively, we might assume that the state space of the quantum system is

$$L^2(\mathbb{R}^{3n}, \mathbb{C}) \equiv \otimes_1^n L^2(\mathbb{R}^3, \mathbb{C}), \qquad (4.28)$$

where the second term is the tensor product of n Hilbert spaces $L^2(\mathbb{R}^3, \mathbb{C})$, defined as the Hilbert space spanned by the products of elements of orthonormal bases in each $L^2(\mathbb{R}^3, \mathbb{C})$ (see Section 25.13 in the mathematical supplement for a definition for abstract Hilbert spaces). However this is not so, as we explain below.

The indistinguishability of the particles means that all probability distributions which can be extracted from an n-particle wave function $\Psi(x_1, \ldots, x_n)$ should be invariant with respect to permutations of the coordinates, x_j. To illustrate this, consider bound states, which can be always taken to be real-valued wave functions. Since $|\Psi(x_1, \ldots, x_n)|^2$ is invariant under permutations, this implies that $\Psi(x_1, \ldots, x_n)$ is invariant under particle permutations, modulo a change of sign.

A *permutation* is a one-to-one and onto map π of $\{1, 2, \ldots, n\}$ into itself, and the collection of all permutations of n indices forms the *symmetric group* S_n. To a permutation π, we associate a transformation of $\mathbb{R}^{3n}$ (also denoted by π)

$$\pi(x_1, \ldots, x_n) = (x_{\pi(1)}, \ldots, x_{\pi(n)}).$$

We consider the map of S_n into unitary (i.e. preserving the inner product) operators on $L^2(\mathbb{R}^{3n})$ defined by

$$(T_\pi \Psi)(x) = \Psi(\pi^{-1}x),$$

with the property $T_{\pi_1}T_{\pi_2} = T_{\pi_1\pi_2}$ (this is why we need π^{-1} on the r.h.s.), called a (unitary) *representation* of S_n.

Let $U(X)$ denote the space of unitary operators acting on a Hilbert space X. For a group G, a representation $T : G \to U(X)$ is called *irreducible* iff X has no non-trivial proper subspace invariant under all T_g, $g \in G$. The *dimension* of T is the dimension of X.

There are exactly two one-dimensional irreducible representations of S_n, and they are of special interest in the spinless case: one with $\Psi(x_1, \ldots, x_n)$ totally invariant under permutations, $T_\pi \Psi = \Psi \; \forall \pi \in S_n$, and one with $\Psi(x_1, \ldots, x_n)$ transforming as

$$T_\pi \Psi = (-1)^{\#(\pi)}\Psi \; \forall \pi \in S_n, \tag{4.29}$$

where $\#(\pi)$ is the number of transpositions making up the permutation π (so $(-1)^{\#(\pi)}$ is the *parity* of $\pi \in S_n$ – see the end of this section for more details on irreducible representations of S_n.)

The particles described by the wave functions of the first type are called *bosons* and by the second type, *fermions*. Correspondingly, we have the following subspaces of $L^2(\mathbb{R}^{3n})$:

$$\mathcal{H}_{\text{bose}}^{\text{total}} := \{\Psi \in L^2(\mathbb{R}^{3n}, \mathbb{C}) \mid T_\pi \Psi = \Psi \; \forall \pi \in S_n\}, \tag{4.30}$$

$$\mathcal{H}_{\text{fermi}}^{\text{total}} := \{\Psi \in L^2(\mathbb{R}^{3n}, \mathbb{C}) \mid T_\pi \Psi = (-1)^{\#(\pi)}\Psi \; \forall \pi \in S_n\} \tag{4.31}$$

Including spin. To complete the picture, we have to take into account particle spins. The one-particle space for a particle of spin r is

$$L^2(\mathbb{R}^3 \times \{-r, \ldots, r\}, \mathbb{C}) = L^2(\mathbb{R}^3, \mathbb{C}^{2r+1}), \tag{4.32}$$

with wave functions of the form $\Psi(x, s) = (\Psi_{-r}(x), \ldots, \Psi_r(x))$. Assume we have n identical particles of spin r. In this case, the state space is

$$L^2(\mathbb{R}^3, \mathbb{C}^{2r+1})^{\otimes n} \equiv L^2(\mathbb{R}^{3n}, \mathbb{C}^{(2r+1)n}) \tag{4.33}$$

(see the remark above, and Section 25.13 about tensor products), consisting of functions which can be written as $\Psi(x_1, s_1, \ldots, x_n, s_n)$. Now, the operator H_n acts on such functions.

The indistinguishability of the particles means that all probability distributions which can be extracted from an n-particle wave function (4.33) should be symmetric with respect to permutations of the coordinates and spins of the identical particles. Since for bound states we can restrict ourselves to real wave functions, this is equivalent to the property that $\Psi(x_1, s_1, \ldots, x_n, s_n)$ is invariant under such permutations, modulo a change of sign.

Recall that all elementary (and composite) particles are divided into two groups: particles with half integer spins, called fermions (e.g. electrons, protons, and neutrons have spin $1/2$), and particles with integer spins, called bosons (particles related to interactions). For bosons, the wave functions, $\Psi(x_1, s_1, \ldots, x_n, s_n)$, should be symmetric with respect to permutations of the coordinates and spins of identical particles, and for fermions, antisymmetric. In particular, the state space for fermions of spin $\frac{1}{2}$ is

$$\mathcal{H}_{fermi} := \{\Psi \in L^2(\mathbb{R}^{3n}, \mathbb{C}^{2n}) \mid T_\pi \Psi = (-1)^{\#(\pi)} \Psi, \, \forall \pi \in S_n\} \qquad (4.34)$$

where, as above, π is a permutation of the n indices, $\pi : (1, 2, \ldots, n) \rightarrow (\pi(1), \pi(2), \ldots, \pi(n))$, $\#(\pi)$ is the number of transpositions making up the permutation π, and

$$(T_\pi \Psi)(x_1, s_1, \ldots, x_n, s_n) = \Psi(x_{\pi(1)}, s_{\pi(1)}, \ldots, x_{\pi(n)}, s_{\pi(n)}).$$

Below, we will write the space (4.31) as $\bigwedge_{i=1}^n L^2(\mathbb{R}^3, \mathbb{C}^2)$.

Since the Hamiltonian H_n does not act on the spin variables, by separation of variables, we may consider it acting on functions $\psi(x_1, \ldots, x_n)$ of the coordinates only, which arise from (4.34) by, say, taking inner products in the spin variables with functions of $s_1, \ldots, s_n$. What are the symmetry properties of these functions with respect to permutations of the coordinates? To answer this question one has to dip into the theory of representations of the symmetric group S_n (the group of permutations of n indices). We do not do so here, but just formulate below the outcome of the theory. Here we summarize:

- for identical particles the state space is not $L^2(\mathbb{R}^{3n})$, but rather a subspace of it, defined by certain symmetry properties with respect to permutations of the particles.

Now, we summarize relevant elements of the theory. Consider n fermions of spin r. Denote by α partitions of the integer n into ordered positive integers $2r+1 \geq \alpha_1 \geq \alpha_2 \geq \cdots \geq \alpha_k \geq 1$, $\alpha_1 + \alpha_2 + \cdots + \alpha_k = n$. Denote the set of such α's by A_r. These can be visualized as arrangements of n squares into k rows containing $\alpha_1, \alpha_2, \ldots, \alpha_k$ squares each, called *Young diagrams*. For example, with $n = 3$:

In particular, for spin one half, $r = \frac{1}{2}$, we have one- and two-column Young diagrams. To a Young diagram, one associates a *Young tableau* by filling in squares with particles.

We associate with a given Young diagram α, the space $\mathcal{H}^\alpha$ of wave functions which are symmetric with respect to permutations of particles in the same row, and antisymmetric with respect to those in the same column, of some tableau T (e.g. a canonical one) associated with α. One can show that the subspaces $\mathcal{H}^\alpha$ are mutually orthogonal and satisfy

$$\mathcal{H}_{\text{fermi}} = \oplus_{\alpha \in A_r} \mathcal{H}^\alpha. \qquad (4.35)$$

In technical terms, irreducible representations of the symmetric group S_n are in one-to-one correspondence with $\alpha \in A_r$, and on the subspace $\mathcal{H}^\alpha$, the representation of S_n is a multiple of the irreducible one labeled by α.

The spaces (4.30) and (4.31) correspond to the irreducible representations of S_n, with the one-row and one-column Young diagrams, respectively. The remaining subspaces $\mathcal{H}^\alpha$ correspond to fermions with higher spins.

Irreducible representations of S_n can be connected, via Weyl's theory of dual pairs of groups, to irreducible representations of $SU(2)$ carried by the spin space $\mathbb{C}^{(2r+1)n}$, and therefore determine the total spin of the corresponding wave functions.

Remark 4.6 *In dimension 2, the state spaces of n identical particles is classified not by irreducible representations of the permutation group S_n, but by irreducible representations of the braid group B_n. See [111, 112, 262] for reviews, and a book with a collection of articles and commentary.*

4.7 Supplement: Hamiltonian Formulation of Classical Mechanics

In this supplement we discuss briefly the hamiltonian formulation of classical mechanics. For more details and extensions see Mathematical Supplement 26.

The starting point here is the *principle of minimal action*: solutions of physical equations minimize (more precisely, make stationary) certain functionals, called *action functionals*. It is one of the basic principles of modern physics. The action functional, $S : \phi \mapsto S(\phi)$, is the integral of the form

$$S(\phi) := \int_0^T L\big(\phi(t), \dot\phi(t)\big)\, dt, \qquad (4.36)$$

where $L : X \times V \to \mathbb{R}$ is a twice differentiable function, called a *Lagrangian function*, or *Lagrangian*, V is a finite-dimensional inner-product vector space, called the space of velocities, X is an open subset of V, called the position, or configuration, space, and $\phi(t)$ is a differentiable path in X.

The functional $S(\phi)$ is defined on the space of paths $\mathcal{P}_{a,b} = \{\phi \in C^1([0, T]; X) \mid \phi(0) = a, \phi(T) = b\}$, for some $a, b \in X$. We can write $\mathcal{P}_{a,b}$ in the form $\mathcal{P}_{a,b} = \{\phi_0 + \phi \mid \phi \in \mathcal{P}_0\} \equiv \phi_0 + \mathcal{P}_0$, where ϕ_0 is a fixed element of $\mathcal{P}_{a,b}$, and $\mathcal{P}_0 := \mathcal{P}_{0,0}$. Now, $\mathcal{P}_0$ is a vector space and consequently $\mathcal{P}_{a,b}$ is an affine space. For classical mechanics, $L(x, v) = \frac{1}{2}mv^2 - V(x) : \mathbb{R}^3 \times \mathbb{R}^3 \to \mathbb{R}$ and $\phi(t) = x(t)$, and therefore the action functional is given by

$$S(\phi) = \int_0^T \big(\frac{m}{2}|\dot\phi|^2 - V(\phi)\big)\, dt. \qquad (4.37)$$

The dynamics is given by critical points of this functional, and the dynamical equation is the equation for critical points

$$S'(\phi) = 0, \tag{4.38}$$

called the Euler-Lagrange equation. Here $S'(\phi)$ is the *differential* or *variational derivative* of S at ϕ, defined as follows. Let V^* be the space dual to V, i.e. the space of bounded linear functions or functionals on V (see Section 25.1). The action of $l \in V^*$ is denoted as $\langle l, v \rangle$. We define $S'(\phi) : [0, T] \to V^*$ by the equation

$$\int_0^T \langle S'(\phi)(t), \xi(t) \rangle dt = \frac{d}{d\lambda} S(\phi_\lambda)|_{\lambda=0}, \tag{4.39}$$

where $\phi_\lambda := \phi + \lambda\xi$, for any $\xi \in \mathcal{P}_0$.

For simplicity, let $V = \mathbb{R}^m$, and denote by $\partial_\phi L$ and $\partial_{\dot\phi} L$ the gradients of $L(\phi, \dot\phi)$ in the variables ϕ and $\dot\phi$, respectively. Using (4.36), we compute $\frac{d}{d\lambda} S(\phi_\lambda)|_{\lambda=0} = \int_0^T \left(\partial_{\dot\phi} L(\phi)\dot\xi + \partial_\phi L(\phi)\xi \right) dt$. Integrating the first term on the r.h.s. by parts and using that $\xi(0) = \xi(T) = 0$, we arrive at $\frac{d}{d\lambda} S(\phi_\lambda)|_{\lambda=0} = \int_0^T \left(-\partial_t(\partial_{\dot\phi} L(\phi)) + \partial_\phi L(\phi) \right) \xi(t) \, dt$. If $\frac{d}{d\lambda} S(\phi_\lambda)|_{\lambda=0} = 0$, for any $\xi \in \mathcal{P}_0$, this implies the equation

$$-\partial_t(\partial_{\dot\phi} L(\phi, \dot\phi)) + \partial_\phi L(\phi, \dot\phi) = 0 \tag{4.40}$$

(see Section 26.2 for more details and generalizations). Applying this to the classical mechanics action functional (4.37), we arrive at Newton's equation of Classical Mechanics:

$$m\ddot\phi = -\nabla V(\phi).$$

Suppose now that the dynamics of a system are determined by the action principle, with a differentiable Lagrangian function/functional $L : X \times V \to \mathbb{R}$ defined on a space $X \times V$. We define the energy of a path ϕ as

$$\text{energy}\,(\phi) := \frac{\partial L}{\partial \dot\phi} \cdot \dot\phi - L.$$

We have

Lemma 4.7 (Conservation of energy) If $\bar\phi$ is a critical path of the action (4.36), then the energy is conserved, energy $(\bar\phi) =$ const.

Proof. We compute

$$\frac{d}{ds}\left(\frac{\partial L}{\partial \dot\phi} \cdot \dot\phi - L \right) = \frac{\partial^2 L}{\partial \dot\phi^2} \ddot\phi \cdot \dot\phi + \frac{\partial^2 L}{\partial \dot\phi \partial \phi} \dot\phi \cdot \dot\phi$$
$$+ \frac{\partial L}{\partial \dot\phi} \cdot \ddot\phi - \frac{\partial L}{\partial \dot\phi} \cdot \ddot\phi - \frac{\partial L}{\partial \phi} \cdot \dot\phi$$
$$= \left(\frac{d}{ds} \frac{\partial L}{\partial \dot\phi} - \frac{\partial L}{\partial \phi} \right) \cdot \dot\phi.$$

Since for $\bar\phi$, the expression on the right hand side vanishes, the result follows.
$\square$

We pass now to the new variables $(x, v) \to (x, k)$, where k, as a function of x and v, is given by

$$k = \partial_v L(x, v). \tag{4.41}$$

Note that k belongs to the space, V^*, dual to V (see Section 25.1). We assume that the equation (4.41) has a unique solution for v (which holds if L is *strictly convex* in the second variable, which means

$$L(x, sv + (1 - s)v') < sL(x, v) + (1 - s)L(x, v'), \ \forall s \in (0, 1), \tag{4.42}$$

$\forall x \in X$, $v, v' \in V$, and it is guaranteed by the inequality $d_v^2 L(x, v) > 0$, $\forall x \in X$, $v \in V$). With this in mind, we express the energy in the new variables, as

$$h(x, k) = \big(\langle k, v \rangle - L(x, v) \big) \big|_{v : \partial_v L(x, v) = k} \tag{4.43}$$

where the notation $\langle \cdot, \cdot \rangle$ stands for the coupling between V^* and V (Section 25.1). This defines the *Hamiltonian function/functional*, $h : X \times V^* \to \mathbb{R}$.

Theorem 4.8 If $L(x, v)$ and $h(x, k)$ are related by (4.43), then the Euler-Lagrange equation (4.40) for the action (4.36) is equivalent to the equations (called Hamilton's equations)

$$\dot{x} = \partial_k h(x, k), \ \dot{k} = -\partial_x h(x, k). \tag{4.44}$$

Proof. Assume (4.40) is satisfied. First, we note that the equations (4.43) and (4.41) imply $\partial_k h(x, k) = v + (k - \partial_v L(x, v)) \partial_k v = v = \dot{x}$ and $\partial_x h(x, k) = -\partial_x L(x, v) + (k - \partial_v L(x, v)) \partial_x v = -\partial_x L(x, v)$. Now, the last equation and the equations (4.40) and (4.41) imply $\dot{k} = -\partial_x h(x, k)$, which gives (4.44). Now, assuming (4.44), we obtain (4.40) from (4.41), $\partial_x h(x, k) = -\partial_x L(x, \dot{x})$ and $\dot{k} = -\partial_x h(x, k)$. $\square$

Applying (4.43) to the classical mechanics Lagrangian, $L(x, v) = \frac{mv^2}{2} - V(x)$, we arrive at the classical Hamiltonian function

$$h(x, k) = \frac{1}{2m} |k|^2 + V(x), \tag{4.45}$$

which leads to Hamilton's equations $\dot{x} = \frac{1}{m} k$, $\dot{k} = -\partial_x V(x)$, which are equivalent to Newton's equations.

Another example of a Lagrangian is that for a classical relativistic particle (in units with speed of light $c = 1$):

$$L(x, \dot{x}) = -m\sqrt{1 - \dot{x}^2}.$$

($ds = \sqrt{1 - \dot{x}^2} \, dt$ is the proper time of the particle.) The generalized momentum in this case is $k = \partial_v L(x, v) = \frac{m\dot{x}}{\sqrt{1 - \dot{x}^2}}$ and the Hamiltonian is $h(x, k) = \sqrt{|k|^2 + m^2}$.

Problem 4.9 Prove this.

Next, we recognize that Hamilton's equations (4.44) can be written as

$$\dot{z}_t = \{z, h\}(z_t), \qquad z_t = (x(t), k(t)) \in Z = X \times V^*, \tag{4.46}$$

where now x_j and k_j together are thought of as a path in the *phase space* Z, and for any pair f, g of differentiable functions on Z, $\{f.g\}$ denotes the function

$$\{f, g\} = \nabla_x f \cdot \nabla_k g - \nabla_k f \cdot \nabla_x g. \tag{4.47}$$

The map $(f, g) \to \{f, g\}$, given by (4.47), is a bilinear map, which has the following properties: for any functions f, g, and h from Z to $\mathbb{R}$,

1. $\{f, g\} = -\{g, f\}$ (skew-symmetry)
2. $\{f, \{g, h\}\} + \{g, \{h, f\}\} + \{h, \{f, g\}\} = 0$ (the Jacobi identity)
3. $\{f, gh\} = \{f, g\}h + g\{f, h\}$. (Leibniz rule)

Bilinear maps, $(f, g) \to \{f, g\}$, having these properties, are called *Poisson brackets*. The map (4.47) also obeys $\{f, g\} = 0 \ \forall \ g \implies f = 0$. Poisson brackets with the latter property are said to be non-degenerate. Note that a space of smooth functions (or functionals), together with a Poisson bracket, has the structure of a *Lie Algebra*.

The space Z together with a Poisson bracket on $C^\infty(Z, \mathbb{R})$ is called a Poisson space. A *Hamiltonian system* is a pair: a Poisson space, $(Z, \{\cdot, \cdot\})$, and a Hamiltonian function, $h : Z \to \mathbb{R}$. In this case Hamilton's equations are given by (4.46). Classical mechanics of one particle is a Hamiltonian system with the phase space $Z = \mathbb{R}^3 \times \mathbb{R}^3$, with bracket (4.47) and the Hamiltonian (4.45).

Remark 4.10 Our definition of a Hamiltonian system differs from the standard one in using the Poisson bracket instead of a symplectic form. The reason for using the Poisson bracket is its direct relation to the commutator.

Definition 4.11 Functions on a Poisson space, $(Z, \{\cdot, \cdot\})$, are called *classical observables*.

The classical evolution of observables is given by $f(z, t) = f(z_t)$, where z_t is the solution of (4.46) with the initial condition z. Note that $f(z, t)$ solves the equation

$$\frac{d}{dt} f(z, t) = \{f, h\}(z, t).$$

with the initial condition $f(z, 0) = f(z)$. Conversely, a solution of this equation with an initial condition $f(z)$ is given by $f(z, t) = f(z_t)$.

Problem 4.12 Prove this.

The equation above implies that an observable $f(z)$ is *conserved* or is a *constant of motion*, i.e. $f(z_t)$ is independent of t, if and only if its Poisson bracket with the Hamiltonian h vanishes: $\{f, h\} = 0$.

The map $t \to \phi_t$, where $\phi_t(z) := z_t$ and z_t is the solution of (4.46) with the initial condition z, is called the *flow* associated with the differential equation (4.46).

Particle coupled to an external electro-magnetic field. As an application of the above machinery, we consider a system of charges interacting with an electro-magnetic field. Of course, if the external field is purely electric, then it is a potential field, and fits within the framework we have considered already.

Suppose, then, that a magnetic field B, and an electric field, E, are present, $B, E : \mathbb{R}^{3+1} \mapsto \mathbb{R}^3$. The law of motion of a classical particle of mass m and electric charge e is given by Newton's equation with the Lorentz force,

$$m\ddot{x}(t) = eE(x(t), t) + \frac{e}{c}\dot{x} \wedge B(x(t), t). \qquad (4.48)$$

To find a hamiltonian formulation of this equation we first derive it from the minimum action principle. We know from the theory of electro-magnetism (Maxwell's equations) that the electric and magnetic fields, E and B, can be expressed in terms of the vector potential $A : \mathbb{R}^{3+1} \mapsto \mathbb{R}^3$, and the scalar potential $\Phi : \mathbb{R}^3 \mapsto \mathbb{R}$ via

$$E = -\nabla\Phi - \frac{1}{c}\partial_t A, \qquad B = \text{curl}A. \qquad (4.49)$$

The action functional which gives (4.48) is given by

$$S(\phi) = \int_0^T \left(\frac{m}{2}|\dot{\phi}|^2 - e\Phi(\phi) + \frac{e}{c}\dot{\phi} \cdot A(\phi)\right) dt. \qquad (4.50)$$

Indeed, we find the Euler-Lagrange equation (see (4.40)) for this functional. Using that $L(\phi, \dot{\phi}) = \frac{m}{2}|\dot{\phi}|^2 - e\Phi(\phi) + \frac{e}{c}\dot{\phi} \cdot A(\phi)$, we compute

$$\partial_{\dot{\phi}}L(\phi, \dot{\phi}) = m\dot{\phi} + \frac{e}{c}A(\phi), \ \partial_\phi L(\phi, \dot{\phi}) = -e\nabla\Phi(\phi) - \frac{e}{c}\nabla_\phi(\dot{\phi} \cdot A(\phi)).$$

Plug this into (4.40) and use the relations $\frac{d}{dt}A(\phi) = \partial_t A(\phi) + (\dot{\phi}\cdot\nabla)A(\phi)$, $\nabla(v \cdot A) - (v \cdot \nabla)A(\phi) = v \wedge \text{curl}A$ and (4.49) to obtain (4.48).

Now, the generalized momentum is $k = m\dot{x} + \frac{e}{c}A(x)$ and the classical Hamiltonian function is $h_{A,\Phi}(x, k) = k \cdot v - L(x, v)|_{m\dot{x} = k - \frac{e}{c}A(x)}$, which gives

$$h_{A,\Phi}(x, k) = \frac{1}{2m}(k - \frac{e}{c}A(x))^2 + e\Phi(x).$$

Defining the Poisson bracket as in (4.47), we arrive at the hamiltonian formulation for a particle of mass m and charge e moving in the external electric and magnetic fields E and B.

5

Uncertainty Principle and Stability of Atoms and Molecules

One of the fundamental implications of quantum theory is the *uncertainty principle*. It states that certain pairs of physical quantities cannot be measured simultaneously with arbitrary accuracy. Mathematically, it follows from the fact that the corresponding observables do not commute. The key example here is provided by the observables of position and momentum. They do not commute, as seen from the commutation relation:

$$\frac{i}{\hbar}[p, x] = 1 \tag{5.1}$$

(this is a matrix equation, meaning $\frac{i}{\hbar}[p_j, x_k] = \delta_{jk}$). In this chapter, we establish precise mathematical statements of the uncertainty principle for the position and momentum observables.

5.1 The Heisenberg Uncertainty Principle

We consider a particle in a state ψ and think of the observables x and p as random variables with probability distributions $|\psi|^2$ and $|\hat{\psi}|^2$ respectively. Recall that the means of x_j and p_j in the state ψ ($\in D(x_j) \cap D(p_j)$) are $\langle x_j \rangle_\psi$ and $\langle p_j \rangle_\psi$, respectively. The *dispersion* of x_j in the state ψ is

$$(\Delta x_j)^2 := \langle (x_j - \langle x_j \rangle_\psi)^2 \rangle_\psi$$

and the dispersion of p_j is

$$(\Delta p_j)^2 := \langle (p_j - \langle p_j \rangle_\psi)^2 \rangle_\psi.$$

Theorem 5.1 (The Heisenberg uncertainty principle) For any state $\psi \in D(x_j) \cap D(p_j)$,

$$\Delta x_j \Delta p_j \geq \frac{\hbar}{2}. \tag{5.2}$$

© Springer-Verlag GmbH Germany, part of Springer Nature 2020
S. J. Gustafson and I. M. Sigal, *Mathematical Concepts of Quantum Mechanics*, Universitext, https://doi.org/10.1007/978-3-030-59562-3_5

Proof. The basic ingredient is commutation relation (5.1). For notational simplicity, we assume $\langle x \rangle_\psi = \langle p \rangle_\psi = 0$. Note that for two self-adjoint operators A, B, and $\psi \in D(A) \cap D(B)$,

$$\langle i[A, B] \rangle_\psi = -2\,\mathrm{Im}\langle A\psi, B\psi \rangle. \tag{5.3}$$

So assuming ψ is normalized ($\|\psi\| = 1$), and $\psi \in D(x_j) \cap D(p_j)$, we obtain

$$1 = \langle \psi, \psi \rangle = \langle \psi, \frac{i}{\hbar}[p_j, x_j]\psi \rangle = -\frac{2}{\hbar}\,\mathrm{Im}\langle p_j\psi, x_j\psi \rangle$$
$$\leq \frac{2}{\hbar}|\langle p_j\psi, x_j\psi \rangle| \leq \frac{2}{\hbar}\|p_j\psi\|\|x_j\psi\| = \frac{2}{\hbar}(\Delta p_j)(\Delta x_j).$$

This does it. $\square$

What are the states which minimize the uncertainty, i.e. the l.h.s. of (5.2)? Clearly, the states for which $-\mathrm{Im}\langle p_j\psi, x_j\psi \rangle = \|p_j\psi\|\|x_j\psi\|$ would do this. This equality is satisfied by states obeying the equation $p_j\psi = i\mu x_j\psi$ for some $\mu > 0$. Solving the latter equation we obtain $\psi_\mu := \prod_j \left(\frac{\mu_j}{\pi\hbar}\right)^{1/4} e^{-\sum \mu_j x_j^2/2\hbar}$ for any $\mu_j > 0$. Of course, shifting these states in coordinate and momentum as

$$\psi_{yq\mu\varphi} := \prod_j \left(\frac{\mu_j}{\pi\hbar}\right)^{1/4} e^{(iq\cdot x + i\varphi - \sum \mu_j (x_j - y_j)^2/2)/\hbar} \tag{5.4}$$

would give again states minimizing the uncertainty principle. These states are called *coherent states*. They are obtained by scaling and translating the Gaussian state $\phi := (\pi\hbar)^{-3/4} e^{-|x|^2/2\hbar}$ and can be written as

$$\psi_{yq\mu\varphi} := e^{i(q\cdot y + \varphi)/\hbar} T_{yq}\phi, \tag{5.5}$$

where T_{yq} is the shift operator in coordinate and momentum: $T_{yq} := e^{i(q\cdot x - p\cdot y)/\hbar}$. Note that

$$\langle \psi_{yq\mu\varphi}, x_j\psi_{yq\mu\varphi} \rangle = y_j, \quad \langle \psi_{yq\mu\varphi}, p_j\psi_{yq\mu\varphi} \rangle = q_j, \tag{5.6}$$

$$\langle \psi_{yq\mu\varphi}, (x - y_j)^2\psi_{yq\mu\varphi} \rangle = \frac{\hbar}{2\mu}, \tag{5.7}$$

$$\langle \psi_{yq\mu\varphi}, (p - q_j)^2\psi_{yq\mu\varphi} \rangle = \frac{\hbar\mu}{2}. \tag{5.8}$$

Problem 5.2 Prove (5.5) - (5.8).

5.2 A Refined Uncertainty Principle

The following result is related to the Heisenberg uncertainty principle.

Theorem 5.3 (Refined uncertainty principle) On $L^2(\mathbb{R}^3)$,

$$-\Delta \geq \frac{1}{4|x|^2}.$$

Recall that for operators A, B, we write $A \geq 0$ if $\langle \psi, A\psi \rangle \geq 0$ for all $\psi \in D(A)$, and we write $A \geq B$ if $A - B \geq 0$. So by the above statement, we mean $\langle \psi, (-\Delta - \frac{1}{4|x|^2})\psi \rangle \geq 0$ for ψ in an appropriate dense subspace of $D(-\Delta)$. We will prove it for $\psi \in C_0^\infty(\mathbb{R}^d)$.

Proof. We will ignore domain questions, leaving these as an exercise for the reader. Compute

$$\sum_{j=1}^d i[|x|^{-1}p_j|x|^{-1}, x_j] = |x|^{-1} \sum_{j=1}^d i[p_j, x_j]|x|^{-1} = \hbar d|x|^{-2}$$

($d =$ space dimension $= 3$). Hence, using (5.3) again,

$$\hbar d \||x|^{-1}\psi\|^2 = -2\sum_{j=1}^d \text{Im}\langle |x|^{-1}p_j|x|^{-1}\psi, x_j\psi \rangle$$

and therefore, using

$$p_j|x|^{-1} = |x|^{-1}p_j + [p_j, |x|^{-1}] = |x|^{-1}p_j + i\hbar\frac{x_j}{|x|^3},$$

we obtain

$$\hbar(d-2)\||x|^{-1}\psi\|^2 = -2\,\text{Im}\sum_{j-1}^d \langle p_j\psi, \frac{x_j}{|x|^2}\psi \rangle.$$

Now the Cauchy-Schwarz inequality implies

$$|\sum_{j=1}^d \langle p_j\psi, \frac{x_j}{|x|^2}\psi \rangle| \leq \langle \psi, |p|^2\psi \rangle^{1/2} \||x|^{-1}\psi\|$$

(prove this!), which together with the previous equality gives

$$\hbar|d-2|\||x|^{-1}\psi\|^2 \leq 2\langle \psi, |p|^2\psi \rangle^{1/2}\||x|^{-1}\psi\|.$$

Squaring this, and observing that $\||x|^{-1}\psi\|^2 = \langle \psi, |x|^{-2}\psi \rangle$ and $\langle \psi, |p|^2\psi \rangle = \hbar^2\langle \psi, -\Delta\psi \rangle$ yields (for $d \geq 3$)

$$\langle \psi, -\Delta\psi \rangle \geq \frac{|d-2|^2}{4}\langle \psi, \frac{1}{|x|^2}\psi \rangle$$

which, for $d = 3$, implies the desired result. $\square$

5.3 Application: Stability of Atoms and Molecules

In classical mechanics, atoms and molecules are unstable: as the electrons orbit the nuclei, they radiate away energy and fall onto the nuclei. The demonstration that this is not so in quantum mechanics was one of the first triumphs of the theory.

The statement that a quantum system (with the particles interacting vial Coulomb potentials) is *stable with respect to collapse* is expressed mathematically by the property that the Hamiltonian H (and therefore the energy), is bounded from below. First we demonstrate the stability for the simplest quantum system – the hydrogen-type ion. The latter is described by the Schrödinger operator

$$H_{hyd} = -\frac{\hbar^2}{2m}\Delta - \frac{e^2 Z}{|x|}$$

on $L^2(\mathbb{R}^3)$ (m and $-e$ are the mass and charge of the electron respectively and eZ is the charge of the nucleus). The refined uncertainty principle gives $H_{hyd} \geq \frac{\hbar^2}{8m|x|^2} - \frac{e^2 Z}{|x|}$. The right hand side reaches its minimum at $|x|^{-1} = 4me^2 Z/\hbar^2$ and so

$$H_{hyd} \geq -\frac{2me^4 Z^2}{\hbar^2}. \tag{5.9}$$

Thus, the energy of the hydrogen atom is bounded from below, and the electron does not collapse onto the nucleus.

Now we show how to extend the argument above to an arbitrary system of electrons and nuclei, by considering for simplicity an atom with N electrons and an infinitely heavy nucleus. According to Section 4.5, the Schrödinger operator of this system is given by

$$H_{at} = \sum_{j=1}^{N}(-\frac{\hbar^2}{2m}\Delta_{x_j} - \frac{e^2 Z}{|x_j|}) + \frac{1}{2}\sum_{i \neq j}\frac{e^2}{|x_i - x_j|}, \tag{5.10}$$

acting on $L^2(\mathbb{R}^{3N})$. Here m and $-e$ are the electron mass and charge, $x = (x_1, \ldots, x_N)$ are the electron coordinates, and the term $\sum_1^N(-\frac{e^2 Z}{|x_j|})$ on the r.h.s. is the sum of Coulomb interaction potentials between the electrons and the nuclei (Ze is the charge of the nucleus) and the last term, between the electrons. For a neutral atom, $Z = N$. For the moment, we ignore the fact that the electrons are fermions.

To prove a lower bound on H_{at} we observe that the electron-electron interaction potential is positive and therefore we have the following lower bound for H_{at}:

$$H_{at} \geq H_{at}^{indep}, \tag{5.11}$$

where $H_{at}^{indep} := \sum_1^N(-\frac{\hbar^2}{2m}\Delta_{x_j} - \frac{e^2 Z}{|x_j|})$. Using, for each term on the r.h.s., the bound (5.9), we obtain

$$H_{at} \geq -N\frac{2m(e^2Z)^2}{\hbar^2}.$$

(5.12)

This bound works but it is rather rough. First of all it ignores the electron-electron repulsion, but most importantly, it ignores that the electrons are fermions (see Section 4.6). In section 14.1, we show how to take into account the second feature and how to improve on the first one.

6

Spectrum and Dynamics

Given a quantum observable (a self-adjoint operator) A, what are the possible values A can take in various states of the system? The interpretation of $\langle \psi, A\psi \rangle$ as mean value of the observable A in a state ψ, which is validated by quantum experiments, leads to the answer. It is the spectrum of A. The most important observable is the energy – the Schrödinger operator, H, of a system. Hence the spectrum of H gives the possible values of the energy.

The goal of this chapter is to develop techniques for finding the spectra of Schrödinger operators. A rough classification of the spectra is into discrete and continuous (also called essential) components. Such a classification is related to the space-time behaviour of solutions of the corresponding Schrödinger equations. Thus our main thrust is toward describing these components. We begin by presenting the general theory, and then proceed to applications. In particular, we explain how the spectrum of a Schrödinger operator gives us important information about the solutions of the Schrödinger equation. Details about the general machinery are presented in Mathematical Supplement 25.

6.1 The Spectrum of an Operator

We begin by giving some key definitions and statements related to the spectrum. More details can be found in Section 25.8. (Note we will often omit the identity operator $\mathbf{1}$ in expressions like $A - z\mathbf{1}$.)

Definition 6.1 The *spectrum* of an operator A on a Hilbert space $\mathcal{H}$ is the subset of $\mathbb{C}$ given by

$$\sigma(A) := \{\lambda \in \mathbb{C} \mid A - \lambda \text{ is not invertible (has no bounded inverse)}\}.$$

The complement of the spectrum of A in $\mathbb{C}$ is called the *resolvent set* of A: $\rho(A) := \mathbb{C}\backslash\sigma(A)$. For $\lambda \in \rho(A)$, the operator $(A - \lambda)^{-1}$, called the *resolvent* of A, is well-defined and bounded.

The following exercise asks for the spectrum of our favorite operators.

© Springer-Verlag GmbH Germany, part of Springer Nature 2020
S. J. Gustafson and I. M. Sigal, *Mathematical Concepts of Quantum Mechanics*, Universitext, https://doi.org/10.1007/978-3-030-59562-3_6

Problem 6.2 Prove that as operators on $L^2(\mathbb{R}^d)$ (with their natural domains),

1. $\sigma(1) = \{1\}$.
2. $\sigma(p_j) = \mathbb{R}$.
3. $\sigma(x_j) = \mathbb{R}$.
4. $\sigma(V) = \overline{\mathrm{Ran}(V)}$, where V is the multiplication operator on $L^2(\mathbb{R}^d)$ by a continuous function $V(x) : \mathbb{R}^d \to \mathbb{C}$.
5. $\sigma(-\Delta) = [0, \infty)$.
6. $\sigma(f(p)) = \overline{\mathrm{Ran}(f)}$, where $f(p) := \mathcal{F}^{-1} f \mathcal{F}$ with $f(k)$ the multiplication operator on $L^2(\mathbb{R}^d)$ by a continuous function $f(k) : \mathbb{R}^d \to \mathbb{C}$.

The following two results state important facts about the spectrum. They are proved in Section 25.8.

Theorem 6.3 The spectrum $\sigma(A) \subset \mathbb{C}$ is a closed set.

Theorem 6.4 The spectrum of a self-adjoint operator is real: A self-adjoint $\implies \sigma(A) \subset \mathbb{R}$.

6.2 Spectrum and Measurement Outcomes

We saw in Section 3.2 that the postulates of quantum mechanics imply that the probability that, in the state ψ, the physical quantity corresponding to an observable A is in a set Ω is given by Born's rule

$$\mathrm{Prob}_\psi(A \in \Omega) = \langle \psi, \chi_\Omega(A)\psi \rangle. \tag{6.1}$$

Here $\chi_\Omega(\lambda)$ is the characteristic function of the set Ω (i.e. $\chi_\Omega(\lambda) = 1$, if $\lambda \in \Omega$ and $\chi_\Omega(\lambda) = 0$, if $\lambda \notin \Omega$) and the operator $\chi_\Omega(A)$ can be defined using an operator calculus (see Section 25.11, Eq (25.43)). We define it in an important special case below. Conclusion (6.1) is validated by quantum experiments.

For A self-adjoint, the operator $\chi_\Omega(A)$ is also self-adjoint. Motivated by (6.1), we call it the probability observable. Mathematically, it is a projection, i.e. it satisfies (see Exercise 6.5 below)

$$\chi_\Omega(A)^2 = \chi_\Omega(A). \tag{6.2}$$

One can show, using the formula (25.43) for functions of operators given in the Mathematical Supplement (we sketch the proof below), that

$$\chi_\Omega(A) = \chi_{\Omega \cap \sigma(A)}(A). \tag{6.3}$$

This relation implies

$$\chi_\Omega(A) \neq 0 \Leftrightarrow \Omega \cap \sigma(A) \neq \emptyset. \tag{6.4}$$

The latter property shows that the spectrum is equal to the set of all the values the physical quantity represented by the observable A could take, or in short

$$\sigma(A) = \{\text{values of the physical quantity represented by } A\}. \tag{6.5}$$

Due to the relation (6.1), the above equation suggests that $\sigma(A)$ can be interpreted as the set of all possible values of the observable A. The most important observable is the energy – the Schrödinger operator, H, of a system. Hence the spectrum of H gives the possible values of the energy.

Now, we sketch a proof of (6.3). First, we recall from Section 25.11, Eq (25.43), that for $\Omega \subset \mathbb{C}$ with $\partial\Omega \subset \rho(A)$, we use that $\chi_\Omega(\lambda) := \frac{1}{2\pi i} \oint_\gamma (z - \lambda)^{-1} dz$ to define $\chi_\Omega(A)$ by the Cauchy integral

$$\chi_\Omega(A) := \frac{1}{2\pi i} \oint_\gamma (z - A)^{-1} dz, \tag{6.6}$$

where $\gamma = \partial\Omega$. The integral here can be understood either as the Bochner integral of vector-valued functions with values in a Banach space, i.e. as the norm limit Riemann sum approximations, or in the weak sense:

$$\langle \psi, \chi_\Omega(A)\phi \rangle := \frac{1}{2\pi i} \oint_\gamma \langle \psi, (A - z)^{-1}\phi \rangle dz,$$

for any $\psi, \phi \in \mathcal{H}$. Here we use the fact that the knowledge of $\langle \psi, T\phi \rangle$ for all ψ and ϕ determines the operator T uniquely. We call $\chi_\Omega(A)$ the characteristic function of the operator A. This definition coincides with all other definitions for $\chi_\Omega(A)$.

Problem 6.5 Prove (6.2), or, more generally, that $(f \cdot g)(A) = f(A)g(A)$.

We claim that with this definition we have (6.3) (assuming for simplicity that $\partial\Omega \subset \rho(A)$). Indeed, if $\Omega \cap \sigma(A) = \emptyset$, then the resolvent $(A - z)^{-1}$ is analytic in Ω, and therefore by Cauchy's theorem $\oint_\gamma (A - z)^{-1} dz = 0$, and so $\chi_\Omega(A) = 0$.

One familiar reason for $A - \lambda$ not to be invertible, is that $(A - \lambda)\psi = 0$ has a non-zero solution, $\psi \in \mathcal{H}$. In this case we say that λ is an *eigenvalue* of A and ψ is called a corresponding *eigenvector*.

If ψ is a normalized eigenvector of A with eigenvalue λ, then the equation $(A - z)\psi = (\lambda - z)\psi$ implies $(A - z)^{-1}\psi = (\lambda - z)^{-1}\psi$ for any z in the resolvent set of A. This and definition (6.6) imply

$$\chi_\Omega(A)\psi = \chi_\Omega(\lambda)\psi,$$

which, together with definition (6.1) gives

$$\text{Prob}_\psi(A \in \Omega) = 1 \text{ if } \lambda \in \Omega \text{ and } = 0 \text{ if } \lambda \notin \Omega. \tag{6.7}$$

Hence measuring the corresponding physical observable in the state ψ will always give the same answer, λ.

Now consider a superposition $\psi = \sum_i a_i \psi_i$ of several normalized eigenvectors ψ_i of A, with eigenvalues λ_i. By Problem 6.6(2) below, these ψ_i can be chosen to be mutually orthogonal. Then by $\chi_\Omega(A)\psi_i = \chi_\Omega(\lambda_i)\psi_i$ and the orthonormality of the ψ_i, we have

$$\mathrm{Prob}_\psi(A \in \Omega) = \sum_i |a_i|^2 \, \mathrm{Prob}_{\psi_i}(A \in \Omega),$$

which is consistent with an interpretation of the the the coefficients a_i as probability ampltudes.

Similar statements, though technically more complicated, could be made about the essential spectra discussed below.

As a partial generalization of (6.7), we have that for any $\psi \in \mathrm{Ran}\chi_{\Omega'}(A)$, we have

$$\mathrm{Prob}_\psi(A \in \Omega) = 1 \text{ if } \Omega' \subset \Omega \text{ and } = 0 \text{ if } \Omega' \cap \Omega = \emptyset. \tag{6.8}$$

6.3 Classification of Spectra

We begin with the simplest type of spectrum. The *discrete spectrum* of an operator A is

$$\sigma_d(A) = \{\lambda \in \mathbb{C} \mid \lambda \text{ is an isolated eigenvalue of } A \text{ with finite multiplicity}\}$$

(isolated meaning some neighbourhood of λ is disjoint from the rest of $\sigma(A)$). Here the *multiplicity* of an eigenvalue λ is the dimension of the *eigenspace*

$$\mathrm{Null}(A - \lambda) := \{v \in \mathcal{H} \mid (A - \lambda)v = 0\}.$$

Problem 6.6 1. Show $\mathrm{Null}(A - \lambda)$ is a vector space.
 2. Show that if A is self-adjoint, eigenvectors of A corresponding to different eigenvalues are orthogonal.

The rest of the spectrum is called the *essential spectrum* of the operator A:

$$\sigma_{ess}(A) := \sigma(A) \backslash \sigma_d(A).$$

Remark 6.7 Some authors may use the terms "point spectrum" and "continuous spectrum" rather than (respectively) "discrete spectrum" and "essential spectrum'.

Problem 6.8 For the following operators on $L^2(\mathbb{R}^d)$ (with their natural domains), show that

1. $\sigma_{ess}(p_j) = \sigma(p_j) = \mathbb{R}$;
2. $\sigma_{ess}(x_j) = \sigma(x_j) = \mathbb{R}$;
3. $\sigma_{ess}(-\Delta) = \sigma(-\Delta) = [0, \infty)$.

Hint: Show that these operators do not have discrete spectrum.

Problem 6.9 Show that if $U : \mathcal{H} \to \mathcal{H}$ is unitary, then $\sigma(U^*AU) = \sigma(A)$, $\sigma_d(U^*AU) = \sigma_d(A)$, and $\sigma_{ess}(U^*AU) = \sigma_{ess}(A)$.

Problem 6.10 Let A be a self-adjoint operator on $\mathcal{H}$. If λ is an accumulation point of $\sigma(A)$, then $\lambda \in \sigma_{ess}(A)$. Hint: use the definition of the essential spectrum, and the fact that the spectrum is a closed set.

For a self-adjoint operator A the sets {span of eigenfunctions of A} and {span of eigenfunctions of A}$^\perp$, where

$$W^\perp := \{\psi \in \mathcal{H} \mid \langle \psi, w \rangle = 0 \ \forall \, w \in W\},$$

are invariant under A in the sense of the definition

Definition 6.11 A subspace $W \subset \mathcal{H}$ of a Hilbert space $\mathcal{H}$ is *invariant* under an operator A if $Aw \in W$ whenever $w \in W \cap D(A)$.

Problem 6.12 Assume A is a self-adjoint operator. Show that

1. If W is invariant under A, then so is $W^\perp$;
2. The span, V, of the eigenfunctions of A, and its orthogonal complement, $V^\perp$, are invariant under A;
3. Suppose further that A has only finitely many eigenvalues, all of them with finite multiplicity. Show that the restricted operator $A|_V$ has a purely discrete spectrum;
4. Show that the restricted operator $A|_{V^\perp}$ has a purely essential spectrum.

The spaces {span of eigenfunctions of A} and {span of eigenfunctions of A}$^\perp$ is said to be the *subspaces of the discrete and essential spectra* of A.

6.4 Bound and Decaying States

We show how the classification of the spectrum introduced in the previous section is related to the space-time behaviour of solutions of the Schrödinger equation

$$i\hbar \frac{\partial \psi}{\partial t} = H\psi$$

with given initial condition

$$\psi|_{t=0} = \psi_0,$$

where H is a self-adjoint Schrödinger operator acting on $L^2(\mathbb{R}^3)$. Naturally, we want to distinguish between states which are localized for all time, and those whose essential support moves off to infinity. We assume all functions below are normalized.

Suppose first that $\psi_0 \in \{$ span of eigenfunctions of $H\}$. Then for any $\epsilon > 0$, there is an R such that

$$\inf_t \int_{|x| \leq R} |\psi|^2 \geq 1 - \epsilon. \tag{6.9}$$

To see this, note that if $H\psi_0 = \lambda\psi_0$, then $e^{-\frac{iHt}{\hbar}}\psi_0 = e^{-\frac{i\lambda t}{\hbar}}\psi_0$, and so

$$\int_{|x| \geq R} |\psi|^2 = \int_{|x| \geq R} |\psi_0|^2 \to 0$$

as $R \to \infty$. Such a ψ is called a *bound state*, as it remains essentially localized in space for all time. A proof of (6.9) in the general case is given at the end of this section.

On the other hand, if

$$\psi_0 \in \{\text{span of eigenfunctions of } H\}^\perp,$$

where $W^\perp := \{\psi \in \mathcal{H} \mid \langle \psi, w \rangle = 0 \ \forall \, w \in W\}$, then for all R,

$$\int_{|x| \leq R} |\psi|^2 \to 0 \tag{6.10}$$

as $t \to \infty$, in the sense of *ergodic mean*. Convergence $f(t) \to 0$ in ergodic mean as $t \to \infty$ means that

$$\frac{1}{T} \int_0^T f(t)dt \to 0$$

as $T \to \infty$. This result is called the *Ruelle theorem*. We sketch the proof below (see, eg, [73, 162] for a complete proof). Such a state, ψ, is called a *decaying state*, as it eventually leaves any fixed ball in space.

We say that a system in a bound state is stable, while in a decaying state, unstable. This notion differs from the notion of stability in dynamical systems. Indeed, solutions of the Schrödinger equation are always dynamically orbitally stable (see Section 14.1 for the relevant definitions). The notion of stability here characterizes the space-time behaviour of the system, whether it stays essentially in a bounded region of the space, or falls apart with fragments departing from each other.

Now, we saw above that solutions of the Schödinger equation with initial conditions in the discrete spectral subspace describe bound states, those in the essential spectral subspace describe decaying states. Hence the spectral classification for a Schrödinger operator H leads to a space-time characterization of the quantum mechanical evolution, $\psi = e^{-iHt/\hbar}\psi_0$. Namely, the classification of the spectrum into discrete and essential parts corresponds to a classification of the dynamics into localized (bound) states and locally-decaying (scattering) states.

Finally, we give proof of (6.9) and (6.10). *Proof of equation (6.9) in the general case*: if

$$\psi_0 \in \{ \text{ span of eigenfunctions of } H\},$$

then ψ_0 can be written as $\psi_0 = \sum_j a_j \psi_j$ where $a_j \in \mathbb{C}$, $\sum_j |a_j|^2 = 1$, (we assume in what follows that $\|\psi_0\| = 1$) and $\{\psi_j\}$ is an orthonormal set ($\langle \psi_j, \psi_k \rangle = \delta_{jk}$) of eigenfunctions of H: $H\psi_j = \lambda_j \psi_j$. We will assume that the above sum has only a finite number of terms, say $\psi_0 = \sum_{j=1}^{N} a_j \psi_j$. Otherwise an additional continuity argument is required below. The solution $\psi = e^{-iHt/\hbar}\psi_0$ can be written as

$$\psi = \sum_{j=1}^{N} e^{-i\lambda_j t/\hbar} a_j \psi_j.$$

Multiplying this equation by the characteristic function of the exterior of the R−ball, $|x| \geq R$, taking the L^2−norm and using the triangle inequality, we obtain

$$\left(\int_{|x|\geq R} |\psi|^2 \right)^{1/2} \leq \sum_{j=1}^{N} |a_j| \left(\int_{|x|\geq R} |\psi_j|^2 \right)^{1/2}. \tag{6.11}$$

To estimate the second factor on the right hand side, for any $\varepsilon > 0$, we choose R such that for all j,

$$\left(\int_{|x|\geq R} |\psi_j|^2 \right)^{1/2} \leq \frac{\varepsilon}{\sqrt{N}}.$$

Using this estimate and applying the Cauchy-Schwarz inequality to the sum on the right hand side of (6.11) and using that $\sum_{j=1}^{N} 1 = N$ and $\sum_j |a_j|^2 = 1$, we obtain

$$\left(\int_{|x|\geq R} |\psi|^2 \right)^{1/2} \leq \varepsilon$$

and so (6.9) follows. $\qquad\square$

Sketch of proof of (6.10): in this proof we display the time dependence as a subindex t. We suppose the potential $V(x)$ is bounded, and so as a multiplication operator $\|V\| = \|V\|_\infty$, and therefore $V \geq -\|V\|\mathbf{1}$. Since also $-\Delta \geq 0$, the operator H is bounded below: $H \geq -\|V\|\mathbf{1}$. Let $\lambda > \|V\| + 1$ so that $H + \lambda > 1$, and let $\phi_t := (H + \lambda)\psi_t = e^{-iHt}(H + \lambda)\psi_0$. Defining

$$B := (-\Delta + \lambda)(H + \lambda)^{-1}, \tag{6.12}$$

and using that $B = \mathbf{1} - V(H + \lambda)^{-1}$, which shows that B is bounded, we see that $\psi_t = (-\Delta + \lambda)^{-1} B \phi_t$. Let χ_Ω denote the characteristic function of a set $\Omega \subset \mathbb{R}^3$ and let $K(x, y)$ be the integral kernel of the operator $\chi_\Omega(-\Delta + \lambda)^{-1}$. Using (6.12) and the notation $K_x(y) := K(x, y)$, we obtain that

$$\chi_\Omega \psi_t = \chi_\Omega (-\Delta + \lambda)^{-1} B\phi_t = \langle K_x, B\phi_t \rangle = \langle B^* K_x, \phi_t \rangle. \qquad (6.13)$$

We use the notation $\| \cdot \|_x$ for the L^2−norm in the variable x and claim now that

$$\frac{1}{T} \int_0^T dt \| \langle B^* K_x, \phi_t \rangle \|_x^2 \to 0, \qquad \text{as} \qquad T \to \infty. \qquad (6.14)$$

To prove this claim we use the fact that $\int |K(x,y)|^2 dx dy < \infty$ to show that

$$|\langle B^* K_x, \phi_t \rangle| \le \| B^* K_x \| \|\phi_t\| \le \| B^* \| \| K_x \| \| (H + \lambda)\psi_0 \| \in L^2(dx)$$

(uniformly in t). Next, we want to prove that

$$\forall x, \ \frac{1}{T} \int_0^T dt |\langle B^* K_x, \phi_t \rangle|^2 \to 0, \text{ as } T \to \infty. \qquad (6.15)$$

Then (6.14) follows from interchange of t− and x−integration on the l.h.s. and the Lebesgue dominated convergence theorem.

The proof of (6.15) is a delicate one. First note that, since $\psi_0 \perp$ the eigenfunctions of H, so is $\phi_0 := (H + \lambda)\psi_0$ (*show this*). Next, we write

$$|\langle f, \phi_t \rangle|^2 = \langle f \otimes \bar{f}, \phi_t \otimes \bar{\phi}_t \rangle = \langle f \otimes \bar{f}, e^{-itL/\hbar} \phi_0 \otimes \bar{\phi}_0 \rangle. \qquad (6.16)$$

where L is an operator acting on $\mathcal{H} \otimes \mathcal{H}$ given by $L := H \otimes 1 - 1 \otimes H$. The relation $\phi_0 \perp$ (the eigenfunctions of H) implies that $\phi_0 \otimes \bar{\phi}_0 \perp$ (the eigenfunctions of $H \otimes 1 - 1 \otimes H$). Hence we can compute the time integral of the r.h.s.:

$$\frac{1}{T} \int_0^T dt \langle f \otimes \bar{f}, e^{-itL/\hbar} \phi_0 \otimes \bar{\phi}_0 \rangle = \frac{1}{T} \langle f \otimes \bar{f}, \frac{e^{-iTL/\hbar} - 1}{-iL/\hbar} \phi_0 \otimes \bar{\phi}_0 \rangle. \qquad (6.17)$$

The delicate point here is to show that, for nice f and ϕ_0, the r.h.s. is well-defined and is bounded by $CT^{-\delta}$, $\delta > 0$. We omit showing this here (this can be done, for example. by using the spectral decomposition theorem, see [244]). Then (6.15) holds, which completes the proof (since this shows that for any bounded set $\Omega \subset \mathbb{R}^3$ we have that $\| \chi_\Omega \psi_t \| \to 0$, in the sense of ergodic mean, as $t \to \infty$, which is equivalent to (6.10)). $\square$

6.5 Spectra of Schrödinger Operators

We now want to address the question of how to characterize the essential spectrum of a self-adjoint operator A. Is there a characterization of $\sigma_{ess}(A)$ similar to that of $\sigma_d(A)$ in terms of some kind of eigenvalue problem? In particular, we address this question for Schrödinger operators. We begin with

Definition 6.13 Let A be an operator on $L^2(\mathbb{R}^d)$. A sequence $\{\psi_n\} \subset D(A)$ is called a *spreading sequence* for A and λ if

1. $\|\psi_n\| = 1$ for all n
2. for any bounded set $B \subset \mathbb{R}^d$, $\text{supp}(\psi_n) \cap B = \emptyset$ for n sufficiently large
3. $\|(A - \lambda)\psi_n\| \to 0$ as $n \to \infty$.

Clearly, if not for the second condition, a sequence consisting of a repeated eigenfunction would fit this definition. The second condition implies that we can choose a subsequence $\{\psi'_n\}$ so that

$$\text{supp}\,\psi'_n \subset \{|x| \geq R_n\},$$

with $R_n \to \infty$. In what follows, we always assume that the sequence $\{\psi_n\}$ has this property.

As an example, a spreading sequence on $L^2(\mathbb{R}^3)$ for $-\frac{\hbar^2}{2m}\Delta$ and any $\lambda \geq 0$ is given by $\psi_n(x) := n^{-3/2}f(|x|/n)e^{ik\cdot x/\hbar}$, where f is a smooth (suitably normalized) function s.t. $f(x) = 0$ for $|x| \leq 1$, and $\frac{1}{2m}|k|^2 = \lambda$.

Problem 6.14 Show that the sequence constructed above is a spreading sequence for $-\frac{\hbar^2}{2m}\Delta$ and $\lambda = \frac{1}{2m}|k|^2$.

Theorem 6.15 (Weyl-Zhislin theorem) If $H = -\frac{\hbar^2}{2m}\Delta + V$ is a Schrödinger operator, with real potential $V(x)$ which is continuous and bounded from below, then

$$\sigma_{ess}(H) = \{\lambda \mid \text{there is a spreading sequence for } H \text{ and } \lambda\}.$$

We sketch a proof of this result later.

Now, we describe the spectra of self-adjoint Schrödinger operators

$$H = -\frac{\hbar^2}{2m}\Delta + V.$$

Our first result covers the case when the potential tends to zero at infinity. Recall that we have proved in Section 2.2 that H is self-adjoint.

Theorem 6.16 Let $V : \mathbb{R}^d \to \mathbb{R}$ be continuous, with $V(x) \to 0$ as $|x| \to \infty$. Then $\sigma_{ess}(H) = [0, \infty)$ (so H can have only negative isolated eigenvalues, possibly accumulating at 0).

Proof. We have, by the triangle inequality,

$$\|(H - \lambda)\psi_n\| - \|V\psi_n\| \leq \|(-\frac{\hbar^2}{2m}\Delta - \lambda)\psi_n\| \leq \|(H - \lambda)\psi_n\| + \|V\psi_n\|.$$

Suppose $\{\psi_n\}$ is a spreading sequence. Then the term $\|V\psi_n\|$ goes to zero as $n \to \infty$ because V goes to zero at infinity and $\{\psi_n\}$ is spreading. So λ is in the essential spectrum of H if and only if $\lambda \in \sigma_{ess}(-\frac{\hbar^2}{2m}\Delta)$. We have (see Problem 6.8) $\sigma_{ess}(-\frac{\hbar^2}{2m}\Delta) = \sigma_{ess}(-\Delta) = [0, \infty)$, and consequently $\sigma_{ess}(H) = [0, \infty)$. $\square$

Problem 6.17 Extend Theorems 6.15 and 6.16 to real potentials $V(x)$ satisfying (2.8).

The bottom of the essential spectrum (0 in the present case), is called the *ionization threshold*, since above this energy the particle is no longer localized, but moves freely.

Our next theorem covers *confining* potentials – that is, potentials which increase to infinity with x. As we have mentioned in Section 2.2, Schrödinger operators with such potentials are self-adjoint.

Theorem 6.18 Let $V(x)$ be a continuous function on $\mathbb{R}^d$ satisfying $V(x) \geq 0$, and $V(x) \to \infty$ as $|x| \to \infty$. Then $\sigma(H)$ consists of isolated eigenvalues $\{\lambda_n\}_{n=1}^{\infty}$ with $\lambda_n \to \infty$ as $n \to \infty$. Moreover, the corresponding eigenfunctions form an (orthonormal) basis in $L^2(\mathbb{R}^d)$.

Proof. Suppose λ is in the essential spectrum of H, and let $\{\psi_n\}$ be a corresponding spreading sequence. Then as $n \to \infty$,

$$0 \leftarrow \langle \psi_n, (H - \lambda)\psi_n \rangle = \langle \psi_n, -\frac{\hbar^2}{2m}\Delta\psi_n \rangle + \langle \psi_n, V\psi_n \rangle - \lambda$$

$$= \frac{\hbar^2}{2m}\int |\nabla\psi_n|^2 + \int V|\psi_n|^2 - \lambda$$

$$\geq \inf_{y \in \text{supp}(\psi_n)} V(y) - \lambda \to \infty$$

(because $\{\psi_n\}$ is spreading), which is a contradiction. Thus the essential spectrum is empty.

Now we show that H has an infinite number of eigenvalues, with finite multiplicities, tending to $+\infty$. Suppose, on the contrary, that H has a finite number of eigenvalues with finite multiplicities, and let f be a non-zero element of L^2 which is orthogonal to all the eigenfunctions of H. Then, for any $z \in \mathbb{C}$, the equation $(H - z)\psi = f$ has a unique solution $\psi = (H - z)^{-1}f$ in $\{\text{span of eigenfunctions of } H\}^{\perp}$. The function $(H - z)^{-1}f$ is analytic in $z \in \mathbb{C}$ and satisfies $\|(H - z)^{-1}f\| \leq |\text{Im } z|^{-1}\|f\|$. Hence, by a straightforward extension of the Liouville theorem, $(H - z)^{-1}f = 0$, a contradiction. Since the essential spectrum is empty, the eigenvalues have finite multiplicities and cannot accumulate, and since $H \geq 0$, they must tend to $+\infty$. The above argument shows also that the eigenfunctions of H form a basis. $\square$

Problem 6.19 What are the essential spectra, and what are the possible locations of the discrete spectra, of the following operators (justify your answer):
(a) $H = -\frac{\hbar^2}{2m}\Delta - 10|x|^3 + |x|^4$,
(b) $H = -\frac{\hbar^2}{2m}\Delta - (1 + |x|)^{-2}$.

Sketch of the proof of Theorem 6.15. We will prove the theorem for Schrödinger operators with H_0-bounded potentials V – that is, potentials satisfying the estimate (2.8).

Let $\{\psi_n\}$ be a spreading sequence for H and λ, and let $\phi_n = \frac{(H-\lambda)\psi_n}{\|(H-\lambda)\psi_n\|}$. Evidently, $\|\phi_n\| = 1$. Since $(H-\lambda)^{-1}\phi_n = \frac{\psi_n}{\|(H-\lambda)\psi_n\|}$, and $\|(H-\lambda)\psi_n\| \to 0$, we obtain that

$$\|(H-\lambda)^{-1}\phi_n\| \to \infty,$$

as $n \to \infty$. Therefore $(H-\lambda)^{-1}$ is unbounded, which implies that $\lambda \in \sigma(H)$. We will prove that $\lambda \notin \sigma_d(H)$. Indeed, suppose on the contrary that $\lambda \in \sigma_d(H)$. Let M denote the eigenspace of λ. Then $(H-\lambda)$ is invertible on $M^\perp$ (show this). Let P and $P^\perp$ be the orthogonal projections on M and $M^\perp$, respectively. (For the definition and properties of projection operators see Section 25.7.) We have $P + P^\perp = 1$. Since the sequence $\{\psi_n\}$ is spreading and the operator P can be written as $P = \sum_j |\phi_j\rangle\langle\phi_j|$, where $\{\phi_n\}$ is an orthonormal basis in $M = \mathrm{Ran}P$, we have $\|P\psi_n\| \to 0$ and therefore $\|P^\perp\psi_n\| \to 1$. Hence for the normalized sequence $\psi_n^\perp := \frac{P^\perp\psi_n}{\|P^\perp\psi_n\|}$, we have that $(H-\lambda)\psi_n^\perp \to 0$ and therefore $(H-\lambda)$ is not invertible on $M^\perp$, a contradiction. Hence $\lambda \notin \sigma_d(H)$ and therefore $\lambda \in \sigma_{ess}(H)$.

Suppose now that $\lambda \in \sigma_{ess}(H)$. Then there is a sequence ϕ_n with $\|\phi_n\| = 1$ and $\|(H-\lambda)^{-1}\phi_n\| \to \infty$. Let $\psi_n = \frac{(H-\lambda)^{-1}\phi_n}{\|(H-\lambda)^{-1}\phi_n\|}$. We claim that for every bounded set Ω we have that $\|\chi_\Omega\psi_n\| \to 0$ as $n \to \infty$. Indeed, we can assume without loss of generality that $V \geq 0$, which implies that $(H+1)$ is invertible. So

$$\chi_\Omega\psi_n = \chi_\Omega(-\Delta+1)^{-1}(-\Delta+1)(H+1)^{-1}(H+1)\psi_n. \tag{6.18}$$

Now, we have

$$\chi_\Omega(-\Delta+1)^{-1}f = \int K(x,y)f(y)dy, \tag{6.19}$$

with $K \in L^2(\mathbb{R}^3 \times \mathbb{R}^3)$.

Problem 6.20 Use properties of the Fourier transform to show that $K(x,y) = \chi_\Omega(x)G(x-y)$, where $G(y) = C\frac{e^{-|y|}}{|y|}$ and C is a constant.

Define $B := (-\Delta+1)(H+1)^{-1}$. The relation

$$B = 1 - V(H+1)^{-1}, \tag{6.20}$$

and the assumption that V is H_0-bounded, imply that B is bounded. Using this definition, and using that

$$(H+1)\psi_n = \frac{\phi_n}{\|(H-\lambda)^{-1}\phi_n\|} + (\lambda+1)\psi_n, \tag{6.21}$$

we obtain that

$$\chi_\Omega\psi_n = \chi_\Omega(-\Delta+1)^{-1}B(\lambda+1)\psi_n + \chi_\Omega(-\Delta+1)^{-1}B\frac{\phi_n}{\|(H-\lambda)^{-1}\phi_n\|}. \tag{6.22}$$

We consider the first term on the r.h.s.. We write this term as $(\lambda+1)\langle K_x, B\psi_n\rangle$, where $K_x(y) := K(x,y)$. Since B is bounded, we have that

$\forall x$, $\langle K_x, B\psi_n \rangle = \langle B^*K_x, \psi_n \rangle$. If the vector B^*K_x were in the domain of $(H - \lambda)^{-1}$, then we would have $\langle K_x, B\psi_n \rangle = \frac{1}{r_n}\langle (H - \lambda)^{-1}B^*K_x, \phi_n \rangle$, where $r_n := \|(H - \lambda)^{-1}\phi_n\| \to \infty$, and therefore

$$\forall x, \ \langle K_x, B\psi_n \rangle \to 0. \tag{6.23}$$

In general, B^*K_x might not be in the domain of $(H - \lambda)^{-1}$, but we can show that the latter domain is dense and therefore B^*K_x can be approximated by vectors from this domain: $\forall \epsilon > 0$, there is f_ϵ such that $\|B^*K_x - f_\epsilon\| \leq \epsilon$. Now, one can modify the argument above to show (6.23). (A different argument showing (6.23) goes along the lines of the proof of Theorem 25.57 in the mathematical supplement.) Furthermore, since $\int |K(x,y)|^2 dx dy < \infty$, we have

$$|\langle B^*K_x, \psi_n \rangle| \leq \|B^*K_x\|\|\psi_n\| \leq \|B^*\|\|K_x\| \in L^2(dx).$$

Hence, by the Lebesgue dominated convergence theorem, we have, using the notation $\|\cdot\|_x$ for the L^2-norm in the variable x, that

$$\|\langle K_x, B(\lambda + 1)\psi_n \rangle\|_x = |\lambda + 1|\|\langle B^*K_x, \psi_n \rangle\|_x \to 0. \tag{6.24}$$

For the second term on the r.h.s. of (6.22), we use that B is bounded, to obtain that

$$\left\|\chi_\Omega(-\Delta + 1)^{-1}B\frac{\phi_n}{\|(H - \lambda)^{-1}\phi_n\|}\right\| \leq \|B\|\left\|\frac{\phi_n}{\|(H - \lambda)^{-1}\phi_n\|}\right\| \to 0. \tag{6.25}$$

Thus we conclude that for any bounded set $\Omega \subset \mathbb{R}^3$,

$$\|\chi_\Omega \psi_n\| \to 0. \tag{6.26}$$

Let $B(R)$ be a ball of radius R centered at the origin and let $R_m \to \infty$ as $m \to \infty$. Since $\|\chi_\Omega \psi_n\| \to 0$ as $n \to \infty$ for any bounded set Ω we have that $\forall m$, $\|\chi_{B(R_m)}\psi_n\| \to 0$ as $n \to \infty$. Hence using a diagonal procedure and passing to a subsequence, if necessary, we obtain that $\|\chi_{B_n}\psi_n\| \to 0$, as $n \to \infty$, for $B_n := B(R_{m(n)})$ and some subsequence $m(n)$, satisfying $m(n) \to \infty$, as $n \to \infty$. Let

$$f_n = \frac{(1 - \chi_{B_n})\psi_n}{\|(1 - \chi_{B_n})\psi_n\|}.$$

Evidently $\|f_n\| = 1$ and $\text{supp}(f_n) \cap \Omega = \emptyset$ for all bounded Ω provided that n is large (depending on Ω, of course). To finish the proof it suffices to show that

$$\|(H - \lambda)f_n\| \to 0. \tag{6.27}$$

To prove this relation, we compute $(H - \lambda)(1 - \chi_{B_n})\psi_n = (1 - \chi_{B_n})(H - \lambda)\psi_n - [H, \chi_{B_n}]\psi_n$ and $[H, \chi_{B_n}] = -((2\nabla\chi_{B_n} \cdot \nabla + \Delta\chi_{B_n})$. Therefore $\|(H - \lambda)(1 - \chi_{B_n})\psi_n\| \leq \|(1 - \chi_{B_n})(H - \lambda)\psi_n\| + \|[H, \chi_{B_n}]\psi_n\| \leq \|(1 - \chi_{B_n})\|\|(H -$

$\lambda)\psi_n\| + \|(2\nabla\chi_{B_n} \cdot \nabla + \Delta\chi_{B_n})\psi_n\| \to 0$, as $n \to \infty$. Since $\|(1 - \chi_{B_n})\psi_n\| \to 1$, (6.27) follows. Hence f_n is a spreading sequence for H and λ. $\square$

To conclude this section, we present a result on the spectra of Schrödinger operators on bounded domains with Dirichlet boundary conditions.

Theorem 6.21 Let Λ be a parallelepiped in $\mathbb{R}^d$, and V a continuous function on Λ. Then the Schrödinger operator $H = -\Delta + V$, acting on the space $L^2(\Lambda)$ with Dirichlet or periodic boundary conditions, has purely discrete spectrum, accumulating at $+\infty$.

To be precise, the operator "H on $L^2(\Lambda)$ with Dirichlet boundary conditions" should be understood as the unique self-adjoint extension of H from $C_0^\infty(\Lambda)$. This theorem is proved in Section 25.10.

6.6 Particle in a Periodic Potential

We consider a particle moving in a periodic potential. A primary example of this situation is an electron moving in the potential created by ions or atoms of a solid crystal lattice. Such a particle is described by a self-adjoint Schrödinger operator

$$H = -\frac{\hbar^2}{2m}\Delta + V$$

on $L^2(\mathbb{R}^3)$, with the potential $V(x)$, having certain periodicity properties which we now explain.

First, we identify the notion of physical crystal lattice with the mathematical *(Bravais) lattice*, $\mathcal{L}$, which is defined as the subset of $\mathbb{R}^3$ given by

$$\mathcal{L} = \{m_1 s_1 + m_2 s_2 + m_3 s_3 \mid m_1, m_2, m_3 \in \mathbb{Z}\}$$

for some three linearly independent vectors $s_1, s_2, s_3 \in \mathbb{R}^3$, called a *basis* of $\mathcal{L}$. A basis in $\mathcal{L}$ is not unique.

We say $V(x)$ is periodic with respect to a lattice $\mathcal{L}$ if $V(x + s) = V(x)$ for any $s \in \mathcal{L}$. This implies that the operator H commutes with the lattice translations,

$$T_s H = H T_s, \ \forall s \in \mathcal{L}, \tag{6.28}$$

where T_s is the translation operator, given by $T_s f(x) = f(x + s)$.

Problem 6.22 Show that T_s are unitary operators satisfying $T_t T_s = T_{t+s}$.

Due to (6.28), the operator H can be decomposed into a direct fiber integral, as follows. First, let Ω be the lattice cell given by $\Omega := \{x_1 s_1 + x_2 s_2 + x_3 s_3 \mid 0 \le x_1, x_2, x_3 \le 1\}$. Let $\mathcal{L}^*$ be the lattice dual to $\mathcal{L}$, i.e. the lattice with the basis s_1^*, s_2^*, s_3^*, satisfying $s_i^* \cdot s_j = \delta_{ij}$, and let Ω^* be a lattice cell in $\mathcal{L}^*$.

Now, we define the Hilbert space $\mathcal{H} = L^2(\Omega^*, dk; L^2(\Omega))$ of vector functions $\psi : \Omega^* \ni k \to \psi_k \in L^2(\Omega)$, with the standard inner product

$\langle \psi, \phi \rangle := \int_{\Omega^*} dk \langle \psi_k, \phi_k \rangle_{L^2(\Omega)}$, where dk is the standard Lebesgue measure on Ω^*, normalized so that $\int_{\Omega^*} dk = 1$. We call $\mathcal{H}$ the direct fiber integral, denoted as

$$\mathcal{H} = \int_{\Omega^*}^{\oplus} \mathcal{H}_k dk,$$

where $\mathcal{H}_k := L^2(\Omega)$, for each $k \in \Omega^*$, while its elements are written as $\psi = \int_{\Omega^*}^{\oplus} \psi_k dk$.

Given operators H_k, $k \in \Omega^*$, acting on $\mathcal{H}_k$, we define the operator $\psi_k(x) \rightarrow H_k \psi_k(x)$ on $\mathcal{H} = \int_{\Omega^*}^{\oplus} \mathcal{H}_k dk$ and denote it $\int_{\Omega^*}^{\oplus} H_k dk$, so that $\int_{\Omega^*}^{\oplus} H_k dk \int_{\Omega^*}^{\oplus} \psi_k dk = \int_{\Omega^*}^{\oplus} H_k \psi_k dk$.

Define the operator $U : L^2(\mathbb{R}^3) \rightarrow \mathcal{H}$ on smooth functions with compact support by the formula

$$(Uv)_k(x) = \sum_{t \in \mathcal{L}} \chi_k^{-1}(t) T_t v(x).$$

where $\chi_k(t) = e^{ik \cdot t}$, the character of $\mathcal{L}$, i.e., a homomorphism from $\mathcal{L} \rightarrow U(1)$ (see the remark below). We now have the following Bloch decomposition result:

Proposition 6.23 The operator U extends uniquely to a unitary operator and

$$UHU^{-1} = \int_{\Omega^*}^{\oplus} H_k dk \qquad (6.29)$$

where H_k, $k \in \Omega^*$, is the restriction of operator H to $\mathcal{H}_k$, with domain consisting of vectors $v \in \mathcal{H}_k \cap H^2$ satisfying the boundary conditions

$$T_t v(x) = \chi_k(t) v(x), \qquad (6.30)$$

for the basis elements $t = s_1, s_2, s_3$.

Proof. We begin by showing that U is an isometry on smooth functions with compact support. Using Fubini's theorem we calculate

$$\|Uv\|_{\mathcal{H}}^2 = \int_{\Omega^*} \|(Uv)_k\|_{\mathcal{H}_k}^2 dk = \int_{\Omega^*} \int_{\Omega} \left| \sum_{t \in \mathcal{L}} \chi_k^{-1}(t) T_y v(x) dx \right|^2 dk$$

$$= \int_{\Omega} \left(\sum_{t,s \in \mathcal{L}} T_t v(x) \overline{\tau_s v(x)} \int_{\Omega^*} \chi_k^{-1}(t) \chi_k(s) dk \right) dx$$

$$= \int_{\Omega} \sum_{t \in \mathcal{L}} |T_t v(x)|^2 dx = \int_{\mathbb{R}^2} |v(x)|^2 dx.$$

Hence $\|Uv\|_{\mathcal{H}} = \|v\|_{\mathcal{H}}$ and U extends to an isometry on all of $\mathcal{H}$. To show that U is in fact a unitary operator we define $U^* : \mathcal{H} \rightarrow \mathcal{H}$ by the formula

$$U^* g(x + t) = \int_{\Omega^*} \chi_k(t) g_k(x) dk,$$

for $t \in \mathcal{L}$ and $x \in \Omega$. Straightforward calculations show that U^* is the adjoint of U and that it too is an isometry, proving that U is a unitary operator.

For (6.29), we need to first show that $(Uv)_k$ is in the domain of H_k. For $(Uv)_k$ we have

$$
\begin{aligned}
T_t(Uv)_k(x) &= \sum_{s \in \mathcal{L}} \chi_k^{-1}(s) T_t T_s v(x) \\
&= \sum_{s \in \mathcal{L}} \chi_k^{-1}(s) T_{t+s} v(x) \\
&= \chi_k(t) \sum_{s \in \mathcal{L}} \chi_k^{-1}(t+s) T_{t+s} v(x) \\
&= \chi_k(t)(Uv)_k(x).
\end{aligned}
$$

Hence if $v \in D(H)$, then $Uv \in D(H_k)$. Next, we have that

$$
\begin{aligned}
(H_k(Uv)_k)(x) &= \sum_{t \in \mathcal{L}} \chi_k^{-1}(t) H T_t v(x) \\
&= \sum_{t \in \mathcal{L}} \chi_k^{-1}(t) T_t H v(x) \\
&= (UHv)_k(x),
\end{aligned}
$$

which establishes (6.29). $\square$

Since the resolvents of the operators H and H_k are related as $U(H - z)^{-1} U^{-1} = \int_{\Omega^*}^{\oplus} (H_k - z)^{-1} dk$, we deduce that

Theorem 6.24

$$
\sigma(H) = \bigcup_{k \in \Omega^*} \sigma(H_k). \tag{6.31}
$$

Since Ω is compact, by Theorem 6.21, the spectra of H_k are purely discrete, say, $\{\lambda_n(k)\}$. This shows that the spectrum of H is the union of the sets $\{\lambda_n(k) \mid k \in \Omega^*\}$, called the *bands*. This is a key result in solid state physics. Mathematically, it reduces the investigation of the essential spectrum of a periodic quantum hamiltonian to the eigenvalue problem (the Bloch eigenvalue problem)

$$
H_k \psi_k = \lambda(k) \psi_k, \quad \psi_k \in \mathcal{H}_k. \tag{6.32}
$$

Remark: eigenvalue problem (6.32) involves boundary conditions (6.30). We can define the Hilbert space $\mathcal{H}_k$ without reference to a specific fundamental cell (which is not unique) and incorporating boundary conditions (6.30) directly. Namely, we define

$$
\mathcal{H}_k := \{v \in L^2_{\mathrm{loc}}(\mathbb{R}^3) \mid T_t v = \chi_k(t) v \; \forall t \in \mathcal{L}\}, \tag{6.33}
$$

with an inner product that of $L^2(\Omega)$ for some fundamental cell Ω.

To connect the above theory with the Bloch theory of solid state physics, we consider eigenvalue problem (6.32) for the hamiltonian H_k on $\mathcal{H}_k$.

Since ψ_k satisfy $T_t \psi_k = \chi_k(t)\psi_k \ \forall t \in \mathcal{L}$, we have that $e^{-ik\cdot x}\psi_k(x)$ is $\mathcal{L}$-periodic. Indeed, by the property of the exponentials, we have $T_s e^{-ik\cdot x}\psi_k(x) = e^{-ik\cdot x - ik\cdot s}T_s \psi_k(x) = e^{-ik\cdot x}e^{-ik\cdot s}\chi_k(t)e^{ik\cdot s}\psi_k(x) = e^{-ik\cdot x}\psi_k(x)$. Hence the eigenfunctions $\psi_k(x)$ have the representation

$$\psi_k(x) = e^{ik\cdot x}\phi_k(x),$$

where $\phi_k(x)$ are $\mathcal{L}$-periodic. This result is the celebrated Bloch theorem of condensed matter physics.

Remark: equation (6.30) is an eigen-equation for the operators T_s.

Remark: the dual group to $\mathcal{L}$ is the group consisting of all characters of $\mathcal{L}$, i.e., all homomorphisms from $\mathcal{L} \to U(1)$. Explicitly, for $k \in \Omega^*$, we have the character χ_k given by

$$\chi_k(t) = e^{ik\cdot t}.$$

Since $\chi_{k+k'}(t) = \chi_k(t), \forall k' \in \mathcal{L}^*$, the dual group of $\mathcal{L}$ can be identified with the fundamental cell, Ω^*, of $\mathcal{L}^*$.

6.7 Angular Momentum

In this section we study the angular momentum operators (or angular momentum observables, or just angular momenta)

$$L_j = (x \times p)_j, \tag{6.34}$$

where $p = -i\hbar\nabla$, as usual. The term is justified in Section 3.4 on conservation laws, by observing that $\frac{1}{i\hbar}L_j$ are the generators of the rotation group $SO(3)$. They satisfy the commutation relations

$$[L_k, L_l] = i\hbar\epsilon^{klm}L_m. \tag{6.35}$$

Here ϵ^{klm} is the Levi-Chivita symbol: $\epsilon^{123} = 1$ and ϵ^{klm} changes sign under the permutation of any two indices.

Problem 6.25 *Prove* (6.35). *(Hint: Use that $[p_j, x_k] = -i\hbar\delta_{ij}$.)*

We define the squared magnitude of the angular momentum, $L^2 = L_1^2 + L_2^2 + L_3^2$. Note that L^2 commutes with L_k:

$$[L^2, L_k] = 0, \ \forall k. \tag{6.36}$$

Problem 6.26 *Prove* (6.36). *(Hint: Use that $[p_j, x_k] = -i\hbar\delta_{ij}$.)*

Furthermore, L^2 is a homogeneous degree zero operator, in the sense that it commutes with the scalings $T_\lambda^{\text{scal}} : \psi(x) \to \lambda^{3/2}\psi(\lambda x)$, $\lambda > 0$, and therefore it does not act on the radial variable $r = |x|$.

Theorem 6.27 *(a) The spectrum of the operator L^2 is purely discrete; (b) it consists of isolated eigenvalues $\lambda = \hbar^2 l(l+1)$, where $l = 0, 1, 2, \ldots$, of the multiplicities $2l + 1$.*

A proof of (a) is somewhat involved and we skip it here (see however the remark after the proof).We will derive the proof of (b) from the following

Theorem 6.28 *Let L_j be operators satisfying the commutation relations (6.35). Assume the operator $L^2 = L_1^2 + L_2^2 + L_3^2$ has purely discrete spectrum. Then this spectrum consists of the eigenvalues $\lambda = \hbar^2 l(l+1)$, where $l = 0, \frac{1}{2}, 1, \frac{3}{2}, 2, \ldots$, of the multiplicities $2l + 1$.*

Proof (of Theorem 6.28). Let λ be an eigenvalue of L^2 with eigenspace V_λ. It follows from (6.36) that the eigenspace V_λ is invariant under L_3, i.e. $L_3 V_\lambda \subset V_\lambda$. (Indeed, if $L^2\phi_\lambda = \lambda\phi_\lambda$, then $L^2 L_3\phi_\lambda = L_3 L^2\phi_\lambda = \lambda L_k\phi_\lambda$.) Hence one can choose a basis in V_λ consisting of common eigenfunctions $\phi_{\lambda,\mu}$ for these operators:

$$L^2\phi_{\lambda,\mu} = \lambda\phi_{\lambda,\mu}, \ L_3\phi_{\lambda,\mu} = \mu\phi_{\lambda,\mu}.$$

(By choosing an arbitrary orthonormal basis in the finite dimensional subspace V_λ, one can reduce this problem to one for matrices.)

We still would like to use the remaining operators L_1, L_2 and so we form the combinations

$$L_\pm := L_1 \pm iL_2. \tag{6.37}$$

The virtue of these new operators is the following commutation relations, which are easily checked:

$$[L^2, L_\pm] = 0, \ [L_+, L_-] = 2\hbar L_3, \ [L_3, L_\pm] = \pm\hbar L_\pm. \tag{6.38}$$

Problem 6.29 *Prove (6.38). (Hint: Use that $[p_j, x_k] = -i\hbar\delta_{ij}$.)*

The last relation can be rewritten as $L_3 L_\pm = L_\pm(L_3 \pm \hbar)$ which shows that $L_\pm$ are raising/lowering operators in the following sense: apply the operators $L_\pm$ to the eigenfunctions $\phi_{\lambda,\mu}$ and use $L_3\phi_{\lambda,\mu} = \mu\phi_{\lambda,\mu}$ and $L_3 L_\pm = L_\pm(L_3 \pm \hbar)$ to obtain

$$L_3 L_\pm\phi_{\lambda,\mu} = L_\pm(L_3 \pm \hbar)\phi_{\lambda,\mu} = (\mu \pm \hbar)\phi_{\lambda,\mu}.$$

Hence we see that the operators $L_\pm$ raise/lower eigenvalues of L_3. More precisely, assuming the eigenfunctions $\phi_{\lambda,\mu}$ are non-degenerate (correspond to different eigenvalues), we have $L_\pm\phi_{\lambda,\mu} = C\phi_{\lambda,\mu\pm\hbar}$, for some constants C.

Next, we claim that $\mu^2 \le \lambda$, for given λ. Indeed, we use $L^2 = L_1^2 + L_2^2 + L_3^2 \ge L_3^2$ to conclude that $\mu^2 = \langle\phi_{\lambda,\mu}, L_3^2\phi_{\lambda,\mu}\rangle \le \langle\phi_{\lambda,\mu}, L^2\phi_{\lambda,\mu}\rangle = \lambda$. Furthermore, since $\bar{L}_3 = -L_3$, if μ is an eigenvalue of L_3 on V_λ, then so is $-\mu$.

Denote $\mu_* := \max\{|\mu| : \text{given } \lambda\}$. Then by the definition of μ_* and the raising/lowering property of the operators $L_\pm$, we find that

$$L_\pm \phi_{\lambda, \pm\mu_*} = 0.$$

Next, we use $L_\pm L_\mp = L_1^2 + L_2^2 \pm \hbar L_3$ to obtain the following relation between the operators L^2, $L_\pm$ and L_3:

$$L^2 = L_\pm L_\mp \mp \hbar L_3 + L_3^2. \tag{6.39}$$

Taking the expectation of this equation in the eigenfunction $\phi_{\lambda, -\mu_*}$, and using that $L_- \phi_{\lambda, -\mu_*} = 0$, we find

$$\lambda = \langle \phi_{\lambda, -\mu_*}, L^2 \phi_{\lambda, -\mu_*} \rangle = \langle \phi_{\lambda, -\mu_*}, (-\hbar L_3 + L_3^2) \phi_{\lambda, -\mu_*} \rangle = \hbar \mu_* + \mu_*^2.$$

This gives $\lambda = \mu_*(\mu_* + \hbar)$. If we denote $\mu_* = \hbar l$, then $\lambda = \hbar^2 l(l+1)$.

Finally, since, using the raising operator L_+, we can go from $\phi_{\lambda, -\mu_*}$ to ϕ_{λ, μ_*} in an integer number of steps, we conclude that $2\mu_*/\hbar$ is a nonnegative integer, $\mu_* = \hbar l$, $l \in \frac{1}{2}\mathbb{Z}^+$. For each l, the index m runs through the values $m = -l, \ldots, l$. Hence the multiplicity of the eigenvalue $\lambda = \hbar^2 l(l+1)$ of the operator L^2 is $2l + 1$.

Proof (of Theorem 6.27(b).). To prove statement (b), it remains to show that only integer l's are realized now. To this end, we note that we can solve the eigenvalue problem for L_3 easily by using spherical coordinates (r, θ, ϕ), in which $L_3 = \hbar \partial_\phi$, to obtain $L_3 e^{im\phi} = \hbar m e^{im\phi}$, so that $\mu = \hbar m$, where m must be an integer. So we have shown that every eigenvalue λ of L^2 is of the form $\lambda = \hbar^2 l(l+1)$, with $l = 0, 1, \ldots$, and has multiplicity $2l + 1$.

This proof allows one to construct all eigenfunctions of L^2, by solving the equation $L_- \phi_{\lambda, -\mu_*} = 0$, for $\lambda = \hbar^2 l(l+1)$ and $\mu_* = \hbar l$, and then using the raising operator L_+ to obtain all other eigenfunctions with the same l. (In spherical co-ordinates, we have $L_\pm = \hbar e^{\pm i\phi}(\pm \partial_\theta + i \cot \theta \partial_\phi)$.)

Thus we determined completely the eigenvalues of the operators L^2 and L_3: $\lambda = \hbar^2 l(l+1)$ and $\mu = \hbar m$, where $m, l \in \mathbb{Z}$ and $m = -l, \ldots, l$. Denote the joint eigenfunction of the operators L^2 and L_3, corresponding to the eigenvalues $\lambda = \hbar^2 l(l+1)$ and $\mu = \hbar m$, by $Y_l^m(\theta, \phi)$:

$$L^2 Y_l^m = \hbar^2 l(l+1) Y_l^m, \qquad L_3 Y_l^m = \hbar m Y_l^m. \tag{6.40}$$

The functions $Y_l^m(\theta, \phi)$ are the celebrated *spherical harmonics*. We know it is of the form $Y_l^m(\theta, \phi) = \tilde{Y}_l^m(\theta) e^{im\phi}$, $m \in \mathbb{Z}$, where $\tilde{Y}_l^m(\theta)$ is an eigenfunction of L^2: $L^2 \tilde{Y}_l^m(\theta) = \hbar^2 l(l+1) \tilde{Y}_l^m(\theta)$. More precisely,

$$Y_l^k(\theta, \phi) = c_{lk} P_l^{|k|}(\cos(\theta)) e^{ik\phi} \tag{6.41}$$

where $l = 0, 1, \ldots$; $k \in \{-l, -l+1, \ldots, l-1, l\}$; c_{lk} is a constant; and the *Legendre function* P_l^k can be written as

$$P_l^k(u) = \frac{(1-u^2)^{k/2}}{2^l l!}(\frac{d}{du})^{l+k}(u^2-1)^l.$$ (6.42)

This completes our excursion into the spectral theory of the operator L^2.

Problem 6.30 Show that in the spherical coordinates (r, θ, ϕ),

$$L^2(f(r)g(\theta, \phi)) = -f(r)\Delta_\Omega g(\theta, \phi),$$

where Δ_Ω is the Laplace-Beltrami operator on $\mathbb{S}^2$, given in spherical coordinates (θ, ϕ), by

$$\Delta_\Omega = \frac{1}{\sin(\theta)}\frac{\partial}{\partial\theta}(\sin(\theta)\frac{\partial}{\partial\theta}) + \frac{1}{\sin^2(\theta)}\frac{\partial^2}{\partial\phi^2}.$$

One could interpret Δ_Ω as a quantum hamiltonian of a 'free' particle on the sphere.

Using the Weyl criterion of essential spectrum and the fact that the sphere $\mathbb{S}^2$ is a compact space, one can show that Δ_Ω has purely discrete spectrum. Above, we found this spectrum explicitly using commutation relations, i.e. purely algebraic methods.

7

Special Cases and Extensions

In this chapter we consider several specific quantum systems which either fall outside of the general theory considered in Chapter 6, or are accessible to explicit computations providing the desired results. The results presented below not only illustrate some of the general arguments presented above, but also form a basis for our intuition about quantum behaviour.

7.1 The Square Well and Torus

The Square Well

We consider a one-dimensional potential well of finite depth V_0, and width a (see Fig. 7.1).

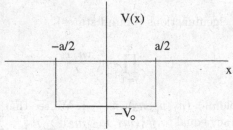

Fig. 7.1. The finite well.

The determination of the discrete and essential spectra is straightforward, and is left as an exercise.

Problem 7.1 Show

1. $\sigma_{ess}(H) = [0, \infty)$
2. $\sigma_d(H) \subset (-V_0, 0)$

© Springer-Verlag GmbH Germany, part of Springer Nature 2020
S. J. Gustafson and I. M. Sigal, *Mathematical Concepts of Quantum Mechanics*, Universitext, https://doi.org/10.1007/978-3-030-59562-3_7

3. the equations for the eigenvalues E, for $-V_0 < E < 0$, are

$$k \tan(ak/2) = K \qquad k \cot(ak/2) = -K$$

where

$$K = \sqrt{-\frac{2mE}{\hbar^2}} \qquad k = \sqrt{\frac{2m(E + V_0)}{\hbar^2}}$$

If we shift the well upward as $H_{V_0} := H + V_0$, then we can take the limit as $V_0 \to \infty$ to obtain the quantum hamiltonian H_∞ of a particle in an infinite well. In three dimensions, the potential of the infinite well can be thought of as

$$V(x) = \begin{cases} 0 & x \in W \\ \infty & x \notin W \end{cases}$$

where $W := [0, L]^3 \subset \mathbb{R}^3$. This means we take $\psi \equiv 0$ outside W, and that we impose Dirichlet boundary conditions

$$\psi|_{\partial W} = 0 \tag{7.1}$$

on the wave function inside W. It is a simple matter to solve the eigenvalue equation

$$-\frac{\hbar^2}{2m} \Delta \psi = E\psi \tag{7.2}$$

in W with the boundary condition (7.1), using the method of separation of variables. Doing so, we obtain eigenvalues (energy levels)

$$E_n = \frac{\hbar^2 \pi^2}{2mL^2} \sum_{j=1}^{3} n_j^2$$

with corresponding eigenfunctions (bound states)

$$\psi_n(x) = \prod_{j=1}^{3} \sin(\frac{\pi n_j x_j}{L})$$

for each integer triple $n = (n_1, n_2, n_3)$, $n_j \geq 1$. We see that the eigenvalue E_n occurs with degeneracy equal to $\#\{(m_1, m_2, m_3)| \sum m_j^2 = \sum n_j^2\}$. We remark that the ground-state (lowest) energy $E_{(1,1,1)} = \frac{3\hbar^2 \pi^2}{2mL^2}$ is non-degenerate.

The Torus

Now we consider a particle on a torus, say $\mathbb{T}^3 = \mathbb{R}^3/L\mathbb{Z}^3$, and with no external potential, $V \equiv 0$. This corresponds to considering the Schrödinger operator $-\frac{\hbar^2}{2m} \Delta$ in the cube $W = [0, L]^3$ (or any other fundamental cell of the lattice $L\mathbb{Z}^3$), with periodic boundary conditions. Solving the eigenvalue equation (7.2) with boundary conditions

$$\left\{ \begin{array}{l} \psi(x)|_{x_j=0} = \psi(x)|_{x_j=L} \\ \partial\psi/\partial x_k|_{x_j=0} = \partial\psi/\partial x_k|_{x_j=L}, \end{array} \right. \tag{7.3}$$

for all j, k, leads to (separation of variables again) eigenfunctions

$$\psi_n(x) = \prod_{j=1}^{3} \left\{ \begin{array}{l} \sin(\frac{2\pi n_j x_j}{L}) \\ \cos(\frac{2\pi n_j x_j}{L}) \end{array} \right\}$$

with eigenvalues

$$E_n = \frac{2\pi^2\hbar^2}{mL^2} \sum_1^3 n_j^2,$$

for $n_j \geq 0$. The ground state energy, $E_{(0,0,0)} = 0$ is non-degenerate, with eigenfunction $\psi_{(0,0,0)} \equiv 1$. The spacing between energy levels is greater than for the infinite well, and the degeneracy is higher.

Problem 7.2 Determine the spectrum of x, and the eigenvalues of p, on $L^2(W)$ with periodic boundary conditions. Challenge: show that p has no essential spectrum (hint – use the Fourier transform for periodic functions).

7.2 Motion in a Spherically Symmetric Potential

We consider a particle moving in a potential which is spherically symmetric, i.e. $V = V(|x|)$ depends only on $r = |x|$. Then the corresponding Schrödinger operator is

$$H = -\frac{\hbar^2}{2m}\Delta + V(|x|), \tag{7.4}$$

acting on the Hilbert space $L^2(\mathbb{R}^3)$.

If $V(r) \to 0$ as $r \to \infty$, then from Section 6.5, we know that H is self-adjoint and has essential spectrum filling in the half-line $[0, \infty)$. Hence the discrete spectrum, if it exists, is negative with only one possible accumulation point at 0. Our goal here is to use the spherical symmetry of V to reduce the eigenvalue problem for (7.4) to a problem in the radial variable $r = |x|$, only; i.e. an ODE problem.

Because the potential is spherically symmetric, the Schrödinger operator H commutes with rotations and, consequently, with their generators, the angular momenta

$$L_j = (x \times p)_j, \tag{7.5}$$

where $p = -i\hbar\nabla$ as usual. Hence the angular momentum is conserved.

The squared magnitude of the angular momentum operator $L^2 = L_1^2 + L_2^2 + L_3^2$ also commutes with H. Furthermore, L^2 is a homogeneous degree zero operator in the sense that it commutes with the scalings $T_\lambda^{scal} : \psi(x) \to \lambda^{3/2}\psi(\lambda x)$, $\lambda > 0$, and therefore it does not act on the radial variable $r = |x|$. Indeed, if we introduce spherical coordinates (r, θ, ϕ), where

$$x_1 = r\sin(\theta)\cos(\phi), \quad x_2 = r\sin(\theta)\sin(\phi), \quad x_3 = r\cos(\theta),$$

$0 \le \theta < \pi$, $0 \le \phi < 2\pi$, then L^2 acts only on the angles (θ, ϕ). A straightforward computation gives

$$-\hbar^2\Delta = -\hbar^2\Delta_r + \frac{1}{r^2}L^2, \qquad (7.6)$$

where Δ_r is the *radial Laplacian*, acting only on the radial variable $r = |x|$, given by

$$\Delta_r = \frac{1}{r^2}\frac{\partial}{\partial r}r^2\frac{\partial}{\partial_r} = \frac{\partial^2}{\partial r^2} + \frac{2}{r}\frac{\partial}{\partial_r}. \qquad (7.7)$$

Problem 7.3 Prove (7.6). (Hint: Compute $L^2 = L_1^2 + L_2^2 + L_3^2$ as follows. $L^2 = (x_2 p_3 - x_3 p_2)^2 +$ cyclic permutations $= (x_2 p_3)^2 + (x_3 p_2)^2 - x_2 p_3 x_3 p_2 - x_3 p_2 x_2 p_3 +$ cyclic permutations $= r^2 p^2 - \sum_j x_j^2 p_j^2 - \sum_{ij} p_i x_i x_j p_j + \sum_j p_j x_j^2 p_j$. This gives $L^2 = r^2 p^2 + (p \cdot x)(x \cdot p)$, from which one can easily derive (7.6).)

Problem 7.4 Show that in spherical coordinates (r, θ, ϕ) the Laplacian becomes

$$\Delta = \Delta_r + \frac{1}{r^2}\Delta_\Omega$$

where Δ_r is the radial Laplacian introduced in (7.7), and Δ_Ω is the Laplace-Beltrami operator on $\mathbb{S}^2$, given in spherical coordinates by

$$\Delta_\Omega = \frac{1}{\sin(\theta)}\frac{\partial}{\partial \theta}\left(\sin(\theta)\frac{\partial}{\partial \theta}\right) + \frac{1}{\sin^2(\theta)}\frac{\partial^2}{\partial \phi^2}.$$

We now return to our original problem - the eigenvalue problem for the spherically symmetric Schrödinger operator, (7.4). In view of (7.6) and the fact that L^2 commutes with Δ_r and H, we seek eigenfunctions of H in the separated-variables form

$$\psi(r, \theta, \phi) = R(r)Y_l^k(\theta, \phi)$$

where Y_l^k is an eigenfunction of L^2 corresponding to the eigenvalue $\hbar l(l+1)$ (see (6.40)), a spherical harmonic. Plugging this into the eigenvalue equation $H\psi = E\psi$, we obtain

$$\left(\frac{\hbar^2}{2m}[-\Delta_r + \frac{l(l+1)}{r^2}] + V(r)\right)R = ER. \qquad (7.8)$$

Usually, one cannot solve this equation explicitly with exception of a very few cases. The Schrödinger operator of the hydrogen atom is one of these cases, to which we now proceed.

7.3 The Hydrogen Atom

A hydrogen atom consists of a proton and an electron, interacting via a Coulomb force law. Let us make the simplifying assumption that the nucleus (the proton) is infinitely heavy, and so does not move. We also consider more generally the hydrogen-type atom, or ion, with nucleus of charge eZ, where e is the charge of the proton, and $-e$ that of the electron. Placing the nucleus at the origin, we have the electron moving under the influence of the external (Coulomb) potential $V(x) = -Ze^2/|x|$. The appropriate Schrödinger operator is therefore

$$H = -\frac{\hbar^2}{2m}\Delta - \frac{e^2 Z}{|x|}, \tag{7.9}$$

acting on the Hilbert space $L^2(\mathbb{R}^3)$. In Section 13.1 we will see how to reduce the problem of the more realistic hydrogen atom - when the nucleus has a finite mass (a two-body problem) - to the problem studied here.

As usual, we want to study the spectrum of H. The first step is to invoke Theorems 2.9 and 6.16 and Problem 6.17 to conclude that H is self-adjoint and has essential spectrum filling in the half-line $[0, \infty)$. Our goal, then, is to find the bound-states (eigenfunctions) and bound-state energies (eigenvalues). It is a remarkable fact that we can find these explicitly. Indeed, aside from the infinite well, the only multi-dimensional potentials for which the Schrödinger eigenvalue problem can be solved explicitly are the harmonic oscillator and the Coulomb potential.

Because the Coulomb potential is spherically symmetric (depends only on $r = |x|$), the results of the previous section can be applied to the Schrödinger operator (7.9). It is a remarkable fact that, in this case, we can solve the equation (7.8) explicitly and find the eigenvalues explicitly. Indeed, aside from the infinite well, the only multi-dimensional potentials for which the Schrödinger eigenvalue problem can be solved explicitly are the harmonic oscillator and the Coulomb potential.

For the Coulomb potential, the equation (7.8) reads

$$\left(\frac{\hbar^2}{2m}[-\Delta_r + \frac{l(l+1)}{r^2}] - e^2 Z/r\right)R = ER. \tag{7.10}$$

Eq. (7.10) is an ordinary differential equation (ODE) and is well-studied (see, eg, [186]). Without going into details, we remark that one can show (by power-series methods) that (7.10) has square-integrable solutions only for

$$n := \frac{e^2 Z}{\hbar}\sqrt{\frac{-m}{2E}} \in \{l+1, l+2, \ldots\}.$$

The corresponding eigenfunctions, R_{nl} are of the form

$$R_{nl}(r) = \rho^l e^{-\rho/2} F_{nl}(\rho)$$

where $\rho = \frac{2me^2Z}{n\hbar^2}r$, and F_{nl} is a polynomial.

In full, then, the solutions of the eigenvalue problem $H\psi = E\psi$ are

$$\psi(r,\theta,\phi) = R_{nl}(r)Y_l^k(\theta,\phi)$$

where

$$l = 0,1,2,\ldots;\quad k \in \{-l,-l+1,\ldots,l\};\quad n \in \{l+1,l+2,\ldots\};$$

and the eigenvalues are

$$E\ (=E_n) = -\left(\frac{me^4Z^2}{2\hbar^2}\right)\frac{1}{n^2}. \tag{7.11}$$

So we see that the hydrogen atom has an infinite number of bound states below the essential spectrum (which starts at zero), which accumulate at zero (this result is obtained in Section 8.3 by a general technique, without solving the eigenvalue problem). The ground state energy, attained when $l = k = 0, n = 1$, is $E_1 = -me^4Z^2/2\hbar^2$. An easy count finds the degeneracy of the energy level E_n to be

$$\sum_{l=0}^{n-1}(2l+1) = n^2. \tag{7.12}$$

Finally, we note that the expression (7.11) is in agreement with the empirical formula ("Balmer series")

$$\Delta E = R(\frac{1}{n_f^2} - \frac{1}{n_i^2}).$$

Here $1 \leq n_f < n_i$ are integers labeling the final and initial states of the atom in a radiation process, R is a constant, and ΔE is the difference of the two energy levels. This formula predates quantum mechanics, and was based on measurements of absorption and emission spectra.

7.4 The Harmonic Oscillator

The Hamiltonian of the quantum *harmonic oscillator* in r dimensions is

$$H_{ho} = -\frac{\hbar^2}{2m}\Delta + \frac{1}{2}m\sum_{i=1}^{r}\omega_i^2 x_i^2,$$

acting on the space $L^2(\mathbb{R}^r)$. By Theorem 6.18, $\sigma(H)$ consists of isolated eigenvalues, increasing to infinity. We will solve the eigenvalue problem explicitly for this operator.

Theorem 7.5 *The spectrum of H_{ho} is*

$$\sigma(H_{ho}) = \{\sum_{i=1}^{r} \hbar\omega_i(n_i + 1/2) \mid n_i = 0, 1, 2, \ldots\}.$$

To prove this result, we derive a representation of the operator H which facilitates its spectral analysis. It also prepares us for a similar technique we will encounter in the more complex situation of second quantization and quantum electrodynamics (see Chapters 20 and 21). We introduce the *creation* and *annihilation* operators

$$a_j := \frac{1}{\sqrt{2m\hbar\omega_j}}(m\omega_j x_j + ip_j) \text{ and } a_j^* := \frac{1}{\sqrt{2m\hbar\omega_j}}(m\omega_j x_j - ip_j). \quad (7.13)$$

These operators are adjoint of each other (for the definition of the operator adjoint to a given operator see Definition 25.23) and they satisfy the commutation relation

$$[a_i, a_j^*] = \delta_{ij}. \quad (7.14)$$

Using that $H = \sum_{i=1}^{r}\left(-\frac{\hbar^2}{2m}\partial_{x_i}^2 + \frac{1}{2}m\omega_i^2 x_i^2\right)$, the Hamiltonian H can be re-written in terms of a_i and a_i^*'s as follows:

$$H_{ho} = \sum_{i=1}^{r} \hbar\omega_i\left(a_i^* a_i + \frac{1}{2}\right). \quad (7.15)$$

We say that this expression is in *normal form* because a^* appears to the left of a. Now, we are ready to prove Theorem 7.5.

Proof (of Theorem 7.5). First we find the ground state and ground state energy of H_{ho}. We define the *particle number operators*

$$N_i := a_i^* a_i,$$

so that, by the expression (7.15),

$$H_{ho} = \sum_{i=1}^{r} \hbar\omega_i(N_i + \frac{1}{2}),$$

Now, because a_i^* is the adjoint of a_1, N_i are non-negative, $N_i \geq 0$:

$$\langle\psi, N_i\psi\rangle = \|a_i\psi\|^2 \geq 0$$

for any ψ. Therefore the ψ with the smallest eigenvalue, $\sum_{i=1}^{r}\frac{1}{2}\hbar\omega_i$ (i.e. the one that minimizes the average energy, $\langle\psi, H_{ho}\psi\rangle$), satisfies $N_i\psi = 0$, $\forall i$, which holds if and only if $a_i\psi = 0$ $\forall i$. These equations,

$$a_i\psi = \frac{1}{\sqrt{2m\hbar\omega_i}}(m\omega_i x_i + \hbar\partial/\partial x_i)\psi = 0,$$

can be easily solved, giving the unique (up to a multiplicative constant which we denote c) family of solutions

$$\psi_{0i}(x_i) := ce^{-m\omega_i x_i^2/2\hbar}$$

We choose c to have ψ_{0i} normalized as $\|\psi_{0i}\| = 1$ which gives $c := (2\pi m\hbar\omega_i)^{-1/4}$. To find the (normalized) ground state of H_{ho}, we have to multiply these functions to obtain

$$\psi_0 := \prod_{i=1}^{r} (2\pi m\hbar\omega_i)^{-1/4} e^{-m\sum_{i=1}^{r}\omega_i x_i^2/2\hbar}. \tag{7.16}$$

This is the ground state of H_{ho}, with the ground state energy $\sum_{i=1}^{r} \frac{1}{2}\hbar\omega_i$.

To find the exited states we observe that (7.14) imply that the operators N_i satisfy the relations

$$N_i a_i = a_i(N_i - 1), \tag{7.17}$$

$$N_i a_i^* = a_i^*(N_i + 1). \tag{7.18}$$

The commutation relation (7.18) implies $N_i(a_i^*)^n = (a_i^*)^n(N_i + n)$, which gives $N_i a^* \psi_0 = a_i^* \psi_0$ and in general

$$N_i(a_i^*)^n \psi_0 = n_i(a_i^*)^n \psi_0.$$

Thus $\phi_{n_i} := (a_i^*)^{n_i} \psi_0$ is an eigenfunction of N_i with eigenvalue n_i and therefore $\phi_{\mathbf{n}} := \prod_{i=1}^{r}(a_i^*)^{n_i}\psi_0$ is an eigenfunction of H_{ho} with eigenvalue $\sum_{i=1}^{r} \hbar\omega_i(n_i + \frac{1}{2})$.

Problem 7.6 Show that $\|\phi_{\mathbf{n}}\|^2 = |c|^2 \prod_{i=1}^{r} n_i!$. Hint: write $\|\phi_{\mathbf{n}}\|^2 = \langle \psi_0, \prod_{i=1}^{r}(a_i)^{n_i}(a_i^*)^{n_i}\psi_0\rangle$, then pull the a_i's through the a_i^*'s (including the necessary commutators) until they hit ψ_0 and annihilate it.

So $\psi_{\mathbf{n}} := \prod_{i=1}^{r} \frac{1}{\sqrt{n_i!}}(a_i^*)^{n_i}\psi_0$, $\mathbf{n} := (n_1, \ldots, n_r)$, is a normalized eigenfunction of H_{ho} with eigenvalue $\sum_{i=1}^{r} \hbar\omega_i(n_i + 1/2)$.

We now show that these are the only eigenfunctions. To simplify the notation we do this in dimension 1, i.e. for $r = 1$. It follows from the commutation relation (7.17) that $Na^n = a^n(N - n)$ (this is adjoint of $N(a^*)^n = (a^*)^n(N+n)$). Hence if ψ is any eigenfunction of N with eigenvalue $\lambda > 0$, then

$$Na^m\psi = (\lambda - m)a^m\psi. \tag{7.19}$$

If we choose m so that $\lambda - m < 0$ we get a contradiction to $N \geq 0$ unless $a^m\psi = 0$. Let j be the largest integer s.t. $a^j\psi \neq 0$ and $a^{j+1}\psi = aa^j\psi = 0$. This implies

$$a^j\psi = c\psi_0 \tag{7.20}$$

where $c \neq 0$ is a constant. Hence by $N\psi_0 = 0$, (7.20) and (7.19), we have $0 = Na^j\psi = (\lambda - j)a^j\psi$ and therefore $\lambda = j$. Thus ψ corresponds to the eigenvalue j. If ψ is not proportional to ψ_j, then we can choose it to satisfy $\langle \psi, \psi_j \rangle = 0$. However, by $a^j\psi = c\psi_0$ and $\psi_j = (a^*)^j\psi_0$, we have that $\langle \psi, \psi_j \rangle = \langle a^j\psi, \psi_0 \rangle = c\|\psi_0\|^2 \neq 0$, a contradiction. So we are done.

In the proof we obtained the following result

Theorem 7.7 *The eigenfunctions of the operator H_{ho} corresponding to the eigenvalues $\sum_{i=1}^{r} \hbar\omega_i(n_i + 1/2) \mid n_i = 0, 1, 2, \ldots$, are*

$$\psi_n = \prod_{i=1}^{r} (1/\sqrt{n_i!})(a_i^*)^{n_i} \psi_0, \quad n := (n_1, \ldots, n_r), \tag{7.21}$$

with ψ_0 given in (7.16) and a_i^* the first order differential operators defined in (7.13).

Remark 7.8 One can extend the last part of the proof above to show that any function $f \in L^2(\mathbb{R}^r)$ can be written as

$$f(x) = \sum_n c_n \psi_n = \sum_n c_n \prod_{i=1}^{r} \frac{1}{\sqrt{n_i!}} (a_i^*)^{n_i} \psi_0.$$

Hence the set $\{\psi_n\}$ form an orthonormal basis in $L^2(\mathbb{R}^r)$ and the space $L^2(\mathbb{R}^r)$ is isometric to the space $\mathcal{F}^{(r)} := \oplus_{n=0}^{\infty} \mathbb{C}_{sym}^n$, where $\mathbb{C}_{sym}^n$ is equal to $\{0\}$ for $n = 0$ and to $\mathbb{C}^n/S_n$ for $n \geq 1$. Here S_n is the symmetric group of permutations of n indices.

Remark 7.9 Since ψ_0 is positive and normalized, $\int \psi_0^2 = 1$, the operator $U:$ $f \to \psi_0^{-1} f$ maps unitarily the space $L^2(\mathbb{R}^r, d^r x)$ into the space $L^2(\mathbb{R}^r, \psi_0^2 d^r x)$. Under this map the operator H_{ho} is mapped into $L := UHU^{-1}$ acting on the space $L^2(\mathbb{R}^r, \psi_0^2 d^r x)$. We compute

$$L = \sum_{j=1}^{r} \hbar(-\frac{\hbar}{2m}\partial_{x_j} + \omega_j x_j)\partial_{x_j}. \tag{7.22}$$

The operator L, with $\hbar = 1$, $2m = 1$, is the generator of the Ornstein-Uhlenbeck stochastic process (see [135]).

Problem 7.10 Find the ground state energy of the ideal quantum gas in $\mathbb{R}^3$, with the Schrödinger operator given by (7.56), with potential $U(y) := |y|^2$, for (a) $n = 27$ identical spinless fermions and (b) $n = 12$ bosons (in terms of the one-particle energies).

7.5 A Particle in a Constant Magnetic Field

Consider a charged quantum particle moving in a constant magnetic field, B, with no electric field present. According to (4.15), the quantum Hamiltonian of such a system is

$$H(A) = \frac{1}{2m}(p - eA)^2, \tag{7.23}$$

acting on $L^2(\mathbb{R}^3)$, where A is the vector potential of the magnetic field B and therefore satisfies

$$\mathrm{curl}A = B. \tag{7.24}$$

Recall that the corresponding Schrödinger equation has the gauge symmetry

$$H(A + \nabla\chi) = e^{ie\chi/\hbar}H(A)e^{-ie\chi/\hbar}. \tag{7.25}$$

We fix the gauge by choosing a special solution of equation (7.24). A possible choice for A is

$$A'(x) = \frac{1}{2}B \times x. \tag{7.26}$$

If, we chose the x_3 axis along B so that $B = (0,0,b)$, then A' becomes $A'(x) = \frac{1}{2}(-x_2, x_1, 0)$. Another possibility, again supposing $B = (0,0,b)$, is

$$A''(x) = b(-x_2, 0, 0). \tag{7.27}$$

Problem 7.11 Check that both (7.26) and (7.27) yield the magnetic field B, and that the two are gauge-equivalent.

Note that if the magnetic field is directed along the x_3 axis, i.e. $B = (0,0,b)$, then $H(A)$ is of the form

$$H(A) = H_A + \frac{1}{2m}p_3^2, \tag{7.28}$$

where the operator H_A acts only on the variables x_1 and x_2 and is of the form

$$H_A := \sum_1^2 \frac{1}{2m}(p_i - eA_i)^2 = -\frac{\hbar^2}{2m}\Delta_A, \tag{7.29}$$

where $\Delta_A := \nabla_A^* \nabla_A$, with $\nabla_A := \nabla - ieA/\hbar$, on $\mathbb{R}^2$. We say that the magnetic field is *perpendicular to the plane* (i.e. along the x_3−direction).

Problem 7.12 Prove (7.28) - (7.29).

The spectrum of the operator $H(A)$ in (7.28) can be found by separation of variables:

$$\sigma(H(A)) = \overline{\sigma(H_A) + [0,\infty)},$$

where $[0,\infty)$ is the spectrum of $\frac{1}{2m}p_3^2$. Thus it suffices to find the spectrum of H_A.

From now on we deal with the operator H_A in 2D, acting only on the variables x_1 and x_2. Let $\omega_b := \frac{eb}{m}$ denote the cyclotron frequency (the frequency of the classical circular motion of a particle of charge e in a constant magnetic field of magnitude b). We have

Theorem 7.13 *The spectrum of H_A consists of the eigenvalues*

$$\hbar\omega_b(n + \frac{1}{2}), \quad n = 0, 1, \ldots . \tag{7.30}$$

with infinite multiplicity. (These eigenvalues are called Landau levels.*)*

Proof. We may introduce the harmonic oscillator annihilation and creation operators, α and α^*, with

$$\alpha := \hbar[(\nabla_A)_1 + i(\nabla_A)_2], \quad \alpha^* := \hbar[-(\nabla_A)_1 + i(\nabla_A)_2], \tag{7.31}$$

where $\nabla_A := \nabla - i\frac{e}{\hbar}A(x)$. These operators satisfy the standard commutation relations:

$$[\alpha, \alpha^*] = 2e\hbar\,\mathrm{curl}\,A = 2be\hbar. \tag{7.32}$$

Indeed, $\frac{1}{\hbar^2}[\alpha, \alpha^*] = [(\nabla_A)_1, i(\nabla_A)_2] + [i(\nabla_A)_2, -(\nabla_A)_1] = 2\frac{e}{\hbar}(\partial_1 A_2 - \partial_2 A_1)$. The operator H_A can be expressed in terms of these operators as

$$H_A = \frac{1}{2m}(\alpha^*\alpha + be\hbar). \tag{7.33}$$

Indeed, $\frac{1}{\hbar^2}\alpha^*\alpha = -(\nabla_A)_1^2 - (\nabla_A)_2^2 - i[(\nabla_A)_1, (\nabla_A)_2] = -\Delta_A - be/\hbar$. The the argument used above for the standard harmonic oscillator shows that the spectrum of H_A consist of the eigenvalues (7.30).

Problem 7.14 Prove (7.32), (7.33) and that (7.30) are the eigenvalues of H_A.

Now, we show that the eigenvalues (7.30) have infinite multiplicities. First, we note that the gauges (7.26) and (7.27) become

$$A'(x) = \frac{1}{2}bx^\perp, \quad x^\perp := (-x_2, x_1), \quad A''(x) = b(-x_2, 0). \tag{7.34}$$

We find the eigenfunctions of the operator $H_{A'}$, with A' given in (7.34). First we write the annihilation operator, α, given in (7.31), explicitly as

$$\alpha = 2\hbar\bar{\partial} + \frac{be}{2}z, \quad \bar{\partial} := \frac{1}{2}(\partial_{x_1} + i\partial_{x_2}), \quad z := x_1 + ix_2. \tag{7.35}$$

Next, a straightforward computation shows that

$$e^{\frac{be}{4\hbar}|z|^2}(\bar{\partial} + \frac{be}{4\hbar}z)e^{-\frac{be}{4\hbar}|z|^2} = \bar{\partial}. \tag{7.36}$$

Problem 7.15 Prove (7.35) and (7.36).

Hence ϕ solves the equation $\alpha\phi = 0$ iff $f(z) := e^{\frac{be}{4\hbar}|z|^2}\phi$ solves the equation $\bar{\partial}f = 0$. The latter equation implies that f is a holomorphic function. Hence the eigenspace of the operator $H_{A'}$ corresponding to the lowest eigenvalue $\frac{\hbar\omega_b}{2}$ is

$$V_0 := \{f(z)e^{-\frac{be}{4\hbar}|z|^2} : f \text{ is a holomorphic function s.t.}$$

$$f(z)e^{-\frac{be}{4\hbar}|z|^2} \in L^2(\mathbb{R}^2)\}. \tag{7.37}$$

V_0 is the eigenspace corresponding to the lowest eigenvalue $\frac{\hbar\omega_b}{2}$. It is called the *zeroth Landau level* subspace. Higher Landau level subspaces are obtained by applying powers of the creation operator α^* to V_0: the eigenspace corresponding to the eigenvalue $\hbar\omega_b(n + \frac{1}{2}), n \geq 1$, the *nth Landau level*, is given by

$$V_n := (\alpha^*)^n V_0.$$

Clearly, all these subspaces are infinite dimensional. This completes the proof of Theorem 7.13

To find a convenient basis in the infinite dimensional space V_0, we use the creation and annihilation operators

$$\beta := \hbar[(\pi_A)_1 - i(\pi_A)_2], \quad \beta^* := \hbar[-(\pi_A)_1 - i(\pi_A)_2], \tag{7.38}$$

where $\pi_A := \nabla + i\frac{e}{\hbar}A(x) = \bar{\nabla}_A$. We can compute as before

$$\beta = 2\hbar\partial + \frac{be}{2}\bar{z}, \quad \beta^* = -2\hbar\bar{\partial} + \frac{be}{2}z. \tag{7.39}$$

Problem 7.16 Prove (7.39), that $[\beta, \beta^*] = 2be\hbar$ and that α, α^* and β, β^* commute mutually.

Now, the equations $\alpha\psi_0 = 0$ and $\beta\psi_0 = 0$ have the solution $\psi_0 := e^{-\frac{be}{4\hbar}|z|^2}$, unique up to multiplicative constant. Applying $(\beta^*)^m$ to this solution, one generates an orthogonal basis in V_0. An orthogonal basis in V_n is given by $(\beta^*)^m(\alpha^*)^n\psi_0, m = 0, 1, \ldots$.

The elements, $\psi_m := (\beta^*)^m\psi_0, m = 0, 1, \ldots$, of the basis in V_0 are eigenfunctions of the angular momentum operator

$$L_3 = x_1 p_2 - x_2 p_1.$$

Problem 7.17 Prove this. Hint: show that (a) $\alpha^*\alpha - \beta^*\beta = -2ebL_3$, (b) in the z variables $L_3 = \hbar(z\partial - \bar{z}\bar{\partial})$, (c) $\psi_m = c_m z^m \psi_0$, for some c_m. The last two properties give $L_3\psi_m = \hbar m\psi_m$.

Problem 7.18 Show that Schrödinger operator H_A in the first gauge in (7.34) is of the form

$$H_{A'} := \frac{1}{2m}[p_1^2 + p_2^2 + \frac{1}{4}e^2b^2(x_1^2 + x_2^2) - ebL_3] \tag{7.40}$$

where $L_3 = x_1p_2 - x_2p_1$ is the third component of the angular momentum operator. For the second gauge, the corresponding Schrödinger operator is

$$H_{A''} := \frac{1}{2m}[(p_1 + ebx_2)^2 + p_2^2]. \tag{7.41}$$

To analyze $H_{A''}$, we apply the Fourier transform to only the first variable ($x_1 \mapsto k_1$). This results in the unitarily equivalent operator

$$\tilde{H}_{A''} = \frac{1}{2m}p_2^2 + \frac{m\omega_b^2}{2}(x_2 + \frac{1}{eb}k_1)^2, \tag{7.42}$$

where, recall, $\omega_b := \frac{eb}{m}$ is the cyclotron frequency, and k_1 acts as a multiplication operator. We see that $\tilde{H}_{A''}$ acts as a harmonic oscillator (centred at $-\frac{1}{eb}k_1$) in the variable x_2, and as a multiplication operator in k_1. In the following problem you are asked to determine the spectrum of this operator.

Problem 7.19

1. Show (7.41) and (7.42), and use (7.42) to prove directly that the energy levels of $H_{A''}$ are given by (7.30).
2. Show that the corresponding eigenfunctions are of the form

$$\psi_{n,\alpha}(x_1, x_2) = (2\pi\hbar)^{-1/2} \int e^{ix_1k_1/\hbar} f(k_1)\phi_n(x_2 + k_1/eb)dk_1$$

where ϕ_n is the nth eigenfunction of the harmonic oscillator and f is an arbitrary function from $L^2(\mathbb{R})$.

Problem 7.20 Let A be a vector potential with zero magnetic field, curl$A = 0$. Show that $H(A)$ is unitarily equivalent to $H(A = 0)$, i.e. there is a unitary operator U s.t. $H(A) = U^*H(A = 0)U$. (Hint: Show that there is a function χ s.t. $A = \nabla\chi$ and that the gauge transformation, $\psi(x) \rightarrow e^{i\chi(x)}\psi(x)$ provides the desired unitary operator.

To conclude this subsection, we consider the simple but instructive case of a particle on a flat torus, say

$$\mathbb{T}^2 = \mathbb{R}^2/\mathcal{L}, \quad \mathcal{L} := \{m_1L_1e_1 + m_2L_2e_2 : m_1, m_2 \in \mathbb{Z}\} \text{ for some } L_1, L_2 > 0,$$

with no external electrostatic potential, $V \equiv 0$, and with a *constant vector potential*, $A = (A_1, A_2)$. The corresponding Schrödinger operator is still (7.23), acting on

$$L^2(\mathbb{T}^2) := \{\psi \in L^2_{\mathrm{loc}}(\mathbb{R}^2) : \psi(x + L_i e_i) = \psi(x),\ i = 1, 2\}$$

(or on $L^2(\Omega)$, where Ω is a fundamental cell of $\mathcal{L}$, with the domain $H^2(\Omega)$, with the boundary conditions (7.3)). Here $e_1 := (1, 0), e_2 := (0, 1)$. Since $B = \mathrm{curl}A = 0$, on $L^2(\mathbb{R}^2)$ the Schrödinger operator (7.23) with such an A is gauge equivalent to the one without the vector potential, i.e. $H(A = 0) = \frac{1}{2m}|p|^2$. But not on the torus. The gauge transformation,

$$\psi(x) \to e^{iA \cdot x}\psi(x),$$

which removes A on $L^2(\mathbb{R}^2)$ (see Problem 7.20), does not leave $L^2(\mathbb{T}^2)$ invariant. Hence, though the magnetic field corresponding to A is zero, $B = \mathrm{curl}A = 0$, the operator $H(A)$ is not unitary equivalent to $H(A = 0)$. In particular, the spectrum of $H(A)$ depends on A: it is straightforward to check that the eigenvalues are

$$E_n(A) = \frac{1}{2m} \sum_1^3 \left(\frac{2\pi\hbar}{L_j}\right)^2 \left(n_j - \Phi_j/\Phi_0\right)^2,$$

for $n_j \geq 0$, where $\Phi_j := A_j L_j$ and $\Phi_0 := \frac{2\pi\hbar}{e}$.

Thus even for zero magnetic field, the vector potential, which is classically irrelevent, affects the Schrödinger operator. This surprising result is due to the subtle topology of the torus. $\mathbb{T}^2$ is not a simply connected space: one could draw two (classes of) loops on $\mathbb{T}^2$ which cannot be contracted to points (or to each other), say, $\gamma_1(t) := (L_1 t, 0)$ and $\gamma_2(t) := (0, L_2 t), 0 \leq t \leq 1$ (remember that for each i, the points and $\gamma_i(0)$, and $\gamma_i(1)$ are identified by taking the quotient by $\mathcal{L}$). Topologically, $\mathbb{T}^2$ is equivalent to a hollow donut in $\mathbb{R}^3$ and the loops above, to the loops around the two 'holes' of the donut. Assuming that the magnetic field B is confined to these holes, we see that though it is zero on the donut's surface, its vector potential cannot be gauged away. But what really matters are the magnetic fluxes through the holes, ' $\int_{D_i} B$ ', where D_i is a cross-section of the i-th hole, which, by the Stokes theorem, are

$$\int_{\partial D_i} A_i = A_j L_j := \Phi_j.$$

Only when these fluxes are integers times the 'flux quantum' Φ_0 does the vector potential A have no effect on the spectrum of $H(A)$.

Remark 7.21 Mathematically, A is a flat connection on a line bundle over the torus $\mathbb{T}^2$ (see e.g. [83]) and $e^{i\Phi_j}$ are holonomies for this connection (cf. Section 7.6).

7.6 Aharonov-Bohm Effect

Consider a charged quantum particle in 2 dimensions moving in external electric and magnetic fields with the potentials V and A, respectively. Let $H(A)$ denote the corresponding Schrödinger operator, i.e.

$$H(A) := -\frac{\hbar^2}{2m}\Delta_A + V,$$

where, recall $-\hbar^2\Delta_A := (-i\hbar\nabla - eA)^2$. Furthermore, we assume that V is radially symmetric, $V(x) = U(|x|)$, and A is of the form[1]

$$A(x) := f(|x|)x^\perp/|x|^2, \qquad x^\perp := (-x_2, x_1). \tag{7.43}$$

Note that $\text{div}A = 0$, and $B = \text{curl}A = 2g + rg'$, where $g(r) := f(r)/r^2$. (Indeed, $\text{curl}A = 2g + (x_1x_1/r + x_2x_2/r)g' = 2g + rg'$.) The operator $H(A)$ with A and V specified above will be denoted by H_f.

The computation above implies that, for $f(r) = b$ (a constant), the magnetic field is $B = 2\pi b\delta_0$, with $2\pi b =$ the total flux of B through $\mathbb{R}^2$: $\int_{\mathbb{R}^2} B = 2\pi b$.

Using that $(p - eA)^2 = |p|^2 - 2eA \cdot p + |eA|^2$, where, recall, $p := -i\hbar\nabla$, and passing to polar co-ordinates, (r, θ), we find that

$$(-i\hbar\nabla - eA)^2 = -\hbar^2\Delta_r + \frac{1}{r^2}\left(-\hbar^2\partial_\theta^2 + 2ef(r)(i\hbar\partial_\theta) + e^2f^2(r)\right)$$

$$= -\hbar^2\Delta_r + \frac{1}{r^2}(i\hbar\partial_\theta + ef(r))^2,$$

where Δ_r is the radial Laplacian, $\Delta_r := \frac{1}{r}\partial_r r\partial_r$. Using this, we obtain

$$H_f = -\frac{\hbar^2}{2m}\Delta_r + \frac{1}{2mr^2}(i\hbar\partial_\theta + ef(r))^2 + U(r).$$

Expanding in a Fourier series, $\psi(r, \theta) = \sum_{m\in\mathbb{Z}}\psi_m(r)e^{im\theta}$ and using separation of variables, we see that

$$H_f = \oplus_{m\in\mathbb{Z}}H_{f,m},$$

where $H_{f,m} := -\frac{\hbar^2}{2m}\Delta_r + \frac{1}{2mr^2}(-\hbar m + ef(r))^2 + U(r)$. This yields

$$\sigma(H_f) = \cup_m\sigma(H_{f,m}).$$

Thus, writing $f(r) \equiv \frac{\hbar}{e}\alpha$, we have $H_\alpha \equiv H_{f=\hbar\alpha/e} = \oplus_{m\in\mathbb{Z}}H_{\alpha,m}$, where

$$H_{\alpha,m} := -\frac{\hbar^2}{2m}\Delta_r + \frac{\hbar^2}{2mr^2}(m - \alpha)^2 + U(r).$$

We see that H_α is periodic in α: at each period, every $H_{\alpha,m}$ in the decomposition $H_\alpha = \oplus_{m\in\mathbb{Z}}H_{\alpha,m}$ moves one step down resulting in

$$H_{\alpha+1} = H_\alpha.$$

---

[1] The vector field $A(x)$ is said to be spherically equivariant in the sense that it satisfies the equation $A(gx) = gA(x)$ for any $g \in O(2)$.

(The corresponding eigenvalues move accordingly, which is referred to in the literature as the spectral flow.)

The general theory developed in Section 6.5 (with the caveat that special care be taken of the singular potential $\frac{\hbar^2}{2mr^2}(m-\alpha)^2$), implies that the spectrum of H_α consists of: the continuum $[0,\infty)$ and possibly, depending on $U(r)$, negative eigenvalues, if $U(r) \to 0$ as $r \to \infty$; and of isolated eigenvalues with finite multiplicities accumulating at ∞, if $U(r) \to \infty$ as $r \to \infty$. Clearly, these eigenvalues depend on the magnetic flux $\int B = 2\pi\hbar\alpha/e$, even though there is no magnetic field in the physical space $\mathbb{R}^2/\{0\}$ in which particle moves. Moreover, one can show that in the first case, the quantities characterizing the essential spectrum, such as the scattering matrix, also depend on α.

Mathematically, this is not surprising as the space $\mathbb{R}^2/\{0\}$ is not simply connected (it has a 'hole' at $x = 0$) and there are vector fields A, with zero magnetic fields, which are not gauge equivalent to 0 (flat connections in mathematical language). Such vector potentials are called topological vector potentials, and the corresponding fluxes (in the ambient space), topological fluxes.

Note that we have already encountered the effect of the topology of the physical space on the properties of Schrödinger operators at the end of Section 7.5, where we considered the movement of a particle on a torus with a constant vector potential having zero magnetic field.

The set-up above can be generalized to several topological fluxes: $H_{\underline{\alpha}} = H(A)$, $\underline{\alpha} := (\alpha_1, \dots, \alpha_r)$ for A given by

$$A(x) := \sum_1^r \frac{\hbar\alpha_i}{e} \frac{(x-a_i)^\perp}{|x-a_i|^2}, \tag{7.44}$$

where, recall, $x^\perp := (-x_2, x_1)$, with the magnetic field given by $B = 2\pi\frac{\hbar}{e}\sum_i \alpha_i \delta_{a_i}$. (One can also take individual vector potentials in different gauges.) In this case, the operator $H_{\underline{\alpha}}$ does not decompose into one-dimensional ones, but the conclusions reached above about the spectrum of H_α still hold for $H_{\underline{\alpha}}$. Furthermore, $H_{\underline{\alpha}}$ is periodic in each α_i:

$$H_{\underline{\alpha}+e_i} = H_{\underline{\alpha}},$$

where $\{e_i\}$ is the standard basis in $\mathbb{R}^r$. In particular, we can think of $\underline{\alpha}$ as varying in the standard flat torus $\mathbb{T}^r := \mathbb{R}^r/\mathbb{Z}^r$.

Remark 7.22 The structure of the Aharonov-Bohm magnetic potential (7.43) with $f(r) \equiv \hbar\alpha/e$, or (7.44), is especially simple in the complex representation. For this, we go from the real vector variable $x = (x_1, x_2)$ to the complex one $z = x_1 + ix_2$ and from the real vector potential $A = (A_1, A_2)$ to the complex function $A^c := A_1 - iA_2$. Then, e.g. (7.43), with $f(r) \equiv \hbar\alpha/e$, becomes

$$A^c(z) = -(\hbar\alpha/e)i/z. \tag{7.45}$$

7.7 Linearized Ginzburg-Landau Equations of Superconductivity

One of the greatest achievements in condensed matter physics is A.A. Abrikosov's discovery of vortex lattice solutions of the Ginzburg-Landau theory of super-conductivity (Abrikosov was awarded the Nobel prize for this discovery). We formulate the key mathematical problem of Abrikosov's paper.

Let $\mathcal{L}$ be a Bravais lattice in $\mathbb{R}^2$; i.e. for some basis $\{\omega_1, \omega_2\}$ in $\mathbb{R}^2$,

$$\mathcal{L} := \{\alpha = \sum_{i=1}^{2} m_i \omega_i : m_i \in \mathbb{Z}\}.$$

Let Δ_A stand for the covariant Laplacian defined as $\Delta_A := (\nabla - iA)^2$. The Abrikósov problem can be formulated as the eigenvalue problem

$$-\Delta_A \psi = \kappa^2 \psi, \tag{7.46}$$

where A is a vector potential of a constant external magnetic field b, $\operatorname{curl} A = b$, $\kappa > 0$ is a material constant (different for different materials) and ψ is a twice differentiable function satisfying

$$\psi(x + s) = e^{ig_s(x)} \psi(x), \tag{7.47}$$

for some functions $g_s \in C^2(\mathbb{R}^2; \mathbb{R})$ and each $s \in \mathcal{L}$. We can fix the gauge as

$$A(x) = \frac{b}{2} x^\perp, \qquad g_s(x) = \frac{b}{2} s^\perp \cdot x + c_s, \tag{7.48}$$

where $x^\perp = (-x_2, x_1)$ for $x = (x_1, x_2)$, and c_s satisfies

$$c_{s+t} - c_s - c_t - \frac{1}{2} bs \wedge t \in 2\pi \mathbb{Z}.$$

(For more details and for the relation of the Abrikosov problem to equations on line bundles see [268].)

Let Ω be a fundamental cell [2] of the lattice $\mathcal{L}$, say

$$\Omega := \left\{ \sum_{i=1}^{2} a_i \omega_i \mid -\frac{1}{2} < a_i \leq \frac{1}{2} \right\}.$$

(7.47)-(7.48) are consistent with the fact that ψ is a single valued function if and only if the magnetic flux, $b|\Omega|$, through the fundamental cell Ω is quantized:

$$b|\Omega| = 2\pi n, \tag{7.49}$$

---

[2] A *fundamental cell* of $\mathcal{L} \subset \mathbb{C}$ is a subset $\Omega \subset \mathbb{C}$ such that (i) $\Omega \cap (\Omega + s) = \emptyset \; \forall s \in \mathcal{L}/\{0\}$ and (ii) $\cup_{s \in \mathcal{L}} \Omega = \mathbb{C}$. (The second property says that Ω tessellates $\mathbb{C}$.)

for some integer n. (Performing translations by s and then by t should lead to the same value as by t and then by s. For more details see Supplement 7.9.) This is the celebrated quantization condition.

To finish the formulation of the eigenvalue problem (7.46)-(7.47), we have to define the space on which Δ_A acts, and its domain. We define $L^2_{\mathrm{per}}(\mathbb{R}^2)$ to be the space of locally L^2 functions satisfying the gauge-periodicity condition (7.47)-(7.48). It becomes a Hilbert space under the inner product of $L^2(\Omega)$ for an arbitrary fundamental cell Ω of the lattice $\mathcal{L}$. We consider the eigenvalue problem (7.46)- (7.47) on the corresponding Sobolev space $H^2_{\mathrm{per}}(\mathbb{R}^2)$ of index 2.

A key step in proving the existence of Abrikosov vortex lattice solutions is the following

Theorem 7.23 The values of b for which the problem (7.46)- (7.47), with the quantization condition $b|\Omega| = 2\pi n$, has a non-trivial solution, are $b = \kappa^2/(2k+1)$, $k = 0, 1, 2, \cdots$. Furthermore, for the largest value $b = \kappa^2$, it has exactly n linearly independent solutions.

Proof. One can show that the operator Δ_A with domain $H^2_{\mathrm{per}}(\mathbb{R}^2)$ is self-adjoint. We denote it by H. By Theorem 7.13, with $\hbar = 1, m = 1/2$ and $e = 1$, the spectrum of H is given by

$$\sigma(H) = \{ (2k+1)b : k = 0, 1, 2, \dots \}.$$

Moreover, the eigenspace of the lowest eigenvalue satisfies

$$\mathrm{Null}(H - b) = \mathrm{Null}\ \alpha. \tag{7.50}$$

Now, identifying $x \in \mathbb{R}^2$ with $z = x_1 + ix_2 \in \mathbb{C}$, the arguments of the proof of Theorem 7.13 imply that

$$\mathrm{Null}\ \alpha = \{\psi(z) := f(z)e^{-\frac{b}{4}|z|^2} \mid f \text{ is holomorphic}, \quad \psi \text{ satisfies (7.47)-(7.48)}\}. \tag{7.51}$$

Hence we look for $\psi(z)$ in the form

$$\psi(z) = e^{\frac{b}{4}(z^2 - |z|^2)}\Theta(z), \tag{7.52}$$

where $\Theta(z)$ is an entire function, and ψ satisfies (7.47)-(7.48).

Viewing $\mathcal{L}$ as a subset of $\mathbb{C}$, and using the rotation symmetry, if necessary, we can assume that $\mathcal{L}$ has a basis of the form $\{r, r\tau\}$, where r is a positive real number and $\tau \in \mathbb{C}, \mathrm{Im}\,\tau > 0$.[3] Setting $r = 1$ (by rescaling) and taking $c_r = c_\tau = 0$, the quasiperiodicity (7.47)-(7.48) of ψ transfers to Θ as follows:

---

[3] We identify $x \in \mathbb{R}^2$ with $z = x_1 + ix_2 \in \mathbb{C}$, and view $\mathcal{L}$ as a subset of $\mathbb{C}$. Then set $\tau = \frac{r'}{r}$ for a basis r, r'. Although the basis is not unique, the value of τ is, and it is used that as a characterization of the shape of the lattice. Using the rotation symmetry we can assume that if $\mathcal{L}$ has as a basis $\{r, r\tau\}$, where r is a positive real number and $\tau \in \mathbb{C}, \mathrm{Im}\,\tau > 0$.

$$\Theta(z+1) = \Theta(z), \tag{7.53a}$$

$$\Theta(z+\tau) = e^{-2\pi i n z}e^{-i\pi n\tau}\Theta(z). \tag{7.53b}$$

To complete the proof, we now need to show that the space of analytic functions which satisfy these relations form a vector space of dimension n. The first relation ensures that Θ has an absolutely convergent Fourier expansion of the form

$$\Theta(z) = \sum_{k=-\infty}^{\infty} c_k e^{2\pi i k z}.$$

The second relation, on the other hand, leads to a relation for the coefficients of the expansion:

$$c_{k+n} = e^{in\pi\tau}e^{2ki\pi\tau}c_k. \tag{7.54}$$

That means such functions are determined solely by the values of $c_0, \ldots, c_{n-1}$, and therefore form an n-dimensional vector space. This completes the proof of Theorem 7.23.

Problem 7.24 Prove (7.53) and (7.54).

Holomorphic functions satisfying (7.53) are called the theta functions. The proof implies also the form of the corresponding eigenfunctions: (7.52).

Historical Remark. Abrikosov found the vortex lattice solutions solutions using the linearized Ginzburg-Landau equations, and taking into account the nonlinearity in the first order of perturbation theory. For this, he considered the eigenvalue problem (7.46)- (7.47) in the case $n = 1$. In this case, the space (7.50) is one-dimensional and spanned by the function (setting $b = 1$ for simplicity)

$$\psi := e^{\frac{i}{2}x_2(x_1+ix_2)} \sum_{k=-\infty}^{\infty} c_k e^{2\pi i k(x_1+ix_2)}, \quad c_k = c e^{ik\pi\tau} \prod_{m=1}^{k-1} e^{i2m\pi\tau}. \tag{7.55}$$

This is the leading approximation to the Abrikosov lattice solution ([6]). The normalization coefficient c cannot be found from the linear theory and is obtained by taking into account nonlinear terms by perturbation theory (see [294] and references therein for a rigorous treatment and [146, 267] for a review).

7.8 Ideal Quantum Gas and Ground States of Atoms

Consider a system of N identical particles not interacting with each other in an external potential $U(x)$. It is called the ideal quantum gas. The Schrödinger operator for such a system is

$$H_{\text{igas}} = \sum_{j=1}^{N}(-\frac{\hbar^2}{2m}\Delta_{x_j} + U(x_j)). \qquad (7.56)$$

Assume the Schrödinger operator $-\frac{\hbar^2}{2m}\Delta_x + U(x)$ has an isolated eigenvalue at the bottom of its spectrum – the ground state energy. In this case, so does H_{igas} and, for bosons, its ground state and ground state energy are given by $\Phi(x_1,\ldots,x_n) = \phi_1(x_1)\ldots\phi_1(x_N)$, and

$$E_{\text{igas}}^{(N)} = Ne_1, \qquad (7.57)$$

respectively, where where $\phi_1(x)$ and e_1 are a ground state and the ground state energy of the Schrödinger operator $-\frac{\hbar^2}{2m}\Delta_x + U(x)$.

Problem 7.25 Prove the above statement.

On the other hand, for spinless fermions, for $N \geq 2$, the ground state of the operator H_{igas} cannot be given by the product $\prod_1^N \phi_1(x_j)$. A little contemplation shows that it is given by the anti-symmetric product,

$$\bigwedge_1^N \phi_j(x_j),$$

of N bound states, $\phi_j(x)$, of the operator H_{igas}, corresponding to the N lowest energies $e_1,\ldots,e_N$, counting multiplicities. In this case, the ground state energy, $E_{\text{igas}}^{(N)}$, of H_{igas} is given by

$$E_{\text{igas}}^{(N)} = \sum_1^N e_j$$

Problem 7.26 Prove the above statement.

This is, in general a much higher number. For example, for non-degenerate eigenvalues converging quickly to 0, this number remains $O(1)$ for large N.

Now we consider the most important case of spin $\frac{1}{2}$. Recall that the state space of a system of N fermions of spin $\frac{1}{2}$ is given by (4.31),

$$\mathcal{H}_{fermi} := \bigwedge_1^N (L^2(\mathbb{R}^3) \otimes \mathbb{C}^2). \qquad (7.58)$$

Hence, after separation of spin variables and for $N > 2$, the ground state of the operator $H_{\text{igas}} := \sum_1^N (-\frac{\hbar^2}{2m}\Delta_{x_j} + U(x_j))$ is given by the anti-symmetric product,

$$\bigwedge_1^{N/2} (\phi_{2j-1}(x_{2j-1})\phi_{2j}(x_{2j})),$$

(if N is even) of $\frac{N}{2}$ bound states, $\phi_j(x)$, corresponding to the $\frac{N}{2}$ lowest energies, say $E_1, \ldots, E_{N/2}$. (The case of N odd requires a slight modification.) In this case, the ground state energy of H_{igas} is given by $E_{\text{igas}}^{(N)} := 2 \sum_1^{N/2} E_j$, so that

$$H_{\text{igas}} \geq 2 \sum_1^{N/2} E_j. \tag{7.59}$$

Now we apply the above analysis to an estimation of the ground state energies of atoms. We have shown in Subsection 5.3 that the Schrödinger operator, H_{at}, of the atom (or ion) with N electrons and an infinitely heavy nucleus of the charge Ze is estimated from below as

$$H_{at} \geq H_{\text{igas}}^{el}, \tag{7.60}$$

where $H_{\text{igas}}^{el} := \sum_1^N \left(-\frac{\hbar^2}{2m} \Delta_{x_j} - \frac{e^2 Z}{|x_j|} \right)$. The latter operator is the quantum Hamiltonian of the ideal gas of N electrons in the external potential $-\frac{e^2 Z}{|x|}$ created by the nucleus. If we ignore the fact that the electrons are fermions, then we have according to (7.60), (7.57) and (7.11),

$$H_{at} \geq -N \frac{2m(e^2 Z)^2}{\hbar^2}. \tag{7.61}$$

Now, we estimate H_{at} from below using the fact that the electrons are fermions. Then by (7.60) and (7.59), $H_{at} \geq H_{\text{igas}}^{el} \geq 2 \sum_{j=1}^{N/2} E_j$, where the E_j are given by (7.11). To estimate the r.h.s. we should take into account the multiplicities of the eigenvalues. We relabel $\{E_j, j = 1, 2, \ldots\}$ as $\{E_{k,m}, m = 1, 2, \ldots, m_k, k = 1, 2, \ldots\}$, with $E_{k,m}$ distinct for different k, and equal for the same k. Thus N independent particles, in the lowest energy state, occupy the K lowest energy levels of the operator $-\frac{\hbar^2}{2m} \Delta_x - \frac{e^2 Z}{|x|}$, where K is defined by $\sum_1^K m_k = N/2$. The corresponding energy of H_{igas} is

$$E_{\text{igas}}^{(N)} = 2 \sum_{k=1}^K m_k E_{k,m_1}, \quad \text{with} \quad \sum_1^K m_k = N/2.$$

We know from (7.11) and (7.12) that $E_{k,m} - \frac{me^4 Z^2}{2\hbar^2} k^{-2}$ with the multiplicities $m_k = k^2$. Then the previous two equations yield $K \sim N^{1/3}$ and

$$E_{\text{igas}}^{(N)} \sim -Z^2 K \sim -Z^2 N^{1/3}.$$

For large N, this gives a much higher number than the one on the r.h.s. of (7.61).

Note that it is easy to give a realistic estimate of the sum $2 \sum_1^{\frac{N}{2}} E_j$, when E_j are the eigenvalues of the operator $-\frac{\hbar^2}{2m} \Delta_x - \frac{e^2 Z}{|x|}$. However, if one wants to obtain a more realistic estimate on the ground state energy of (5.10), then one should not throw away the electron repulsion. To keep the problem manageable, one replaces it by some mean field term (see Section 14.1). In this case we want to bound $\sum_1^{\frac{N}{2}} E_j$, where E_j are eigenvalues of the effective one-particle operator containing the mean-field potential. This is substantially more difficult ,and to do this, one uses the *Lieb-Thirring inequalities*.

Problem 7.27 Find the ground state energy of H_{igas} (in terms of the one-particle energies) for (a) 11 identical fermions of spin 3/2 on the subspace corresponding to a Young diagram with with columns of lengths $4, 4, 2, 1$; (b) 17 identical fermions of spin 1/2 on the subspace corresponding to a Young diagram with with columns of lengths $10, 7$.

7.9 Supplement: $\mathcal{L}$–equivariant functions

Consider the two dimensional Bravais lattice $\mathcal{L} = \{m\omega_1 + n\omega_2 : m, n \in \mathbb{Z}\}$, where (ω_1, ω_2) is some basis in $\mathbb{R}^2$. A function Ψ and a vector field A are said to be $\mathcal{L}$–*equivariant* if and only if there are differentiable functions, $g_s : \mathbb{R}^2 \to \mathbb{R}$, $s \in \mathcal{L}$, s.t. [4]

$$\Psi(x + s) = e^{ig_s(x)}\Psi(x) \text{ and } A(x + s) = A(x) + \nabla g_s(x), \ \forall s \in \mathcal{L}. \quad (7.62)$$

Since T_s^{trans} is a group, the family of functions g_s should satisfy

$$g_{s+t}(x) - g_s(x + t) - g_t(x) \in 2\pi\mathbb{Z}, \qquad s, t \in \mathcal{L}. \quad (7.63)$$

This can be seen by evaluating the effect of translation by $s+t$ in two different ways (as $\Psi((x + t) + s)$ and $\Psi(x + (t + s))$).

We list some important properties of g_s:

- If (Ψ, A) satisfy (7.62) with $g_s(x)$, then $T_\chi^{gauge}(\Psi, A)$ satisfies (7.62) with $g_s(x) \to g_s'(x)$, where

$$g_s'(x) = g_s(x) + \chi(x + s) - \chi(x). \quad (7.64)$$

- The following functions satisfy (7.63):

$$g_s^b(x) = \frac{b}{2}s \wedge x + c_s, \quad (7.65)$$

  where b satisfies the flux quantization relation $b|\Omega| \in 2\pi\mathbb{Z}$ and c_s are numbers satisfying

$$c_{s+t} - c_s - c_t - \frac{1}{2}bs \wedge t \in 2\pi\mathbb{Z}. \quad (7.66)$$

---

[4] Eq. (7.62) implies that $g_s(x) = \int_{x_0}^x A_s(y)dy$, for some x_0, where $A_s(x) := A(x + s) - A(x)$. Since $\text{curl}A_s(x) = \text{curl}A(x + s) - \text{curl}A(x) = 0 \ \forall s \in \mathcal{L}$, the integral $\int_{x_0}^x A_s(y)dy$ is independent of the path between x_0 and x. For a constant magnetic field vector potential $A^b(x) = \frac{b}{2}x^\perp$, where $x^\perp = (-x_2, x_1)$, this gives $g_s^b(x) = \frac{b}{2}s \wedge x + c_s$, where c_s is a constant of integration (cf. (7.65) below).

- Every exponential g_s satisfying the cocycle condition (7.63) is equivalent to the exponential (7.65) with (7.66).
- The exponentials g_s satisfying the cocycle condition (7.63) are classified by the irreducible representation of the group of lattice translations.

Indeed, the first and second statements are straightforward. For the third property, which was shown by A. Weil and generalized by R. Gunning, we refer to [144].

We show that under (7.62) the flux of the magnetic field is quantized:

$$\int_\Omega \text{curl} A \in 2\pi \mathbb{Z}. \tag{7.67}$$

Indeed, let, as above, Ω be a fundamental cell of the lattice $\mathcal{L}$. Take $\Omega :=$ $\{\sum_1^2 a_i \omega_i : -\frac{1}{2} < a_i \leq \frac{1}{2}\}$. We compute, by Stokes' theorem

$$\int_\Omega \text{curl} A = \int_{\partial\Omega} A.$$

Breaking the integral on the r.h.s. into the integrals over the four sides of the parallelogram Ω and using that due to (7.62), the sum of integrals over parallel sides give $g_{\omega_2}(\omega_1) - g_{\omega_2}(0)$ and $g_{\omega_1}(0) - g_{\omega_1}(\omega_2)$, which yields $\int_{\partial\Omega} A = g_{\omega_2}(\omega_1) - g_{\omega_2}(0) - (g_{\omega_1}(\omega_2) - g_{\omega_1}(0))$. By (7.63), this implies (7.67).

If we plug $A(x)$ from (7.48) into this formula, we find the flux quantization relation

$$b|\Omega| = 2\pi n, \quad n \in \mathbb{Z}.$$

Bound States and Variational Principle

In this chapter we develop powerful techniques for proving existence of bound states (eigenfunctions) corresponding to isolated eigenvalues. We also give estimates of their number.

8.1 Variational Characterization of Eigenvalues

We consider, for the moment, a self-adjoint operator H, acting on a Hilbert space $\mathcal{H}$. The main result of this chapter is the following important characterization of eigenvalues of H in terms of the minimization problem for the "energy" functional $\langle \psi, H\psi \rangle$.

Theorem 8.1 (1) The *Ritz variational principle*:

$$\inf_{\psi \in D(H), \, \|\psi\|=1} \langle \psi, H\psi \rangle = \inf \sigma(H).$$

(2) The left hand side has a minimizer ψ if and only if $H\psi = \lambda\psi$, with $\lambda := \inf \sigma(H)$.

(3) If there is a $\psi \in D(H)$ (called a *test function*) with $\|\psi\| = 1$ and

$$\langle \psi, H\psi \rangle < \inf \sigma_{ess}(H),$$

then H has at least one eigenvalue below its essential spectrum.

As an example application of this theorem, we consider the bound state problem for the Hydrogen atom. The Schrödinger operator for the Hydrogen atom is

$$H = -\frac{\hbar^2}{2m}\Delta - \frac{e^2}{|x|}$$

acting on the Hilbert space $L^2(\mathbb{R}^3)$ (see Section 5.3). Take the (normalized) test function $\psi(x) = \sqrt{\mu^3/\pi}e^{-\mu|x|}$ for some $\mu > 0$ to be specified later and compute (passing to spherical coordinates)

© Springer-Verlag GmbH Germany, part of Springer Nature 2020
S. J. Gustafson and I. M. Sigal, *Mathematical Concepts of Quantum Mechanics*, Universitext, https://doi.org/10.1007/978-3-030-59562-3_8

$$\langle \psi, H\psi \rangle = \frac{\hbar^2}{2m} \int |\nabla \psi|^2 - e^2 \int \frac{1}{|x|} |\psi|^2$$

$$= \frac{\hbar^2}{2m} \frac{\mu^3}{\pi} 4\pi \int_0^\infty e^{-2\mu r} r^2 dr - e^2 \frac{\mu^3}{\pi} 4\pi \int_0^\infty e^{-2\mu r} r dr.$$

To compute the above integrals, note

$$\int_0^\infty e^{-\alpha r} r^2 dr = \frac{d^2}{d\alpha^2} \int_0^\infty e^{-\alpha r} dr$$

and

$$\int_0^\infty e^{-\alpha r} r dr = -\frac{d}{d\alpha} \int_0^\infty e^{-\alpha r} dr.$$

Since $\int_0^\infty e^{-\alpha r} dr = \alpha^{-1}$, we find $\int_0^\infty e^{-\alpha r} r^2 dr = 2\alpha^{-3}$, and $\int_0^\infty e^{-\alpha r} r dr = \alpha^{-2}$. Substituting these expressions with $\alpha = 2\mu$ into the formula for $\langle \psi, H\psi \rangle$, we obtain

$$\langle \psi, H\psi \rangle = \frac{\hbar^2}{2m} \mu^2 - e^2 \mu.$$

The right hand side has a minimum at $\mu = me^2/\hbar^2$, which is equal to

$$\langle \psi, H\psi \rangle|_{\mu=me^2/\hbar^2} = -\frac{me^4}{2\hbar^2}.$$

Since $\sigma_{ess}(H) = [0, \infty)$ (according to Theorem 6.16, which can be extended to cover singular potentials like the Coulomb potential – see Problem 6.17 in Section 6.5) we conclude that H has negative eigenvalues, and the lowest negative eigenvalue, λ_1, satisfies the estimate

$$\lambda_1 \leq -\frac{me^4}{2\hbar^2}.$$

This should be compared with the lower bound

$$\lambda_1 > -\frac{2me^4}{\hbar^2}$$

found in Section 5.3.

The rest of this section is devoted to the proof of this theorem and a generalization. We begin with some useful characterizations of operators in terms of their spectra.

Theorem 8.2 Let H be a self-adjoint operator with $\sigma(H) \subset [a, \infty)$. Then $H \geq a$ (i.e. $\langle u, Hu \rangle \geq a\|u\|^2$ for all $u \in D(H)$).

Proof. Without loss of generality, we can assume $a = 0$ (otherwise we can consider $H - a\mathbf{1}$ instead of H). First we suppose H is bounded (we will pass to the unbounded case later). If $b > 2\|H\|$, then the operator $(H + b)^{-1}$ is positive, as follows from the Neumann series

$$(H+b)^{-1} = b^{-1}(1+b^{-1}H)^{-1} = b^{-1}\sum_{j=0}^{\infty}(-b^{-1}H)^j$$

(see (25.42)). For all $\lambda > 0$, the operator $(H+\lambda)^{-1}$ is bounded, self-adjoint, and differentiable in λ (in fact it is analytic – see Section 25.9). Compute

$$\frac{\partial}{\partial\lambda}(H+\lambda)^{-1} = -(H+\lambda)^{-2} < 0,$$

Hence for any $0 < c < b$,

$$(H+c)^{-1} = (H+b)^{-1} + \int_c^b (H+\lambda)^{-2}d\lambda$$

and therefore $(H+c)^{-1} > (H+b)^{-1} > 0$. Now any $u \in D(H)$ can be written in the form $u = (H+c)^{-1}v$ for some $v \in \mathcal{H}$ (show this). Since

$$\langle u, (H+c)u\rangle = \langle v, (H+c)^{-1}v\rangle > 0$$

for any $c > 0$, we conclude that $\langle u, Hu\rangle \geq 0$ for all $u \in D(H)$, as claimed.

In order to pass to unbounded operators, we proceed as follows. Let $c > 0$ and $A := (H+c)^{-1}$, a bounded operator since $-c \notin \sigma(H)$. For any $\lambda \neq 0$, we have

$$A + \lambda = (H+c)^{-1} + \lambda = \lambda(H+c)^{-1}(H+c+\lambda^{-1}). \tag{8.1}$$

Hence for $\lambda > 0$, the operator $A + \lambda$ is invertible, and so $\sigma(A) \subset [0, \infty)$ (in fact, one can see from (8.1) that $A + \lambda$ is also invertible if $\lambda < -c^{-1}$, and so $\sigma(A) \subset [0, c^{-1}]$). By the proof above, $A := (H+c)^{-1} \geq 0$. Repeating the argument at the end of this proof, we find that $H + c \geq 0$. Since the latter is true for any $c > 0$, we conclude that $H \geq 0$. $\square$

Theorem 8.3 Let $S(\psi) := \langle\psi, H\psi\rangle$ for $\psi \in D(H)$ with $\|\psi\| = 1$. Then $\inf \sigma(H) = \inf S$. Moreover, $\lambda := \inf \sigma(H)$ is an eigenvalue of H if and only if there is a minimizer for $S(\psi)$ among $\psi \in D(H)$ with the constraint $\|\psi\| = 1$.

Proof. As in the proof of Lemma 25.21, we compute, for $\psi \in D(H)$ and $z \in \mathbb{R}$ satisfying $z < \inf S =: \mu$,

$$\|(H - z\mathbf{1})\psi\|^2 = \langle(H - z\mathbf{1})\psi, (H - z\mathbf{1})\psi\rangle$$
$$= \|(H - \mu)\psi\|^2 + 2(\mu - z)\langle\psi, (H - \mu)\psi\rangle + |\mu - z|^2\|\psi\|^2. \tag{8.2}$$

Since $\langle\psi, (H - \mu)\psi\rangle \geq 0$, this gives

$$\|(H - z\mathbf{1})\psi\| \geq |\mu - z|\|\psi\|. \tag{8.3}$$

As in the proof of Lemma 25.21, this shows that the operator $H - z\mathbf{1}$ is invertible and therefore $z \notin \sigma(H)$. Therefore $\inf S \leq \inf \sigma(H)$. Now let $\lambda :=$

$\inf \sigma(H)$. By Theorem 8.2, $\langle \psi, H\psi \rangle \geq \lambda \|\psi\|^2$ for any $\psi \in D(H)$. Hence $\inf S \geq \lambda = \inf \sigma(H)$, and therefore $\inf S = \inf \sigma(H)$ as required.

Now if $\lambda = \inf \sigma(H)$ is an eigenvalue of H, with normalized eigenvector ψ_0, then

$$S(\psi_0) = \langle \psi_0, H\psi_0 \rangle = \lambda = \inf S,$$

and therefore ψ_0 is a minimizer of S. On the other hand, if ψ_0 is a minimizer of S, among $\psi \in D(H)$, $\|\psi\| = 1$, then it satisfies the Euler-Lagrange equation (see Section 26.5)

$$S'(\psi_0) = 2\lambda \psi_0$$

for some λ. Since $S'(\psi) = 2H\psi$ (see again Section 26.5), this means that ψ_0 is an eigenvector of H with eigenvalue λ. Moreover,

$$S(\psi_0) = \langle \psi_0, H\psi_0 \rangle = \lambda \|\psi_0\|^2 = \lambda.$$

Since $S(\psi_0) = \inf S = \inf \sigma(H)$, we conclude that $\lambda = \inf \sigma(H)$ is an eigenvalue of H (with eigenvector ψ_0). $\square$

Proof. of Theorem 8.1: Theorem 8.3 gives the proof of the first two parts of the theorem – the Ritz variational principle: for any $\psi \in D(H)$,

$$\langle \psi, H\psi \rangle \geq \lambda = \inf \sigma(H)$$

and equality holds iff $H\psi = \lambda\psi$.

To obtain the third part of the theorem, stating that if we can find ψ with $\|\psi\| = 1$ and $\langle \psi, H\psi \rangle < \inf \sigma_{ess}(H)$ then we know that H has at least one eigenvalue below its essential spectrum, we note that, by Theorem 8.3,

$$\inf \sigma(H) = \inf S < \langle \psi, H\psi \rangle < \inf \sigma_{ess}(H),$$

so $\lambda = \inf \sigma(H)$ must be an (isolated) eigenvalue of H. $\square$

The variational principle above can be extended to higher eigenvalues.

Theorem 8.4 (Min-max principle) The operator H has at least n eigenvalues (counting multiplicities) below $\inf \sigma_{ess}(H)$ if and only if $\lambda_n < \inf \sigma_{ess}(H)$, where the number λ_n is given by

$$\lambda_n = \inf_{\{X \subset D(H) \,|\, \dim X = n\}} \max_{\{\psi \in X \,|\, \|\psi\| = 1\}} \langle \psi, H\psi \rangle. \tag{8.4}$$

In this case, the n-th eigenvalue (labeled in non-decreasing order) is exactly λ_n.

Sketch of proof. We prove only the "if" part of the theorem. The easier "only if" part is left as an exercise. We proceed by induction. For $n = 1$, the statement coincides with that of Theorem 8.3. Now assume that the "if" statement holds for $n \leq m-1$, and we will prove it for $n = m$. By the induction assumption, the operator H has at least $m - 1$ eigenvalues, $\lambda_1, \ldots, \lambda_{m-1}$ (counting multiplicities), all $< \inf \sigma_{ess}(H)$. We show that H has at least m eigenvalues. Let V_{m-1} denote the subspace spanned by the (normalized) eigenvectors $\psi_1, \ldots, \psi_{m-1}$, corresponding to $\lambda_1, \ldots, \lambda_{m-1}$. Then

1. the subspace V_{m-1}, and its orthogonal complement, $V_{m-1}^{\perp}$, are invariant under the operator H, and
2. the spectrum of the restriction, $H|_{V_{m-1}^{\perp}}$, of H to the invariant subspace $V_{m-1}^{\perp}$ is $\sigma(H)\backslash\{\lambda_1,\ldots,\lambda_{m-1}\}$.

Problem 8.5 Prove statements 1 and 2.

Now apply Theorem 8.3 to the operator $H|_{V_{m-1}^{\perp}}$ to obtain

$$\inf_{\{\psi \in V_{m-1}^{\perp} \cap D(H)\ |\ \|\psi\|=1\}} \langle \psi, H\psi \rangle = \inf\{\sigma(H)\backslash\{\lambda_1,\ldots,\lambda_{m-1}\}\}. \tag{8.5}$$

On the other hand, let X be any m-dimensional subspace of $D(H)$. There exists $\phi \in X$ such that $\phi \perp V_{m-1}$, and $\|\phi\| = 1$. We have

$$\langle \phi, H\phi \rangle \geq \inf_{\{\psi \in V_{m-1}^{\perp} \cap D(H)\ |\ \|\psi\|=1\}} \langle \psi, H\psi \rangle.$$

Hence for any such X

$$\max_{\{\psi \in X\ |\ \|\psi\|=1\}} \langle \psi, H\psi \rangle \geq \inf_{\{\psi \in V_{m-1}^{\perp} \cap D(H)\ |\ \|\psi\|=1\}} \langle \psi, H\psi \rangle$$

and therefore λ_m defined by (8.4) obeys

$$\lambda_m \geq \inf_{\{\psi \in V_{m-1}^{\perp} \cap D(H)\ ||\ \|\psi\|=1\}} \langle \psi, H\psi \rangle. \tag{8.6}$$

Since $\lambda_m < \inf \sigma_{ess}(H)$ by assumption, and due to (8.5), we have

$$\lambda_m' := \inf\{\sigma(H)\backslash\{\lambda_1,\ldots,\lambda_{m-1}\}\} < \inf \sigma_{ess}(H) \tag{8.7}$$

is the m-th eigenvalue of H. Moreover, Equations (8.5)-(8.7) imply that $\lambda_m \geq \lambda_m'$.

Now we show that $\lambda_m \leq \lambda_m'$. Let ψ_m be a normalized eigenvector corresponding to λ_m', and let $V_m = span\{\psi_1,\ldots,\psi_m\}$. Then

$$\lambda_m \leq \max_{\psi \in V_m, \|\psi\|=1} \langle \psi, H\psi \rangle = \lambda_m'.$$

Hence $\lambda_m = \lambda_m'$. Thus we have shown that H has at least m eigenvalues (counting multiplicities) $< \inf \sigma_{ess}(H)$, and these eigenvalues are given by (8.4). $\square$

This theorem implies that if we find an n-dimensional subspace X, such that

$$\mu_n := \sup_{\psi \in X, \|\psi\|=1} \langle \psi, H\psi \rangle < \inf \sigma_{ess}(H),$$

then H has at least n eigenvalues less than $\inf \sigma_{ess}(H)$, and the largest of these eigenvalues satisfies the bound

$$\lambda_n \leq \mu_n.$$

This result will be used in Section 8.3.

There is another formulation of the min-max (more precisely sup-inf) principle, in which Equation 8.4 is replaced by the equation

$$\lambda_n = \sup_{\{X \subset D(H) \,|\, \dim X = n\}} \inf_{\{\psi \in X^\perp \,|\, \|\psi\|=1\}} \langle \psi, H\psi \rangle.$$

A proof of this theorem is similar to the proof above.

The following useful statement is a simple consequence of the min-max principle. Suppose A and B are self-adjoint operators with $A \leq B$. Denote the j-th eigenvalue of A below its essential spectrum (if it exists) by $\lambda_j(A)$ (and similarly for B). Suppose also that the eigenvectors of B corresponding to the eigenvalues $\lambda_1(B), \ldots, \lambda_j(B)$ lie in $D(A)$. Then $\lambda_j(A) \leq \lambda_j(B)$. To see this, let V_j denote the span of the first j eigenvectors of B, and observe that $\max_{\{\psi \in V_j \,|\, \|\psi\|=1\}} \langle \psi, B\psi \rangle = \lambda_j(B)$. Since $V_j \subset D(A)$, (8.4) gives

$$\lambda_j(A) \leq \max_{\{\psi \in V_j \,|\, \|\psi\|=1\}} \langle \psi, A\psi \rangle$$

$$\leq \max_{\{\psi \in V_j \,|\, \|\psi\|=1\}} \langle \psi, B\psi \rangle = \lambda_j(B).$$

Another useful criterion for finding eigenvalues of self-adjoint operators goes as follows. If for some $\lambda \in \mathbb{R}$ and $\epsilon > 0$, there is a function $\psi \in D(A)$ such that

$$\|(A - \lambda)\psi\| \leq \epsilon \|\psi\|, \tag{8.8}$$

then the operator A has spectrum in the interval $[\lambda - \epsilon, \, \lambda + \epsilon]$:

$$\sigma(A) \cap [\lambda - \epsilon, \, \lambda + \epsilon] \neq \emptyset.$$

To prove this statement we use the inequality

$$\|(A - z)^{-1}\| \leq [\text{dist}(z, \sigma(A))]^{-1}$$

for $z \in \rho(A)$, which extends the inequality (25.11). For a proof of this inequality see [162, 244]. Now if A has no spectrum in $[\lambda - \epsilon, \, \lambda + \epsilon]$, i.e. if $[\lambda - \epsilon, \, \lambda + \epsilon] \subset \rho(A)$, then

$$\|(A - \lambda)^{-1}\| < \frac{1}{\epsilon}$$

which contradicts (8.8), since (8.8) implies $\|\phi\| \leq \epsilon \|(A - \lambda)^{-1}\phi\|$ for $\phi := (A - \lambda)\psi$.

8.2 Exponential Decay of Bound States

Consider the Schrödinger operator $H = -\frac{\hbar^2}{2m}\Delta + V$ with the potential $V :$ $\mathbb{R}^d \to \mathbb{R}$ which is continuous and decays $V(x) \to 0$ as $|x| \to \infty$. Recall that H is self-adjoint and its essential spectrum, $\sigma_{ess}(H)$, fills in the semi-axis $[0, \infty)$. hence the spectrum on the negative axis consists of isolated eigenvalues of finite multiplicities.

Theorem 8.6 If H has a bound state, $\psi(x)$, with an energy $E < 0$ (i.e. below the ionization threshold 0), then $\psi(x)$ satisfies the exponential bound

$$\int |\psi(x)|^2 e^{2\alpha|x|} dx < \infty, \ \forall \alpha < \sqrt{-E}. \tag{8.9}$$

Proof. Let J be a real, bounded, smooth function supported in $\{|x| \geq R\}$. By the condition on the potential V, there is $\epsilon = \epsilon(R) \to 0$, as $R \to \infty$, s.t.

$$JHJ \geq -\epsilon J^2. \tag{8.10}$$

We assume now that ∇J is supported in $\{R \leq |x| \leq 2R\}$. Let f be a bounded twice differentiable, positive function and define $H_f := e^f H e^{-f}$. We compute

$$H_f = H - \frac{\hbar^2}{2m}[|\nabla f|^2 - \nabla f \cdot \nabla - \nabla \cdot \nabla f]. \tag{8.11}$$

Then $(H_f - E)\Phi = 0$, where $\Phi := e^f \Psi$, and therefore $(H_f - E)J\Phi = [H_f, J]\Phi$. On the other hand, by (8.11) and (8.10) and the fact that the operator $\nabla f \cdot \nabla + \nabla \cdot \nabla f$ is anti-self-adjoint, we have

$$\mathrm{Re}\langle J\Phi, (H_f - E)J\Phi \rangle = \langle J\Phi, (H - \frac{\hbar^2}{2m}|\nabla f|^2 - E)J\Phi \rangle$$
$$\geq \delta \|J\Phi\|^2.$$

where $\delta := -\epsilon - E - \frac{\hbar^2}{2m}\sup_{x \in \mathrm{supp}\,J} |\nabla f|^2$. Then the last two equations imply,

$$\delta \|J\Phi\|^2 \leq \mathrm{Re}\langle J\Phi, (H_f - E)J\Phi \rangle \leq \|J\Phi\| \|[H_f, J]\Phi\|.$$

Now we take for f a sequence of bounded functions approximating $\alpha(1 + |x|^2)^{1/2}$, with $\alpha < \sqrt{-E}$. Taking the limit in the last inequality gives (8.9). $\square$

8.3 Number of Bound States

Let $H = -\frac{\hbar^2}{2m}\Delta + V(x)$ be a Schrödinger operator acting on $L^2(\mathbb{R}^3)$. Assume $V(x) \to 0$ as $|x| \to \infty$. In this section, we address the questions

- Does H have any bound states?
- If so, how many bound states does it have?

We begin with a very simple example. Suppose that $V(x) \geq 0$. Clearly $H \geq 0$, and so H certainly has no negative eigenvalues. On the other hand, $\sigma_{ess}(H) = [0, \infty)$ by Theorem 6.16. So H has no isolated eigenvalues.

The next example is physically clear, but mathematically more subtle:

If $x \cdot \nabla V(x) \leq 0$,   i.e. $V(x)$ is *repulsive*,

then H has no eigenvalues.

Let us assume here that V is twice differentiable, and that $V(x)$, together with $x \cdot \nabla V(x)$, vanish as $|x| \to \infty$. The proof of the above statement is based on the following *virial relation*: define the self-adjoint operator

$$A := \frac{1}{2}(x \cdot p + p \cdot x).$$

This operator is the generator of a one-parameter group of unitary transformations called *dilations*:

$$\psi(x) \mapsto e^{3\theta/2}\psi(e^{\theta}x)$$

for $\theta \in \mathbb{R}$. Let us show formally (i.e. ignoring domain issues) that if ψ is an eigenfunction of H, then

$$\langle \psi, i[H, A]\psi \rangle = 0. \tag{8.12}$$

Letting λ be the eigenvalue corresponding to ψ, and using (5.3), we have

$$\langle \psi, i[H, A]\psi \rangle = \langle \psi, i[H - \lambda, A]\psi \rangle = -2Im\langle (H - \lambda)\psi, A\psi \rangle.$$

Since $(H - \lambda)\psi = 0$, (8.12) follows.

On the other hand, a simple computation (left as an exercise) yields

$$i[H, A] = -\frac{\hbar^2}{m}\Delta - x \cdot \nabla V(x). \tag{8.13}$$

Therefore $i[H, A] \geq -\frac{\hbar^2}{m}\Delta$ by the repulsivity condition on $V(x)$, which implies

$$\langle \psi, i[H, A]\psi \rangle \geq \frac{\hbar^2}{m}\|\nabla\psi\|^2 > 0,$$

a contradiction to (8.12). Thus the operator H, with $V(x)$ satisfying $x \cdot \nabla V(x) \leq 0$ has no eigenvalues.

Problem 8.7 Show (8.13).

We turn our attention now from a situation where there are no bound states, to one where there are many. In particular, we will prove that the Schrödinger operator

$$H = -\frac{\hbar^2}{2m}\Delta - \frac{q}{|x|} \tag{8.14}$$

with $q > 0$, has an infinite number of bound states. The potential $V(x) = -q/|x|$ is called the (attractive) *Coulomb potential*. For appropriate q, the operator H describes either the the Hydrogen atom, or a one-electron ion. In Section 7.3, we went further, solving the eigenvalue problem for H exactly, finding explicit expressions for the eigenfunctions and (infinitely many) eigenvalues. Nevertheless, it is useful to be able to prove the existence of infinitely many bound states using an argument that does not rely on the explicit solvability of the eigenvalue problem.

We would first like to apply Theorem 6.16 to locate the essential spectrum. However, the fact that the Coulomb potential is singular at the origin is a possible obstacle. In fact, Theorem 6.16 can be extended to cover this case (see Problem 6.17 and, eg, [73]), and we may conclude that H is self-adjoint, with essential spectrum equal to the half-line $[0, \infty)$.

To prove that the operator (8.14) has an infinite number of negative eigenvalues, we will construct an infinite sequence of normalized, mutually orthogonal test functions, $u_n(x)$, such that

$$\langle u_n, H u_n \rangle < 0. \tag{8.15}$$

The "min-max principle" (described in Section 8.1) then implies that H has an infinite number of eigenvalues.

We begin by choosing a single function, $u(x)$, which is smooth, and which satisfies

$$\|u\| = 1 \quad \text{and} \quad \text{supp}(u) \subset \{ x \in \mathbb{R}^3 \mid 1 < |x| < 2 \}.$$

Then we set $u_n(x) := n^{-3/2} u(n^{-1}x)$ for $n = 1, 2, 4, 8, \ldots$.

Problem 8.8 Show that $\langle u_m, u_n \rangle = \delta_{mn}$, and that $\langle u_m, H u_n \rangle = 0$ if $m \neq n$.

Given the results of the exercise, it remains to show (8.15) for n sufficiently large. Indeed, changing variables to $y = n^{-1}x$, we compute

$$\langle u_n, H u_n \rangle = \frac{\hbar^2}{2m} \int |\nabla u_n(x)|^2 dx - q \int \frac{1}{|x|} |u_n(x)|^2 dx$$

$$= n^{-2} \frac{\hbar^2}{2m} \int |\nabla u(y)|^2 dy - n^{-1} q \int \frac{1}{|y|} |u(y)|^2 dy$$

$$< 0$$

for n sufficiently large, as the second term – the potential term – prevails for large n. Thus we have proved that the operator (8.14) has an infinite number of negative eigenvalues.

With this example in mind, we address the question of whether the Schrödinger operator

$$H = -\frac{\hbar^2}{2m} \Delta + V(x)$$

has a finite (including possibly 0) or infinite number of negative eigenvalues. Assume $V(x)$ behaves at infinity as

$$V(x) = c|x|^{-\alpha} \quad \text{for } |x| \text{ sufficiently large} \tag{8.16}$$

for some constant c. We test the operator H on the functions $u_n(x)$ constructed above. It is left as an exercise to show that, as above, $\langle u_m, Hu_n \rangle = 0$ for $m \neq n$, and

$$\langle u_n, Hu_n \rangle \quad \begin{cases} < 0 & \text{if } \alpha < 2 \text{ and } c < 0 \\ > 0 & \text{otherwise} \end{cases}$$

for n sufficiently large.

Problem 8.9 Prove these last two statements.

Let X_n be the n-dimensional subspace spanned by the functions $u_m, \ldots, u_{m+n}$, for m sufficiently large. The results of Problems 8.8 and 8.9 imply that

$$\sup_{\psi \in X_n, \|\psi\|=1} \langle \psi, H\psi \rangle < 0$$

for all n, provided $\alpha < 2$ and $c < 0$. Invoking the min-max principle again, we see that if $\alpha < 2$ and $c < 0$, then the operator H has an infinite number of bound states.

It is shown below that H has only a finite number of bound states if $\alpha > 2$. The borderline for the question of the number of eigenvalues is given by the inequality

$$-\Delta \geq \frac{1}{4|x|^2}$$

(the Uncertainty Principle – see Section 5.2). This inequality shows that the kinetic term $-\frac{\hbar^2}{2m}\Delta$ dominates at ∞ if $\alpha > 2$, or if $\alpha = 2$ and $\frac{2m}{\hbar^2}c > -\frac{1}{4}$. Otherwise, the potential term favouring eigenvalues wins out. This simple intuition notwithstanding, there is presently no physically motivated proof of the finiteness of the discrete spectrum for $\alpha > 2$. The proof presented below uses mathematical ingenuity rather than physical intuition.

We now prove finiteness of the number of eigenvalues for $\alpha > 2$. To simplify the argument slightly, we assume the potential $V(x)$ is non-positive and denote $U(x) := -V(x) \geq 0$. Let $\lambda < 0$ be an eigenvalue of H with eigenfunction ϕ. The eigenvalue equation $(H - \lambda)\phi = 0$ can be re-written as

$$(-\frac{\hbar^2}{2m}\Delta - \lambda)\phi = U\phi.$$

Since $\lambda < 0$ is in the resolvent set of the operator $-\frac{\hbar^2}{2m}\Delta$, we can invert $(-\frac{\hbar^2}{2m}\Delta - \lambda)$ to obtain

$$\phi = (-\frac{\hbar^2}{2m}\Delta - \lambda)^{-1}U\phi,$$

the *homogeneous Lippmann-Schwinger equation*. By introducing the new function $v := U^{1/2}\phi$, this equation can be further re-written to read $v = K(\lambda)v$ where the operator $K(\lambda)$ is defined as

$$K(\lambda) := U^{1/2}(-\frac{\hbar^2}{2m}\Delta - \lambda)^{-1}U^{1/2}.$$

We can summarize the above derivation as follows:

$$\lambda < 0 \ EV \ H \quad \leftrightarrow \quad 1 \ EV \ K(\lambda)$$

and so

$$\#\{\lambda < 0 \mid \lambda \ EV \ H\} = \#\{\lambda < 0 \mid 1 \ EV \ K(\lambda)\}. \tag{8.17}$$

The next step is to prove that

$$\#\{\lambda < 0 \mid 1 \ EV \ K(\lambda)\} = \#\{\nu > 1 \mid \nu \ EV \ K(0)\}. \tag{8.18}$$

To prove (8.18), we begin by showing that

$$\frac{\partial}{\partial\lambda}K(\lambda) > 0 \ \forall \ \lambda \leq 0 \tag{8.19}$$

and

$$K(\lambda) \to 0 \ \text{as} \ \lambda \to -\infty. \tag{8.20}$$

Writing

$$\langle\phi, K(\lambda)\phi\rangle = \langle U^{1/2}\phi, (-\frac{\hbar^2}{2m}\Delta - \lambda)^{-1}U^{1/2}\phi\rangle$$

and differentiating with respect to λ, we obtain

$$\frac{\partial}{\partial\lambda}\langle\phi, K(\lambda)\phi\rangle = \langle U^{1/2}\phi, (-\frac{\hbar^2}{2m}\Delta - \lambda)^{-2}U^{1/2}\phi\rangle$$

$$= \|(-\frac{\hbar^2}{2m}\Delta - \lambda)^{-1}U^{1/2}\phi\|^2 > 0$$

which proves (8.19). To establish (8.20), we need to derive the integral kernel of the operator $K(\lambda)$. Using the fact that the operator $(-\frac{\hbar^2}{2m}\Delta - \lambda)^{-1}$ has integral kernel

$$(-\frac{\hbar^2}{2m}\Delta - \lambda)^{-1}(x, y) = \frac{m}{2\pi\hbar^2|x - y|}e^{-\sqrt{\frac{2m|\lambda|}{\hbar^2}}|x-y|}$$

(see Equation (25.57)), we find that the integral kernel for $K(\lambda)$ is

$$K(\lambda)(x, y) = U(x)^{1/2}\frac{m}{2\pi\hbar^2|x - y|}e^{-\sqrt{\frac{2m|\lambda|}{\hbar^2}}|x-y|}U(y)^{1/2}$$

(see Section 25.3). Using the estimate

$$\|K\| \le \left(\int_{\mathbb{R}^3 \times \mathbb{R}^3} |K(x,y)|^2 dxdy\right)^{1/2}$$

(see again Section 25.3), we obtain

$$\|K(\lambda)\| \le \frac{m}{2\pi\hbar^2} \left(\int_{\mathbb{R}^3 \times \mathbb{R}^3} \frac{U(x)U(y)}{|x-y|^2} e^{-2\sqrt{\frac{2m|\lambda|}{\hbar^2}}|x-y|} dxdy\right)^{1/2}.$$

Since $\exp(-\sqrt{\frac{2m|\lambda|}{\hbar^2}}|x-y|) \to 0$ as $\lambda \to -\infty$, Equation (8.20) follows.

Now we show that the relations (8.19) and (8.20) imply (8.18). By (8.20), for all λ sufficiently negative, all of the eigenvalues of $K(\lambda)$ are less than 1. By (8.19), the eigenvalues, $\nu_m(\lambda)$, of $K(\lambda)$ increase monotonically with λ. Hence if $\nu_m(\lambda_m) = 1$ for some $\lambda_m < 0$, then $\nu_m(0) > \nu_m(\lambda_m) = 1$. Similarly, if $\nu_m(0) > 1$, then there is a $\lambda_m < 0$ such that $\nu_m(\lambda_m) = 1$. In other words, there is a one-to-one correspondence between the eigenvalues $\nu_m(0)$ of $K(0)$ which are greater than 1, and the points λ_m at which some eigenvalue $\nu_m(\lambda)$ crosses 1 (see Fig. 8.1).

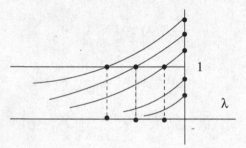

Fig. 8.1. Eigenvalues of $K(\lambda)$.

Thus (8.18) follows.

The relations (8.17) and (8.18) imply

$$\#\{\lambda < 0 \mid \lambda \text{ EV } H\} = \#\{\nu > 1 \mid \nu \text{ EV } K(0)\}. \tag{8.21}$$

The quantity on the left hand side of (8.21) is what we would like to estimate, while the quantity on the right hand side is what we *can* estimate. Indeed, we have

$$\#\{\nu > 1 \mid \nu \text{ EV } K(0)\} = \sum_{\nu_m>1, \nu_m EV K(0)} 1 \le \left(\sum_{\nu_m>1, \nu_m EV K(0)} \nu_m^2\right)^{1/2}$$

$$\le \left(\sum_{\nu_m EV K(0)} \nu_m^2\right)^{1/2} = \left(tr(K(0)^2)\right)^{1/2}$$

$$\tag{8.22}$$

(see Section 25.11 for the definition of tr, the *trace*). On the other hand (see Section 25.3),

$$tr(K(0)^2) = \int |K(0)(x,y)|^2 dxdy = \left(\frac{m}{2\pi\hbar^2}\right)^2 \int \frac{U(x)U(y)}{|x-y|^2} dxdy \qquad (8.23)$$

Collecting equations (8.21)- (8.23) and recalling that $V(x) = -U(x)$, we obtain

$$\#\{\lambda < 0 \mid \lambda \text{ EV } H\} \le \frac{m}{2\pi\hbar^2} \left(\int \frac{|V(x)V(y)|}{|x-y|^2} dxdy\right)^{1/2}.$$

Under our assumption (8.16) on the potential $V(x)$, with $\alpha > 2$,

$$\int \frac{|V(x)V(y)|}{|x-y|^2} dxdy < \infty,$$

so that the number of negative eigenvalues of the operator H is finite. This is the fact we set out to prove.

The argument used above (Equation (8.21) in particular) is called the *Birman-Schwinger principle*, and the operator $K(\lambda)$ is the *Birman-Schwinger operator*.

Scattering States

In this chapter we study scattering states in a little more detail. As we saw in Section 6.4, scattering states are solutions of the time-dependent Schrödinger equation

$$ i\hbar \frac{\partial \psi}{\partial t} = H\psi \tag{9.1} $$

with initial condition orthogonal to all eigenfunctions of H:

$$ \psi|_{t=0} = \psi_0 \in \mathcal{H}_b^{\perp} \tag{9.2} $$

where $\mathcal{H}_b := span\{$ eigenfunctions of $H\}$ is the subspace of bound states of H.

We will have to make a more precise assumption on the potential $V(x)$ entering the Schrödinger operator $H = -\frac{\hbar^2}{2m}\Delta + V(x)$: we assume, for simplicity, that

$$ |\partial_x^\alpha V(x)| \leq C(1 + |x|)^{-\mu - |\alpha|} \tag{9.3} $$

for $|\alpha| \leq 2$ and for some $\mu > 0$ (and C a constant). The notation needs a little explanation: α is a multi-index $\alpha = (\alpha_1, \alpha_2, \alpha_3)$ with each α_j a non-negative integer, $|\alpha| = \sum_{j=1}^3 \alpha_j$, and

$$ \partial_x^\alpha = \prod_{j=1}^3 \partial_{x_j}^{\alpha_j}. $$

The question we want to address is what is the asymptotic behaviour of the solution $\psi = e^{-iHt/\hbar}\psi_0$ of (9.1)-(9.2) as $t \to \infty$. First observe that (6.10) shows that ψ moves away from any bounded region of space as $t \to \infty$. Hence, since

$$ V(x) \to 0 \qquad \text{as} \qquad |x| \to \infty, \tag{9.4} $$

we expect that the influence of the potential $V(x)$ diminishes as $t \to \infty$. Thus the following question arises: does the evolution ψ approach a free evolution,

© Springer-Verlag GmbH Germany, part of Springer Nature 2020
S. J. Gustafson and I. M. Sigal, *Mathematical Concepts of Quantum Mechanics*, Universitext, https://doi.org/10.1007/978-3-030-59562-3_9

say $\phi = e^{-iH_0 t/\hbar}\phi_0$, where $H_0 = -\frac{\hbar^2}{2m}\Delta$, for some $\phi_0 \in L^2(\mathbb{R}^3)$, as $t \to \infty$? Put differently, given $\psi_0 \in \mathcal{H}_b^\perp$, is there $\phi_0 \in L^2(\mathbb{R}^3)$ such that

$$\|e^{-iHt/\hbar}\psi_0 - e^{-iH_0 t/\hbar}\phi_0\| \to 0 \tag{9.5}$$

as $t \to \infty$? This (or a modification of this question discussed below) is the problem of *asymptotic completeness*. It is the central problem of mathematical scattering theory. It conjectures that all the possible free motions, together with the bound state motions, form a complete set of possibilities for the asymptotic behaviour of solutions of time-dependent Schrödinger equations.

There are two principal cases depending on the decay rate, μ, of the potential $V(x)$ (more precisely, the decay rate, $\mu+1$, of the force $-\nabla V(x)$), and it is only in the first of these cases that the asymptotic completeness property formulated above holds. We now discuss these cases in turn.

9.1 Short-range Interactions: $\mu > 1$

In this case, asymptotic completeness can be proved under condition (9.3). We can reformulate the asymptotic completeness property by defining the *wave operator*, Ω^+:

$$\Omega^+ \phi := \lim_{t \to \infty} e^{iHt/\hbar} e^{-iH_0 t/\hbar}\phi. \tag{9.6}$$

Below we will show that under the condition (9.3) with $\mu > 1$, the limit exists for any $\phi \in L^2(\mathbb{R}^3)$. Further, the operator Ω^+ is an isometry:

$$\|\Omega^+ \phi\| = \|\phi\|. \tag{9.7}$$

Indeed, (9.6) implies

$$\|\Omega^+ \phi\| = \lim_{t \to \infty} \|e^{iHt/\hbar} e^{-iH_0 t/\hbar}\phi\| = \|\phi\|$$

since the operators $e^{iHt/\hbar}$ and $e^{-iH_0 t/\hbar}$ are isometries.

The existence of the wave operator Ω^+ means that given a free evolution $e^{-iH_0 t/\hbar}\phi_0$, there is a full evolution $e^{-iHt/\hbar}\psi_0$ such that (9.5) holds. To see this, note that since $e^{iHt/\hbar}$ is an isometry, (9.5) can be re-written as

$$\|\psi_0 - e^{iHt/\hbar} e^{-iH_0 t/\hbar}\phi_0\| \to 0$$

as $t \to \infty$, which is equivalent to the relation

$$\psi_0 = \Omega^+ \phi_0. \tag{9.8}$$

Thus the existence of the wave operator Ω^+ is equivalent to the existence of scattering states – i.e. states $e^{-iHt/\hbar}\psi_0$ for which (9.5) holds for some ϕ_0.

We have

$$\mathrm{Ran}(\Omega^+) \subset \mathcal{H}_b^\perp.$$

Indeed, for any $\phi_0 \in L^2(\mathbb{R}^3)$ and $g \in \mathcal{H}_b$ such that $Hg = \lambda g$, we have

$$\langle g, \Omega^+ \phi_0 \rangle = \lim_{t \to \infty} \langle g, e^{iHt/\hbar} e^{-iH_0 t/\hbar} \phi_0 \rangle = \lim_{t \to \infty} \langle e^{-iHt/\hbar} g, e^{-iH_0 t/\hbar} \phi_0 \rangle$$
$$= \lim_{t \to \infty} e^{i\lambda t/\hbar} \langle g, e^{-iH_0 t/\hbar} \phi_0 \rangle. \tag{9.9}$$

There is a general theorem implying that the right hand side here is zero, but we will prove it directly. Recalling expression (2.23) for the action of the free evolution operator $e^{-iH_0 t/\hbar}$, we find

$$\langle g, e^{-iH_0 t/\hbar} \phi_0 \rangle = \left(\frac{2\pi i \hbar t}{m} \right)^{-3/2} \int_{\mathbb{R}^3} \int_{\mathbb{R}^3} \bar{g}(x) e^{im|x-y|^2/(2\hbar t)} \phi_0(y) dy dx$$

and therefore, if ϕ_0 and g are integrable functions, i.e.,

$$\int_{\mathbb{R}^3} |\phi_0(x)| dx < \infty \qquad \text{and} \qquad \int_{\mathbb{R}^3} |g(x)| dx < \infty,$$

we have

$$|\langle g, e^{-iH_0 t/\hbar} \phi_0 \rangle| \leq \left(\frac{2\pi \hbar t}{m} \right)^{-3/2} \int_{\mathbb{R}^3} |g| \int_{\mathbb{R}^3} |\phi_0|$$

and thus the right hand side of (9.9) vanishes. For general ϕ_0 and g the result is obtained by continuity. This argument can also be extended to cover the case where g is a linear combination of eigenfunctions.

We can similarly define the wave operator Ω^- describing the asymptotic behaviour as $t \to -\infty$:

$$\Omega^- \phi := \lim_{t \to -\infty} e^{iHt/\hbar} e^{-iH_0 t/\hbar} \phi.$$

This operator maps free states $e^{-iH_0 t/\hbar} \phi_0$ into states $e^{-iHt/\hbar} \psi_0$ which approach these free states as $t \to -\infty$. We have (in an appropriate sense)

$$H\Omega^\pm = \Omega^\pm H_0. \tag{9.10}$$

Indeed, by changing variables, we can obtain the following *intertwining relations*

$$e^{-iHt/\hbar} \Omega^\pm = \Omega^\pm e^{-iH_0 t/\hbar}. \tag{9.11}$$

Differentiating these relations at $t = 0$, we obtain 9.10.

Let $\mathcal{H}_{sc} := Ran\Omega^+$, the set of scattering states. We have shown that $\mathcal{H}_{sc} \subset \mathcal{H}_b^\perp$. The property of *asymptotic completeness* states that

$$\mathcal{H}_{sc} = \mathcal{H}_b^\perp \qquad \text{or} \qquad \mathcal{H}_b \oplus Ran\mathcal{H}_{sc} = L^2(\mathbb{R}^3)$$

i.e., that the scattering states and bound states span the entire state space $L^2(\mathbb{R}^3)$. We will prove this property under some restrictive conditions below.

One can also define the *scattering operator*

$$S := {\Omega^+}^* \Omega^-$$

which maps asymptotic states at $t = -\infty$ into asymptotic states at $t = \infty$:

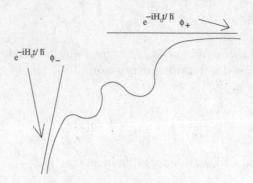

Fig.9.1. $S : \phi_- \to \phi_+$

Equations (9.11) and (9.10) imply $e^{-iH_0t/\hbar}S = Se^{-iH_0t/\hbar}$ and

$$H_0S = SH_0.$$

The property of asymptotic completeness implies that the Hamiltonian H restricted to the invariant subspace $\mathcal{H}_b^\perp$ is unitarily equivalent to the free Hamiltonian H_0:

$$H = \Omega^\pm H_0 {\Omega^\pm}^* \text{ on } \mathcal{H}_b^\perp.$$

Let $\Omega^\pm(x, y)$ be the integral kernels of the operators $\Omega^\pm$, and let $\psi^\pm(x, k)$ denote their Fourier transforms with respect to the second variable, y. Equation (9.10) implies that

$$H\psi^\pm(x, k) = \frac{|k|^2}{2m}\psi^\pm(x, k). \tag{9.12}$$

In other words, $\psi^\pm(x, k)$ are generalized eigenfunctions of the operator H, with eigenvalue $|k|^2/2m$. These generalized eigenfunctions are of the form

$$\psi^\pm(x, k) = e^{\pm ik \cdot x} + O\left(\frac{1}{|x|}\right)$$

as $|x| \to \infty$ (see, eg., [302, 246]). They are called *scattering eigenfunctions* of H.

Problem 9.1 Prove relation (9.12).

9.2 Long-range Interactions: $\mu \leq 1$

In this case, the relation (9.5) must be modified to read

$$\|e^{iHt/\hbar}\psi_0 - e^{-iS(t)}\phi_0\| \to 0$$

as $t \to \infty$, where for $\frac{1}{2} < \mu \le 1$, the operator-family $S(t)$ is defined to be $S(t) = S(p,t)$, where the function $S(k,t)$ satisfies the equation

$$\frac{\partial}{\partial t}S(k,t) = H_0(k) + V\left(\frac{1}{m}kt\right).$$

Here $H_0(k) = \frac{1}{2m}|k|^2$ is the classical free Hamiltonian function. The operator $S(p,t)$ is defined according to the rules described in Section 4.2. Note that in this equation, the coordinate x in the potential $V(x)$ is replaced by its free, classical expression, $\frac{1}{m}kt$. With this modification, the asymptotic completeness property can again be established under assumption (9.3) with $\frac{1}{2} < \mu \le 1$. For $0 < \mu \le \frac{1}{2}$, the expression for the operator-family $S(t)$ is more complicated (see [77]). As in the short-range case, one can introduce modified wave operators and investigate their properties (see the references given in Chapter 27).

9.3 Wave Operators

As promised, we prove the existence and completeness of the wave operators $\Omega^{\pm}$. To simplify the proof, we impose a somewhat stronger condition on the potential: $V(x) \in L^2(\mathbb{R}^3)$. Below, we also make use of the space

$$L^1(\mathbb{R}^3) := \{\psi : \mathbb{R}^3 \to \mathbb{C} \mid \int_{\mathbb{R}^3} |\psi(x)|dx < \infty\}.$$

Theorem 9.2 If $V \in L^2(\mathbb{R}^3)$, then the wave operators $\Omega^{\pm}$, introduced formally above, exist.

Proof. Denote $\Omega^t := e^{iHt}e^{-iH_0t}$ (we drop the $\hbar$ in what follows, just to simplify the notation). Since $\|\Omega^t\| \le const$, uniformly in t (in fact, $\|\Omega^t\| = 1$), it suffices to prove the existence of the limit (9.6) (and similarly for $t \to -\infty$) on functions $\phi \in L^2(\mathbb{R}^3) \cap L^1(\mathbb{R}^3)$ (the existence of the limit for $\phi \in L^2$ will then follow by approximating ϕ by elements of the dense subspace $L^2 \cap L^1$). For $t \ge t'$, we write the vector-function $\Omega^t\phi - \Omega^{t'}\phi$ as the integral of its derivative:

$$\Omega^t\phi - \Omega^{t'}\phi = \int_{t'}^t \frac{d}{ds}\Omega^s\phi ds.$$

Using the relation $\frac{d}{ds}e^{iHs} = iHe^{iHs}$, and similarly for e^{-iH_0s} (see Chapter 2), we find

$$\frac{d}{ds}\Omega^s = iHe^{iHs}e^{-iH_0s} + e^{iHs}(-iH_0)e^{-iH_0s} = ie^{iHs}Ve^{-iH_0s},$$

as $H - H_0 = V$. Since $\|e^{iH\hat{s}}\| = 1$, and using the Minkowski inequality $\|\int \phi(s)ds\| \leq \int \|\phi(s)\|ds$ (see eg. [106]), we have

$$\|\Omega^t\phi - \Omega^{t'}\phi\| \leq \int_t^{t'} \|Ve^{-iH_0s}\phi\|ds. \tag{9.13}$$

We estimate the integrand as follows:

$$\|Ve^{-iH_0s}\phi\|^2 = \int_{\mathbb{R}^3} |V(x)\left(e^{-iH_0s}\phi\right)(x)|^2 dx$$

$$\leq \sup_{x'\in\mathbb{R}^3} |\left(e^{-iH_0s}\phi\right)(x')|^2 \int_{\mathbb{R}^3} |V(x)|^2 dx,$$

yielding

$$\|Ve^{-iH_0s}\phi\| \leq \|V\|_{L^2}\sup_{x'}|(e^{-iH_0s}\phi)(x')|.$$

Finally, recalling the bound (see (2.24))

$$\left|\left(e^{-iH_0s/\hbar}\phi\right)(x)\right| \leq \left(\frac{2\pi\hbar s}{m}\right)^{-3/2}\int_{\mathbb{R}^3}|\phi(x)|dx \tag{9.14}$$

from Section 2.4, we obtain

$$\|Ve^{-iH_0s}\phi\| \leq (const)s^{-3/2}\|V\|_{L^2}\|\phi\|_{L^1}$$

and so

$$\int_1^\infty \|Ve^{-iH_0s}\phi\|ds \leq (const)\|V\|_{L^2}\|\phi\|_{L^1}.$$

This shows that the right hand side in (9.13) vanishes as $t', t \to \infty$. In other words, for any sequence $t_j \to \infty$, $\{\Omega^{t_j}\phi\}$ is a Cauchy sequence, and so $\{\Omega^t\phi\}$ converges as $t \to \infty$. Convergence for $t \to -\infty$ is proved in the same way. $\square$

Finally, as promised, we prove asymptotic completeness in a special case.

Theorem 9.3 Assume $V \in L^\infty \cap L^1$ with $\|V\|_{L^\infty}$ and $\|V\|_{L^1}$ sufficiently small. Then $\text{Ran}\Omega_\pm = \mathcal{H}$. Consequently, $\mathcal{H}_b$ is empty, and asymptotic completeness holds.

Proof. We begin by proving that for any $\psi \in L^1 \cap L^2$, $e^{-itH}\psi$ has some decay in time, as measured in the $\|\cdot\|_{L^2+L^\infty}$ norm, which is defined by

$$\|f\|_{L^2+L^\infty} := \inf_{f=g+h}(\|g\|_{L^2} + \|h\|_{L^\infty}).$$

In fact, we claim

$$\|e^{-itH}\psi\|_{L^2+L^\infty} \leq (const)(1+|t|)^{-3/2}. \tag{9.15}$$

To show this, set $M(t) := \sup_{0 \le s \le t}(1+|s|)^{3/2}\|e^{-isH}\psi\|_{L^2+L^\infty}$. Writing, as above,

$$e^{isH_0}e^{-isH}\psi - \psi = \int_0^s \frac{d}{d\tau}e^{i\tau H_0}e^{-i\tau H}\psi d\tau$$

and applying e^{-isH_0}, we arrive at the *Duhamel formula*

$$e^{-isH}\psi = e^{-isH_0}\psi - i\int_0^s e^{-i(s-\tau)H_0}Ve^{-isH}\psi ds.$$

We estimate as follows:

$$\|e^{-isH}\psi\|_{L^2+L^\infty} \le \|e^{-isH_0}\psi\|_{L^2+L^\infty} + \int_0^{s-1}\|e^{-i(s-\tau)H_0}Ve^{-i\tau H}\psi\|_{L^\infty}d\tau$$
$$+ \int_{s-1}^s \|e^{-i(s-\tau)H_0}Ve^{-i\tau H}\psi\|_{L^2}d\tau.$$

Using (9.14) as above, together with Hölder's inequality, we find

$$\|e^{-isH}\psi\|_{L^2+L^\infty} \le (const)[(1+|s|)^{-3/2}\|\psi\|_{L^1\cap L^2}$$
$$+ \int_0^{s-1}|s-\tau|^{-3/2}\|V\|_{L^1\cap L^2}\|e^{-i\tau H}\psi\|_{L^2+L^\infty}d\tau$$
$$+ \int_{s-1}^s \|V\|_{L^2\cap L^\infty}\|e^{-i\tau H}\psi\|_{L^2+L^\infty}d\tau]$$
$$\le (const)[(1+|s|)^{-3/2}\|\psi\|_{L^1\cap L^2} + \|V\|_{L^1\cap L^\infty}M(s)I(s)],$$

where $I(s) := \int_0^{s-1}|s-\tau|^{-3/2}(1+|\tau|)^{-3/2}d\tau + \int_{s-1}^s(1+|\tau|)^{-3/2}d\tau$, and so multiplying through by $(1+|s|)^{3/2}$, and taking supremum over $s \le t$ and using $\sup_{0 \le s \le t}I(s) \le (const)$, we find

$$M(t) \le (const)[\|\psi\|_{L^1\cap L^2} + \|V\|_{L^1\cap L^\infty}M(t)]. \tag{9.16}$$

Thus if $\|V\|_{L^1\cap L^\infty}$ is sufficiently small (so that $(const)\|V\|_{L^1\cap L^\infty} \le 1/2$, say), (9.16) implies $M(t) \le (const)$ for all t, and so (9.15) holds.

Now denote $W(t) := e^{iH_0 t}e^{-iHt}$, and, as above, write

$$W(t) - W(t') = i\int_{t'}^t e^{iH_0 s}Ve^{-iHs}ds.$$

Using Hölder again, for $\psi \in L^1 \cap L^2$ we find

$$\|(W(t)-W(t'))\psi\|_{L^2} \le \int_{t'}^t \|V\|_{L^2\cap L^\infty}\|e^{-iHs}\psi\|_{L^2+L^\infty}ds$$
$$\le (const)\int_{t'}^t(1+|s|)^{-3/2}ds \to 0$$

as $t > t' \to \infty$. Hence $\{W(t)\psi\}$ is a Cauchy sequence in L^2, and therefore $\lim_{t\to\infty} W(t)\psi$ exists for all $\psi \in L^1 \cap L^2$. Since the operator $W(t)$ is bounded uniformly in t, this limit exists for all $\psi \in L^2$.

Denote $W_\pm \psi := \lim_{t\to\pm\infty} W(t)\psi$. We want to show that $\Omega_\pm W_\pm = 1$, $W_\pm \Omega_\pm = 1$ and $W_\pm = \Omega_\pm^*$. Indeed, since $\Omega_\pm = s - \lim e^{iHt} e^{-iH_0 t}$ and $W_\pm = s - \lim e^{iH_0 t} e^{-iHt}$ we have that $\Omega_\pm W_\pm = s - \lim e^{iHt} e^{-iH_0 t} e^{iH_0 t} e^{-iHt} = 1$ and similarly for the other relations. The relation $\Omega_\pm W_\pm = 1$ implies that $\mathrm{Ran}\,\Omega_\pm = \mathcal{H}$, i.e., $\Omega_\pm$ is unitary. This implies the asymptotic completeness.

9.4 Appendix: The Potential Step and Square Well

Potential step. In the one-dimensional case one can say much more about scattering process. In particular, one can introduce very useful reflection and transmission coefficients as illustrated in the example of the one-dimensional potential

$$V(x) = \begin{cases} V_0 & x > 0 \\ 0 & x \le 0 \end{cases}$$

with $V_0 > 0$. Consider the "eigenvalue" problem

$$-\frac{\hbar^2}{2m}\psi'' + V\psi = E\psi. \tag{9.17}$$

We put the term eigenvalue in quotation marks because we will allow solutions, ψ, which are not L^2-functions. Solving this eigenvalue problem separately in the two different regions gives us a general solution of the form

$$\psi = \begin{cases} Ae^{ik_0 x} + Be^{-ik_0 x} & (\frac{\hbar^2 k_0^2}{2m} = E) & x < 0 \\ Ce^{ik_1 x} + De^{-ik_1 x} & (\frac{\hbar^2 k_1^2}{2m} = E - V_0) & x > 0 \end{cases}.$$

There are no bound states (L^2 solutions), but we can say something about the scattering states.

Suppose $0 < E < V_0$, and take $k_1 = iK$ where $K = \sqrt{(2m/\hbar^2)(V_0 - E)} > 0$. Then for a bounded solution, we require $D = 0$. Imposing the condition that ψ be continuously differentiable at 0, that is

$$\psi|_{0-} = \psi|_{0+}, \qquad \partial\psi/\partial x|_{0-} = \partial\psi/\partial x|_{0+},$$

leads to the equations

$$A + B = C, \qquad ik_0(A - B) = -KC.$$

After some manipulation, we find that

$$\frac{B}{A} = \frac{k_0 - iK}{k_0 + iK},$$

or making the dependence on the energy E explicit,

$$\frac{B}{A} = \frac{\sqrt{E} - i\sqrt{V_0 - E}}{\sqrt{E} + i\sqrt{V_0 - E}}.$$

Similarly, if $E > V_0$, we obtain

$$\frac{B}{A} = \frac{\sqrt{E} - \sqrt{E - V_0}}{\sqrt{E} + \sqrt{E - V_0}},$$

and

$$R(E) := \left|\frac{B}{A}\right|^2 = \frac{1}{V_0^2}[2E - V_0 - 2\sqrt{E}\sqrt{E - V_0}]^2$$

($R(E)$ is called the *reflection coefficient*). In particular, if $\frac{E - V_0}{V_0} \ll 1$, then

$$R(E) \approx 1 - 4\sqrt{\frac{E - V_0}{V_0}}$$

and almost all of the wave is reflected. This is in spite of the fact that the energy of the particle lies above the barrier. In classical mechanics, the particle would pass over the barrier.

The square well. We consider the "eigenvalue" problem, $H\psi = E\psi$, for the square well potential defined in Section 7.1, and for positive energies $E > 0$. The "eigenvalue" equation is given by the same expression (9.17), but we do not expect this equation to have L^2 solutions, i.e. bound states. So we look for bounded solutions which converge to plane waves as $x \to \pm\infty$. Such solutions were called *scattering states* above.

Consider the situation where a plane wave $\psi_{inc}(x) = Ae^{ikx}$ is incoming from the left. Then on the left of the well, ψ is a superposition of incoming and reflected plane waves:

$$\psi = Ae^{ikx} + AB(E)e^{-ikx} \qquad x < -a/2$$

while on the right of the well, ψ is an outgoing (transmitted) plane wave:

$$\psi = AC(E)e^{ik(x-a)} \qquad x > a/2.$$

Problem 9.4 Show that

$$C(E) = [\cos(\bar{k}a) - i\frac{\bar{k}^2 + k^2}{2\bar{k}k}\sin(\bar{k}a)]^{-1}$$

where

$$\bar{k} = \sqrt{\frac{2m(E + V_0)}{\hbar^2}} \quad \text{and} \quad k = \sqrt{\frac{2mE}{\hbar^2}}.$$

This implies

$$T(E) := |C(E)|^2 = \left[1 + \frac{V_0^2}{4E(E + V_0)} \sin^2(a\bar{k})\right]^{-1}$$

($T(E)$ is the *transmission coefficient*) which is sketched in Fig. 9.2.

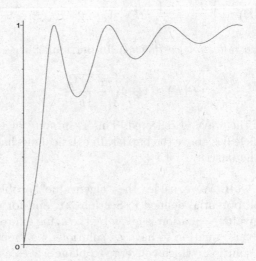

Fig. 9.2. Transmission coefficient.

We see that at the energies satisfying $\sin(\bar{k}a) = 0$, i.e.

$$E = -V_0 + \frac{n^2\pi^2\hbar^2}{2ma^2} > 0, \qquad n = 1, 2, \ldots$$

$T(E)$ has maxima ($T(E) = 1$) which are called *resonances*. The corresponding values of E are the *resonance energies*. We remark that for large n these are approximately equal to the energy levels of the infinite well of the same width.

Existence of Atoms and Molecules

In this chapter we prove existence of stationary, well localized and stable states of atoms and, in a certain approximation, molecules. These are the lowest energy states, and their existence means that our quantum systems exist as well-localized objects, and do not disintegrate into fragments under sufficiently small perturbations.

General many-body systems are considered in Chapter 13.

10.1 Essential Spectra of Atoms and Molecules

Recall from Section 4.5 that a molecule with N electrons of mass m and charge $-e$, and M nuclei of masses m_j and charges $Z_j e$, $j = 1, \dots, M$, is described by the Schrödinger operator

$$H_{mol} = -\sum_{1}^{N} \frac{\hbar^2}{2m} \Delta_{x_j} - \sum_{1}^{M} \frac{\hbar^2}{2m_j} \Delta_{y_j} + V(x, y) \qquad (10.1)$$

acting on $L^2_{sym}(\mathbb{R}^{3(N+M)})$. Here $L^2_{sym}(\mathbb{R}^{3(N+M)})$ is a symmetry subspace of $L^2(\mathbb{R}^{3(N+M)})$ reflecting the fact that electrons and some of the nuclei are identical particles, $x = (x_1, \dots, x_N)$ and $y = (y_1, \dots y_M)$ are the electron and nucleus coordinates, respectively, and

$$V(x, y) = \frac{1}{2} \sum_{i \neq j} \frac{e^2}{|x_i - x_j|} - \sum_{i,j} \frac{e^2 Z_j}{|x_i - y_j|} + \frac{1}{2} \sum_{i \neq j} \frac{e^2 Z_i Z_j}{|y_i - y_j|}, \qquad (10.2)$$

the sum of Coulomb interaction potentials between the electrons (the first term on the r.h.s.), between the electrons and the nuclei (the second term), and between the nuclei (the third term). (See Section 4.5.) For a neutral molecule, we have

$$\sum_{j=1}^{M} Z_j = N.$$

© Springer-Verlag GmbH Germany, part of Springer Nature 2020
S. J. Gustafson and I. M. Sigal, *Mathematical Concepts of Quantum Mechanics*, Universitext, https://doi.org/10.1007/978-3-030-59562-3_10

If $M = 1$, the resulting system is called an atom, or Z-atom ($Z = Z_1$). In the case of atoms, the last term in (10.2) is absent.

Since nuclei are much heavier than electrons, in the leading approximation one can suppose that the nuclei are frozen at their positions. (For a more precise statement and discussion of this point see Section 12.1 below.) One then considers, instead of (10.1), the Schrödinger operator

$$H_N^{BO}(y) = -\sum_1^N \frac{\hbar^2}{2m} \Delta_{x_j} + V(x, y) \qquad (10.3)$$

acting in $L_{sym}^2(\mathbb{R}^{3N})$, the positions $y \in \mathbb{R}^{3M}$ of the nuclei appearing as parameters. This is called the Born-Oppenheimer approximation. It plays a fundamental role in quantum chemistry, where most computations are done with the operator $H_N^{BO}(y)$. We discuss the justification of this approximation in Section 12.2.

Theorem 10.1 (Kato theorem) The operators H_{mol} and $H_N^{BO}(y)$ are self-adjoint and bounded below.

Proof. The fact that the operators H_{mol} and $H_N^{BO}(y)$ are bounded below was shown in Section 5.3. Next, to fix ideas we prove the self-adjointness for H_{mol} only. The self-adjointness for $H_N^{BO}(y)$ is proven similarly. As in Problem 2.8 we have

$$\left\| \frac{1}{|x_i - x_j|} \psi \right\| \leq a \left\| \frac{1}{2m} \Delta_{x_i} \psi \right\| + b \|\psi\|,$$
$$\left\| \frac{1}{|x_i - y_j|} \psi \right\| \leq a \left\| \frac{1}{2m} \Delta_{x_i} \psi \right\| + b \|\psi\|, \qquad (10.4)$$

with $a > 0$ arbitrary and b depending on a and for all $\psi \in \mathcal{D}(H_0)$, where $H_0 := -\sum_1^N \frac{1}{2m} \Delta_{x_j} - \sum_1^M \frac{1}{2m_j} \Delta_{y_j}$. *(Prove this. Hint: Pick the variable x_i and follow the instructions in Problem 2.8, then integrate over the rest of the variables.)* Let V be given in (10.2). Then the last two estimates imply

$$\|V\psi\| \leq a\|H_0\psi\| + b\|\psi\| \qquad (10.5)$$

with $a > 0$ arbitrary and b depending on a. By Theorem 2.9, this implies that the operator H_{mol} is self-adjoint. $\square$

To simplify the exposition, in what follows we consider only atoms and molecules with fixed (infinitely heavy) nuclei and denote the corresponding quantum Hamiltonians by H_N. Thus H_N is equal to $H_N^{BO}(y)$, where in the case of atoms ($M = 1$) $y \equiv y_1$ is set to 0. For atoms the assumption of fixed nuclei is a minor one and can be easily removed; for molecules, it is a crucial one. A rigorous existence theory for molecules, apart from hydrogen, still does not exist.

We begin with the HVZ theorem describing the essential spectra of atoms and Born-Oppenheimer (BO) molecules.

Theorem 10.2 (HVZ theorem for atoms and BO molecules) $\sigma_{ess}(H_N)$ $= [\Sigma_N, \infty)$, where $\Sigma_N = \inf \sigma(H_{N-1})$.

This theorem is a special case of the general HVZ theorem proven in Chapter 13, so we omit its proof. The energy $\Sigma_N = \inf \sigma(H_{N-1})$ is called the ionization threshold. To obtain $\sigma_{ess}(H_N)$ we take one of the electrons to infinity and let it move freely there. The rest of the atom is placed in the ground state, so that the energy of the atom is

$$Energy = \Sigma_N + \frac{1}{2m}|k|^2 \quad \forall k \tag{10.6}$$

where k is the momentum of the electron which is placed at infinity. Varying $|k|$ from 0 to ∞ we see that (10.6) ranges over $[\Sigma_N, \infty)$. For molecules with mobile nuclei, the bottom of the essential spectrum is likely below its ionization threshold. However, the theorem above would imply the existence of bound states of H_{mol}, with $m/\min m_j$ sufficiently small, likely smaller then it is for real molecules (see Section 12.1).

10.2 Bound States of Atoms and BO Molecules

Are atoms or BO molecules stable? To answer this question we have to determine whether H_N has at least one bound state.

Theorem 10.3 For $N < Z + 1$, H_N has infinite number of eigenvalues below its ionization threshold $\Sigma_N = \inf \sigma(H_{N-1})$. Bound states, $\Psi_N^{(i)}(x_1, x_2, ..., x_N)$, of H_N, with energies $E_N^{(i)} < \Sigma_N$, satisfy the exponential bound

$$\int |\Psi_N^{(i)}(x)|^2 e^{2\alpha|x|} dx < \infty, \quad \forall \alpha < \sqrt{\Sigma_N - E_N^{(i)}}. \tag{10.7}$$

Proof. To simplify the exposition, we assume the ground states, if they exist, are unique. In the case without statistics, i.e. on the entire $L^2(\mathbb{R}^{3N})$, this is not hard to prove (see [247]). In the case with statistics, this is not known and is, probably, not true. However, a generalization of the proof below to the case of multiple ground states is straightforward.

We prove this theorem by induction in N. (This is strictly speaking not necessary, but is convenient.) We have shown already that it holds for the hydrogen atom, i.e. $N = 1$. Assume now it holds for $k \leq N - 1$ and prove it for $k = N$. Let $\Psi_{N-1}(x_1, x_2, ..., x_{N-1})$ be the normalized ground state of H_{N-1} with the ground state energy $E_{N-1} < E_{N-2}$.

First, for simplicity, we *ignore the statistics* and consider H_N on the entire space $L^2(\mathbb{R}^{3N})$. We use the variational principle with the test function

$$\phi = \Psi_{N-1}(x_1, ..., x_{N-1}) f(x_N), \tag{10.8}$$

where $f \in L^2(\mathbb{R}^3)$, $\|f\| = 1$. Using that $H_N = H_{N-1} - \frac{\hbar^2}{2m}\Delta_{x_N} + I_N$, with

$$I_N(x) := \sum_{i=1}^{N-1} \frac{e^2}{|x_i - x_N|} - \frac{e^2 Z}{|x_N|}, \qquad (10.9)$$

and $H_{N-1}\Psi_{N-1} = E_{N-1}\Psi_{N-1}$, we obtain

$$H_N \phi = (E_{N-1} + I_N)\phi + \Psi_{N-1}(-\frac{\hbar^2}{2m}\Delta_{x_N})f. \qquad (10.10)$$

This implies that

$$\langle \phi, H_N \phi \rangle = E_{N-1} + \langle f, -\frac{\hbar^2}{2m}\Delta f \rangle + \langle \phi, I_N \phi \rangle. \qquad (10.11)$$

By the exponential bound (10.7) for Ψ_{N-1}, i.e. $\int |\Psi_{N-1}(x)|^2 e^{2\alpha|x|} dx < \infty$, for any $\alpha < E_{N-2} - E_{N-1}$, it follows that

$$|\langle \Psi_{N-1}, (I_N(x) - I_N(x_N))\Psi_{N-1}\rangle_{L^2(\mathbb{R}^{N-1})}| \le (\text{const})|x_N|^{-2},$$

where $I_N(x_N) := I_N(x)|_{x_i=0} \; \forall i$. Observe that $I_N(x_N) = -q/|x_N|$, where $q := (Z - N + 1)e^2$, which implies

$$\langle \Psi_{N-1}, I_N(x)\Psi_{N-1}\rangle_{L^2(\mathbb{R}^{N-1})} \le -\frac{q}{|x_N|} + \frac{\text{const}}{|x_N|^2}, \qquad (10.12)$$

which, in turn, together with (10.11) and $\|\phi\| = 1$, gives

$$\langle \phi, (H_N - \Sigma_N)\phi \rangle \le \left\langle f, \left(-\frac{\hbar^2}{2m}\Delta_{x_N} - \frac{q}{|x_N|} + \frac{\text{const}}{|x_N|^2}\right) f \right\rangle_{L^2(\mathbb{R}^3)}.$$

Let $f \in C_0^\infty(\mathbb{R}^3)$ and satisfy $\|f\| = 1$ and

$$\text{supp}(f) \subset \{x_N \in \mathbb{R}^3 \mid 1 < |x_N| < 2\}.$$

Then the functions

$$f_n(x_N) = n^{-3/2} f(n^{-1}x_N), \qquad n = 1, 2, 4, 8, \ldots,$$

are orthonormal, and have disjoint supports. Thus the corresponding trial states $\phi_n(x) = \Psi_{N-1}(x_1, ..., x_{N-1})f_n(x_N)$ satisfy $\langle \phi_n, H_N \phi_m \rangle = 0$ for $n \ne m$, and

$$\langle \phi_n, (H_N - \Sigma_N)\phi_n \rangle \le -c_1 \frac{q}{n} + c_2 \frac{1}{n^2} < 0,$$

for $c_1 = \int \frac{1}{|x|}|f(x)|^2 dx$ and some positive constant c_2, if n is sufficiently large. Using this one can show that H_N possesses infinitely many discrete eigenvalues below the threshold Σ_N.

Taking into account statistics. Now we show how to modify the proof above to the case of spinless fermions, i.e. for the state space $L_{sym}^2(\mathbb{R}^{3N}) :=$

$\wedge_1^N L^2(\mathbb{R}^3)$. (The fact that electrons have spin $\frac{1}{2}$ can be accommodated similarly.) As before, let S_N be the group of permutations of N indices and define the anti-symmetrization projection

$$P_N^A \Psi = \frac{1}{N!} \sum_{\pi \in S_N} (-1)^{\#(\pi)} \Psi(x_{\pi(1)}, ..., x_{\pi(N)})$$

where, recall, $\#(\pi)$ is the number of transpositions making up the permutation π ($(-1)^{\#(\pi)}$ is the parity of $\pi \in S_N$). We replace the test function (10.8) by the function $\phi = P_N^A(\Psi_{N-1} \otimes f) / \|P_N^A(\Psi_{N-1} \otimes f)\|$. This gives $\phi := \sum_{j=1}^N \phi^{(j)}$ where, with the normalization constant c,

$$\phi^{(j)} := \pm c \Psi_{N-1}(x_1, ..., x_{j-1}, x_{j+1}, ..., x_N) f(x_j). \tag{10.13}$$

Let $I_j := \sum_{i:i \neq j} \frac{e^2}{|x_i - x_j|} - \frac{e^2 Z}{|x_j|}$. We take $f_\alpha(x) = \alpha^{\frac{3}{2}} f(\alpha x)$ with $\|f\| = 1$ and denote by ϕ_α and $\phi_\alpha^{(i)}$ the corresponding test functions. Then we have, for $i \neq j$,

$$|\langle \phi_\alpha^{(i)}, \phi_\alpha^{(j)} \rangle| \lesssim \frac{\alpha^3}{N}, \tag{10.14}$$

and, similarly,

$$|\langle \phi_\alpha^{(i)}, (-\frac{\hbar^2}{2m} \Delta_{x_j} + I_j) \phi_\alpha^{(j)} \rangle| \lesssim \frac{\alpha^3}{N}. \tag{10.15}$$

For $\alpha \to 0$, (10.14) implies that the normalization constant c in (10.13) is $c \asymp O(\frac{1}{\sqrt{N}})$. Then the equation (10.15) implies

$$\langle \phi_\alpha, H_N \phi_\alpha \rangle \leq E_{N-1} + \langle f_\alpha, (-\frac{\hbar^2}{2m} \Delta_x - \frac{q}{|x|}) f_\alpha \rangle + O(\alpha^3 N).$$

Hence $\langle \phi_\alpha, H_N \phi_\alpha \rangle < E_{N-1}$, if $\alpha \ll \frac{1}{\sqrt{N}}$ and $\alpha < \langle f, \frac{q}{|x|} f \rangle / \langle f, -\frac{\hbar^2}{2m} \Delta f \rangle$. This proves the existence of the ground state energy for H_N $\forall N$.

Finally, the bound (10.7) is a special case of the general exponential bound for many-body bound states given in Theorem 13.7, Section 13.6 (see also Theorem 8.6 of Section 8.2). $\square$

Problem 10.4 Go over the proof above and, where necessary, fill in the details.

A more refined estimate of $\langle \phi, I_N \phi \rangle$ (replacing (10.12)). Consider the one-electron density

$$\rho_N(y) := \int |\Psi_N(y, x_2, ..., x_N)|^2 dx_2 ... dx_N. \tag{10.16}$$

Assume that for any N, $\rho_N(y)$ is spherically symmetric. Then we have a bound considerably stronger than (10.12):

$$\langle \phi, I_N \phi \rangle \leq -\frac{(Z-N+1)e^2}{|x_N|}. \tag{10.17}$$

Indeed, we write $\langle \phi, I_N \phi \rangle = \langle f, Wf \rangle$ where

$$W(x_N) = \int I_N |\Psi_{N-1}(x_1,...,x_{N-1})|^2 d^{N-1}x. \tag{10.18}$$

We compute

$$W(x_N) = (N-1)e^2 \int \frac{\rho_{N-1}(y)}{|x-x_N|}dy - \frac{e^2 Z}{|x_N|}, \tag{10.19}$$

where $\rho_{N-1}(x_1) = \int |\Psi_{N-1}|^2 dx_2...dx_{N-1}$. It is not hard to show that ρ_{N-1} is spherically symmetric. Hence we have by Newton's theorem

$$\int_{\mathbb{R}^3} \frac{\rho_{N-1}(y)dy}{|y-x_N|} = \frac{1}{|x_N|}\int_{|y|\leq|x_N|} \rho_{N-1}(y)dy + \int_{|y|\geq|x_N|}\frac{1}{|y|}\rho_{N-1}(y)dy. \tag{10.20}$$

Using that $\int \rho_{N-1}(y)dy = 1$, this can be estimated as

$$\int_{\mathbb{R}^3} \frac{\rho_{N-1}(y)dy}{|y-x_N|} \leq \frac{1}{|x_N|}\int_{\mathbb{R}^3}\rho_{N-1}(y)dy = \frac{1}{|x_N|}. \tag{10.21}$$

(Moreover, $\rho_{N-1}(x) = O(e^{-\delta|x|}) \implies \int_{|y|\leq|x_N|}\rho_{N-1}(y)dy = 1 + O(e^{-\delta|x_N|})$.)
The equations (10.18) - (10.21) imply (10.17).

Remark 10.5 The accumulation of eigenvalues at Σ_N can be studied by similar arguments. Consider trial wave functions, ϕ_{nm}, constructed as above, with $f_{nm}(x_N)$ a hydrogen atom eigenfunction of energy $-n^{-2}$ (in suitable units). Then one can show

$$\|(H_N - E_n)\phi_{nm}\| \leq (\text{const}) \, n^{-\alpha}$$

for some $\alpha > 3$, where $E_n = \Sigma - n^{-2}$. This implies that H_N has groups of eigenvalues close to E_n compared to the spacing $E_{n+1} - E_n$ as $n \to \infty$ (Rydberg states). This analysis can be easily extended to take into account the particle statistics.

Problem 10.6 Show that the Schrödinger operator describing the Helium atom with infinitely heavy nucleus has at least one discrete eigenvalue (isolated eigenvalue of finite multiplicity).

10.3 Open Problems

Though atoms and molecules have been studied since the advent of quantum mechanics, there are many open problems in their rigorous theory. We mention here three of these problems:

1. Existence of molecules.
2. Non-existence of negative atomic ions with more than a few (two?) extra electrons.
3. Uniqueness or non-uniqueness of the ground states.

It is easy to prove the uniqueness of the ground states on the entire space, say $L^2(\mathbb{R}^{3N})$, however there are no techniques available to deal with the fermionic subspace (see Section 4.5).

Perturbation Theory: Feshbach-Schur Method

As we have seen, many basic questions of quantum dynamics can be reduced to finding and characterizing the spectrum of the appropriate Schrödinger operator. Though this task, known as *spectral analysis*, is much simpler than the task of analyzing the dynamics directly, it is far from trivial. The problem can be greatly simplified if the Schrödinger operator H under consideration is very close an operator H_0 whose spectrum we already know. In other words, the operator H is of the form $H = H_\kappa$, where

$$H_\kappa = H_0 + \kappa W, \tag{11.1}$$

H_0 is an operator at least part of whose spectrum is well understood, κ is a small parameter called the *coupling constant*, and W is an operator, called the *perturbation*. (All the operators here are assumed to be self-adjoint.)

If the operator W is bounded relative to H_0, say in the sense that $D(H_0) \subset D(W)$ and

$$\begin{aligned} \|Wu\| \leq c\|H_0 u\| + c'\|u\| \\ \text{for some } c, c' \geq 0, \quad \text{for all } u \in D(H_0), \end{aligned} \tag{11.2}$$

then the "standard" perturbation theory applies, and allows us to find or estimate eigenvalues of H_κ (see [247, 176, 162]).

In this chapter, we describe a powerful technique which allows us to estimate eigenvalues of H_κ even in cases where W is not bounded in terms of H_0. This is important in applications. We consider several examples of applications of this method, one of which is the hydrogen atom in a weak constant magnetic field B. Combining the expressions derived in Sections 7.3 and 7.5, we see that the Schrödinger operator for such an atom is

$$H_B := \frac{1}{2m}(p - eA)^2 - \frac{e^2}{|x|}, \tag{11.3}$$

where $e < 0$ denotes the electron charge, and we have kept units in which the speed of light is $c = 1$. Recall that the vector potential A is related to

© Springer-Verlag GmbH Germany, part of Springer Nature 2020
S. J. Gustafson and I. M. Sigal, *Mathematical Concepts of Quantum Mechanics*, Universitext, https://doi.org/10.1007/978-3-030-59562-3_11

the magnetic field B by $B = \nabla \times A$. In the notation of Section 7.5, $H_B = H(A, -e^2/|x|)$. Expanding the square in (11.3), we find

$$H_B = H_0 + W_B \tag{11.4}$$

where $H_0 = \frac{1}{2m}p^2 - e^2/|x|$ is the Schrödinger operator of the hydrogen atom (see Section 7.3), and

$$W_B = \frac{|e|}{m}A \cdot p + \frac{e^2}{2m}|A|^2. \tag{11.5}$$

Here we have assumed the gauge condition $\nabla \cdot A \equiv 0$. The small parameter κ here is the strength, $|B|$, of the magnetic field.

Now we know from Section 7.3 that the operator H_0 has a series of eigenvalues

$$E_n := -\left(\frac{me^4}{2\hbar^2}\right)\frac{1}{n^2}, \quad n = 1, 2, \ldots$$

as well as continuous spectrum in $[0, \infty)$. One would expect that for weak magnetic fields B, the operator H_B has eigenvalues $E_{B,n}$ close to E_n, at least for the few smallest E_n's. This is not so obvious as it might seem at first glance, since the perturbation W_B is not bounded relative to H_0 – i.e., (11.2) does not hold for any c and c', small or large. (The perturbation (11.5) grows in x: take for example $A := \frac{B}{2}(-x_2, x_1, 0)$.) The method we present below does show rigorously that such eigenvalues exist, though we will make only formal computations of $E_{B,n}$.

Two other examples, one we present below and a third, in the next chapter, display different physical phenomena, which conceptually and technically are considerably more complicated.

11.1 The Feshbach-Schur Method

Before returning to our perturbation problem, we state a general result used below, which allows us to reduce a perturbation problem on a large space to one on a small space. (For some motivating discussion see Appendix 12.7.) Let P and $\bar{P}$ be orthogonal projections (i.e. $P, \bar{P}$ are self-adjoint, $P^2 = P$, and $\bar{P}^2 = \bar{P}$) on a separable Hilbert space X, satisfying $P + \bar{P} = 1$. Let H be a self-adjoint operator on X. We assume that $\mathrm{Ran}P \subset D(H)$, that $H_{\bar{P}} := \bar{P}H\bar{P} \upharpoonright_{\mathrm{Ran}\bar{P}}$ is invertible, and

$$\|R_{\bar{P}}\| < \infty, \qquad \|PHR_{\bar{P}}\| < \infty \quad \text{and} \quad \|R_{\bar{P}}HP\| < \infty, \tag{11.6}$$

where $R_{\bar{P}} = \bar{P}H_{\bar{P}}^{-1}\bar{P}$. We define the operator

$$F_P(H) := P(H - HR_{\bar{P}}H)P \upharpoonright_{\mathrm{Ran}\,P}. \tag{11.7}$$

We call F_P the *Feshbach-Schur map*. The key result for us is the following:

Theorem 11.1 Assume (11.6) hold. Then the operators H and $F_P(H)$ are isospectral at 0, in the sense that

(a) $0 \in \sigma(H) \iff 0 \in \sigma(F_P(H))$,
(b) $H\psi = 0 \iff F_P(H)\phi = 0$

where ψ and ϕ are related by $\phi = P\psi$ and $\psi = Q\phi$, with the (bounded) operator Q given by

$$Q = Q(H) := P - R_{\bar{P}}HP. \tag{11.8}$$

A proof of this theorem is given in an appendix, Section 11.4. Moreover, under the conditions above,

$$H \text{ is self-adjoint} \implies F_P(H) \text{ is self-adjoint}. \tag{11.9}$$

The latter property is an example of transmission properties of the Feshbach-Schur map.

Now we return to our general perturbation problem. Thus we consider a family, $H_\kappa, \kappa \geq 0$, of self-adjoint operators of the form (11.1). Assume the operator H_0 has an isolated eigenvalue λ_0 of finite multiplicity. Let P be the orthogonal projection onto the eigenspace $\text{Null}(H_0 - \lambda_0)$ spanned by all the eigenfunctions of H_0 corresponding to the eigenvalue λ_0, and let $\bar{P} := 1 - P$.

We apply Theorem 11.1 to the family of operators $H = H_\kappa - \lambda$ for some λ close to λ_0, with the projections P and $\bar{P}$ defined as above. Observe that

$$\lambda \in \sigma_d(H_\kappa) \iff 0 \in \sigma_d(H_\kappa - \lambda), \tag{11.10}$$

and that $F_P(H_\kappa - \lambda)$, if it is well-defined, is a family of $m \times m$ matrices, where m is the multiplicity of the eigenvalue λ_0 of the operator H_0 (indeed, P is a rank-m projection – i.e., $\dim \text{Ran}P = m$). Thus the perturbation problem reduces the problem of finding an eigenvalue (and an eigenfunction) of an (infinite-dimensional) operator H, to the problem of finding the values λ for which

$$0 \in \sigma(F_P(H_\kappa - \lambda)).$$

Such values are called *singular values* of the family $F_P(H_\kappa - \lambda)$.

We need to discuss the problem of finding singular values of the family $F_P(H_\kappa - \lambda)$ of matrices, but first we address the issue of defining $F_P(H_\kappa - \lambda)$. Write

$$H_{\bar{P}} = H_{0\bar{P}} + \kappa W_{\bar{P}},$$

where we are using the notation $A_Q := QAQ \upharpoonright_{\text{Ran}Q}$. Assume the operators W and H_0 satisfy the conditions

$$WP \quad \text{and} \quad PW \quad \text{are bounded}, \tag{11.11}$$

$$\bar{P}H_\kappa\bar{P} - \lambda_0 \quad \text{is invertible on} \quad \text{Ran}\bar{P} \quad \text{for} \quad |\kappa| < \kappa_0. \tag{11.12}$$

(If H_0 and W are self-adjoint and if WP is bounded, then so is PW.) Since $\bar{P}H_\kappa P = \kappa\bar{P}WP$, the condition (11.6) is satisfied for $H_\kappa - \lambda$, with λ sufficiently close to λ_0, and therefore $F_P(H_\kappa - \lambda)$ is well-defined. (In this case, $R_{\bar{P}} = \bar{P}(H_{\bar{P}} - \lambda)^{-1}\bar{P}$, $H_{\bar{P}} := \bar{P}H_\kappa\bar{P}\restriction_{\text{Ran}\bar{P}}$.)

The operator $H_{0\bar{P}} := \bar{P}H_0\bar{P}\restriction_{\text{Ran}\bar{P}}$ has no spectrum near λ_0. The same is true, as can be readily verified, for the operator $H_{\bar{P}} = H_{0\bar{P}} + \kappa\bar{P}W\bar{P}$ for $|\kappa|$ sufficiently small, if the operator W is bounded relative to H_0 in the sense (11.2). Otherwise, justifying the assumption (11.12) is a delicate matter, and is done on a case-by-case basis. For the examples of the hydrogen atom in a weak constant magnetic field, and of a particle system in a weak time-periodic (electric) field, we address this question below. Under the assumptions above, conditions (11.6) hold. Consequently, the operator $F_P(H_\kappa - \lambda)$ is well-defined for λ close to λ_0.

Now let's compute the operator $F_P(H_\kappa - \lambda)$. We write it as the sum of three terms

$$F_P(H_\kappa - \lambda) = H_P - \kappa^2 U(\lambda) - \lambda \tag{11.13}$$

where $H_P := PH_\kappa P\restriction_{\text{Ran}P}$ and

$$U(\lambda) := PW\bar{R}(\lambda)WP\restriction_{\text{Ran}P}, \quad \text{with} \quad \bar{R}(\lambda) = \bar{P}(\bar{P}H_\kappa\bar{P} - \lambda)^{-1}\bar{P}. \tag{11.14}$$

The matrix family $U(\lambda)$ is called the *level shift* operator. Since $PH_0 = H_0P = \lambda_0 P$ and $P\bar{P} = 0$, we have

$$PH_0P = \lambda_0 P, \quad PH\bar{P} = \kappa PW\bar{P}, \quad \bar{P}HP = \kappa\bar{P}WP.$$

These relations yield

$$F_P(H_\kappa - \lambda) = \kappa W_P - \kappa^2 U(\lambda) + \lambda_0 - \lambda. \tag{11.15}$$

Theorem 11.1 and equations (11.10) and (11.15) imply the relation

$$\lambda \in \sigma_d(H_\kappa) \iff \lambda - \lambda_0 \in \sigma_d(\kappa W_P - \kappa^2 U(\lambda)).$$

Note that the operator on the right itself depends on the spectral parameter λ. Expanding the resolvent

$$\bar{R}(\lambda) = \bar{P}(H_{0\bar{P}} - \lambda_0 + \kappa W_{\bar{P}} + \lambda_0 - \lambda)^{-1}\bar{P}$$

in a Neumann series in $\kappa W_P + \lambda_0 - \lambda$ we obtain

$$\bar{R}(\lambda) = \bar{R}_0 + O(|\kappa| + |\lambda - \lambda_0|)$$

where $\bar{R}_0 = \bar{P}(H_{0\bar{P}} - \lambda_0)^{-1}\bar{P}$. Consequently,

$$U(\lambda) = U_0 + O(|\kappa| + |\lambda - \lambda_0|)$$

where $U_0 := PW\bar{R}_0WP\restriction_{\text{Ran}P}$. Therefore any eigenvalue λ_κ of H_κ sufficiently close to λ_0 has the form

$$\lambda_\kappa = \lambda_0 + \kappa\mu_k + O(|\kappa|^3)$$

where $\mu_\kappa \in \sigma_d(W_P - \kappa U_0)$. Similarly, we can obtain expressions for λ_κ to any order in κ. Observe that if λ_0 is a simple eigenvalue (i.e. of multiplicity one) of H_0, with normalized eigenfunction ψ_0, then W_P and U_0 are just the real numbers $\langle \psi_0, W\psi_0 \rangle$ and $\langle W\psi_0, \bar{R}_0 W\psi_0 \rangle$, and we have

$$\lambda_\kappa = \lambda_0 + \kappa\langle \psi_0, W\psi_0 \rangle + \kappa^2\langle W\psi_0, \bar{R}_0 W\psi_0 \rangle + O(|\kappa|^3). \qquad (11.16)$$

Our next step is to examine the structure of the quadratic term in more detail. Suppose that the operator H_0 is self-adjoint and has isolated eigenvalues λ_j, $j = 0, 1, \ldots$, counting multiplicities, with

$$\lambda_j < \inf \sigma_{ess}(H_0).$$

Let $\{\psi_j\}$ be corresponding normalized eigenfunctions, and let P_{ψ_j} be the rank-one orthogonal projections onto these eigenfunctions:

$$P_{\psi_j} = |\psi_j\rangle\langle\psi_j|, \quad \text{or} \quad P_{\psi_j}f = \langle \psi_j, f \rangle\psi_j.$$

Then we have

$$\bar{P} = \sum_{j\neq 0} P_{\psi_j} + P_{ess}$$

where $P_{ess} := \mathbf{1} - \sum_j P_{\psi_j}$ is the projection onto the essential spectral subspace of H_0. Then we can write

$$\langle \bar{P}W\psi_0, \bar{R}_0\bar{P}W\psi_0 \rangle = \sum_{j\neq 0} |\langle \psi_j, W\psi_0 \rangle|^2 (\lambda_j - \lambda_0)^{-1}$$
$$+ \langle P_{ess}W\psi_0, (H_{0,ess} - \lambda_0)^{-1} P_{ess}W\psi_0 \rangle \qquad (11.17)$$

where $H_{0,ess} := P_{ess}H_0P_{ess}$ is the essential spectral part of the operator H_0. Now we can interpret the coefficient $\langle \psi_0, W\psi_0 \rangle$ in (11.16) as due to a direct interaction of the bound state ψ_0 with itself, the term $|\langle \psi_j, W\psi_0 \rangle|^2 (\lambda_j - \lambda_0)^{-1}$ as due to the interaction of ψ_0 with itself via the bound state ψ_j, and the last term in (11.17) as due to the interaction of ψ_0 with itself via the essential spectral states of H_0.

In conclusion of this section we sketch a proof of an extension of the central theorem of perturbation theory

Theorem 11.2 Assume (11.11) hold, the operator H_0 has an isolated eigenvalue, λ_0, of a finite multiplicity, m, and there is $\kappa_0 > 0$ s.t. (11.12) holds. Then for $|\kappa|$ sufficiently small the operator H_κ has eigenvalues $\lambda_\kappa^{(i)}$ near λ_0 of the total multiplicity equal to m. Moreover, if H_κ is self-adjoint, then the eigenvalues $\lambda_\kappa^{(i)}$ have the expansions of the form (11.16) - (11.17).

Sketch of proof. For $|\lambda - \lambda_0| < \|(\bar{P}H_\kappa\bar{P} - \lambda_0)^{-1}\|^{-1}$, the operator $\bar{P}H_\kappa\bar{P} - \lambda$ is invertible on $\mathrm{Ran}\bar{P}$ and therefore, due to (11.11), $F_P(\lambda)$ exist. Moreover, the eigenvalues, $\lambda_\kappa^{(i)}$, of H_κ near λ_0, if they exist, are also eigenvalues of

$F_P(\lambda_\kappa^{(i)})$. Let $\nu_j(\lambda, \kappa)$ be the eigenvalues of $F_P(\lambda)$ (remember that, though we do not display this, the latter operator depends on κ). To simplify the exposition, assume they are simple. The general case is treated similarly. Then the eigenvalues $\lambda_\kappa^{(i)}$ must solve the equation

$$\lambda = \nu_i(\lambda, \kappa).$$

To show that this equation has a unique solution for κ sufficiently small and this solution is close to λ_0, we observe that, due to (11.15), $\nu_i(\lambda, \kappa)$ is a differentiable function, $\lambda_0 = \nu_i(\lambda_0, 0)$, $|\nu_i(\lambda, \kappa) - \lambda_0| \lesssim \kappa$ and $\partial_\lambda \nu_i(\lambda, \kappa) = O(\kappa)$. Therefore by the implicit function theorem, the equation $\lambda = \nu_i(\lambda, \kappa)$ has a unique solution, $\lambda_\kappa^{(i)}$, and this solution satisfies $\lambda_\kappa^{(i)} = \lambda_0 + O(\kappa)$. Proving the expansions of the form (11.16) - (11.17) follows the arguments leading to this expansion. $\square$

As was mentioned above, the condition (11.12) is satisfied if W is H_0-bounded (*show this*). It also holds if H_0 is self-adjoint and W is non-negative and either λ_0 is the ground state energy or we take for P the orthogonal projection on the eigenspace corresponding to the eigenvalues $\leq \lambda_0$. In the next chapter we consider the celebrated Born-Oppenheimer approximation, for which no condition of the theorem above holds, but which still can be handled by the present technique. This little discussion indicates the power of the method.

11.2 The Zeeman Effect

We can apply the theory developed above to compute the energy levels of the Schrödinger operator of a hydrogen atom in a weak homogeneous magnetic field (the *Zeeman effect*). Recall that for any magnetic field this operator is given by the expression (11.4)- (11.5). For a constant magnetic field B, we can choose the vector potential to be $A(x) = \frac{1}{2}B \times x$ (see Equation (7.26)). Then the perturbation W_B can be re-written as

$$W_B = \kappa \hat{B} \cdot L + \frac{m}{2}\kappa^2(\hat{B} \times x)^2$$

where $L := x \times p$ is the operator of angular momentum (see Section (7.2)), $\kappa = \frac{|e|}{2m}|B|$, and $\hat{B} := |B|^{-1}B$. Thus κ is a small parameter in our problem.

Let $\lambda_0 = E_n$ be the n-th energy level of the Hydrogen atom Schrödinger operator H_0. One can show that if λ is close to λ_0, then the operator $H_{B\bar{P}}$ has no spectrum near λ_0, and therefore the Feshbach-Schur operator $F_P(H_B - \lambda)$ is well-defined. We will not do this here, but mention only that this follows from the inequality $(H_{0\bar{P}} - \lambda_0)^{-1}W_B(H_{0\bar{P}} - \lambda_0)^{-1} \geq -c\kappa$ on $\mathrm{Ran}\bar{P}$, for some uniform $c > 0$ and for κ sufficiently small. The idea of a proof of the latter inequality is that $(H_{0\bar{P}} - \lambda_0)^{-1}$ controls p entering W_B through L and x entering W_B through L is controlled by the positive term $\frac{m}{2}\kappa^2(\hat{B} \times x)^2$.

We compute formally the perturbation expansion for the eigenvalue $\lambda_\kappa = E_{n,B}$ of the operator H_B near the eigenvalue $\lambda_0 = E_n$ of the operator H_0. This computation is slightly more complicated than the corresponding computation in the abstract case considered above, since our perturbation is of the form

$$W_B = \kappa W_1 + \kappa^2 W_2$$

where $W_1 := \hat{B} \cdot L$ and $W_2 := \frac{m}{2}(\hat{B} \times x)^2$, rather than of the form κW considered above.

First, we take the x_3-axis to be in the direction of the magnetic field vector B, so that $W_1 = L_3$ (see Section 7.2) and $W_2 = \frac{m}{2}r_\perp^2$, where $r_\perp^2 = x_1^2 + x_2^2$.

Second, since the operator H_0 is invariant under rotations, we know that L commutes with P:

$$[L, P] = 0,$$

and therefore

$$P W_B \bar{P} = \kappa^2 P W_2 \bar{P}, \quad \text{and} \quad \bar{P} W_B P = \kappa^2 \bar{P} W_2 P.$$

Third, by the results of Section 7.3, the projection P on the eigenspace corresponding to the eigenvalue E_n is given by

$$P = \sum_{l=0}^{n-1} \sum_{k=-l}^{l} P_{nlk}$$

where $P_{nlk} = |\psi_{nlk}\rangle\langle\psi_{nlk}|$, and $\psi_{nlk}(x) = R_{nl}(r)Y_l^k(\theta, \phi)$ is the normalized eigenfunction derived in Section 7.3. Since $L_3 \psi_{nlk} = \hbar k \psi_{nlk}$, this gives

$$PW_1 P = \sum_{l=0}^{n-1} \sum_{k=-l}^{l} \hbar k P_{nlk} \tag{11.18}$$

and

$$PW_2 P = \sum_{l,l'=0}^{n-1} \sum_{k=-\min(l,l')}^{\min(l,l')} |\psi_{nlk}\rangle\langle\psi_{nl'k}|\langle\psi_{nlk}, \frac{m}{2}r_\perp^2 \psi_{nl'k}\rangle. \tag{11.19}$$

Proceeding as in the abstract case, we find that $E_{n,B} - E_n$ is an eigenvalue of

$$\kappa PW_1 P + \kappa^2(PW_2 P + -PW_1 \bar{R}_0 W_1 P) + O(|\kappa|^3). \tag{11.20}$$

Thus for $n \neq 1$, we have in the leading order

$$E_{n,B} = E_n + \mu|B|k + O(|B|^2),$$

where $\mu := |e|\hbar/2m$ is the *Bohr magneton*, for $k = -l, \ldots, l$ and $l = 0, \ldots, n-1$. Thus the magnetic field lifts the degeneracy of the energy levels in the direction of the angular momentum. This is called the *Zeeman effect*.

For the ground state energy $E_{n=1}$, the term linear in κ vanishes and the expansion yields

$$E_{1,B} = E_1 + \frac{me^2}{3}|B|^2\bar{r}^2 + a(e|B|)^4 + O(|B|^5)$$

where

$$\bar{r}^2 = \int |x|^2 |\psi_{100}|^2 \quad \text{and} \quad a = \frac{m^2}{4}\langle r_\perp^2 \psi_{100}, \bar{P}(H_{0\bar{P}} - E_0)^{-1}\bar{P}r_\perp^2 \psi_{100}\rangle.$$

Here we have used spherical coordinates to simplify the first integral.

11.3 Time-Dependent Perturbations

Our second example is an atom placed in a spatially localized but time-periodic electric field. We write the total Schrödinger operator as in (11.1):

$$H_\kappa = H_0 + \kappa W,$$

where H_0 is the Schrödinger operator of an atom, or, to fix ideas, a one-particle Schrödinger operator

$$H_0 = -\frac{\hbar^2}{2m}\Delta + V(x)$$

with $V(x)$ continuous and decaying to zero as $|x| \to \infty$ as in Theorem 6.16, so that $\sigma_{ess}(H_0) = [0, \infty)$. (Of course, for an atom, the potential $V(x)$ has singularities, but the analysis below can be easily generalized to this case.) We also suppose H_0 has discrete eigenvalues $E_j < 0$ with corresponding normalized eigenfunctions $\phi_j(x)$, $j = 0, 1, \ldots$. The perturbation $W = W(x,t)$ is assumed to be smooth, vanishing as $|x| \to \infty$, and time-periodic with period $T = \frac{2\pi}{\omega}$.

Since the perturbation W depends on time, so does the operator $H_\kappa = H_\kappa(t)$, and the spectrum of H_κ (at each moment of time) does not tell us much about the dynamics – that is, the solution of the time-dependent Schrödinger equation

$$i\hbar\frac{\partial\psi}{\partial t} = H_\kappa(t)\psi. \tag{11.21}$$

So we have to deal with this equation directly. The question we address is what happens to the bound states $\psi_j(x,t) := \phi_j(x)e^{-iE_jt/\hbar}$ of the unperturbed equation

$$i\hbar\frac{\partial\psi}{\partial t} = H_0\psi$$

when the perturbation $\kappa W(x,t)$ is "switched on"? More precisely, are there time-(quasi) periodic solutions $\psi_{\kappa,j}(x,t)$ of (11.21) which are L^2 functions of x, and such that $\psi_{\kappa,j} \to \psi_j$ as $\kappa \to 0$ in, say, the $L^2(\mathbb{R}^3)$-norm? If so, then we

would like to find an approximate expression for such solutions. If not, what
are the descendants of $\psi_j(x,t)$?

We fix j and consider the perturbation theory for the j-th state $\psi_j(x,t)$.
We look for a solution to (11.21) of the form

$$\psi_{\kappa,j}(x,t) = \phi_{\kappa,j}(x,t)e^{-iE_{\kappa,j}t/\hbar} \tag{11.22}$$

where the function $\phi_{\kappa,j}$ is time-periodic with period T, and L^2 in x, and
with $\phi_{\kappa,j} \to \phi_j$ (in the L^2-sense) and $E_{\kappa,j} \to E_j$ as $\kappa \to 0$. Plugging the
expression (11.22) into (11.21), we find the equation for $\phi_{\kappa,j}$:

$$(H_\kappa(t) - i\hbar\frac{\partial}{\partial t} - E_{\kappa,j})\phi = 0. \tag{11.23}$$

We look for solutions of this equation in the space $L^2(\mathbb{R}^n \times S_T)$ where S_T
is the circle of circumference T – i.e., we assume that the function $\phi(x,t)$ is
periodic in t with period T, and satisfies

$$\int_0^T \int_{\mathbb{R}^n} |\phi(x,t)|^2 dx dt < \infty.$$

Thus we can treat Equation (11.23) as an eigenvalue equation for the operator

$$K_\kappa := H_\kappa(t) - i\hbar\frac{\partial}{\partial t} \quad \text{on} \quad L^2(\mathbb{R}^n \times S_T).$$

We call this operator the *Bloch-Floquet Hamiltonian*, since the general approach we describe here, of reducing the time-dependent problem (11.21) to
an eigenvalue problem on a larger space, follows the outlines of the theory
laid out in parallel by F. Bloch in solid state physics and by Floquet in mathematics.

Thus our task is to find out whether the operator K_κ on $L^2(\mathbb{R}^n \times S_T)$ has
an eigenvalue $E_{\kappa,j}$ close to the eigenvalue E_j of the operator H_0. Since κ is
assumed to be small, we treat this as a perturbation problem:

$$K_\kappa = K_0 + \kappa W.$$

We begin by examining the spectrum of the operator $K_0 := H_0 - i\hbar\frac{\partial}{\partial t}$ on
$L^2(\mathbb{R}^n \times S_T)$. Using the facts that

$$\sigma_{ess}(H_0) = [0,\infty) \quad \text{and} \quad \sigma_d(H_0) = \{E_j\}_{j=0}^N$$

and, on $L^2(S_T)$,

$$\sigma\left(i\hbar\frac{\partial}{\partial t}\right) = \sigma_d\left(i\hbar\frac{\partial}{\partial t}\right) = \{\hbar\omega n\}_{n=-\infty}^\infty,$$

where, recall, $\omega = 2\pi/T$ (see Section 7.1), and separation of variables, we find
that

$$\sigma_{ess}(K_0) = \bigcup_{n=-\infty}^{\infty} [\hbar\omega n, \infty)$$

and

$$\sigma_{pp}(K_0) = \{\hbar\omega n + E_m \mid n \in \mathbb{Z},\ m = 0, 1, \ldots\}.$$

Here σ_{pp} denotes the full set of eigenvalues (including non-isolated ones). Thus the essential spectrum of K_0 fills the entire real axis $\mathbb{R}$, and the eigenvalues of K_0 (which are infinite in number) lie on top of the essential spectrum, or, as it is said, are *embedded* in the essential spectrum. The eigenfunctions corresponding to the eigenvalues $E_{mn} = E_m + \hbar\omega n$ are given by

$$\psi_{mn}(x, t) := \phi_m(x)e^{-i\omega nt}.$$

Note that the E_j's themselves are eigenvalues of K_0, with eigenfunctions $\psi_{j0}(x) = \phi_j(x)$:

$$K_0 \psi_{j0} = E_j \psi_{j0}.$$

Now we apply the Feshbach projection method to the operator $K_0 + \kappa W$ in order to find its eigenvalues near E_j. Let us assume, for simplicity, that E_j is a simple eigenvalue of H_0 (for example, $j = 0$ and E_0 is the ground state energy), and that

$$E_m + \hbar\omega n \neq E_j \quad \forall\, (m, n) \neq (j, 0)$$

(in other words, that there is no "accidental" degeneracy in the spectrum of K_0). As the projection P we use the orthogonal rank-one projection onto the eigenspace of K_0, spanned by the eigenfunction $\psi_{j0} = \phi_j$, corresponding to the eigenvalue E_j, that is $P = |\psi_{j0}\rangle\langle\psi_{j0}|$. Then we will find from (11.16) the expansion for the desired eigenvalue $E_{\kappa,j}$:

$$E_{\kappa,j} = E_j + \kappa E_{j1} + \kappa^2 E_{j2} + O(\kappa^3) \qquad (11.24)$$

where

$$E_{j1} := \langle\langle \psi_{j0}, W\psi_{j0} \rangle\rangle$$

and

$$E_{j2} := \langle\langle \bar{P}W\psi_{j0}, \bar{R}_0\bar{P}W\psi_{j0} \rangle\rangle.$$

Here, recall that $\bar{P} = \mathbb{1} - P$ and $\bar{R}_0 = \bar{P}(\bar{P}K_0\bar{P} - E_j - i0)^{-1}\bar{P}$, and $\langle\langle\cdot\rangle\rangle$ is used to denote the inner product in the space $L^2(\mathbb{R}^n \times S_T)$:

$$\langle\langle u, v \rangle\rangle := \int_0^T \int_{\mathbb{R}^n} \bar{u}v\, dx\, dt.$$

Notice that we inserted $-i0$ into the resolvent above. This was not there in our previous discussion of the Feshbach perturbation theory, and the reason for it will become apparent shortly.

Let $P_{mn} = P_{\psi_{mn}}$ be the rank-one orthogonal projections onto the eigenfunctions ψ_{mn} of the operator K_0. Then

$$P_{ess} := 1 - \sum_{mn} P_{mn}$$

defines the projection onto the essential spectral subspace of K_0. Thus

$$P = P_{j0} \quad \text{and} \quad \bar{P} = \sum_{(m,n)\neq(j,0)} P_{mn} + P_{ess}.$$

By the definition of the inner product we have

$$E_{j1} = \langle \phi_j, W_0 \phi_j \rangle$$

where $W_0(x) := \int_0^T W(x,t)dt$, and the inner product on the right is the $L^2(\mathbb{R}^n)$ inner product.

Now we compute the third term on the right hand side of (11.24) – that is, the coefficient of κ^2. It can be written as (see (11.17))

$$E_{j2} = \sum_{(m,n)\neq(j,0)} \frac{|\langle\langle \psi_{mn}, W\psi_{j0}\rangle\rangle|^2}{E_{mn} - E_j} + \langle\langle P_{ess}W\psi_{j0}, \bar{R}_0 P_{ess}W\psi_{j0}\rangle\rangle. \quad (11.25)$$

By the definition of the ψ_{mn}'s, we have

$$\langle\langle \psi_{mn}, W\psi_{j0}\rangle\rangle = \int_0^T \langle \phi_m, W(t)\phi_j\rangle e^{i\omega nt}dt = \langle \phi_m, W_n\phi_j \rangle \quad (11.26)$$

where W_n is the n-th Fourier coefficient of W:

$$W_n := \int_0^T W(t)e^{i\omega nt}dt.$$

This simplifies the expression of the first term on the right hand side of (11.25). Now we analyze the second term. By separation of variables,

$$P_{ess} = \sum_n P_{ess}^{H_0} \otimes P_n$$

where $P_{ess}^{H_0}$ is the projection onto the essential spectral subspace of the operator H_0,

$$P_{ess}^{H_0} := 1 - \sum_m P_{\psi_m},$$

and P_n is the projection onto the eigenfunction $e^{-i\omega nt}$ of the operator $i\partial/\partial t$. Inserting this into the last term in (11.25), and using

$$P_{ess}^{H_0} \otimes P_n W\psi_{j0} = \int_0^T P_{ess}^{H_0} W(t)\phi_j e^{i\omega nt}dt = P_{ess}^{H_0} W_n\phi_j$$

and

$$\langle\langle P_{ess}^{H_0} f \otimes e^{-i\omega nt}, \bar{R}_0(P_{ess}^{H_0} f \otimes e^{-i\omega nt})\rangle\rangle$$
$$= \langle P_{ess}^{H_0} f, (H_{0ess} - \hbar\omega n - E_j - i0)^{-1} P_{ess}^{H_0} f\rangle,$$

where $H_{0ess} := H_0 P_{ess}^{H_0}$, we find

$$\langle\langle P_{ess} W\psi_{j0}, \bar{R}_0 P_{ess} W\psi_{j0}\rangle\rangle$$
$$= \sum_n \langle P_{ess}^{H_0} W_n\phi_j, (H_{0ess} - \hbar\omega n - E_j - i0)^{-1} P_{ess}^{H_0} W_n\phi_j\rangle.$$

Substituting (11.26) and (11.27) into (11.25), we obtain

$$E_{j2} = \sum_{(m,n)\neq(j,0)} \frac{|\langle\phi_m, W_n\phi_j\rangle|^2}{E_m + \hbar\omega n - E_j}$$
$$+ \sum_n \langle P_{ess}^{H_0} W_n\phi_j, (H_{0ess} - E_j - \hbar\omega n - i0)^{-1} P_{ess}^{H_0} W_n\phi_j\rangle. \tag{11.27}$$

Thus we have obtained detailed expressions for the first and second order coefficients in the expansion of $E_{\kappa,j}$ in κ. Now let's analyze the expression for $E_{j,2}$ a little further. Start with the well-known formula

$$\int_{-\infty}^{\infty} \frac{f(\lambda)}{\lambda + i0} d\lambda := \lim_{\epsilon\to 0^+} \int_{-\infty}^{\infty} \frac{f(\lambda)}{\lambda + i\epsilon} d\lambda$$
$$= PV \int_{-\infty}^{\infty} \frac{f(\lambda)}{\lambda} d\lambda - 2\pi i \int_{-\infty}^{\infty} f(\lambda)\delta(\lambda) d\lambda$$

where $PV \int_{-\infty}^{\infty}$ denotes the principal value of the singular integral, defined by

$$PV \int_{-\infty}^{\infty} \frac{f(\lambda)}{\lambda} d\lambda := \lim_{\epsilon\to 0^+} \left(\int_{-\infty}^{-\epsilon} + \int_{\epsilon}^{\infty}\right) \frac{f(\lambda)}{\lambda} d\lambda,$$

and $\delta(\lambda)$ denotes the Dirac delta function (centred at zero). Thus we find that E_{j2} is a complex number of the form

$$E_{j2} = E_{j2}^{Re} - i\Gamma_j,$$

whose real part, E_{j2}^{Re}, is the sum of the first term on the right hand side of (11.27) and the real part (or principal value) of the second term. The imaginary part is $-\Gamma_j$, where

$$\Gamma_j = 2\pi \sum_n \langle P_{ess}^{H_0} W_n\phi_j, \delta(H_{0ess} - \hbar\omega n - E_j) P_{ess}^{H_0} W_n\phi_j\rangle.$$

The last expression is known as the *Fermi golden rule*. Since $\delta(\lambda) \geq 0$, we have $\Gamma \geq 0$. Then for "generic" perturbations (that is, barring an accident), we expect

$$\Gamma_j > 0.$$

This means that for κ sufficiently small, $E_{\kappa,j}$ is a complex number of the form

$$E_{\kappa,j} = E_{\kappa,j}^{Re} - i\kappa^2 \Gamma_j + O(\kappa^3)$$

where $E_{\kappa,j}^{Re} := \kappa E_{j1} + \kappa^2 E_{j,2}^{Re}$. This suggests that for generic perturbations, and for κ sufficiently small, the operator K_κ has no eigenvalues near E_j. So what happened to the eigenvalue E_j of K_0 (or H_0)? It is apparently unstable under small generic perturbations. But does it just disappear without a trace, or is something left behind? It turns out that the eigenvalue E_j of H_0 (or K_0) gives rise to a *resonance* of H_0 (or K_0). A theory of resonances is briefly described in Chapter 17. The method of complex deformations indicated there, together with the Feshbach method described in this section, can be used to establish the existence of resonance 'eigenvalues' $E_{\kappa,j}$ born out of E_j (see [73] and references therein). The number $-\operatorname{Im} E_{\kappa,j}$ is called the *width* of the resonance at $E_{\kappa,j}$, and $T_{\kappa,j} := \hbar(-\operatorname{Im} E_{\kappa,j})^{-1}$ gives the *lifetime* of the resonance. The Fermi golden rule gives the leading order of the resonance width

$$-\operatorname{Im} E_{\kappa,j} = \kappa^2 \Gamma_j + O(\kappa^3)$$

(in fact $O(\kappa^3)$ can be replaced by $O(\kappa^4)$).

A physical interpretation of the phenomenon of instability of bound states under time-periodic perturbations is that an atom in a photon field becomes unstable, as photons of sufficiently high energy can break it up. This photoelectric effect was predicted by Einstein in 1905. To develop a consistent theory of the photoelectric effect, one has to use the quantized electro-magnetic field (or Maxwell equations) given in Chapter 21.

11.4 Appendix: Proof of Theorem 11.1

First, in addition to (11.8), we define the operator

$$Q^\# = Q^\#(H) := P - PHR_{\bar{P}}. \tag{11.28}$$

The operators P, Q and $Q^\#$ have the following **properties:**

$$\operatorname{Null}Q \cap \operatorname{Null}H' = \{0\}, \qquad \operatorname{Null}P \cap \operatorname{Null}H = \{0\}, \tag{11.29}$$

$$HQ = PH', \tag{11.30}$$

$$Q^\#H = H'P, \tag{11.31}$$

where $H' = F_P(H)$. We prove relations (11.29) - (11.31). The second relation in (11.29) is shown in Proposition 11.3, while the first one follows from the inequality

$$\|Qu\|^2 = \|Pu\|^2 + \|R_{\bar{P}}HPu\|^2 \geq \|Pu\|^2.$$

In the first equality, we used the fact that the projections P and $\bar{P}$ are orthogonal.

Now we prove relations (11.30) - (11.31). Using the definition of $Q(H)$, we transform

$$
\begin{aligned}
HQ &= HP - H\bar{P}H_{\bar{P}}^{-1}\bar{P}HP \\
&= PHP + \bar{P}HP - PH\bar{P}H_{\bar{P}}^{-1}\bar{P}HP - \bar{P}H\bar{P}H_{\bar{P}}^{-1}\bar{P}HP \\
&= PHP - PH\bar{P}H_{\bar{P}}^{-1}\bar{P}HP \\
&= PF_P(H).
\end{aligned}
\tag{11.32}
$$

Next, we have

$$
\begin{aligned}
Q^{\#}H &= PH - PH\bar{P}H_{\bar{P}}^{-1}\bar{P}H \\
&= PHP + PH\bar{P} - PH\bar{P}H_{\bar{P}}^{-1}\bar{P}HP - PH\bar{P}H_{\bar{P}}^{-1}\bar{P}H\bar{P} \\
&= PHP - PH\bar{P}H_{\bar{P}}^{-1}\bar{P}HP \\
&= F_P(H)P.
\end{aligned}
$$

This completes the proof of (11.29) - (11.31).

Proposition 11.3 Assume conditions (11.6) are satisfied. Then (11.29) - (11.31) imply that $0 \in \sigma(H) \Rightarrow 0 \in \sigma(H')$ (the part of property (a) which is crucial for us). Moreover, we have $\mathrm{Null}P \cap \mathrm{Null}H = \{0\}$.

Proof. Let $0 \in \rho(H')$. Then we can solve the equation $H'P = Q^{\#}H$ for P to obtain

$$
P = H'^{-1}Q^{\#}H .
\tag{11.33}
$$

The equation $P + \bar{P} = 1$ and the definition $H_{\bar{P}} = \bar{P}H\bar{P}$ imply

$$
\bar{P} = \bar{P}H_{\bar{P}}^{-1}\bar{P}H\bar{P} = \bar{P}H_{\bar{P}}^{-1}(\bar{P}H - \bar{P}HP) .
\tag{11.34}
$$

Substituting expression (11.33) for P into the r.h.s., we find

$$
\bar{P} = \bar{P}H_{\bar{P}}^{-1}(\bar{P} - \bar{P}HPH'^{-1}Q^{\#})H .
$$

Adding this to Equation (11.33) multiplied from the left by P, and using $P + \bar{P} = 1$, yields

$$
1 = \left[\bar{P}H_{\bar{P}}^{-1}\bar{P} - \bar{P}H_{\bar{P}}^{-1}\bar{P}HPH'^{-1}Q^{\#} + PH'^{-1}Q^{\#}\right]H.
$$

Since by our conditions $\bar{P}H_{\bar{P}}^{-1}\bar{P}HP$ is bounded, the expression in the square brackets represents a bounded operator. Hence H has a bounded inverse. So $0 \in \rho(H)$. This proves the first statement.

The second statement follows from the relation

$$
1 = QP + R_{\bar{P}}H ,
\tag{11.35}
$$

which, in turn, is implied by Equation (11.34) and the relation $P + \bar{P} = 1$. Indeed, applying (11.35) to a vector $\phi \in \mathrm{Null}P \cap \mathrm{Null}H$, we obtain $\phi = QP\phi + R_{\bar{P}}H\phi = 0$. $\square$

Now we proceed directly to the proof of Theorem 11.1. Statement (b) follows from relations (11.29)-(11.30). Proposition 11.3 implies that $0 \in \rho(H)$ if $0 \in \rho(H')$, where $H' := F_P(H)$, which is half of statement (a). Conversely, suppose $0 \in \rho(H)$. The fact that $0 \in \rho(H')$ follows from the relation .

$$H'^{-1} = PH^{-1}P \qquad\qquad (11.36)$$

which we set out to prove now. We have by definition

$$H'PH^{-1}P = PHPH^{-1}P - PH\bar{P}H_{\bar{P}}^{-1}\bar{P}HPH^{-1}P$$
$$= PH(1-\bar{P})H^{-1}P - PH\bar{P}H_{\bar{P}}^{-1}\bar{P}H(1-\bar{P})H^{-1}P$$
$$= P .$$

Similarly one shows that $PH^{-1}PH' = P$. Hence H' has the bounded inverse $PH^{-1}P$.

So we have shown that $0 \in \rho(H) \Leftrightarrow 0 \in \rho(F_P(H))$, which is equivalent to $0 \in \sigma(H) \Leftrightarrow 0 \in \sigma(F_P(H))$. $\square$

12

Born-Oppenheimer Approximation and Adiabatic Dynamics

In this chapter, we discuss the Born-Oppenheimer approximation of molecular physics. This approximation lies at the foundation of quantum chemistry. It gave rise to adiabatic theory (see Section 12.5) in quantum mechanics and to multiscale analysis in mathematics and physics. It also led to the theory of geometrical phases discussed in Section 12.6. We begin with the stationary theory and then present informally the time-dependent one.

12.1 Problem and Heuristics

We consider a molecule with N electrons and M nuclei of masses $m_1, \ldots, m_M$. Its state space is a symmetry subspace $L^2_{sym}(\mathbb{R}^{3(N+M)})$ of $L^2(\mathbb{R}^{3(N+M)})$ and its Schrödinger operator, acting on $L^2_{sym}(\mathbb{R}^{3(N+M)})$, is given, in units such that $\hbar = 1$ and $c = 1$, by

$$H_{mol} = -\sum_1^N \frac{1}{2m}\Delta_{x_j} - \sum_1^M \frac{1}{2m_j}\Delta_{y_j} + V(x,y). \tag{12.1}$$

Here $x = (x_1, \ldots, x_N)$ and $y = (y_1, \ldots, y_M)$ are the electron and nuclear coordinates, and $V(x,y)$ is the sum of Coulomb interaction potentials between the electrons, between the electrons and the nuclei, and between the nuclei (see (10.2)).

As we know, the evolution of the molecule is described by the time-dependent Schrödinger equation (SE)

$$i\partial_t \Psi = H_{mol}\Psi, \tag{12.2}$$

with $\Psi(x,y,t) \in H^2(\mathbb{R}^{3(N+M)}), \forall t \in \mathbb{R}$.

A key fact here is that nuclei are much heavier than electrons and therefore much slower. Consequently, the electrons adjust almost instantaneously to the positions of the nuclei ('slave' to the nuclei) and prefer to be in the state of

© Springer-Verlag GmbH Germany, part of Springer Nature 2020
S. J. Gustafson and I. M. Sigal, *Mathematical Concepts of Quantum Mechanics*, Universitext, https://doi.org/10.1007/978-3-030-59562-3_12

lowest possible energy, i.e. the ground state, of the electronic Schrödinger operator.

Thus, in the first step, one supposes that the nuclei are frozen at their positions and considers, instead of (12.1), the electronic Schrödinger operator

$$H_{el}(y) := -\sum_1^N \frac{1}{2m} \Delta_{x_j} + V(x, y). \tag{12.3}$$

This is the adiabatic regime. The operator $H_{el}(y)$, called the *Born-Oppenheimer Hamiltonian*, acts on a symmetry subspace $L^2_{sym}(\mathbb{R}^{3N})$ of $L^2(\mathbb{R}^{3N})$ and depends on the coordinates, y, of the nuclei, as parameters, and therefore its eigenvalues are functions of y, as well.

In the second step, the ground state energy, $E(y)$, of $H_{el}(y)$ is considered as the potential (interaction) energy of the nuclear motion. Replacing $H_{el}(y)$ in (12.1) by its ground state energy, $E(y)$, leads to the nuclear Hamiltonian

$$H_{nucl} := -\sum_1^M \frac{1}{2m_j} \Delta_{y_j} + E(y). \tag{12.4}$$

where $E = E(y)$ is the operator of multiplication by the function $E(y)$. One expects that the eigenvalues of H_{nucl} give a good approximation to eigenvalues of H_{mol} and that the dynamics generated by (12.4) approximates the true dynamics of nuclei under exact evolution (12.2).

Minimizing the ground state energy, $E(y)$, of the operator $H_{el}(y)$ with respect to y gives the equilibrium positions of the nuclei, i.e. the shape of the molecule. This provides key information for quantum chemistry.

To proceed to our analysis, we use a rescaling to pass to dimensionless variables so that m_j become the ratios of nuclear masses to the electronic one. The small parameter in our problem is

$$\kappa := 1/\min_j m_j,$$

which depending on the nuclei varies from $\approx 1/1836$ to $\approx 1/367000$.

12.2 Stationary Born-Oppenheimer Approximation

In the stationary Born-Oppenheimer approximation, eigenvalues of molecular Hamiltonian (12.1) are approximated by eigenvalues of the nuclear Hamiltonian (12.4). To justify this, we use the Feshbach - Schur method (see Chapter 11).

Assume for simplicity that the ground state energy, $E(y)$, of the electronic Hamiltonian, $H_{el}(y)$, is non-degenerate (which is not hard to prove on $L^2(\mathbb{R}^{3N})$ but is not known and might be, in certain cases, false on

$L^2_{sym}(\mathbb{R}^{3N}))$, and denote the corresponding normalized ground state by $\psi_y(x)$: $H_{el}(y)\psi_y = E(y)\psi_y$.

Also, for technical reasons, we modify the original Hamiltonian by replacing the point nuclear charges by smeared ones. (This is often done in physical and computational literature. In technical jargon, the smeared charges are called form-factors.) This replaces the singular Coulomb interaction potential $V(x, y)$ in (12.3) by a potential which is differentiable as many times as the form-factors are. This will be used in the proof. We keep the same notation $H(y) = H_{el}(y)$ for the modified Hamiltonian.

The main result of this section is the following

Theorem 12.1 *[Born-Oppenheimer approximation] To second order, $O(\kappa^2)$, the ground state energy, E_0, of H_{mol} is the ground state energy of the operator*

$$H_{eff} := H_{nucl} + v, \tag{12.5}$$

acting on $L^2(\mathbb{R}^{3M})$, where H_{nucl} is given in (12.4) and $v = O(\kappa)$ is an operator given by

$$v := \sum_1^M \frac{1}{2m_j} \int |\nabla_{y_j}\psi_y|^2 dx, \tag{12.6}$$

if ψ_y is real, as for (12.3), and by (12.29) below, if ψ_y is complex, as in the case when a magnetic field is present. In the latter case, the operator H_{eff} in (12.5) can be written as

$$H_{eff} = -\sum_1^M \frac{1}{2m_j}(\nabla_{y_j} - iA_j)^2 + E(y) + \tilde{v}, \tag{12.7}$$

where $A_j := i\langle\psi_y, \nabla_{y_j}\psi_y\rangle$, with, recall, $\langle\cdot,\cdot\rangle$ standing for the L^2-inner product in the x variable, and $\tilde{v}$ is given by $\tilde{v} = \sum_1^M \frac{1}{2m_j}\|[P, \nabla_{y_j}]\psi_y\|^2$.

Proof. As was mentioned above, we use the Feshbach - Schur method to prove this result. Recall that in this method, given a quantum Hamiltonian, H, we pick a projection P so that the Feshbach-Schur map F_P is defined on $H - \lambda$ and maps the latter operator to a simpler one.

Let P be an orthogonal projection and $P^\perp = 1 - P$. Introduce the notation $H^\perp = P^\perp H P^\perp$. Assume

(a) The operator $H^\perp - \lambda$ is invertible;
(b) The operator

$$U(\lambda) := PHP^\perp(H^\perp - \lambda)^{-1}P^\perp HP \tag{12.8}$$

is well defined.

The Feshbach-Schur method, as applied to the quantum Hamiltonian H, states that if Conditions (a) and (b) are satisfied, then the Feshbach-Schur map

$$F_P(H - \lambda) = (PHP - U(\lambda))|_{\mathrm{Ran}P} \tag{12.9}$$

is well defined and

$$\lambda \text{ eigenvalue of } H \iff \lambda \text{ eigenvalue of } F_P(H - \lambda). \tag{12.10}$$

Moreover, the eigenfunctions of H and $F_P(H - \lambda)$ corresponding to the eigenvalue λ are connected as

$$H\psi = \lambda\psi \quad \Leftrightarrow \quad F_P(H - \lambda)\phi = \lambda\phi, \tag{12.11}$$

where ϕ, ψ are related by the following equations $\phi = P\psi$, $\psi = Q(\lambda)\phi$. Here the family of operators $Q(\lambda)$ is defined as $Q(\lambda) = P - P^\perp(H^\perp - \lambda)^{-1}P^\perp HP$.

Recall that $\psi_y(x)$ denotes the non-degenerate ground state, $H_{\mathrm{el}}(y)\psi_y = E(y)\psi_y$, of $H_{\mathrm{el}}(y)$, normalized as $\int |\psi_y(x)|^2 dx = 1$. We define P to be the orthogonal projection

$$(P\Psi)(x, y) = \psi_y(x) \int \overline{\psi_y(x)}\Psi(x, y)dx. \tag{12.12}$$

Problem 12.2 Check that P is an orthogonal projection. Note that it has infinite rank.

Remark 12.3 In the previous applications of the Feshbach-Schur method, the projection P was related to an isolated eigenvalue of some 'unperturbed' operator H_0. In this section, P is the spectral projection for the band of the spectrum $\{E(y) : y \in \mathbb{R}^{3M}\}$, which might be only partially isolated.

The analysis of Chapter 10 shows that the function $E(y)$ increases as $|y| \to \infty$. Hence it has a minimum $\lambda_* := \min_y E(y)$. (The corresponding minimum points - there could be several of those - determine the shape of the molecule.)

We apply the Feshbach - Schur map with projection (12.12) to $H_{mol} - \lambda$, with the spectral parameter λ close to λ_*.

Before we proceed, we introduce a useful concept and notation. For a family, $\{A(y)\}$, of operators, $A(y)$, acting on the space $L^2(\mathbb{R}^{3N})$ (or a symmetry subspace thereof), we define the fiber integral, $\int^\oplus A(y)dy$, as the operator on $L^2(\mathbb{R}^{3(N+M)})$ acting as [1]

---

[1] We identify vectors $\Psi \in L^2(\mathbb{R}^{3(N+M)})$ as vector-functions $y \in \mathbb{R}^{3M} \to \psi(y) \in L^2(\mathbb{R}^{3N})$, where $\psi(y)(x) := \Psi(x, y)$, and denote $\Psi = \int^\oplus \psi(y)dy$: $\int^\oplus A(y)dy \int^\oplus \psi(y)dy = \int^\oplus A(y)\psi(y)dy$.

$$\left(\int^{\oplus} A(y)dy \ \Psi \right)(x,y) = (A(y)\Psi(\cdot,y))(x).$$

In particular, the operator P defined in (12.12) can be written as the fiber integral $\int^{\oplus} P(y)dy$, where $P(y)$ are the orthogonal projections on the ground state eigenspaces for the operators $H_{el}(y)$. Moreover, using the fiber integral we can define the electronic Hamiltonian on the entire space $L^2(\mathbb{R}^{3(N+M)})$ as

$$H_{el} := \int^{\oplus} H_{el}(y)dy. \tag{12.13}$$

With this notation and by (12.1), the molecular Hamiltonian can be written as

$$H_{mol} = H_{el} + T_{nucl}, \tag{12.14}$$

where, recall, H_{el} is given in (12.13) and $T_{nucl} := -\sum_1^M \frac{1}{2m_j}\Delta_{y_j}$.

Furthermore, given two fiber integral operators $\int^{\oplus} A(y)dy$ and $\int^{\oplus} B(y)dy$, their product is also a fiber integral given by

$$\int^{\oplus} A(y)dy \int^{\oplus} B(y)dy = \int^{\oplus} A(y)B(y)dy. \tag{12.15}$$

Below, we use that for twice differentiable form-factors, i.e. for twice differentiable potentials $V(x,y)$ in (12.3), we have the estimate

$$\|\partial_y^\alpha \psi_y\| \le C, \qquad |\alpha| \le 2, \tag{12.16}$$

for some constant C. This can be shown by using perturbation theory, which we omit here.

Let $E'(y)$ be the first excited state energy of $H_{el}(y)$ above $E(y)$ (i.e. $E'(y) > E(y) \ \forall y$) and let $\bar{P}(y) := 1 - P(y)$. Then $\min_y E'(y) > \min_y E(y) =: \lambda_*$. Next, Eq (12.14) and the relations $T_{nucl} \ge 0$ and (12.15) imply

$$\bar{P}H_{mol}\bar{P} \ge H_{el}\bar{P} \ge \int^{\oplus} H_{el}(y)\bar{P}(y)dy$$

$$\ge \int^{\oplus} E'(y)\bar{P}(y)dy > \lambda_*\bar{P}. \tag{12.17}$$

Therefore, for λ close to λ_*, $H_{\bar{P}} := \bar{P}(H_{mol} - \lambda)\bar{P} \restriction_{\mathrm{Ran}\bar{P}}$ is invertible, giving condition (a).

To show that condition (b) above also holds, we first observe that, by the definition of $R_{\bar{P}}$, we have $PH_{mol}R_{\bar{P}} = E(y)PR_{\bar{P}} = 0$. Together with the above, this gives

$$PH_{mol}R_{\bar{P}} = PT_{nucl}R_{\bar{P}}. \tag{12.18}$$

To show that $PT_{nucl}R_{\bar{P}}$ is a bounded operator, we use a refined version of inequality (12.17) in which one keeps the term T_{nucl} to obtain

$$\bar{P}H_{mol}\bar{P} \geq H_{el}\bar{P} > \lambda_*\bar{P} + \bar{P}T_{nucl}\bar{P}. \tag{12.19}$$

Next, since $PR_{\bar{P}} = 0$, we have $PT_{nucl}R_{\bar{P}} = [P, T_{nucl}]R_{\bar{P}}$. Furthermore, to simplify the notation, we set $2m_j = 1$ for all j, so that $T_{nucl} = -\Delta_y$. With this, we compute $[P, T_{nucl}] = 2(\nabla_y P)\nabla_y + (\Delta_y P)$. The operators $(\nabla_y P)$ and $(\Delta_y P)$ are bounded, so it remains to show that the operator $\nabla_y R_{\bar{P}}$ is bounded. To this end, we write

$$\|\nabla_y R_{\bar{P}}\Psi\|^2 = \langle R_{\bar{P}}\Psi, -\Delta_y R_{\bar{P}}\Psi \rangle \leq \|R_{\bar{P}}\Psi\|\|\bar{P}\Delta_y\bar{P}R_{\bar{P}}\Psi\|$$

and use that, by inequality (12.19) and our convention $-\Delta_y = T_{nucl}$, the operator $\Delta_y\bar{P}R_{\bar{P}}$ is bounded, which implies that so is $\nabla_y R_{\bar{P}}$. Hence by the above, the operator (12.18) is bounded. The latter implies that so is $R_{\bar{P}}H_{mol}P = (PH_{mol}R_{\bar{P}})^*$. This proves condition (b) above.

Problem 12.4 Fill in details in the arguments above.

Thus for λ close to λ_*, $F_P(H_{mol} - \lambda)$ is well defined and according to (12.10),

$$\lambda \in \sigma_d(H_{mol}) \iff 0 \in \sigma_d(F_P(H_{mol} - \lambda)), \tag{12.20}$$

with the corresponding eigenfunctions related accordingly.

Now, remember by the decomposition (12.9) of the Feshbach-Schur operator we have to compute the terms $H_P := PHP \upharpoonright_{\mathrm{Ran}P}$ and $U(\lambda)$, defined in (23.8). We begin with PHP. Using decomposition (12.14) and the eigenequation $H_{el}(y)\psi_y = E(y)\psi_y$, we compute for $\Psi(x, y) \in H^2(\mathbb{R}^{3(N+M)})$

$$(H_{mol}P\Psi)(x, y) = \big[\psi_y(x)H_{nucl} + W\big]f(y) \tag{12.21}$$

where, recall, $H_{nucl} := T_{nucl} + E(y)$ (see (12.4)), $f(y) := \int \overline{\psi_y(x)}\Psi(x, y)dx$, so that $P\Psi = \psi_y f$, and, recall, W is a family of differential operators in y, with coefficients depending on x and y, given by

$$W := [T_{nucl}, \psi_y(x)]. \tag{12.22}$$

Problem 12.5 Show (12.21). Hint: use the decomposition $H_{mol} = H_{el} + T_{nucl}$ of the molecular Hamiltonian (12.1), where $(H_{el}\Psi)(x, y) = (H_{el}(y)\Psi(\cdot, y))(x)$, and the eigen-equation $H_{el}(y)\psi_y = E(y)\psi_y$.

Eqn (12.21) implies for any $\Psi \in H^2(\mathbb{R}^{3(N+M)})$,

$$PH_{mol}P\Psi = \psi_y(H_{nucl} + v)f, \tag{12.23}$$

where f is as above and v the differential operator in the variable y given by

$$v := \mathrm{Re} \int \overline{\psi_y(x)}W dx = \int \overline{\psi_y(x)}[T_{nucl}, \psi_y(x)]dx. \tag{12.24}$$

To compute the operator v, we first evaluate the commutator in (12.22) to find

$$W = -\sum_1^M \frac{1}{2m_j}((\nabla_{y_j}\psi_y) \cdot \nabla_{y_j} + \nabla_{y_j} \cdot (\nabla_{y_j}\psi_y)). \qquad (12.25)$$

Using that $\psi_y\nabla_{y_j} = \nabla_{y_j}\psi_y - (\nabla_{y_j}\psi_y)$, find furthermore that

$$v = -\sum_1^M \frac{1}{2m_j}(a_j\nabla_{y_j} + \nabla_{y_j}a_j - \int |\nabla_{y_j}\psi_y|^2 dx), \qquad (12.26)$$

where $a_j := \langle \psi_y, \nabla_{y_j}\psi_y \rangle$, with $\langle \cdot, \cdot \rangle$ standing for the L^2-inner product in the x variable. We see that $v = O(\kappa)$.

Note that $\mathrm{Ran}P$ can be identified with $L^2(\mathbb{R}^3)$ and therefore the operator $PH_{mol}P$ on $\mathrm{Ran}P$, with $H_{\mathrm{nucl}} + v$ on $L^2(\mathbb{R}^3)$.

Now we consider the term $U(\lambda)$, defined in (12.8). Using (12.21) and $\bar{P} = 1 - P$, we compute $\bar{P}(H_{mol} - \lambda)P\Psi = \bar{P}WP\Psi = Kf$, where K is the first order differential operator given by $K := \bar{P}W = W - v$. This expression and definition (23.8) show that the $U-$term is formally of the order $O(\kappa^2)$ and is of the form $U(\lambda)\Psi = \psi_y w(\lambda)f$, where the operator $w(\lambda)$ acts on $L^2(\mathbb{R}^{3M})$ and is given by

$$w(\lambda) := \langle (W - v)\psi_y, P^\perp(H_{mol}^\perp - \lambda)^{-1}P^\perp(W - v)\psi_y \rangle, \qquad (12.27)$$

where the inner product is taken in the variable x. (Note that due (12.25), $w(\lambda)$ involves derivatives in y up to the second order, it is a second order integro-differential operator.) Hence we obtain that the ground state energy, E_0, of H_{mol} is the ground state energy of the operator

$$H_{\mathrm{nucl}} + v + w$$

with $w := w(E_0)$, acting on $L^2(\mathbb{R}^{3M})$. Since $w(E_0) = O(\kappa^2)$, this gives Theorem 12.1, with v given in (12.26).

If ψ_y real, then $a_j = \nabla_{y_j}\langle \psi_y, \psi_y \rangle$ and therefore $a_j = 0$, due to the normalization $\int |\psi_y|^2 dx = 1$, which gives

$$v = \sum_1^M \frac{1}{2m_j} \int |\nabla_{y_j}\psi_y|^2 dx. \qquad (12.28)$$

This gives the first part of Theorem 12.1.

In general, for ψ_y complex, (12.7) is equivalent to (12.5) with v given by

$$v = -\sum_1^M \frac{1}{2m_j}(a_j\nabla_{y_j} + \nabla_{y_j}a_j + a_j^2 + \tilde{v}), \qquad (12.29)$$

where, recall, $a_j := \langle \psi_y, \nabla_{y_j}\psi_y \rangle$ and $\tilde{v}$ is given by

$$\tilde{v} = \sum_{1}^{M} \frac{1}{2m_j} \| [P, \nabla_{y_j}] \psi_y \|^2. \tag{12.30}$$

To prove (12.29), we note that (12.26) can be rewritten as (12.29), with $\tilde{v}$ is given by

$$\tilde{v} = \sum_{1}^{M} \frac{1}{2m_j} \left(\int |\nabla_{y_j} \psi_y|^2 dx - \left(\int \bar{\psi}_y \nabla_{y_j} \psi_y dx \right)^2 \right). \tag{12.31}$$

Now, we show that this $\tilde{v}$ is equal the one in (12.30). To this end, we insert the partition of unity $P + \bar{P} = 1$, where $P = \langle \psi_y, \cdot \rangle \psi_y$, into $\int |\nabla_{y_j} \psi_y|^2 dx = \langle \nabla_{y_j} \psi_y, \nabla_{y_j} \psi_y \rangle$ to obtain

$$\int |\nabla_{y_j} \psi_y|^2 dx = \langle \nabla_{y_j} \psi_y, (P + \bar{P}) \nabla_{y_j} \psi_y \rangle. \tag{12.32}$$

Using the definition of P, we find

$$\langle \nabla_{y_j} \psi_y, P \nabla_{y_j} \psi_y \rangle = \langle \nabla_{y_j} \psi_y, \psi_y \rangle \langle \psi_y, \nabla_{y_j} \psi_y \rangle. \tag{12.33}$$

Since $\langle \nabla_{y_j} \psi_y, \psi_y \rangle = \nabla_{y_j} \langle \psi_y, \psi_y \rangle - \langle \psi_y, \nabla_{y_j} \psi_y \rangle$ and $\nabla_{y_j} \langle \psi_y, \psi_y \rangle = 0$, this gives

$$\langle \nabla_{y_j} \psi_y, P \nabla_{y_j} \psi_y \rangle = -a_j^2. \tag{12.34}$$

Using that $\bar{P} \psi_y = 0$ and therefore $\bar{P} \nabla_{y_j} \psi_y = [P, \nabla_{y_j}] \psi_y$, the $\bar{P}$-part of can rewritten as

$$\langle \nabla_{y_j} \psi_y, \bar{P} \nabla_{y_j} \psi_y \rangle = \| [P, \nabla_{y_j}] \psi_y \|^2.$$

This equation together with (12.32), shows that the expressions on the r.h.s. of (12.30) and (12.31) are equal, which proves (12.7).

The excited states of H_{mol} come from excited states of $H_{\text{nucl}} + v$ and from bound states of the operator obtained by projecting onto the exited states of $H_{\text{el}}(y)$.

Remark 12.6 In the previous applications of the Feshbach-Schur method, the projection P was related to an isolated eigenvalue of some 'unperturbed' operator H_0. In this section, P is the spectral projection for the band of the spectrum $\{E(y) : y \in \mathbb{R}^{3M}\}$, which might be only partially isolated.

Beyond the Born-Oppenheimer Approximation. We recall (12.20) and define

$$H_{\text{nucl}}(\lambda) := F_P(H_{\text{mol}} - \lambda) + \lambda P.$$

Then

$$\lambda \in \sigma_d(H_{\text{mol}}) \leftrightarrow \lambda \in \sigma_d(H_{\text{nucl}}(\lambda)),$$

Let δ be the gap between the ground state energy, $E(y)$, of $H_N(y)$ and the rest of its spectrum. Then, for $\lambda \leq E(y) - \delta/2$, the operator $H_{\text{nucl}}(\lambda)$ acts on $L^2(\mathbb{R}^{3M})$ and is of the form

$$H_{\text{nucl}}(\lambda) := -\sum_{j=1}^{M} \frac{1}{2m_j} \Delta_{y_j} + E_\kappa(y, \lambda),$$

where, recall, $\kappa := 1/\min_j m_j$ and $E_\kappa(y, \lambda)$ can be computed to any order of κ, e.g. in the second order we have

$$E_\kappa(y, \lambda) = E(y) + \sum_{1}^{M} \frac{1}{2m_j} \int |\nabla_{y_j} \psi_y|^2 dx + O(\kappa^2),$$

with $O(\kappa^2)$ standing for a non-local operator of the indicated order in κ, differentiable in λ, with derivatives of the same order in κ (but maybe non-uniform in y). The leading term in the energy $E_\kappa(y, \lambda)$ gives the operator H_{nucl} defined in (12.4). The energy $E_\kappa(y, \lambda)$ can be used to define the interaction energy in all orders as

$$W_\kappa(y) := E_\kappa(y) - E(\infty), \tag{12.35}$$

where $E_\kappa(y)$ solves the equation $\lambda = E_\kappa(y, \lambda)$. It is worth remembering however that $E_\kappa(y, \lambda)$, and therefore $W_\kappa(y)$, are not potentials anymore.

12.3 Complex ψ_y and Gauge Fields

If ψ_y is not real, as is the case when magnetic fields are present, then we have shown that the operator H_{eff} in (12.5) can be written as (12.7), which gives the effective time-independent Schrödinger equation

$$H_{\text{eff}} f = E f, \qquad H_{\text{eff}} = -\sum_{1}^{M} \frac{1}{2m_j} (\nabla_{y_j} - iA_j)^2 + E(y) + \tilde{v}, \tag{12.36}$$

where, recall, $A_j := i\langle \psi_y, \nabla_{y_j} \psi_y \rangle$, with $\langle \cdot, \cdot \rangle$ standing for the L^2-inner product in the x variable, and $\tilde{v} = \sum_{1}^{M} \frac{1}{2m_j} \|[P, \nabla_{y_j}]\psi_y\|^2$.

Gauge invariance. Under the transformation $f(y, t) \to e^{i\chi(y)} f(y, t), \psi_y(x) \to e^{-i\chi(y)} \psi_y(x)$, the function $\Psi_{\text{BO}}(x, y, t) := \psi_y(x)f(y, t)$ is not changed. However, the vector potential $A_y := i\langle \psi_y, \nabla_y \psi_y \rangle_x$ changes as

$$A_y \to A_y + \nabla \chi(y).$$

Thus an effective equation for f should be invariant under the gauge transformation

$$f(y) \to e^{i\chi(y)} f(y), \quad A_y \to A_y + \nabla \chi(y). \tag{12.37}$$

As can be easily checked (12.36) is indeed invariant under (12.37).

Problem 12.7 Prove that (12.36) is invariant under the gauge transformations (12.37).

Remark 12.8 (Degenerate eigenvalues) If the eigenvalue $E(y)$ is degenerate, then $A(y) := i\langle \psi_y, d_y\psi_y \rangle_x$ takes values in the $k \times k$-matrices, where k is the degeneracy of $E(y)$.[2] In this case, Eq (12.36) is invariant under the gauge transformation

$$f(y) \rightarrow g(y)f(y), \quad A(y) \rightarrow g(y)A(y)g(y)^{-1} + dg(y)g(y)^{-1} \tag{12.38}$$

where $g(y) \in U(k), \forall y \in \mathbb{R}^k$. Hence $A(y)$ transforms as a Yang-Mills gauge field, or a $U(k)$−connection.

12.4 Time-dependent Born-Oppenheimer Approximation

We present a heuristic discussion of the time-dependent Born-Oppenheimer approximation. We look for the solution Ψ to the time-dependent Schrödinger equation (12.2) in the form

$$\Psi(x, y, t) \approx \Psi_{\text{BO}}(x, y, t) := \psi_y(x)f(y, t).$$

We call the functions $\Psi_{\text{BO}}(x, y, t), f \in L^2(\mathbb{R}^{3M})$, the Born-Oppenheimer states. Now, as in the Dirac-Frenkel theory (see e.g. [216] and references therein), projecting the time-dependent Schrödinger equation (12.2) on the manifold of the Born-Oppenheimer states, we arrive the effective time-dependent Schrödinger equation

$$i\partial_t \Psi_{\text{BO}} = PH_{\text{mol}}\Psi_{\text{BO}}, \tag{12.39}$$

where P is the projection defined in (12.12). Using that $\Psi_{\text{BO}} = P\Psi_{\text{BO}}$, so that $PH_{mol}\Psi_{\text{BO}} = PH_{mol}P\Psi_{\text{BO}}$ and relation (12.23), we find the equation for f:

$$i\partial_t f = H_{\text{eff}}f, \quad H_{\text{eff}} := T_{\text{nucl}} + E + v, \tag{12.40}$$

where v is the operator given by (12.26), or (12.29). Moreover, if ψ_y is real, then v is the operator of multiplication by the function (12.6), and, if ψ_y is complex,   then the effective Hamiltonian H_{eff} in (12.40) is given by (12.7), i.e.

$$H_{\text{eff}} := -\sum_1^M \frac{1}{2m_j}(\nabla_{y_j} - iA_j)^2 + E(y) + \tilde{v}. \tag{12.41}$$

Using a time-dependent version of the Feshbach-Schur method of Section 12.2, as in [225], would give the exact effective quantum hamiltonian

---

[2] We can also consider P to be a projection onto several eigenspaces. In this case, k is the total dimension of the eigenspaces considered.

$$H'_{\text{eff}} := H_{\text{nucl}} + v + w, \tag{12.42}$$

with w a time-dependent differential operator, acting on $L^2(\mathbb{R}^{3M})$, of the order $O(\kappa^2)$, so that the original time-dependent Schrödinger equation would be equivalent to the effective one.

Remark 12.9 (The effective action) The (effective) Lagrangian for the effective quantum Hamiltonian (12.41), with $j = 1$ and $m = 1$, is $L_{\text{eff}} = \frac{1}{2}|\dot{y}|^2 + \dot{y} \cdot A(y) - E$. (Indeed, we compute $p \cdot \dot{y} - L_{\text{eff}}|_{p=\dot{y}+A} = [p \cdot \dot{y} - \frac{1}{2}|\dot{y}|^2 - \dot{y} \cdot A(y) + E]|_{p=\dot{y}+A} = \frac{1}{2}|p - A|^2 + E$.) Let $L_{\text{geom}} := \dot{y} \cdot A(y)$. Using this in the path integral representation of the propagator, we obtain the factor

$$e^{iS_{geom}} = e^{i \int A} = e^{i\gamma}.$$

12.5 Adiabatic Motion

As was mentioned in the introduction to this chapter, heavy nuclei move relatively slowly and in the Born-Oppenheimer approximation we fix them in their positions. The next step would be to let them move but take advantage of their slow motion. In this section, we explore the slow, or adiabatic, motion in an abstract context and at the end discuss application of the results to molecular dynamics.

In its original context of thermodynamics, the notion of adiabatic evolution describes a slow dynamics for which at every moment of time the evolving state is approximately a stationary one, but with a possibly different set of parameters. Below, we formulate this statement rigorously in the context of quantum mechanics.

Let $H_s, s \in \mathbb{R}$, be a one-parameter family of Schrödinger operators and consider the adiabatic (slow) evolution

$$i\epsilon \partial_s \psi^\epsilon_s = H_s \psi^\epsilon_s, \quad \psi^\epsilon_{s=0} = \phi_0, \tag{12.43}$$

with a small parameter ϵ (incorporating the Planck constant $\hbar$) quantifying the adiabaticity.

Assume H_s have isolated, simple eigenvalues E_s. We expect that, if the initial condition ϕ_0 is an eigenvector of H_0 with the eigenvalue E_0, then, for ϵ sufficiently small, the solution ψ^ϵ_s is close to an instantaneous eigenvector ϕ_s with the eigenvalue E_s. The next result justifies this. Since even simple eigenvectors are defined up to a phase, we pick normalized eigenvectors:

$$\phi_s \in \text{Null}(H_s - E_s). \tag{12.44}$$

A family H_s of operators is said to be *differentiable* in the resolvent sense iff its resolvent $(H_s - z)^{-1}$ is differentiable, as a family of bounded operators, for some z in each connected component of the resolvent set.

Theorem 12.10 *[Adiabatic theorem] Assume H_s is $(k+1)$-times differentiable in the resolvent sense. Then there are phases γ_s s.t.*

$$\psi_s^\epsilon = e^{i(\gamma_s - \int^s E_r dr/\epsilon)}\phi_s + O(\epsilon^k) \quad as \quad \epsilon \to 0. \tag{12.45}$$

We discuss the phases $e^{i\gamma_s}$ in the next section. Another formulation of the adiabatic regime is to consider the normal evolution for a slowly varying family of quantum Hamiltonians, say, $H_{t/\tau}$, with τ large:

$$i\hbar\partial_t \Psi_t^\tau = H_{t/\tau}\Psi_t^\tau, \ \Psi_{t=0}^\tau = \phi_0. \tag{12.46}$$

Rescaling (12.46) by setting $s := t/\tau$ and $\psi_s^\epsilon := \Psi_{\tau s}^\tau$, where $\epsilon := \hbar/\tau$, we obtain (12.43). Then Theorem 12.10 implies

$$\Psi_{\tau s}^\tau \to \phi_s \quad as \quad \tau \to \infty.$$

Next, we reformulate Theorem 12.10 in a convenient and important way. Let $U_\epsilon(s,r)$ be the propagator generated by H_s/ϵ (see Section 25.6, Theorem 25.32) and $U_\epsilon(s) \equiv U_\epsilon(s,0)$, so that $\psi_s^\epsilon = U_\epsilon(s)\psi_{s=0}^\epsilon$. It satisfies the equation

$$i\epsilon\partial_s U_\epsilon(s) = H_s U_\epsilon(s), \ U_\epsilon(s=0) = \mathbf{1}. \tag{12.47}$$

Let P_s be the orthogonal projection onto $\mathrm{Null}(H_s - E_s)$. Note that, unlike eigenvectors, eigenprojections are defined uniquely. We define the adiabatic hamiltonian

$$H_s^A := i\epsilon[\dot{P}_s, P_s] + E_s \tag{12.48}$$

and let $U_\epsilon^A(s,r)$ be the propagator generated by the hamiltonian H_s^A/ϵ i.e. the two-parameter family of bounded operators satisfying the initial value problem (see (25.30) - (25.31))

$$i\epsilon\frac{\partial}{\partial s}U_\epsilon^A(s,r) = H_s^A U_\epsilon^A(s,r), \quad U(r,r) = \mathbf{1}. \tag{12.49}$$

Let $U_\epsilon^A(s) \equiv U_\epsilon^A(s,0)$. The following (adiabatic) theorem implies Theorem 12.10:

Theorem 12.11 *Assume H_s is $(k+1)$-times differentiable in the resolvent sense. Then we have*

$$U_\epsilon(s)P_0 = U_\epsilon^A(s)P_0 + O(\epsilon^k). \tag{12.50}$$

A proof of Theorem 12.11 is given in Section 12.8. It is based on the Feshbach - Schur approach described in Section 11.1 (with the orthogonal projection P_s). However, unlike Chapter 11, we apply to it to a time-dependent problem, specifically, to Eq. (12.47). Using this approach we derive the equation

$$F_{P_s}(K_{s,\epsilon})P_s U_\epsilon(s) = 0, \tag{12.51}$$

where $F_{P_s}(K_{s,\epsilon})$ is the Feshbach - Schur map, with the projection P_s (cf. (11.7)), applied to the operator $K_{s,\epsilon} := H_s - i\epsilon\partial_s$ acting on s-dependent vectors. Using the Duhamel principle and the notation $\bar{P}_s := 1 - P_s$, we rewrite equation (12.51) as (see (12.72)-(12.73) of Section 12.8)

$$P_s U_\epsilon(s) = U_\epsilon^A(s, 0) P_0 \tag{12.52}$$

$$+ \int_0^s dr\, U_\epsilon^A(s, r) P_r \dot{P}_r \bar{P}_r \int_0^r dt\, \bar{U}_\epsilon^A(r, t) \bar{P}_t \dot{P}_t P_t U_\epsilon(t), \tag{12.53}$$

where $\bar{U}_\epsilon^A(s, r)$ is the propagator generated by the hamiltonian $\bar{H}_s^A :=$ $i\epsilon[\dot{P}_s, P_s] + \bar{H}_s$, with $\bar{H}_s := \bar{P}_s H_s \bar{P}_s$, and estimate the second term on the r.h.s., (12.53), using integration by parts. (In Section 11.3, we came close to this but then used Floquet theory (and the Fourier transform) to convert the time-dependent problem to a time-independent one.)

The next proposition reveals a key property of the adiabatic propagator $U_\epsilon^A(s, r)$:

Proposition 12.12 *The (adiabatic) evolution $U_\epsilon^A(s, r)$ has the intertwining property*

$$U_\epsilon^A(s, r) P_r = P_s U_\epsilon^A(s, r). \tag{12.54}$$

Proof. Using that the propagator $U_\epsilon^A(s, r)$ satisfies the differential equation (12.49) and using the similar equation for $U_\epsilon^A(s, r)^{-1}$, we find

$$\frac{\partial}{\partial s}(U_\epsilon^A(s, r)^{-1} P_s U_\epsilon^A(s, r)) = U_\epsilon^A(s, r)^{-1} B_s U_\epsilon^A(s, r), \tag{12.55}$$

where $B_s := -[\dot{P}_s, P_s]P_s + P_s[\dot{P}_s, P_s] + \dot{P}_s$. Next, relations $P_s \dot{P}_s P_s = 0$ and

$$\dot{P}_s = \dot{P}_s P_s + P_s \dot{P}_s$$

(by differentiating $P_s = P_s^2$) imply that $B_s = 0$. Using this in (12.55) and integrating the latter equation in s from r to s, we find $U_\epsilon^A(s, r)^{-1} P_s U_\epsilon^A(s, r) = P_r$, which gives (12.54). □

Remark 12.13 Because of the Coulomb singularity, in the Born-Oppenheimer case, the family $H(y) = H_{\text{el}}(y)$ is not differentiable, but only Hölder continuous with an arbitrary exponent less than 1. One can extend Theorems 12.10 and 12.11 to fractional derivatives and show that in the Born-Oppenheimer case, Theorems 12.10 and 12.11 hold for any $k < 1$. (Presumably, for curves whose velocity vector fields do not move the Coulomb singularities (the Hunziker vector fields) Theorems 12.10 and 12.11 hold for any k.)

For physical and computational reasons, one often replaces the point nuclear charges by smeared ones (in technical jargon, introduces form-factors). Then the corresponding potential, which is differentiable as many times as the form-factors are, replaces the interaction potential $V(x, y)$ in (12.3). In this case, the family $H(y) = H_{\text{el}}(y)$ is differentiable the same number of times.

12.6 Geometrical Phases

We begin with the adiabatic dynamics given by the time-dependent Schrödinger equation (12.43), which we reproduce here

$$i\epsilon\partial_s\psi_s^\epsilon = H_s\psi_s^\epsilon, \quad \psi_{s=0}^\epsilon = \phi_0. \tag{12.56}$$

Then the adiabatic theorem of Section 12.5 (see Theorem 12.10) says that

$$\psi_s^\epsilon \to \psi_s \in \text{Null}(H_s), \quad \text{as} \quad \epsilon \to 0. \tag{12.57}$$

Recall that we have chosen normalized basis vectors ϕ_s in the eigenspaces $\text{Null}(H_s - E_s)$, see (12.44).

Proposition 12.14 γ_s *of Theorem 12.10 and* ψ_s *of (12.57) are given by*

$$\psi_s = e^{i(\gamma_s - \int^s E_r dr/\epsilon)}\phi_s, \quad \partial_s\gamma_s = i\langle\phi_s, \dot\phi_s\rangle. \tag{12.58}$$

Proof. Introduce $\Psi_s := e^{i\int^s E_r dr/\epsilon}\psi_s$ and $\Psi_s^\epsilon := e^{i\int^s E_r dr/\epsilon}\psi_s^\epsilon$. We show first that $\langle\Psi_s, \dot\Psi_s\rangle = 0$. Indeed,

$$\langle\Psi_s, \dot\Psi_s\rangle = \lim_{\epsilon\to0}\langle\Psi_s, \dot\Psi_s^\epsilon\rangle = \lim_{\epsilon\to0}\langle\psi_s, -i\epsilon^{-1}(H_s - E_s)\psi_s^\epsilon\rangle$$

$$= \lim_{\epsilon\to0}\langle(H_s - E_s)\psi_s, -i\epsilon^{-1}\psi_s^\epsilon\rangle = 0.$$

Since ψ_s and ϕ_s belong to $\text{Null}(H_s - E_s)$ and are normalized, $\Psi_s = U_s\phi_s$, where U_s is a complex number of unit modulus ($U(1)$-factor) defined by this expression. We compute

$$0 = \langle\Psi_s, \dot\Psi_s\rangle = \dot U_s U_s^* + \langle\phi_s, \dot\phi_s\rangle.$$

Since U_s is unitary ($U_s^* = U_s^{-1}$), this equation can be rewritten as

$$\dot U_s = -\langle\phi_s, \dot\phi_s\rangle U_s, \tag{12.59}$$

which can be solved as $U_s = e^{i\gamma_s}$ with γ_s given by (12.58). $\qquad\square$

Let Y be a parameter space ($\mathbb{R}^{3m}$, or a surface therein for the nuclear coordinates in the Born-Oppenheimer approximation, or a torus if y describes magnetic fluxes as in the integer quantum Hall effect, see [25]). We consider a family $H(y), y \in Y$, of Schrödinger operators. Assume $\forall y \in Y$, the operator $H(y)$ has an isolated, simple eigenvalue $E(y)$ and fix a corresponding normalized eigenfunction, $\phi(y)$.

We consider a curve $C \subset Y$ parametrized by $y(s), s \in [0,1]$ and the time-dependent Schrödinger equation (12.56), with

$$H_s := H(y(s)), \tag{12.60}$$

with the initial condition $\psi_{s=0} = \phi(0) \in \text{Null}(H(0) - E(0))$. The adiabatic theorem implies that $\psi_s^\epsilon \to \psi_s \in \text{Null}(H_s - E_s)$, $E_s = E(y(s))$, as $\epsilon \to 0$. By replacing $H(y)$ with $H(y) - E(y)$, we may assume $E(y) \equiv 0$. Then by Proposition 12.14, $\psi_s = e^{i\gamma_{y(s)}}\phi(y(s))$, for $\gamma_{y(s)} \equiv \gamma_s$ given by (12.58). We call ψ_1 the adiabatic transport of $\phi(0) \in \text{Null}(H(0) - E(0))$ along C.

If C is a loop, then we have $y(0) = y(1)$. Since $E(y)$ and therefore E_s are simple eigenvalues, this gives

$$\psi_1 = e^{i\gamma_C}\psi_0, \tag{12.61}$$

where $\gamma_C = \gamma_{y(1)}$. By (12.58), we have

$$\partial_s \gamma_{y(s)} = i\langle \phi(y(s)), \dot\phi(y(s)) \rangle = i\langle \phi(y(s)), \nabla_y \phi(y(s)) \rangle \cdot \dot y(s).$$

We introduce the vector field

$$A_y := i\langle \phi(y), \nabla_y \phi(y) \rangle. \tag{12.62}$$

Then $\partial_s \gamma_{y(s)} = A_{y(s)} \cdot \dot y(s)$. For a loop C, we can write $\gamma_{y(1)} = \int_0^1 A_{y(s)} \cdot \dot y(s)ds = \int_C A_y$, and therefore

$$\gamma_C = \int_C A_y. \tag{12.63}$$

Geometric interpretation. A better way to think about A_y is as a co-vector, or a one-form, i.e.

$$A_y := i\langle \phi(y), d_y \phi(y) \rangle,$$

where d is the differential (exterior derivative) in y. Note that if $\phi(y) \to e^{-i\chi(y)}\phi(y)$, then $A_y \to A_y + d\chi(y)$. Hence A_y could be viewed as a connection one-form on the vector (line) bundle, E, over Y, with the fibers being the eigenspaces

$$F_y := \text{Null}(H(y) - E(y))$$

(or the associated frame bundle over Y with the fibers $F_y := \{\psi \in \text{Null}(H(y) - E(y)) : \|\psi\| = 1\}$). In the physics literature it is called the Berry (hermitian) connection.

The curvature ('magnetic field') of A is given by $F_A := dA$. By Stokes' theorem, we have

$$\gamma_C = \int_S dA_y = \int_S F_A,$$

where S is a surface spanning C, i.e. $\partial S = C$.

If S is a 2 dimensional torus, as in the case of the integer quantum Hall effect (see [20]), then the flux of the curvature F_A over S is quantized:

$$\int_S F_A = 4\pi c_1,$$

where c_1 is the first Chern number of the bundle F over S.[3] It turns out F_A is proportional to the Hall conductivity in the integer quantum Hall effect (IQHE), see [20, 25]. There the surface S is a 2-torus spanned by the magnetic fluxes through the 'holes' (see the last paragraph of Subsection 7.5 and Subsection 7.6). Hence the above relation states that the Hall conductivity averaged over all fluxes is quantized (and proportional the first Chern number).

If the eigenvalues $E(y)$ are degenerate, then A_y is a matrix-valued (connection) one-form on the vector bundle, E, over Y with the fibers $F_y = \text{Null}(H(y) - E(y))$ (or the associated frame bundle, F, over Y with the fibers F_y consisting of orthonormal k-frames (ordered bases) $(\phi_1(y), ..., \phi_r(y))$ in $\text{Null}(H(y) - E(y))$, where $r := \dim \text{Null}(H(y) - E(y)))$.

Considering the adiabatic evolution along the path $y(s)$ parametrizing a curve C and letting $H_s := H(y(s))$, we see from Eq (12.54) that $U_\epsilon^A(s, r)$ can be viewed as a parallel transport on the frame bundle F along C. Moreover, A is the connection related to this parallel transport.

Furthermore, solving Schrödinger equation (12.56) with quantum Hamiltonian (12.60) along a loop gives the map $\psi_0 \to \psi_1$ of the fiber $F_y = \text{Null}(H(y) - E(y))$ into itself. This is the holonomy, U_C, of the connection A. If the eigenvalues $E(y)$ are non-degenerate, then $U_C = e^{i \int_C A_y}$. In the degenerate case, we have, instead,

$$U_C = P e^{i \int_C A_y}, \tag{12.64}$$

where $P e^{i \int_C A_y}$ stands for the 'time-ordered exponential. (This is the Wilson loop, see [300] and the paragraph around (12.38) for more discussion.)

Remark 12.15 If $P(y)$ is the orthogonal projection onto $\text{Null}(H(y) - E(y))$, then the covariant derivative, or connection, on the vector bundle E is given by

$$\nabla := P(y)d_y = d_y - [d_y P(y), P(y)],$$

where d_y is the exterior derivative on Y. If $\{\phi_j(y)\}$ is an orthonormal basis in $\text{Ran}P(y)$, so that $P(y) = \sum_j |\phi_j(y)\rangle\langle\phi_j(y)|$ and $\sum_j f_j(y)\phi_j(y) \in \text{Ran}P(y)$, where $f_j(y)$ are functions of just y, then, due to the relation $[d_y P(y), P(y)] = d_y P(y)P(y)$, the last formula becomes (cf. (12.62))

$$\nabla \sum_j f_j(y)\phi_j(y) = \sum_{ij}(d_y f_j(y)\delta_{ij} + \langle\phi_i(y), d_y\phi_j(y)\rangle f_j(y))\phi_i(y).$$

---

[3] This is a special case of the Chern-Weil theory. A 2 dimensional torus can be identified with a fundamental domain of a lattice $\mathcal{L}$. In this case A is defined on $\mathbb{R}^2$ as an $\mathcal{L}$-equivariant one-form in the sense of (7.62), i.e. it satisfies

$$A(x + s) = A(x) + \nabla\chi_s(x),$$

for any $s \in \mathcal{L}$ and for some functions $\chi_s(x)$. For more details see Section 7.9.

12.7 Appendix: Projecting-out Procedure

Let $\mathcal{H}$ be a Hilbert space, and P a projection on $\mathcal{H}$ (i.e. an operator satisfying $P^2 = P$). Let A be an operator acting on $\mathcal{H}$. Then PAP is an operator acting on $\mathrm{Ran}(P)$: i.e. $PAP : \mathcal{H} \to \mathrm{Ran}(P)$ and consequently $PAP : \mathrm{Ran}(P) \to \mathrm{Ran}(P)$.

The simplest example is when P is a rank-one projection: $P = |\phi\rangle\langle\phi|$ for some $\phi \in \mathcal{H}$, $\|\phi\| = 1$. Here we have used Dirac's bra-ket notation

$$|\phi\rangle\langle\phi| : \psi \mapsto \phi\langle\phi, \psi\rangle.$$

Then $PAP = \langle\phi, A\phi\rangle P$ acts on $\mathrm{Ran}(P) = \{z\phi \mid z \in \mathbb{C}\}$ as multiplication by the complex number $\langle\phi, A\phi\rangle$. This example can be easily generalized to a finite-rank projection, say $P = \sum_{i=1}^{m} |\phi_i\rangle\langle\phi_i|$ where $\{\phi_j\}$ is an orthonormal set in $\mathcal{H}$. In this case the number $\langle\phi, A\phi\rangle$ is replaced by the matrix $(\langle\phi_i, A\phi_j\rangle)_{ij}$.

For another typical example, take $\mathcal{H} = L^2(\mathbb{R})$ and take P to be multiplication of the Fourier transform by the characteristic function, χ_I, of an interval $I \subset \mathbb{R}$:

$$(Pf)(x) = (2\pi)^{-1/2} \int_{-\infty}^{\infty} e^{ikx}\chi_I(k)\hat{f}(k)dk = (2\pi)^{-1/2} \int_I e^{ikx}\hat{f}(k)dk,$$

where $\hat{f}(k) = (2\pi)^{-1/2} \int_{-\infty}^{\infty} e^{-ikx}f(x)dx$ is the Fourier transform of f. P is a projection since $\chi_I^2 = \chi_I$. For a given operator A on $\mathcal{H}$, the operator PAP (assuming it is well-defined) maps functions whose Fourier transforms are supported on I into functions of the same type.

Of interest to us is the following situation: the projection P acts on a space $\mathcal{H} = \mathcal{H}_1 \otimes \mathcal{H}_2$ as $P = P_1 \otimes \mathbf{1}$, where P_1 is a rank-one projection on $\mathcal{H}_1$, say $P_1 = |\phi_1\rangle\langle\phi_1|$, with $\phi_1 \in \mathcal{H}_1$. If A is an operator on $\mathcal{H}$, then the operator PAP is of the form

$$PAP = \mathbf{1} \otimes A_2$$

where A_2 is an operator on $\mathcal{H}_2$ which can formally be written as $A_2 = \langle\phi_1, A\phi_1\rangle_{\mathcal{H}_1}$.

12.8 Appendix: Proof of Theorem 12.11

Applying the projections P_s and $\bar{P}_s := \mathbf{1} - P_s$ to equation (12.47) and using $P_s i\epsilon\partial_s \bar{P}_s = -i\epsilon P_s \dot{P}_s \bar{P}_s$ and $\bar{P}_s i\epsilon\partial_s P_s = i\epsilon\bar{P}_s \dot{P}_s P_s$, we find (the Feshbach - Schur splitting)

$$P_s K_{s,\epsilon} P_s U_\epsilon(s) = -i\epsilon P_s \dot{P}_s \bar{P}_s U_\epsilon(s), \qquad (12.65)$$

$$\bar{P}_s K_{s,\epsilon} \bar{P}_s U_\epsilon(s) = i\epsilon\bar{P}_s \dot{P}_s P_s U_\epsilon(s). \qquad (12.66)$$

We compute the operators $P_s K_{s,\epsilon} P_s$ and $\bar{P}_s K_{s,\epsilon} \bar{P}_s$. First, we differentiate the relation $P_s = P_s^2$ and using the notation $\dot{P}_s := \partial_s P_s$ to obtain

$\dot{P}_s P_s + P_s \dot{P}_s = \dot{P}_s$ and similarly for $\bar{P}_s$. These relations imply $P_s \dot{P}_s P_s = 0$ and $\bar{P}_s \dot{P}_s \bar{P}_s = 0$, which give

$$P_s \partial_s P_s = (\partial_s - [\dot{P}_s, P_s])P_s, \qquad \bar{P}_s \partial_s \bar{P}_s = (\partial_s - [\dot{P}_s, P_s])\bar{P}_s.$$

Observe that the operator $i[\dot{P}_s, P_s]$ is self-adjoint. Using these relations, we compute

$$P_s K_{s,\epsilon} P_s = (i\epsilon\partial_s - H_s^A)P_s, \qquad (12.67)$$

$$\bar{P}_s K_{s,\epsilon} \bar{P}_s = (i\epsilon\partial_s - \bar{H}_s^A)\bar{P}_s, \qquad (12.68)$$

where, recall, $H_s^A := i\epsilon[\dot{P}_s, P_s] + H_s$ and $\bar{H}_s^A := i\epsilon[\dot{P}_s, P_s] + \bar{H}_s$, with $\bar{H}_s := \bar{P}_s H_s \bar{P}_s$. Combining Eqs (12.65) and (12.67), we find

$$(i\epsilon\partial_s - H_s^A)P_s U_\epsilon(s) = -i\epsilon P_s \dot{P}_s \bar{P}_s U_\epsilon(s). \qquad (12.69)$$

Using now the Duhamel formula, we find[4]

$$P_s U_\epsilon(s) = U_\epsilon^A(s,0)P_0 - \int_0^s U_\epsilon^A(s,r)P_r \dot{P}_r \bar{P}_r U_\epsilon(r)dr, \qquad (12.70)$$

where, recall, $U_\epsilon^A(s,r)$ is the propagator generated by the hamiltonian H_s^A. Treating (12.66) similarly, we arrive at

$$\bar{P}_s U_\epsilon(s) = \bar{U}_\epsilon^A(s,0)\bar{P}_0 - \int_0^s \bar{U}_\epsilon^A(s,r)\bar{P}_r \dot{P}_r P_r U_\epsilon(r)dr, \qquad (12.71)$$

where $\bar{U}_\epsilon^A(s,r)$ is the propagator generated by the hamiltonian $\bar{H}_s^A$. Plugging (12.71) into (12.70), we find

$$P_s U_\epsilon(s) = U_\epsilon^A(s,0)P_0 + R_s, \qquad (12.72)$$

where R_s is given by

$$R_s = \int_0^s dr U_\epsilon^A(s,r)P_r \dot{P}_r \bar{P}_r \int_0^r dt \bar{U}_\epsilon^A(r,t)\bar{P}_t \dot{P}_t P_t U_\epsilon(t). \qquad (12.73)$$

---

[4] The Duhamel formula could be rephrased as $P_s U_\epsilon(s) = U_\epsilon^A(s,0)P_0 + (i\epsilon\partial_s - H_s^A)^{-1}W P_s U_\epsilon(s)$ (the sum of homogeneous and inhomogeneous solutions), with $W := -i\epsilon P_s \dot{P}_s$ and

$$(i\epsilon\partial_s - H_s^A)^{-1}f(s) = (i\epsilon)^{-1}\int_0^s U_\epsilon^A(s,r)f(r)dr.$$

To prove the last relation, we use that $U_\epsilon^A(s,r)$ is defined by the initial value problem $(i\epsilon\partial_s - H_s^A)U_\epsilon^A(s,r) = 0$ and $U_\epsilon^A(r,r) = \mathbf{1}$. Hence

$$(i\epsilon\partial_s - H_s^A)(i\epsilon)^{-1}\int_0^s \bar{U}_\epsilon^A(s,r)f(r)dr = (\partial_s \int_0^s)\bar{U}_\epsilon^A(s,r)f(r)dr = f(s).$$

Lemma 12.16

$$R_s = O(\epsilon). \tag{12.74}$$

Proof. Using definition (12.73), the fact that $U_\epsilon^A(s,r)$ commutes with $\dot{P}_r \bar{P}_r = -[\dot{P}_s, P_s]\bar{P}_r$ and the notation $G_\epsilon(t) := \bar{P}_t \dot{P}_t P_t U_\epsilon(t)$, we rewrite the expression for R_s as

$$R_s = \int_0^s dr \, \dot{P}_r \bar{P}_r \int_0^r dt \, V_\epsilon^A(s,r,t) G_\epsilon(t), \tag{12.75}$$

where $V_\epsilon^A(s,r,t) = U_\epsilon^A(s,r) \bar{U}_\epsilon^A(r,t)$. To extract e from this expression, we integrate by parts. First, we notice that because of the gap condition, E_r is in the resolvent set of $\bar{H}_r^A$ and the operator $\bar{H}_r^A - E_r$ is invertible. Next, using the equation $(i\epsilon \partial_r - \bar{H}_r^A + E_r) V_\epsilon^A(s,r,t) = 0$, we derive the formula

$$V_\epsilon^A(s,r,t) = (\bar{H}_r^A - E_r)^{-1} i\epsilon \partial_r V_\epsilon^A(s,r,t), \tag{12.76}$$

Since the operator family $\bar{H}_r^A$ is differentiable in the resolvent sense, the projections P_r and the resolvent $(\bar{H}_r^A - E_r)^{-1}$ are twice differentiable. Using the formula above and integrating by parts, we find

$$R_s = \int_0^s dr \, \dot{P}_r \bar{P}_r \int_0^r dt \, (\bar{H}_r^A - E_r)^{-1} i\epsilon \partial_r V_\epsilon^A(s,r,t) G_\epsilon(t)$$

$$= i\epsilon [\dot{P}_r \bar{P}_r (\bar{H}_r^A - E_r)^{-1} \int_0^r dt \, V_\epsilon^A(s,r,t) G_\epsilon(t)]\big|_{r=0}^{r=s} \tag{12.77}$$

$$- i\epsilon \int_0^s dr \, \partial_r (\dot{P}_r \bar{P}_r (\bar{H}_r^A - E_r)^{-1}) \int_0^r dt \, V_\epsilon^A(s,r,t) G_\epsilon(t) \tag{12.78}$$

$$- i\epsilon \int_0^s dr \, \dot{P}_r \bar{P}_r (\bar{H}_r^A - E_r)^{-1} V_\epsilon^A(s,r,r) G_\epsilon(r). \tag{12.79}$$

Now, we use that $(\bar{H}_r^A - E_r)^{-1}$ is uniformly bounded. Hence estimating the expressions on the r.h.s. of the above relation gives (12.74).

Relations (12.72) and (12.74) imply

$$P_s U_\epsilon(s) = U_\epsilon^A(s,0) P_0 + O(\epsilon). \tag{12.80}$$

Next, we claim that

$$\bar{P}_s U_\epsilon(s) = \bar{U}_\epsilon^A(s,r) \bar{P}_0 + O(\epsilon). \tag{12.81}$$

Indeed, combining (12.71) with (12.80) gives

$$\bar{P}_s U_\epsilon(s) = \bar{U}_\epsilon^A(s,0) \bar{P}_0 - R'_s + O(\epsilon), \tag{12.82}$$

where $R'_s := \int_0^s \bar{U}_\epsilon^A(s,r) \bar{P}_r \dot{P}_r P_r U_\epsilon^A(r,0) P_0 \, dr$. Using that $U_\epsilon^A(s,r)$ commutes with $\bar{P}_r \dot{P}_r P_r = [\dot{P}_s, P_s] P_r$, we write R'_s as

$$R'_s = \int_0^s \bar{V}_\epsilon^A(s,r)\bar{P}_r\dot{P}_r P_0 dr, \tag{12.83}$$

where $\bar{V}_\epsilon^A(s,r) = \bar{U}_\epsilon^A(s,r)U_\epsilon^A(r,0)$. Next, using the equation $i\epsilon\partial_r\bar{V}_\epsilon^A(s,r) = \bar{V}_\epsilon^A(s,r)(\bar{H}_r^A - E_r)$, we derive the formula

$$\bar{V}_\epsilon^A(s,r) = i\epsilon(\partial_r\bar{V}_\epsilon^A(s,r))(\bar{H}_r^A - E_r)^{-1}. \tag{12.84}$$

Using the formula above and integrating the r.h.s. of (12.83) by parts, we find

$$\begin{aligned}R'_s &= i\epsilon\int_0^s (\partial_r\bar{V}_\epsilon^A(s,r))(\bar{H}_r^A - E_r)^{-1}\bar{P}_r\dot{P}_r P_0 dr\\ &= i\epsilon[\bar{V}_\epsilon^A(s,r)(\bar{H}_r^A - E_r)^{-1}\bar{P}_r\dot{P}_r P_0]\big|_{r=0}^{r=s}\\ &\quad - i\epsilon\int_0^s \bar{V}_\epsilon^A(s,r)\partial_r((\bar{H}_r^A - E_r)^{-1}\bar{P}_r\dot{P}_r)P_0 dr.\end{aligned} \tag{12.85}$$

Again, using that $\partial_r^m(\bar{H}_r^A - E_r)^{-1}, m = 0,1$, is uniformly bounded and $\partial_r^n P_r, n = 0,1,2$, are bounded, we estimate the expressions on the r.h.s. of the above relation to obtain $R'_s = O(\epsilon)$, which yields (12.81). Combining (12.80) and (12.81) gives (12.50) for $k = 1$. For a general $k \geq 1$, one integrates (12.83) by parts k times.   □

Discussion. Consider the operator $K_{s,\epsilon} := i\epsilon\partial_s - H_s$ acting on s-dependent functions. Eq (12.69), with (12.71), can be rewritten

$$F_{P_s}(K_{s,\epsilon})P_s U_\epsilon(s) = 0, \tag{12.86}$$

where $F_{P_s}(K_{s,\epsilon})$ is the Feshbach - Schur map, with the projection P_s (cf. (11.7)), applied to the operator $K_{s,\epsilon}$. Indeed, according to (11.13) - (11.14), we have

$$F_{P_s}(K_{s,\epsilon}) = [P_s K_{s,\epsilon}P_s - WP_s]\,\lceil_{\mathrm{Ran}P_s}, \tag{12.87}$$

$$W := P_s(i\epsilon\partial_s)\bar{R}(i\epsilon\partial_s)P_s, \quad \bar{R} = \bar{P}_s(\bar{P}_s K_{s,\epsilon}\bar{P}_s)^{-1}\bar{P}_s. \tag{12.88}$$

This exactly (12.69), with (12.71) as can be seen from (12.67), (12.68) and the relation

$$(i\epsilon\partial_s - \bar{H}_s^A)^{-1}f(s) = (i\epsilon)^{-1}\int_0^s \bar{U}_\epsilon^A(s,r)f(r)dr. \tag{12.89}$$

13

General Theory of Many-particle Systems

In this chapter, we extend the concepts developed in the previous chapters to many-particle systems. Specifically, we consider a physical system consisting of N particles of masses $m_1, \ldots, m_N$ which interact pairwise via the potentials $V_{ij}(x_i - x_j)$, where x_j is the position of the j-th particle. Examples of such systems include atoms or molecules – i.e., systems consisting of electrons and nuclei interacting via Coulomb forces. They were considered in Chapter 10.

13.1 Many-particle Schrödinger Operators

Recall from Section 4.1 that quantizing a system of n particles of masses $m_1, \ldots, m_N$ interacting via pair potentials $V_{ij}(x_i - x_j)$, we arrive at the Schrödinger operator (quantum Hamiltonian)

$$H_n = \sum_{j=1}^{n} \frac{1}{2m_j} p_j^2 + V(x) \tag{13.1}$$

acting on $L^2(\mathbb{R}^{3n})$. Here $p_j := -i\hbar \nabla_{x_j}$ and V is the total potential of the system, given in this case by

$$V(x) = \frac{1}{2} \sum_{i \neq j} V_{ij}(x_i - x_j). \tag{13.2}$$

A key example of a many-body Schrödinger operator, that of a molecule, was given in (4.26), Section 4.1.

As in Theorem 10.1, one can show that if the pair potentials V_{ij} satisfy the condition

$$V_{ij} \in L^2(\mathbb{R}^3) + L^\infty(\mathbb{R}^3)$$

(i.e. each V_{ij} can be represented as the sum of an L^2 function and an L^∞ function) then the operator H_n is self-adjoint and bounded below. Indeed, the

© Springer-Verlag GmbH Germany, part of Springer Nature 2020
S. J. Gustafson and I. M. Sigal, *Mathematical Concepts of Quantum Mechanics*, Universitext, https://doi.org/10.1007/978-3-030-59562-3_13

above conditions imply that the potential $V(x)$ satisfies the inequality (2.8), with $H_0 := \sum_{j=1}^{n} \frac{1}{2m_j} p_j^2$ and $a < 1$. Observe that the Coulomb potentials $V_{ij}(y) = e_i e_j |y|^{-1}$ satisfy this condition:

$$|y|^{-1} = |y|^{-1} e^{-|y|} + |y|^{-1}(1 - e^{-|y|}),$$

for example, with $|y|^{-1} e^{-|y|} \in L^2(\mathbb{R}^3)$ and $|y|^{-1}(1 - e^{-|y|}) \in L^\infty(\mathbb{R}^3)$. After this one proceeds exactly as in the proof of Theorem 10.1.

The spectral analysis of the operator H_n for $n \geq 3$ is much more delicate than that of the one-body Hamiltonians we have mostly considered so far. (We will show shortly that for $n - 2$, the operator H_n is reduced to a one-body Hamiltonian.) We are faced with the following issues:

- identical particle symmetries
- separation of the center-of-mass motion
- complicated behaviour of the potential $V(x)$ at infinity in the configuration space $\mathbb{R}^{3n}$.

In Section 4.5 we commented on the first issue. In the subsequent sections, we will deal with the last two issues in some detail. In the remainder of this section, we comment briefly on them. In what follows, we shall assume $V_{ij}(x) \to 0$ as $|x| \to \infty$, though for many considerations this condition is not required.

Separation of the centre-of-mass motion. The Schrödinger operator (13.1) commutes with the operator of total translation of the system

$$T_h : \psi(x_1, \ldots, x_n) \mapsto \psi(x_1 + h, \ldots, x_n + h)$$

and one can show that, as a result, its spectrum is purely essential. So in order to obtain interesting spectral information about our system, we have to remove this translational invariance ("break" it). One way of doing this is by fixing the centre of mass of the system at, say, the origin:

$$\sum_{j=1}^{n} m_j x_j = 0.$$

We will describe a general mathematical procedure for fixing the centre of mass below, but first will show how to do it in the case of two particles ($n = 2$). In this case, we change the particle variables as follows:

$$x_1, x_2 \mapsto y = x_1 - x_2, z = \frac{m_1 x_1 + m_2 x_2}{m_1 + m_2}. \tag{13.3}$$

Here y is the coordinate of the relative position of the two particles, and z is the coordinate of their centre of mass. Using this change of variables in the two-particle Schrödinger operator

$$H_2 = \frac{1}{2m_1}p_1^2 + \frac{1}{2m_2}p_2^2 + V(x_1 - x_2)$$

acting on $L^2(\mathbb{R}^6)$, we arrive easily at the operator

$$\tilde{H}_2 = \frac{1}{2\mu}p^2 + \frac{1}{2M}P^2 + V(y)$$

where $p = -i\hbar\nabla_y$, $P = -i\hbar\nabla_z$, $\mu = \frac{m_1 m_2}{m_1+m_2}$ (the *reduced mass*), and $M = m_1+m_2$ (the total mass). In fact, it can be shown that H_2 and $\tilde{H}_2$ are unitarily equivalent, with the equivalence given by a unitary realization of the change of coordinates (13.3).

The point now is that one can separate variables in the operator $\tilde{H}_2$. In formal language, this means that $\tilde{H}_2$ can be written in the form

$$\tilde{H}_2 = H \otimes 1 + 1 \otimes H_{CM}$$

on $L^2(\mathbb{R}^6) = L^2(\mathbb{R}_y^3) \otimes L^2(\mathbb{R}_z^3)$ where

$$H = \frac{1}{2\mu}p^2 + V(y)$$

acts on $L^2(\mathbb{R}_y^3)$, and

$$H_{CM} = \frac{1}{2M}P^2$$

acts on $L^2(\mathbb{R}_z^3)$ (see the mathematical supplement, Section 25.13, for a description of the tensor product, $\otimes$). Clearly H and H_{CM} are the Schrödinger operators of the *relative motion* of the particles, and of their *centre of mass motion*, respectively. It is equally clear that of interest for us is H, and not H_{CM}. Note that H has the form of a one-particle Schrödinger operator with external potential $V(y)$. All the analysis we developed for such operators is applicable now to H.

Behaviour of $V(x)$ at infinity. The second issue mentioned above arises from the geometry of the potential (13.2). The point is that $V(x)$ does not vanish as $x \to \infty$ in certain directions, namely in those directions where $x_i = x_j$ for at least one pair $i \neq j$ (we assume here that $V_{ij}(0) \neq 0$). This property is responsible for most of the peculiarities of many-body behaviour. In particular, the spectral analysis of Chapter 6 does not work in the many-body case, and must be modified in significant ways by taking into account the geometry of many-body systems.

13.2 Separation of the Centre-of-mass Motion

This section is devoted to a description of a general method for separating the centre-of-mass motion of a many-body system. After applying this method, one is left with a many-body system whose centre-of-mass is fixed at the origin.

We begin by equipping the n-body configuration space $\mathbb{R}^{3n}$ with the inner-product

$$\langle x, y \rangle := \sum_{i=1}^{n} m_i x_i \cdot y_i \qquad (13.4)$$

where $m_1, \ldots, m_n$ are the masses of the n particles, and $x_i \cdot y_i$ is the usual dot-product in $\mathbb{R}^3$. Next, we introduce the orthogonal subspaces

$$X := \{x \in \mathbb{R}^{3n} \mid \sum_{i=1}^{n} m_i x_i - 0\} \qquad (13.5)$$

and

$$X^{\perp} := \{x \in \mathbb{R}^{3n} \mid x_i = x_j \quad \forall \ i, j\} \qquad (13.6)$$

of $\mathbb{R}^{3n}$. Recall that orthogonality $(X \perp X^{\perp})$ means $\langle x, y \rangle = 0$ for any $x \in X$, $y \in X^{\perp}$. To see that $X \perp X^{\perp}$, suppose $x \in X$, and $y \in X^{\perp}$. Since $y_j = y_1$ for all j, we have

$$\langle x, y \rangle = \sum_{y=1}^{n} m_i x_i \cdot y_i = y_1 \cdot \sum_{y=1}^{n} m_i x_i = 0$$

by the definition of X. We recall here the definition of the direct sum of subspaces.

Definition 13.1 If V_1, V_2 are orthogonal subspaces of a vector space, V, with an inner product, then $V_1 \oplus V_2$ denotes the subspace

$$V_1 \oplus V_2 := \{v_1 + v_2 \mid v_1 \in V_1, v_2 \in V_2\}.$$

Problem 13.2 Show that

$$\mathbb{R}^{3n} = X \oplus X^{\perp}. \qquad (13.7)$$

X is the configuration space of *internal motion* of the n-particle system, and $X^{\perp}$ is the configuration space of the centre-of-mass motion of this system. The relation (13.7) implies that

$$L^2(\mathbb{R}^{3n}) = L^2(X) \otimes L^2(X^{\perp}) \qquad (13.8)$$

(see Section 25.13 for an explanation of the tensor product, $\otimes$).

Let Δ denote the Laplacian on $\mathbb{R}^{3N}$ in the metric determined by (13.4), i.e.

$$\Delta = \sum_{j=1}^{n} \frac{1}{m_j} \Delta_{x_j}.$$

Under the decomposition (13.8), the Laplacian decomposes as

$$\Delta = \Delta_X \otimes 1_{X^{\perp}} + 1_X \otimes \Delta_{X^{\perp}} \qquad (13.9)$$

where 1_X and $1_{X_{\perp}}$ are the identity operators on $L^2(X)$ and $L^2(X^{\perp})$ respectively (see again Section 25.13).

Problem 13.3 Show that $\Delta_{X^\perp} = \frac{1}{M} \Delta_{x_{CM}}$ where $M = \sum_{j=1}^n m_j$ and $x_{CM} = \frac{1}{M} \sum_{j=1}^n m_j x_j$.

Let π_X be the orthogonal projection operator from $\mathbb{R}^{3n}$ to X (see Section 25.7 for background on projections). Explicitly,

$$(\pi_X x)_i = x_i - \frac{1}{\sum_{j=1}^N m_j} \sum_{j=1}^n m_j x_j. \tag{13.10}$$

Problem 13.4 Show that (13.10) is the orthogonal projection operator from $\mathbb{R}^{3n}$ to X. Find the orthogonal projection operator, $\pi_{X^\perp}$, from $\mathbb{R}^{3n}$ to $X^\perp$.

Equation (13.10) implies that

$$(\pi_X x)_i - (\pi_X x)_j = x_i - x_j.$$

Hence the many-body potential (4.24) satisfies

$$V(x) = V(\pi_X x). \tag{13.11}$$

Equations (13.9) and (13.11) imply that the operator H_n given in (13.1) can be decomposed as

$$H_n = H \otimes 1_{X^\perp} + 1_X \otimes T_{CM} \tag{13.12}$$

where

$$H = -\frac{\hbar^2}{2} \Delta_X + V(x)$$

is the Hamiltonian of the internal motion of the system, and

$$T_{CM} = -\frac{\hbar^2}{2} \Delta_{X^\perp}$$

is the Hamiltonian of the motion of its centre-of-mass. Equation (13.12) is the centre-of-mass separation formula, and H is called the Hamiltonian in the centre-of-mass frame. It is a self-adjoint operator under the assumptions on the potentials mentioned above. It is the main object of study in many-body theory.

13.3 Break-ups

Here we describe the kinematics of the break-up of an n-body system into non-interacting systems. First we introduce the notion of a cluster decomposition

$$a = \{C_1 \ldots C_s\}$$

for some $s \leq n$. The C_j are non-empty, disjoint subsets of the set $\{1, \ldots, n\}$, whose union yields the whole set:

$$\bigcup_{j=1}^{s} C_j = \{1,\ldots,n\}.$$

The subsets C_j are called *clusters*. An example of a cluster decomposition for $n = 3$ is $a = \{(12),(3)\}$. The number of clusters, s, in the decomposition a will be denoted by $\#(a)$.

There is only one cluster decomposition with $\#(a) = 1$, and one with $\#(a) = n$. In the first case, the decomposition consists of a single cluster

$$\underline{a} = \{(1\ldots n)\},$$

and in the second case the clusters are single particles:

$$\bar{a} = \{(1)\ldots(n)\}.$$

In the first case the system is not broken up at all, while in the second case it is broken into the smallest possible fragments.

To each cluster decomposition, a, we associate the *intercluster potential*

$$I_a(x) := \sum_{(ij)\not\subset a} V_{ij}(x_i - x_j),$$

where the notation $(ij) \not\subset a$ signifies that the indices i and j belong to different clusters in the decomposition a. Similarly, we associate to a the *intracluster potential*

$$V_a(x) := \sum_{(ij)\subset a} V_{ij}(x_i - x_j)$$

where $(ij) \subset a$ signifies that i and j belong to the same cluster in a. Thus $I_a(x)$ (resp. $V_a(x)$) is the sum of the potentials between particles from different (resp. the same) clusters of a.

The Hamiltonian of a decoupled system (in the total centre-of-mass frame) corresponding to a cluster decomposition, a, is

$$H_a := -\frac{\hbar^2}{2}\Delta_X + V_a(x)$$

acting on $L^2(X)$. For $a = \{C_1,\ldots,C_s\}$, the Hamiltonian H_a describes s non-interacting sub-systems $C_1,\ldots,C_s$. For $s = 1$, $H_a = H$, and for $s = N$, $H_a = -\frac{\hbar^2}{2}\Delta_X$. The operators H_a are also self-adjoint. Note that

$$H = H_a + I_a. \tag{13.13}$$

If $s > 1$, then the system commutes with relative translations of the clusters $C_1\ldots C_s$:

$$T_h H_a = H_a T_h,$$

for $h \in X$ satisfying

$$h_i = h_j \quad \text{if} \quad (ij) \subset a.$$

Here

$$T_h : \psi(x_1, \ldots, x_n) \to \psi(x_1 + h_1, \ldots, x_N + h_n)$$

for $h = (h_1, \ldots, h_n)$. As a result, one can show that the spectrum of H_a is purely essential. This is due to the fact that the clusters in a move freely. One can separate the centre-of-mass motions of the clusters $C_1 \ldots C_s$ in a and establish a decomposition for H_a similar to that for H_n (equation (13.12)); we will do this later.

13.4 The HVZ Theorem

In this section we formulate and prove the key theorem in the mathematical theory of n-body systems – the HVZ theorem. The letters here are the initials of W. Hunziker, C. van Winter, and G.M. Zhislin. This theorem identifies the location of the essential spectrum of the many-body Hamiltonian, H.

Theorem 13.5 (HVZ Theorem) We have

$$\sigma_{ess}(H) = [\Sigma, \infty),$$

where

$$\Sigma := \min_{\#(a)>1} \Sigma_a, \quad \text{with} \quad \Sigma_a := \min(\sigma(H_a)).$$

Note that Σ is the minimal energy needed to break the system into independent parts.

Proof. We begin by showing that $\sigma(H_a) \subset \sigma(H)$ for $\#(a) > 1$. Suppose $\lambda \in \sigma(H_a)$. Then for any $\varepsilon > 0$, there is $\psi \in L^2(X)$ with $\|\psi\| = 1$, such that $\|(H_a - \lambda)\psi\| < \varepsilon$. Let $h \in X$ satisfy $h_i = h_j$ if $(ij) \subset a$, and $h_i \neq h_j$ otherwise. For $s > 0$, let T_{sh} be the operator of coordinate translation by sh. Note that T_{sh} is an isometry. As remarked in the previous section, T_{sh} commutes with H_a, and so

$$\|(H_a - \lambda)T_{sh}\psi\| = \|T_{sh}(H_a - \lambda)\psi\| < \varepsilon.$$

On the other hand, $HT_{sh}\psi \to H_a T_{sh}\psi$ as $s \to \infty$, because the translation T_{sh} separates the clusters in a as $s \to \infty$. So for s sufficiently large, $\|(H - \lambda)T_{sh}\psi\| < \varepsilon$. Since $\varepsilon > 0$ is arbitrary, and $\|T_{sh}\psi\| = \|\psi\| = 1$, we see $\lambda \in \sigma(H)$.

So we have shown $\sigma(H_a) \subset \sigma(H)$. As we remarked earlier, H_a has purely essential spectrum. In other words, $\sigma(H_a) = [\Sigma_a, \infty)$. Thus we have shown $[\Sigma, \infty) \subset \sigma(H)$.

It remains to prove that $\sigma_{ess}(H) \subset [\Sigma, \infty)$. To do this, we introduce a "partition of unity", i.e., a family $\{j_a\}$ of smooth functions on X, indexed by all cluster decompositions, a, with $\#(a) > 1$, such that

$$\sum_{\#(a)>1} j_a^2(x) \equiv 1. \tag{13.14}$$

We can use $\{j_a\}$ to decompose H into pieces which are localized in the supports of the j_a, plus an error term:

$$H = \sum_{\#(a)>1} j_a H j_a - \sum_{\#(a)>1} \frac{\hbar^2}{2} |\nabla j_a|^2. \tag{13.15}$$

Indeed, summing the identity

$$[j_a, [j_a, H]] = j_a^2 H + H j_a^2 - 2 j_a H j_a$$

over a, and using (13.14), we obtain the relation known as the IMS formula (see [73])

$$H = \sum_{\#(a)>1} (j_a H j_a + \frac{1}{2}[j_a [j_a, H]]).$$

Computing $[j_a, [j_a, H]] = -\hbar^2 |\nabla j_a|^2$ finishes the proof of (13.15).

Now we construct an appropriate partition of unity, namely one satisfying

$$\min_{(jk) \not\subset a} |x_j - x_k| \geq \varepsilon |x| \quad \text{for} \quad |x| \geq 1, x \in \text{supp}(j_a) \tag{13.16}$$

for some $\varepsilon > 0$. Indeed, the sets

$$S_a := \{x \in X \mid |x| = 1; \; |x_j - x_k| > 0 \; \forall \; (jk) \not\subset a\}$$

form an open cover of the unit sphere, S, of X (i.e., S_a are open sets and $S \subset \bigcup_a S_a$). For each a, let χ_a be a smooth function supported in S_a, and equal to 1 in a slightly smaller set (such that these smaller sets still cover S). Then the functions $j_a := \chi_a/(\sum \chi_a^2)^{1/2}$ form a partition of unity on S, with $\text{supp}(j_a) \subset S_a$. In fact, (13.16) holds (with $|x| = 1$), because $\text{supp}(j_a)$ is compact. We extend $j_a(x)$ to all of X by setting $j_a(x) := j_a(x/|x|)$ for $|x| > 1$, and for $|x| < 1$, choosing any smooth extension of $j_a(x)$ which preserves (13.15). Thus the partition $\{j_a\}$ satisfies

$$j_a(\lambda x) = j_a(x) \quad \text{for} \quad |x| \geq 1, \lambda \geq 1 \tag{13.17}$$

as well as (13.16). By (13.17),

$$|\nabla j_a(x)|^2 \to 0 \quad \text{as} \quad |x| \to \infty.$$

By (13.16),

$$j_a(H - H_a)j_a = j_a I_a j_a \to 0 \quad \text{as} \quad |x| \to \infty.$$

Returning to (13.15), we conclude that

$$H = \sum_{\#(a)>1} j_a H_a j_a + K$$

where K is multiplication operator vanishing at infinity. An argument similar to the proof of the second part Theorem 6.16 shows that

$$\sigma_{ess}(H) = \sigma_{ess}\left(\sum_{\#(a)>1} j_a H_a j_a\right).$$

Since $H \geq (\inf \sigma(H))\mathbf{1}$ for any self-adjoint operator, H, we see

$$\langle \psi, j_a H_a j_a \psi \rangle = \langle j_a \psi, H j_a \psi \rangle$$
$$\geq \Sigma_a \|j_a \psi\|^2 \geq \Sigma_a \|\psi\|^2$$

for any $\psi \in X$ (note $\Sigma_a \leq 0$). Thus $j_a H_a j_a \geq \Sigma_a$, and therefore $\sum_a j_a H_a j_a \geq \Sigma$, yielding $\sigma_{ess}(\sum_a j_a H_a j_a) \subset [\Sigma, \infty)$, and consequently $\sigma_{ess}(H) \subset [\Sigma, \infty)$. This completes the proof of the HVZ theorem. $\square$

13.5 Intra- vs. Inter-cluster Motion

As was mentioned in Section 13.3, the Hamiltonians H_a describing the system broken up into non-interacting clusters have purely essential spectra. This is due to the fact that the clusters in the decomposition move freely. To understand the finer structure of many-body spectra, we have to separate the centre-of-mass motion of the clusters, as we did with the centre-of-mass motion of the entire system. Proceeding as in Section 13.2, we define the subspaces

$$X^a := \{x \in X \mid \sum_{j \in C_i} m_j x_j = 0 \quad \forall\, i\}$$

and

$$X_a := \{x \in X \mid x_i = x_j \text{ if } (ij) \subset a\}.$$

Problem 13.6 Show that these subspaces are mutually orthogonal ($X^a \perp X_a$) and span X:

$$X = X^a \oplus X_a. \tag{13.18}$$

X^a is the subspace of internal motion of the particles within the clusters of the decomposition a, and X_a is the subspace of the centre-of-mass motion of the clusters of a. As before, (13.18) leads to the decomposition

$$L^2(X) = L^2(X^a) \otimes L^2(X_a)$$

of $L^2(X)$ and the related decomposition of the Laplacian on X:

$$\Delta_X = \Delta_{X^a} \otimes \mathbf{1}_{X_a} + \mathbf{1}_{X^a} \otimes \Delta_{X_a}$$

where Δ_{X^a} and Δ_{X_a} are the Laplacians on the spaces X^a and X_a (or $L^2(X^a)$ and $L^2(X_a)$) in the metric (13.4). Again, if π_{X^a} is the orthogonal projection from X to X^a, then

$$V_a(x) = V_a(\pi_{X^a} x)$$

and consequently we have the decomposition

$$H_a = H^a \otimes 1_{X_a} + 1_{X^a} \otimes T_a, \tag{13.19}$$

where

$$H^a = -\frac{\hbar^2}{2}\Delta_{X^a} + V_a(x)$$

is the Hamiltonian of the internal motion of the particles in the clusters of a, and $T_a = -\frac{\hbar^2}{2}\Delta_{X^a}$ is the Hamiltonian of the centre-of-mass motion of the clusters in a.

Applying the HVZ theorem inductively, we arrive at the following representation of the essential spectrum of H:

$$\sigma_{ess}(H) = \bigcup_{\lambda \in \tau(H)} [\lambda, \infty),$$

where the discrete set $\tau(H)$, called the *threshold set* of H, is defined as

$$\tau(H) := \bigcup_{\#(a)>1} \sigma_d(H^a) \bigcup \{0\},$$

the union of the discrete spectra of the break-up Hamiltonians and zero. The points of $\tau(H)$ are called the *thresholds*. Thus one can think of the essential spectrum of a many-body Hamiltonian H (in the centre-of-mass frame) as a union of branches starting at its thresholds and extending to infinity.

13.6 Exponential Decay of Bound States

Theorem 13.7 If H has a bound state, $\Psi(x)$, with an energy $E < \Sigma$ (i.e. below the ionization threshold Σ, see Theorem 13.5), then $\Psi(x)$ satisfies the exponential bound

$$\int |\Psi(x)|^2 e^{2\alpha|x|} dx < \infty, \ \forall \alpha < \sqrt{\Sigma - E}. \tag{13.20}$$

Proof. Let J be a real, bounded, smooth function supported in $\{|x| \geq R\}$. Proceeding as in the prove of the HVZ theorem above, we show that there is $\epsilon = \epsilon(R) \to 0$, as $R \to \infty$, s.t.

$$JHJ \geq (\Sigma - \epsilon)J^2. \tag{13.21}$$

Denote by ∇_X the gradient on the space (13.5) with the metric (13.4). We assume now that $\nabla_X J$ is supported in $\{R \leq |x| \leq 2R\}$. Let f be a bounded twice differentiable, positive function and define $H_f := e^f H e^{-f}$. We compute

$$H_f = H - \frac{\hbar^2}{2}[|\nabla_X f|^2 - \nabla_X f \cdot \nabla_X - \nabla_X \cdot \nabla_X f]. \tag{13.22}$$

Then $(H_f - E)\Phi = 0$, where $\Phi := e^f \Psi$, and therefore $(H_f - E)J\Phi = [H_f, J]\Phi$. On the other hand, by (13.22) and (13.21) and the fact that the operator $\nabla_X f \cdot \nabla_X + \nabla_X \cdot \nabla_X f$ is anti-self-adjoint, we have

$$\mathrm{Re}\langle J\Phi, (H_f - E)J\Phi\rangle = \langle J\Phi, (H - \frac{\hbar^2}{2}|\nabla_X f|^2 - E)J\Phi\rangle$$
$$\geq \delta\|J\Phi\|^2.$$

where $\delta := \Sigma - \epsilon - E - \sup_{x \in \mathrm{supp}\, J} \frac{\hbar^2}{2m}|\nabla_X f|^2$. Then the last two equations imply

$$\delta\|J\Phi\|^2 \leq \mathrm{Re}\langle J\Phi, (H_f - E)J\Phi\rangle \leq \|J\Phi\|\|[H_f, J]\Phi\|.$$

Now we take for f a sequence of bounded functions approximating $\alpha(1 + |x|^2)^{1/2}$, with $\alpha < \sqrt{\Sigma - E}$. Taking the limit in the last inequality gives (13.20). $\square$

13.7 Remarks on Discrete Spectrum

For $n = 2$, the results in Section 8.3, show that the discrete spectrum of H is finite if the potential $V(x)$ is "short-range", whereas a "long-range" attractive potential produces an infinite number of bound states. The borderline between short- and long-range potentials is marked by the asymptotic behaviour $V(x) \sim |x|^{-2}$ as $|x| \to \infty$ (which is different than the borderline asymptotic behaviour of $|x|^{-1}$ which we encountered in scattering theory in Chapter 9). For $n > 2$, however, the question of whether $\sigma_d(H)$ is finite or infinite cannot be answered solely in terms of the asymptotic fall-off of the intercluster potentials, $I_a(x)$; the nature of the threshold Σ at the bottom of the essential spectrum plays a decisive role.

Now we comment on bound states of molecules with mobile nuclei. We restrict our attention to the case where Σ is a *two-cluster threshold*. This means that for energy Σ and slightly above, the system can only disintegrate into two bound clusters, C_1 and C_2. This fits the case of molecules with dynamic nuclei which break up into atoms or stable ions. However, our next assumption, that the lowest energy break-ups originate from charged clusters (having, of course, opposite total charges), is not realistic.

Ignoring particle statistics, the disintegration of the system into the clusters corresponding to a partition a can be represented by a product wave function

$$\phi(x) = \Psi(x^a) f(x_a), \tag{13.23}$$

where Ψ is the eigenfunction of H^a with eigenvalue Σ, $H^a \Psi = \Sigma \Psi$, and x_a and x^a denote the components of x along the subspaces X_a and X^a respectively. Here f is chosen so that $\langle f, T_a f \rangle$ is arbitrarily small. The condition that Σ is a two-cluster threshold means that Σ is a discrete eigenvalue of H^a, and as a consequence it can be shown that $\Psi(x^a)$ decays exponentially as $|x^a| \to \infty$:

$$|\Psi(x^a)| \leq (const) e^{-\alpha |x^a|} \tag{13.24}$$

for some $\alpha > 0$. Using states of the form (13.23) as trial states to make $\langle \phi, H\phi \rangle < \Sigma$, we can show the existence of an infinite number of bound states, provided that the lowest energy break-ups originate from charged clusters (of opposite total charges). This means that the intercluster potential

$$I_a(x) = \sum_{i \in C_1, k \in C_2} \frac{e_i e_k}{|x_i - x_k|},$$

satisfies $(\sum_{i \in C_1} e_i)(\sum_{k \in C_2} e_k) < 0$. Using the exponential bound (13.24), it follows that

$$|\langle \phi, (I_a(x) - I_a(x_a))\phi \rangle_{L^2(X^a)}| \leq (const)|x_a|^{-2}.$$

Observe that $I_a(x_a) = -q/|x_a|$ with $q > 0$. Since $H = H^a \otimes 1_a + 1^a \otimes T_a + I_a$ (see (13.13) and (13.19)) and $H^a \phi = \Sigma \phi$, the last inequality implies

$$\langle \phi, (H - \Sigma)\phi \rangle \leq \left\langle u, \left(T_a - \frac{q}{|x_a|} + (const)|x_a|^{-2} \right) f \right\rangle_{L^2(X_a)}.$$

As in Section 8.3, we let $f \in C_0^\infty(\mathbb{R}^3)$ satisfy $\|f\| = 1$ and

$$\mathrm{supp}(f) \subset \{x_a \mid 1 < |x_a| < 2\}.$$

Then the functions

$$f_k(x_a) = k^{-3/2} f(k^{-1} x_a), \qquad k = 1, 2, 4, 8, \ldots$$

are orthonormal, and have disjoint supports. Thus the corresponding trial states $\phi_k(x) = \Psi(x^a) f_k(x_a)$ satisfy $\langle \phi_k, H\phi_m \rangle = 0$ for $k \neq m$, and

$$\langle \phi_k, (H - \Sigma)\phi_k \rangle \leq -(const)\frac{q}{k} + (const)\frac{1}{k^2} < 0,$$

for some positive constants, if n is sufficiently large. We can now apply the min-max principle (see Section 8.1) to conclude that H possesses infinitely many discrete eigenvalues below the threshold Σ.

13.8 Scattering States

Unlike in the one-body case, the many-body evolution $\psi = e^{-iHt/\hbar}\psi_0$ behaves asymptotically as a superposition of several (possibly infinitely many) free evolutions, corresponding to different scenarios of the scattering problem. These scenarios, called *scattering channels*, can be described as follows. For a given scattering channel, the system is broken into non-interacting clusters $C_1, \ldots, C_s$ corresponding to some cluster decomposition a, and its motion in each cluster is restricted to a cluster bound state. (If a given decomposition contains a cluster for which the cluster Hamiltonian has no bound state, then this cluster decomposition does not participate in the formation of scattering channels.) In other words, a scattering channel is specified by a pair (a, m), where a is a cluster decomposition such that the operator H^a has some discrete spectrum, and m labels the the eigenfunction $\psi^{a,m}$ of H^a (we suppose that for fixed a, the $\psi^{a,m}$ are chosen orthonormal). The evolution in the channel (a, m) is determined by the pair

$$(\mathcal{H}_{a,m}, H_{a,m})$$

where $\mathcal{H}_{a,m} := \psi^{a,m} \otimes L^2(X_a)$ is the channel Hilbert space, and

$$H_{a,m} := H_a \!\restriction_{\mathcal{H}_{a,m}} = E^{a,m} + T_a$$

where $E^{a,m}$ is the eigenvalue of H^a corresponding to the eigenfunction $\psi^{a,m}$. Thus the channel evolution is given by

$$e^{-iH_{a,m}t/\hbar}(\psi^{a,m} \otimes f)$$

for $f \in L^2(X_a)$.

As in the one-body case (see Chapter 9), the existence and asymptotic form of scattering states for $t \to \infty$ depends crucially of the rate at which the potentials tend to zero for large separations. We suppose that the intercluster potentials satisfy

$$\partial^{(\alpha)} I_a(x) \leq (\text{const})|x|_a^{-\mu - |\alpha|} \tag{13.25}$$

as $|x|_a \to \infty$, for $|\alpha| \leq 2$. Here $|x|_a$ is the intercluster distance, i.e.

$$|x|_a := \min_{(jk) \not\subset a} |x_j - x_k|.$$

As in the one-body case, $\mu = 1$ marks the borderline between short-range and long-range systems, for which the scattering theory is quite different.

Short-range systems: $\mu > 1$. In this case, for a given scattering channel, (a, m), the wave operator

$$\Omega^+_{(a,m)} = \text{s-lim}_{t \to \infty} e^{iHt/\hbar} e^{-iH_{a,m}t/\hbar}$$

can be shown to exist on $\mathcal{H}_{a,m}$. The proof is similar to that for the one-body case (see Chapter 9). This wave operator maps free channel evolutions to full evolutions: setting

$$\psi := \Omega^+_{(a,m)}(\psi^{a,m} \otimes f),$$

we have

$$e^{-iHt/\hbar}\psi \rightarrow e^{-iH_{a,m}t/\hbar}(\psi^{a,m} \otimes f)$$

as $t \to \infty$. The wave operator $\Omega^+_{a,m}$ is an isometry from $\mathcal{H}_{a,m}$ to $\mathcal{H}$. Moreover, the ranges $\mathcal{H}^+_{a,m} := \mathrm{Ran}(\Omega^+_{a,m})$ satisfy

$$\mathcal{H}^+_{a,m} \perp \mathcal{H}^+_{b,n} \quad \text{if} \quad a \neq b \quad \text{or} \quad m \neq n$$

which follows from the fact that

$$\lim_{t \to \infty} \langle e^{-iH_{a,m}t/\hbar}(\psi^{a,m} \otimes f), e^{-iH_{b,n}t/\hbar}(\psi^{b,n} \otimes g) \rangle = 0$$

if $a \neq b$ or $m \neq n$. Therefore, the outgoing scattering states form a closed subspace

$$\mathcal{H}^+ := \bigoplus_{a,m} \mathcal{H}^+_{a,m} \subset L^2(X).$$

Under the condition (13.25) with $\mu > 1$, it has been proved that the property of *asymptotic completeness* holds – i.e., that $\mathcal{H}^+ = L^2(X)$ (see Chapter 27 for references). In other words, as $t \to \infty$, every state approaches a superposition of channel evolutions and bound states (the bound states are the channel with $\#(a) = 1$).

Long-range systems: $\mu \leq 1$. As in the one-body case, it is necessary to modify the form of the channel evolutions in the long-range case. The evolution $e^{-iH_{a,m}t/\hbar}$ with $H_{a,m} = \lambda^{a,m} + T_a$ is replaced by

$$e^{-iH_{a,m}t/\hbar - i\alpha_{a,t}(p_a)}$$

where $p_a = -i\hbar \nabla_{X^a}$. Here $\alpha_{a,t}(p_a)$ is an adiabatic phase, arising from the fact that classically, the clusters are located at $x_a = p_a t(1 + O(t^{-\mu}))$ as $t \to \infty$. The modification $\alpha_{a,t}(p_a)$, whose precise form we will not give here, is similar to that for the one-body case (see Chapter 9). We refer the interested reader to the references listed in Chapter 27 for further details. We remark only that with this modification in place, the existence of the (modified) wave operators, and asymptotic completeness, have been proved for μ not too small.

14

Self-consistent Approximations

Even for a few particles the Schrödinger equation is prohibitively difficult
to solve. Hence it is important to have approximations which work in various regimes. One such approximation, which has a nice unifying theme and
connects to large areas of physics and mathematics, is the one approximating solutions of n-particle Schrödinger equations by products of n one-particle
functions (i.e. functions of 3 variables). This results in a single nonlinear equation in 3 variables, or several coupled such equations. The trade-off here is
the number of dimensions for the nonlinearity. This method, which goes under different names, e.g. the mean-field or self-consistent approximation, is
especially effective when the number of particles, n, is sufficiently large.

14.1 Hartree, Hartree-Fock and Gross-Pitaevski equations

For simplicity we consider a system of n identical, spinless bosons. It is
straightforward to include spin. To extend our treatment to fermions requires
a simple additional step (see discussion below). The Hamiltonian of the system of n bosons of mass m, interacting with each other and moving in an
external potential V is

$$H_n := \sum_{j=1}^{n} h_{x_i} + \frac{1}{2} \sum_{i \neq j} v(x_i - x_j), \tag{14.1}$$

where $h_x = -\frac{\hbar^2}{2m}\Delta_x + V(x)$, acting on the state space $\circledS_1^n L^2(\mathbb{R}^d)$, $d = 1, 2, 3$.
Here v is the interaction potential, and $\circledS$ is the symmetric tensor product.
As we know, the quantum evolution is given by the Schrödinger equation

$$i\hbar \frac{\partial \Psi}{\partial t} = H_n \Psi.$$

© Springer-Verlag GmbH Germany, part of Springer Nature 2020
S. J. Gustafson and I. M. Sigal, *Mathematical Concepts of Quantum
Mechanics*, Universitext, https://doi.org/10.1007/978-3-030-59562-3_14

This is an equation in $3n + 1$ variables, $x_1, ..., x_n$ and t, and it is not a simple matter to understand properties of its solutions. We give a heuristic derivation of the mean-field approximation for this equation. A rigorous derivation is sketched in Section 20.6. First, we observe that the potential experienced by the i-th particle is

$$W(x_i) := V(x_i) + \sum_{j \neq i} v(x_i - x_j).$$

Assuming $v(0)$ is finite, it can be re-written, modulo the constant term $v(0)$, which we neglect, as $W(x_i) = V(x_i) + (v * \rho^{micro})(x_i)$. Here, recall, $f * g$ denotes the convolution of the functions f and g, and ρ^{micro} stands for the (operator of) microscopic density of the n particles, defined by

$$\rho^{micro}(x, t) := \sum_j \delta(x - x_j).$$

Note that the average quantum-mechanical (QM) density in the state Ψ is

$$\langle \Psi, \rho^{micro}(x, t)\Psi \rangle = \rho^{QM}(x, t)$$

where $\rho^{QM}(x, t) := n \int | \Psi(x, x_2, ..., x_n, t) |^2 \, dx_2...dx_n$, the one-particle density in the quantum state Ψ.

In the *mean-field theory*, we replace $\rho^{micro}(x, t)$ with a continuous function, $\rho^{MF}(x, t)$, which is supposed to be close to the average quantum-mechanical density, $\rho^{QM}(x, t)$, and which is to be determined later. Consequently, it is assumed that the potential experienced by the i-th particle is

$$W^{MF}(x_i) := V(x_i) + (v * \rho^{MF})(x_i).$$

Thus, in this approximation, the state $\psi(x, t)$ of the i-th particle is a solution of the following one-particle Schrödinger equation $i\hbar \frac{\partial \psi}{\partial t} = (h + v * \rho^{MF})\psi$ where, recall, $h = -\frac{\hbar^2}{2m}\Delta_x + V(x)$. Of course, the integral of $\rho^{micro}(x, t)$ is equal to the total number of particles, $\int_{\mathbb{R}^3} \rho^{micro}(x, t)dx = n$. We require that the same should be true for $\rho^{MF}(x, t)$: $\int_{\mathbb{R}^3} \rho^{MF}(x, t)dx = n$. We normalize the one-particle state, $\psi(x, t)$, in the same way

$$\int_{\mathbb{R}^3} |\psi(x, t)|^2 dx = n. \tag{14.2}$$

Consider a situation in which we expect all the particles to be in the same state ψ. Then it is natural to take $\rho^{MF}(x, t) = |\psi(x, t)|^2$. In this case ψ solves the *nonlinear* equation

$$i\hbar \frac{\partial \psi}{\partial t} = (h + v * |\psi|^2)\psi. \tag{14.3}$$

This nonlinear evolution equation is called the *Hartree equation* (HE).

If the inter-particle interaction, v, is significant only at very short distances (one says that v is very short range, which technically can be quantified by assuming that the "particle scattering length" a is small), we can replace $v(x) \to 4\pi a \delta(x)$, and Equation (14.3) becomes

$$i\hbar \frac{\partial \psi}{\partial t} = h\psi + 4\pi a |\psi|^2 \psi \tag{14.4}$$

(with the normalization (14.2)). This equation is called the *Gross-Pitaevski equation* (GPE) or *nonlinear Schrödinger equation*. It is a mean-field approximation to the original quantum problem for a system of n bosons. The Gross-Pitaevski equation is widely used in the theory of superfluidity, and in the theory of Bose-Einstein condensation (see Appendix 14.2).

Reconstruction of solutions to the n-particle Schrödinger equation,

$$i\hbar \frac{\partial \Psi}{\partial t} = H_n \Psi, \qquad \Psi|_{t=0} = \otimes_1^n \psi_0. \tag{14.5}$$

How do solutions of (HE) or (GPE) relate to solutions of the original many-body Schrödinger equation (14.5)? It is shown rigorously (see [90, 91, 238]) that the solution of equation (14.5) satisfies, in some weak sense and and in an appropriate regime of $n \to \infty$ and $a \to 0$ with $n4\pi a =: \lambda$ fixed,

$$\Psi - \otimes_1^n \psi \to 0$$

where ψ satisfies, depending on the limiting regime, either (HE) or (GPE) with initial condition ψ_0. For the mean-field regime (replacing for the moment v by gv) of $n \to \infty$ and $g \to 0$, with ng fixed, we have (HE) (see [159, 134, 117, 123, 33, 34]).

It is not obvious how to extend the mean-field argument above to *fermions*. To do this, we explore *another way* to derive formally the Hartree equation from the n-particle Schrödinger one. It goes as follows. For complex ψ, we define $d_{\bar\psi}\mathcal{E}(\psi) = (d_{\psi_1} + id_{\psi_2})\mathcal{E}(\psi)$, where $\psi = \psi_1 + i\psi_2$ and call the critical point equation $d_{\bar\psi}\mathcal{E}(u_*) = 0$ the Euler-Lagrange equation. Now, observe

Proposition 14.1 *The Schrödinger equation is the Euler-Lagrange equation for stationary points of the action functional*

$$S(\Psi) := \int \left\{ -\hbar \operatorname{Im}\langle \Psi, \partial_t \Psi \rangle - \langle \Psi, H_n \Psi \rangle \right\} dt, \tag{14.6}$$

The proof is an exercise in the standard variational calculus – see Chapter 26 for material on variational calculus. We sketch it here:

Proof (Sketch of proof). We write $S(\Psi) := -\int \int \bar\Psi A\Psi \, dx dt$, where $A := i\hbar \partial_t + H_n$. We consider $S(\Psi)$ as a functional of Ψ and $\bar\Psi$, $S(\Psi) \equiv S(\Psi, \bar\Psi)$ and differentiate it in $\bar\Psi$ along $\xi \in C_0^\infty(\mathbb{R}^d \times \mathbb{R})$. We obtain

$$\frac{\partial}{\partial \lambda}\Big|_{\lambda=0} \int \int (\bar\Psi + \lambda\xi) A\Psi = \int \int \xi A\Psi$$

Since $\xi \in C_0^\infty(\mathbb{R}^d \times \mathbb{R})$ is arbitrary, this gives the Schrödinger equation (14.5).

Problem 14.2 Review the basic facts of variational calculus.

Now, for bosons, we consider the the action functional (14.6) the space (not linear!)

$$\{\Psi := \otimes_1^n \psi \mid \psi \in H^1(\mathbb{R}^3)\}, \tag{14.7}$$

where $(\otimes_1^n \psi)$ is the function of $3n$ variables defined by $(\otimes_1^n \psi)(x_1, ..., x_n) := \psi(x_1)...\psi(x_n)$. A simple computation gives

Proposition 14.3 Let $S_H(\psi) := n^{-1}S(\otimes_1^n \psi)$ ('H' stands for Hartree), with $\|\psi\| = 1$, and $w := (n-1)v$. Then we have

$$S_H(\psi) = \int \int_{\mathbb{R}^3} \Big\{ -\hbar \,\mathrm{Im}\langle \psi, \partial_t \psi \rangle - |\nabla \psi|^2 - V|\psi|^2$$

$$- \frac{1}{2}|\psi|^2 w * |\psi|^2 \Big\} dx. \tag{14.8}$$

Recall that the regime in which $w := (n-1)v = O(1)$ is called the mean-field regime.

Proof. First, we compute $\langle \Psi, \partial_t \Psi \rangle$ for $\Psi := \otimes_1^n \psi$. Denote $\psi_j := \psi(x_j)$. The relation $\partial_t \Psi = \sum_i (\prod_{j \neq i} \psi_j) \partial_t \psi_i$ gives

$$\langle \Psi, \partial_t \Psi \rangle = \sum_i \int \prod_j \bar{\psi}_j (\prod_{j \neq i} \psi_j) \partial_t \psi_i = \sum_i \|\psi\|^{2(n-1)} \int \bar{\psi}_i \partial_t \psi_i$$

$$= n\|\psi\|^{2(n-1)} \int \bar{\psi} \partial_t \psi$$

Similarly, for $\Psi := \otimes_1^n \psi$, we compute

$$\langle \Psi, \sum_{i=1}^n (-\Delta_{x_i} + V(x_i))\Psi \rangle = \sum_{i=1}^n \int (|\nabla_{x_i} \Psi|^2 + V(x_i)|\Psi|^2)$$

$$= \sum_{i=1}^n \|\psi\|^{2(n-1)} \int (|\nabla_{x_i} \psi_i|^2 + V(x_i)|\psi_i|^2)$$

$$= n\|\psi\|^{2(n-1)} \int (|\nabla \psi|^2 + V(x)|\psi|^2)$$

Finally, we compute the particle pair interaction terms

$$\langle \Psi, (\frac{1}{2} \sum_{i \neq j} v(x_i - x_j))\Psi \rangle = \frac{1}{2} \sum_{i \neq j} \int (\prod_{k \neq i,j} |\psi_k|^2)|\psi_i|^2 v(x_i - x_j)|\psi_j|^2$$

$$= \frac{1}{2} \sum_{i \neq j} \|\psi\|^{2(n-2)} \int \int |\psi_i|^2 v(x_i - x_j)|\psi_j|^2$$

$$= \frac{1}{2} n(n-1)\|\psi\|^{2(n-2)} \int \int |\psi(x)|^2 v(x - x')|\psi(x')|^2$$

Collecting the terms above and using that $v = (n-1)w$, we arrive at (14.8).

The Euler-Lagrange equation for stationary points of the action functional (14.8) considered on the first set of functions is

$$ i\hbar \frac{\partial \psi}{\partial t} = (h + w * |\psi|^2)\psi. \tag{14.9} $$

Here we assume the normalization $\|\psi\| = 1$. This nonlinear evolution equation is called the *Hartree equation* (HE). It is convenient to pass to the normalization $\|\psi\| = n - 1 \approx n$, which leads to the equation

$$ i\hbar \frac{\partial \psi}{\partial t} = (h + v * |\psi|^2)\psi, \tag{14.10} $$

where $h_x = -\Delta_x + V(x)$.

It is relatively easy to extend this approach to fermions. For (spinless) fermions, we consider the action $S(\Psi)$ on the following function space

$$ \{\Psi := \wedge_1^n \psi_j \mid \psi_i \in H^1(\mathbb{R}^3) \; \forall i = 1, ..., n\}, \tag{14.11} $$

where $\wedge_1^n \psi_j := \det[\psi_i(x_j)]$ is the determinant of the $n \times n$ matrix $[\psi_i(x_j)]$. It is called the *Slater determinant*. Then the Euler-Lagrange equation for $S(\Psi)$ on the latter set gives a system of nonlinear, coupled evolution equations

$$ i\hbar \frac{\partial \psi_j}{\partial t} = (h + v * \sum_i |\psi_i|^2)\psi_j - \sum_i (v * \psi_i \bar{\psi}_j)\psi_i, \tag{14.12} $$

for the unknowns $\psi_1, ..., \psi_n$. This systems plays the same role for fermions as the Hartree equation does for bosons. It is called the *Hartree-Fock equations* (HFE).

Finally, we mention another closely related nonlinear equation: the Ginzburg-Landau equations of superconductivity.

Properties of (HE), (HFE) and (GPE). The Hartree and Gross-Pitaevski equations have the following general features

1. For space dimensions $d = 2, 3$ and assuming v is positive definite, (HE) and (HFE) have solutions globally in time; for (GPE) solutions exist globally (in time) if $a > 0$, but blow-up for certain initial conditions in finite time if $a < 0$.
2. (HE), (HFE) and (GPE) are Hamiltonian systems (see Section 20.6).
3. (HE), (HFE) and (GPE) are invariant under the gauge transformations,

$$ \psi(x) \to e^{i\alpha}\psi(x), \; \alpha \in \mathbb{R}, $$

and, for $v = 0$, the translations, $\psi(x) \to \psi(x+y)$, $y \in \mathbb{R}$, and the Galilean transformations, $v \in \mathbb{R}^3$,

$$ \psi(x) \to e^{i(mv \cdot x + \frac{mv^2 t}{2})/\hbar^2}\psi(x - vt). $$

4. (HFE) is invariant under time-independent unitary transformations of $\{\psi_1, ..., \psi_n\}$.

5. The energy, $E(\psi)$, and the number of particles, $N(\psi)$, (see below) are conserved quantities. Moreover, (HFE) conserves the inner products, $\langle \psi_i, \psi_j \rangle$, $\forall i, j$.

The fourth item shows that the natural object for (HFE) is the subspace spanned by $\{\psi_i\}$, or the corresponding projection $\gamma := \sum_i |\psi_i\rangle\langle\psi_i|$. Then (HFE) can be rewritten as an equation for γ:

$$i\frac{\partial \gamma}{\partial t} = [h(\gamma), \gamma] \tag{14.13}$$

where $h(\gamma) := h + v * \rho_\gamma - ex(\gamma)$, with $\rho_\gamma(x, t) := \gamma(x, x, t) = \sum_i |\psi_i(x, t)|^2$ and $ex(\gamma)$ is the operator with the integral kernel $ex(\gamma)(x, y, t) := v(x - y)\gamma(x, y, t) = \sum_i (\bar{\psi}_i v * \psi_i)$. (We write $ex(\gamma)$ as $ex(\gamma) =: v \# \gamma$.) Here $A(x, y)$ stands for the integral kernel of an operator A.

This can be extended to arbitrary non-negative, trace-class operators (i.e. density operators γ, see Section 18.1). We address this in Section 18.6.

To fix ideas, we will hereafter discuss mainly (GPE). For (HE) and (HFE) the results should be appropriately modified. For (GPE) the energy functional is

$$E(\psi) := \int_{\mathbb{R}^3} \left\{ \frac{\hbar^2}{2m} |\nabla\psi|^2 + V|\psi|^2 + 2\pi a|\psi|^4 \right\} dx.$$

The number of particles for (GPE) and (HE) is given by

$$N(\psi) := \int_{\mathbb{R}^3} |\psi|^2 dx$$

while for (HFE), by $N(\psi) := \sum_i \int_{\mathbb{R}^3} |\psi_i|^2 dx$. Note that the energy and number of particle conservation laws are related to the time-translational and gauge symmetries of the equations, respectively.

The above notions of the energy and number of particles are related to corresponding notions in the original microscopic system. Indeed, let $\Psi := \frac{1}{\sqrt{n}} \otimes_1^n \psi$. Then

$$\langle \Psi, H_n \Psi \rangle = E(\psi) + O(\frac{1}{n})$$

where $E(\psi)$ is the energy for (HE) and

$$n \int |\Psi(x_1, ..., x_n)|^2 \, dx_1...dx_n = \int |\psi(x)|^2 \, dx.$$

The notion of bound state can be extended to the nonlinear setting as follows. The *bound states* are stationary solutions of (HE) or (GPE) of the form

$$\psi(x, t) = \phi_\mu(x)e^{i\mu t}$$

where the profile $\phi_\mu(x)$ is in $H^2(\mathbb{R}^3)$. Note that the profile $\phi_\mu(x)$ satisfies the stationary Gross-Pitaevski equation:

$$h\phi + 4\pi a|\phi|^2\phi = -\hbar\mu\phi \qquad (14.14)$$

(we consider here (GPE) only). Thus we can think of the parameter $-\mu$ as a nonlinear eigenvalue.

A *ground state* is a bound state such that the profile $\phi_\mu(x)$ minimizes the energy for a fixed number of particles:

$$\phi_\mu \text{ minimizes } E(\psi) \quad \text{under } N(\psi) = n$$

(see Chapter 26 which deals with variational, and in particular minimization problems). Thus the nonlinear eigenvalue μ arises as a Lagrange multiplier from this constrained minimization problem. In Statistical Mechanics μ is called the chemical potential (the energy needed to add one more particle/atom, see Section 20.7).

Remark 14.4 1. Mathematically, the ground state can be also defined as a stationary solution with a positive (up to a constant phase factor) profile, $\psi(x,t) = \phi_\mu(x)e^{i\mu t}$ with $\phi_\mu(x) > 0$. Let $\delta(\mu) := \|\phi_\mu\|^2$. Then we have (see [142])

$$\delta'(\mu) > 0 \implies \phi_\mu \text{ minimizes } E(\psi) \text{ under } N(\psi) = n.$$

2. The Lagrange multiplier theorem in Section 26.5 implies that the ground state profile ϕ_μ is a critical point of the functional

$$E_\mu(\psi) := E(\psi) + \hbar\mu N(\psi).$$

In fact, ϕ_μ is a minimizer of this functional under the condition $N(\psi) = n$.

If ϕ_μ is the ground state of (GPE), then $\otimes_1^n \phi_\mu$ is close to the ground state of the $n-$body Hamiltonian describing the Bose-Einstein condensate (see [74] for a review, and [209, 210] and the Appendix below for rigorous results).

It is known that for natural classes of nonlinearities and potentials $V(x)$ there is a ground state. Three cases of special interest are

1. $h := -\frac{\hbar^2}{2m}\Delta + V(x)$ has a ground state, and $\frac{2m}{\hbar^2}n|a| \ll 1$
2. V has a minimum, $\frac{2m}{\hbar^2}n|a| \gg 1$, and $a < 0$
3. $V(x) \to \infty$ as $|x| \to \infty$ (i.e. $V(x)$ is confining) and $a > 0$.

(The first and third cases are straightforward and the second case requires some work [105, 234, 11].)

Stability. We discuss now the important issue of stability of stationary solutions under small perturbations. Namely, we want to know how solutions of our equation with initial conditions close to a stationary state (i.e. small perturbations of $\phi_\mu(x)$) behave. Do these solutions stay close to the stationary state in question, do they converge to it, or do they depart from it? This is obviously a central question. This issue appeared implicitly in Section 6.4 (and in a stronger formulation in Chapter 9) but has not been explicitly articulated yet. This is because the situation in the linear case that we have dealt with so far is rather straightforward. On the other hand, in the nonlinear case, stability questions are subtle and difficult, and play a central role.

We say that a stationary solution, $\phi_\mu(x)e^{i\mu t}$, is *orbitally* (respectively, *asymptotically*) *stable* if for all initial conditions sufficiently close to $\phi_\mu(x)e^{i\alpha}$ (for some constant $\alpha \in \mathbb{R}$), the solutions of the evolution equation under consideration stay close (respectively, converge in an appropriate norm) to a nearby stationary solution (times a phase factor), $\phi_{\mu'}(x)e^{i(\mu' t+\beta(t))}$. Here μ' is usually close to μ, and the phase β depends on time, t. The phase factors come from the fact that our equations have gauge symmetry: if $\psi(x, t)$ is a solution, then so is $e^{i\alpha}\psi(x, t)$ for any constant $\alpha \in \mathbb{R}$. One should modify the statement above if other symmetries are present. The notion of orbital stability generalizes the classical notion of *Lyapunov stability*, well-known in the theory of dynamical systems, to systems with symmetries.

For the linear Schrödinger equation, all bound states, as well as stationary states corresponding to embedded eigenvalues, are orbitally stable. But they are *not* asymptotically stable in general. For most nonlinear evolution equations in unbounded domains, the majority of states are not even Lyapunov/orbitally stable.

For (GPE), if $V \to \infty$ as $|x| \to \infty$ (i.e. V is confining), the ground states are orbitally stable, but not asymptotically stable. If $V \to 0$ as $|x| \to \infty$, the ground states can be proved to be asymptotically stable in some cases (see [277, 293, 145, 127, 128, 72] and references therein).

14.2 Appendix: BEC at T=0

In this appendix we consider briefly the phenomenon of Bose-Einstein condensation, predicted by Einstein in 1925 on the basis of analysis of ideal bose gases and experimentally discovered 70 years later in 1995 in real gases by two groups, one led by Wieman and Cornell at Boulder, and another by Ketterle at MIT. The Gross-Pitaevskii equation arises in the description of this phenomenon. We concentrate on zero temperature.

First we consider a system of n non-interacting bosons in an external potential V. The state space of such a system is the Hilbert space $\mathbb{S}_1^n L^2(\mathbb{R}^d)$, $d = 1, 2, 3$, and the Hamiltonian operator is

$$H_n := \sum_{j=1}^n \left(-\frac{\hbar^2}{2m}\Delta_{x_i} + V(x_i)\right).$$

(acting on $\mathbb{S}_1^n L^2(\mathbb{R}^d)$). By separation of variables, the lowest energy for this Hamiltonian is ne_0, where e_0 is the lowest energy for the one-particle

Schrödinger operator $-\frac{\hbar^2}{2m}\Delta_x + V(x)$. The corresponding eigenfunction – the ground state of H_n – is given by $\mathbb{S}_1^n\phi_0$, where ϕ_0 is the ground state of the operator $-\frac{\hbar^2}{2m}\Delta_x + V(x)$. And that's it – at zero temperature, and in the ground state, all particles are in the same state ϕ_0!

Now we consider a system of n interacting bosons subject to an external potential V, which is described by the Hamiltonian

$$H_n := \sum_{j=1}^{n}\left(-\frac{\hbar^2}{2m}\Delta_{x_i} + V(x_i)\right) + \frac{1}{2}\sum_{i\neq j}v(x_i - x_j),$$

acting on the Hilbert space $\mathbb{S}_1^n L^2(\mathbb{R}^d)$. Here v is the potential of interaction between the particles. Let $\Phi_{n,0}(x_1,\ldots,x_n)$ be a ground state of H_n. How do we tell if in this state all (or the majority of) the particles are individually in some one- particle state, say ϕ_0?

To begin with we would like to describe, say, the coordinate or momentum distributions for a single particle. To extract one-particle information from $\Phi_{n,0}$, we use the information reduction principle elucidated in Section 19.1: we pass to density operators ($\Phi_{n,0} \to P_{\Phi_{n,0}}$ = the rank 1 orthogonal projection on the state $\Phi_{n,0}$) and contract $(n-1)-$particle degrees of freedom. This leads to the *one-particle density operator*:

$$\gamma_1^n := Tr_{n-1}P_{\Phi_{n,0}}$$

where Tr_{n-1} is the trace over $n-1$ of the 3 dimensional coordinates (see Section 19.1). The one-particle density matrix satisfies

- $0 \leq \gamma_1^n \leq 1$
- $Tr\gamma_1^n = 1$
- $\Phi_{n,0} = \otimes_1^n\phi_0 \Leftrightarrow \gamma_1^n = P_{\phi_0}$

Let $\lambda_1 \geq \lambda_2 \geq \cdots \lambda_j \geq \cdots$ be the eigenvalues of γ_n, counting their multiplicities, and let $\chi_1, \chi_2, \cdots \chi_j, \cdots$ be corresponding eigenfunctions. We have the spectral decomposition

$$\gamma_1^n = \sum_{j=1}^{\infty}\lambda_j P_{\chi_j}, \qquad \sum_{j=1}^{\infty}\lambda_j = 1$$

(see Mathematical Supplement, Section 25.11).The eigenvalues λ_j are interpreted as the probabilities for a single particle to be in the states χ_j.

The *Penrose-Onsager criterion* of BEC says that the property of the ground state $\Phi_{n,0}$ of the Hamiltonian H_n

$$\text{max eigenvalue of } \gamma_1^n \to 1 \text{ as } n \to \infty \qquad (POC)$$

corresponds to 100% condensation. The criterion (POC) implies

$$\gamma_1^n - P_{\chi_1} \to 0 \quad \text{as} \quad n \to \infty$$

so that for large n, almost all the particles are in the single state χ_1.

A rigorous proof of BEC in the Gross-Pitaevski limit, in which the number of particles $n \to \infty$ and the scattering length $a \to 0$, so that $na =: \lambda/(4\pi)$ is fixed, is given in [207] (see also [209]). Moreover, they show that in the trace norm,

$$\gamma_1^n - P_{\phi_n^{GP}} \to 0$$

where ϕ_n^{GP} is the Gross-Pitaevski ground state. They also prove convergence for the ground state energies.

15

The Feynman Path Integral

In this chapter, we derive a convenient representation for the integral kernel of the Schrödinger evolution operator, $e^{-itH/\hbar}$. This representation, the "Feynman path integral", will provide us with a heuristic but effective tool for investigating the connection between quantum and classical mechanics. This investigation will be undertaken in the next section.

15.1 The Feynman Path Integral

Consider a particle in $\mathbb{R}^d$ described by a self-adjoint Schrödinger operator

$$H = -\frac{\hbar^2}{2m}\Delta + V(x).$$

Recall that the dynamics of such a particle is given by the Schrödinger equation

$$i\hbar\frac{\partial\psi}{\partial t} = H\psi.$$

Recall also that the solution to this equation, with the initial condition $\psi|_{t=0} = \psi_0$, is given in terms of the evolution operator $U(t) := e^{-iHt/\hbar}$ as

$$\psi = U(t)\psi_0.$$

Our goal in this section is to understand the evolution operator $U(t) = e^{-iHt/\hbar}$ by finding a convenient representation of its integral kernel. We denote the integral kernel of $U(t)$ by $U_t(y,x)$ (also called the *propagator* from x to y).

A representation of the exponential of a sum of operators is provided by the *Trotter product formula* (Theorem 15.2) which is explained in Section 15.3 at the end of this chapter. The Trotter product formula says that

$$e^{-iHt/\hbar} = e^{i(\frac{\hbar^2 t}{2m}\Delta - Vt)/\hbar} = \text{s-}\lim_{n\to\infty} K_n^n$$

© Springer-Verlag GmbH Germany, part of Springer Nature 2020
S. J. Gustafson and I. M. Sigal, *Mathematical Concepts of Quantum Mechanics*, Universitext, https://doi.org/10.1007/978-3-030-59562-3_15

where $K_n := e^{\frac{i\hbar t}{2mn}\Delta}e^{-\frac{iVt}{\hbar n}}$. Let $K_n(x, y)$ be the integral kernel of the operator K_n. Then by Proposition 25.12,

$$U_t(y, x) = \lim_{n \to \infty} \int \cdots \int K_n(y, x_{n-1}) \cdots K_n(x_2, x_1) K_n(x_1, x) dx_{n-1} \cdots dx_1.$$
(15.1)

Now (see Section 25.3)

$$K_n(y, x) = e^{\frac{i\hbar t \Delta}{2mn}}(y, x)e^{-\frac{iV(y)t}{\hbar n}}$$

since V, and hence $e^{-iVt/n\hbar}$, is a multiplication operator (check this).

Using the expression (2.23), and plugging into (15.1) gives us

$$U_t(y, x) = \lim_{n \to \infty} \int \cdots \int e^{iS_n/\hbar} \left(\frac{2\pi i \hbar t}{mn}\right)^{-nd/2} dx_1 \cdots dx_{n-1}$$

where

$$S_n := \sum_{k=0}^{n-1} (mn|x_{k+1} - x_k|^2/2t - V(x_{k+1})t/n)$$

with $x_0 = x$, $x_n = y$. Define the piecewise linear function ϕ_n such that $\phi_n(0) = x$, $\phi_n(t/n) = x_1, \cdots, \phi_n(t) = y$ (see Fig. 15.1).

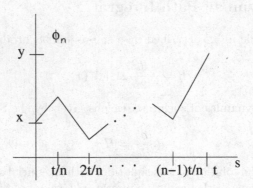

Fig. 15.1. Piecewise linear function.

Then

$$S_n = \sum_{k=0}^{n-1} \left\{ m\frac{|\phi_n((k+1)t/n) - \phi_n(kt/n)|^2}{2(t/n)^2} - V(\phi_n((k+1)t/n)) \right\} t/n.$$

Note that S_n is a Riemann sum for the classical action

$$S(\phi, t) = \int_0^t \left\{ \frac{m}{2}|\dot\phi(s)|^2 - V(\phi(s)) \right\} ds$$

of the path ϕ_n. So we have shown

$$U_t(y,x) = \lim_{n\to\infty} \int_{P^n_{x,y,t}} e^{iS_n/\hbar} D\phi_n \tag{15.2}$$

where $P^n_{x,y,t}$ is the $(n-1)$-dimensional space of paths ϕ_n with $\phi_n(0) = x$, $\phi_n(t) = y$, and which are linear on the intervals $(kt/n, (k+1)t/n)$ for $k = 0,1,\ldots,n-1$, and $D\phi_n = (\frac{2\pi i\hbar t}{nm})^{-nd/2} d\phi_n(t/n)\cdots d\phi_n((n-1)t/n)$.

Heuristically, as $n\to\infty$ ϕ_n approaches a general path, ϕ, from x to y (in time t), and $S_n \to S(\phi)$. Thus we write

$$\boxed{U_t(y,x) = \int_{P_{x,y,t}} e^{iS(\phi,t)/\hbar} D\phi.} \tag{15.3}$$

Here $P_{x,y,t}$ is a space of paths from x to y, defined as

$$P_{x,y,t} := \{\phi : [0,t] \to \mathbb{R}^d | \int_0^t |\dot\phi|^2 < \infty, \quad \phi(0) = x, \quad \phi(t) = y\}.$$

This is the *Feynman path integral*. It is not really an integral, but a formal expression whose meaning is given by (15.2). Useful results are obtained non-rigorously by treating it formally as an integral. Answers we get this way are intelligent guesses which must be justified by rigorous tools.

Note that $P^n_{x,y,t}$ is an $(n-1)$-dimensional sub-family of the infinite-dimensional space $P_{x,y,t}$. It satisfies $P^n_{x,y,t} \subset P^{2n}_{x,y,t}$ and $\lim_{n\to\infty} P^n_{x,y,t} = P_{x,y,t}$ in some sense. We call such subspaces *finite dimensional approximations* of $P_{x,y,t}$. In (non-rigorous) computations, it is often useful to use finite-dimensional approximations to the path space other than the polygonal one above.

We can construct more general finite-dimensional approximations as follows. Fix a function $\phi_{xy} \in P_{x,y,t}$. Then

$$P_{x,y,t} = \phi_{xy} + P_{0,0,t}.$$

Note $P_{0,0,t}$ is a Hilbert space. Choose an orthonormal basis $\{\xi_j\}$ in $P_{0,0,t}$ and define

$$P^n_{0,0,t} := \text{span } \{\xi_j\}^n_1$$

and

$$P^n_{x,y,t} := \phi_{xy} + P^n_{0,0,t}.$$

Then $P^n_{x,y,t}$ is a finite dimensional approximation of $P_{x,y,t}$. Typical choices of ϕ_{xy} and $\{\xi_j\}$ are

1. ϕ_{xy} is piecewise linear and $\{\xi_j\}$ are "splines". This gives the polygonal approximation introduced above.

2. ϕ_{xy} is a classical path (a critical point of the action functional $S(\phi)$) and $\{\xi_j\}$ are eigenfunctions of the Hessian of S at ϕ_{xy} (these notions are described in the Supplement on the calculus of variations, Chapter 26). In this case, if $\eta \in P^n_{0,0,t}$, then we can represent it as

$$\eta = \sum_{j=1}^{n} a_j \xi_j,$$

and we have

$$D\eta = \left(\frac{2\pi it\hbar}{m}\right)^{-d/2} \left(\frac{2\pi n}{t}\sqrt{\frac{m}{\hbar}}\right)^n \prod_{j=1}^{n} da_j.$$

It is reasonable to expect that if

$$\lim_{n\to\infty} \int_{P^n_{x,y,t}} e^{iS(\phi,t)/\hbar} D\phi$$

exists, then it is independent of the finite-dimensional approximation, $P^n_{x,y,t}$, that we choose.

Problem 15.1

1. Compute (using (15.3) and a finite-dimensional approximation of the path space) U_t for
   a) $V(x) = 0$ (free particle)
   b) $V(x) = \frac{m\omega^2}{2}x^2$ (harmonic oscillator in dimension $d = 1$).
2. Derive a path integral representation for the integral kernel of $e^{-\beta H/\hbar}$.
3. Use the previous result to find a path integral representation for $Z(\beta) :=$ tr $e^{-\beta H/\hbar}$ (hint: you should arrive at the expression (16.11)).

15.2 Generalizations of the Path Integral

Here we mention briefly two extensions of the Feynman path integral we have just introduced.

1. Phase-space path integral:

$$U_t(y, x) = \int_{P_{x,y,t} \times \text{ anything}} e^{i\int_0^t (\dot{\phi}\pi - H(\phi,\pi))/\hbar} D\phi D\!\!\!/\pi$$

where $D\!\!\!/\pi$ is the *path measure*, normalized as

$$\int e^{-\frac{i}{2}\int_0^t \|\pi\|^2} D\!\!\!/\pi = 1$$

(in QM, $d^3p = d^3p/(2\pi)^{3/2}$). To derive this representation, we use the Trotter product formula, the expression $e^{-i\epsilon H} \approx 1 - i\epsilon H$ for ϵ small, and the symbolic (pseudodifferential) composition formula. Unlike the representation $\int e^{iS/\hbar} D\phi$, this formula holds also for more complicated H, which are not quadratic in p !

2. A particle in a vector potential $A(x)$. In this case, the Hamiltonian is

$$H(x,p) = \frac{1}{2m}(p - eA(x))^2 + V(x)$$

and the Lagrangian is

$$L(x,\dot{x}) = \frac{m}{2}\dot{x}^2 - V(x) + e\dot{x} \cdot A(x).$$

The propagator still has the representation

$$U_t(y,x) = \int_{P_{x,y,t}} e^{iS(\phi)/\hbar} D\phi,$$

but with

$$S(\phi) = \int_0^t L(\phi,\dot\phi)ds = \int_0^t (\frac{m}{2}\dot\phi^2 - V(\phi))ds + e\int_0^t A(\phi) \cdot \dot\phi ds.$$

Since $A(x)$ does not commute with ∇ in general, care should be exercised in computing a finite-dimensional approximation: one should take

$$\sum A(\frac{1}{2}(x_i + x_{i+1})) \cdot (x_{i+1} - x_i)$$

or

$$\sum \frac{1}{2}(A(x_i) + A(x_{i+1})) \cdot (x_{i+1} - x_i)$$

and not

$$\sum A(x_i) \cdot (x_{i+1} - x_i) \text{ or } \sum A(x_{i+1}) \cdot (x_{i+1} - x_i).$$

15.3 Mathematical Supplement: the Trotter Product Formula

Let A, B, and $A+B$ be self-adjoint operators on a Hilbert space $\mathcal{H}$. If $[A,B] \neq 0$, then $e^{i(A+B)} \neq e^{iA}e^{iB}$ in general. But we do have the following.

Theorem 15.2 (Trotter product formula) Let either A and B be bounded, or A, B, and $A + B$ be self-adjoint and bounded from below. Then for $Re(\lambda) \leq 0$,

$$e^{\lambda(A+B)} = s - \lim_{n\to\infty} (e^{\lambda\frac{A}{n}} e^{\lambda\frac{B}{n}})^n.$$

Remark 15.3 The convergence here is in the sense of the *strong operator topology*. For operators A_n and A on a Hilbert space $\mathcal{H}$, such that $D(A_n) = D(A)$, $A_n \to A$ in the strong operator topology (written s-$\lim_{n\to\infty} A_n = A$) iff $\|A_n\psi - A\psi\| \to 0$ for all $\psi \in D(A)$. For bounded operators, we can take norm convergence. In the formula above, we used a uniform decomposition of the interval $[0, 1]$. The formula still holds for a non-uniform decomposition.

Proof for A,B bounded: We can assume $\lambda = 1$. Let $S_n = e^{(A+B)/n}$ and $T_n = e^{A/n}e^{B/n}$. Now by "telescoping",

$$S_n^n - T_n^n = S_n^n - T_nS_n^{n-1} + T_nS_n^{n-1} + \cdots - T_n^n$$

$$= \sum_{k=0}^{n-1} T_n^k(S_n - T_n)S_n^{n-k-1}$$

so

$$\|S_n^n - T_n^n\| \le \sum_{k=0}^{n-1} \|T_n\|^k\|S_n - T_n\|\|S_n\|^{n-k-1}$$

$$\le \sum_{k=0}^{n-1} (\max(\|T_n\|, \|S_n\|))^{n-1}\|S_n - T_n\|$$

$$\le ne^{\|A\|+\|B\|}\|S_n - T_n\|.$$

Using a power series expansion, we see $\|S_n - T_n\| = O(1/n^2)$ and so $\|S_n^n - T_n^n\| \to 0$ as $n \to \infty$. $\square$

A proof for unbounded operators can be found in [244].

Semi-classical Analysis

In this chapter we investigate some key quantum quantities – such as quantum energy levels – as $\hbar/$(typical classical action)$\to 0$. We hope that asymptotics of these quantities can be expressed in terms of relevant classical quantities. This is called semi-classical (or semi-classical) analysis. To do this, we use the Feynman path integral representation of the evolution operator (propagator) $e^{-iHt/\hbar}$. This representation provides a non-rigorous but highly effective tool, as the path integral is expressed directly in terms of the key classical quantity – the classical action.

The heuristic power of path integrals is that when treated as usual convergent integrals, they lead to meaningful and, as it turns out, correct answers. Thus to obtain a "semi-classical approximation", we apply the method of stationary phase. Recall that the (ordinary) method of stationary phase expands the integral in question in terms of the values of the integrand at the critical points of the phase, divided by the square root of the determinant of the Hessian of the phase at those critical points. The difference here is that the phase – the classical action – is not a function of several variables, but rather a "functional", which (roughly speaking) is a function of an infinite number of variables, or a function on paths. Critical points of the classical action are the classical paths (solutions of Newton's equation) and the Hessians are differential operators. Thus we need some new pieces of mathematics: determinants of operators and elements of the calculus of variations. These are presented in supplementary Section 25.12 and Chapter 26 respectively.

Below we consider a particle in $\mathbb{R}^d$ described by a Schrödinger operator

$$H = -\frac{\hbar^2}{2m}\Delta_x + V(x). \tag{16.1}$$

We want to pass to physical units in which a typical classical action in our system is 1, so that $\hbar$ is now the ratio of the Planck constant to the classical action, so that the regime we are interested in is the one for which $\hbar \to 0$. Let L be a length scale for the potential, and g its size. So roughly, $g = \sup_x |V(x)|$ and $L = g(\sup_x |\nabla V(x)|)^{-1}$. Re-scaling the variable as $x \to x' = x/L$, we find

© Springer-Verlag GmbH Germany, part of Springer Nature 2020
S. J. Gustafson and I. M. Sigal, *Mathematical Concepts of Quantum Mechanics*, Universitext, https://doi.org/10.1007/978-3-030-59562-3_16

$H = gH'$ where

$$H' = -\frac{\hbar'^2}{2m'}\Delta_{x'} + V'(x')$$

where $V'(x') = g^{-1}V(Lx')$ and $\hbar'/\sqrt{m'} = \hbar/(L\sqrt{mg})$. Now the potential $V'(x')$ is essentially of unit size and varies on a unit length scale. The parameter $\hbar'$ is dimensionless. If $\hbar'/\sqrt{m'} \ll 1$, we can consider it a small parameter. As an example suppose $V(x)$ is of the order $100me^4/\hbar^2$, where $me^4/\hbar^2$ is twice the ionization energy of the ground state of the hydrogen atom, and varies on the scale of the Bohr radius (of the hydrogen atom) $L = \hbar^2/(me^2)$. Then $\hbar'/\sqrt{m'} = 1/10$. In the expansions we carry out below, we always have in mind the operator H' and the dimensionless parameter $\hbar'$ with the primes omitted; that is, we think of (16.1) in dimensionless variables.

16.1 Semi-classical Asymptotics of the Propagator

The path integral (15.3) has the form of oscillatory integrals extensively studied in physics and mathematics. One uses the method of stationary phase in order to derive asymptotic expressions for such integrals. It is natural, then, to apply (formally) this method – with small parameter $\hbar$ – to the path integral, in order to derive a semi-classical expression for the Schrödinger propagator $e^{-itH/\hbar}(y, x)$. We do this below. But first, we quickly review the basics of the method of stationary phase (in the finite-dimensional setting, of course).

The stationary phase method. We would like to determine the asymptotics of oscillatory integrals of the form

$$\int_{\mathbb{R}^d} e^{iS(\phi)/\hbar}d\phi$$

as $\hbar \to 0$ (here ϕ is a finite dimensional variable). The basic idea is that as $\hbar \to 0$, the integrand is highly oscillating and yields a small contribution, except where $\nabla S(\phi) = 0$ (i.e., critical points). We now make this idea more precise. Set

$$I(\hbar) := \int_{\mathbb{R}^d} f(\phi)e^{iS(\phi)/\hbar}d\phi$$

where $f \in C_0^\infty(\mathbb{R}^d)$ and S is smooth, and consider two cases:

Theorem 16.1 (stationary phase method) 1. If supp(f) contains no critical points of S, then

$$I(\hbar) = O(\hbar^N) \quad \forall\ N \quad \text{as} \quad \hbar \to 0.$$

2. If supp(f) contains precisely one non-degenerate critical point of S, i.e. $\nabla S(\bar\phi) = 0$ and the matrix of second derivatives $S''(\bar\phi)$ is invertible, then as $\hbar \to 0$,

$$I(\hbar) = (2\pi\hbar)^{d/2}|\det S''(\bar{\phi})|^{-1/2}e^{i\frac{\pi}{4}}sgn(S''(\bar{\phi}))f(\bar{\phi})e^{iS(\bar{\phi})/\hbar}[1 + O(\hbar)]$$
$$(16.2)$$

where for a matrix A, $sgn(A)$ denotes the number of positive eigenvalues minus the number of negative ones.

Proof. To prove the first statement, define the operator

$$L := \frac{\hbar}{i}\frac{\nabla S(\phi)}{|\nabla S(\phi)|^2}\cdot\nabla.$$

Note that $Le^{iS(\phi)/\hbar} = e^{iS(\phi)/\hbar}$, and that for smooth functions f and g,

$$\int_{\mathbb{R}^d} fLg\,dx = \int_{\mathbb{R}^d} (L^T f)g$$

where

$$L^T f := -\frac{\hbar}{i}\nabla \cdot \left[\frac{\nabla S}{|\nabla S|^2}f\right].$$

So for any positive integer N,

$$|I(\hbar)| = |\int_{\mathbb{R}^d} f(\phi)L^N e^{iS(\phi)/\hbar}d\phi| = |\int_{\mathbb{R}^d}[(L^T)^N f(\phi)]e^{iS(\phi)/\hbar}d\phi|$$
$$\leq (const)\hbar^N,$$

establishing the first statement.

Turning to the second statement, suppose supp(f) contains only one critical point, $\bar{\phi}$ of S, which is non-degenerate. We begin with a formal calculation, and then explain how to make it rigorous. Writing $\phi - \bar{\phi} = \sqrt{\hbar}\alpha$, we obtain

$$S(\phi)/\hbar = S(\bar{\phi})/\hbar + \frac{1}{2}\alpha^T S''(\bar{\phi})\alpha + O(\sqrt{\hbar}|\alpha|^3).\qquad (16.3)$$

So

$$I(\hbar) = \hbar^{d/2}e^{iS(\bar{\phi})/\hbar}\int_{\mathbb{R}^d} f(\bar{\phi} + \sqrt{\hbar}\alpha)e^{i\alpha^T S''(\bar{\phi})\alpha/2}e^{iO(\sqrt{\hbar}|\alpha|^3)}d\alpha.$$

Now we use the formula

$$\lim_{R\to\infty}\int_{|\alpha|\leq R} e^{i\alpha^T S''(\bar{\phi})\alpha/2}d\alpha = (2\pi i)^{d/2}[\det S''(\bar{\phi})]^{-1/2}.\qquad (16.4)$$

We can derive this expression by analytically continuing $\int e^{-a\alpha^T S''\alpha}$ from $Re(a) > 0$, though the integral is not absolutely convergent. Some care is needed in choosing the right branch of the square root function. An unambiguous expression for the right hand side is

$$(2\pi)^{d/2}|\det S''(\bar{\phi})|^{-1/2}e^{i\pi\cdot sgn(S''(\bar{\phi}))/4}$$

Using $f(\bar{\phi} + \sqrt{\hbar}\alpha) = f(\bar{\phi}) + O(\sqrt{\hbar})$, yields (16.2). Though this computation shows what is going on, it is not rigorous, since the integral in the remainder diverges.

A more careful computation is based on the Fourier transform, and proceeds as follows. First, as a replacement for (16.3), we use the fact, known as the *Morse Lemma*, that there is a change of variables which makes $S(\phi)$ quadratic near $\bar{\phi}$. More precisely, there exists a smooth function $\Phi : \mathbb{R}^d \to \mathbb{R}^d$ with $\Phi(\bar{\phi}) = \bar{\phi}$, $D\Phi(\bar{\phi}) = 1$, and

$$S(\Phi(\phi)) = S(\bar{\phi}) + \frac{1}{2}(\phi - \bar{\phi})^T S''(\bar{\phi})(\phi - \bar{\phi})$$

for ϕ in a sufficiently small ball, $B_\epsilon(\bar{\phi})$, around $\bar{\phi}$. A proof of this can be found in [92], for example. By the first statement of Theorem 16.1, we can assume that $(\mathrm{supp}\, f) \subset B_\delta(\bar{\phi})$ with δ small enough so that $\Phi^{-1}(B_\delta(\bar{\phi})) \subset B_\epsilon(\bar{\phi})$. Then we have

$$I(\hbar) = \int_{B_\delta(\bar{\phi})} f(\phi)e^{iS(\phi)/\hbar}d\phi = \int_{\Phi^{-1}(B_\delta(\bar{\phi}))} f(\Phi(y))e^{iS(\Phi(y)/\hbar)}|\det D\Phi(y)|dy$$

$$= e^{iS(\bar{\phi})/\hbar} \int_{\Phi^{-1}(B_\delta(\bar{\phi}))} e^{i(y-\bar{\phi})^T S''(\bar{\phi})(y-\bar{\phi})/(2\hbar)} f(\Phi(y))|\det D\Phi(y)|dy$$

$$= e^{iS(\bar{\phi})/\hbar} \int_{\Phi^{-1}(B_\delta(0))} e^{ix^T S''(\bar{\phi})x/(2\hbar)} f(\Phi(\bar{\phi} + x))|\det D\Phi(\bar{\phi} + x)|dx.$$

Now we use the Plancherel formula

$$\int_{\mathbb{R}^d} a(x)b(x)dx = \int_{\mathbb{R}^d} \hat{a}(\xi)\hat{b}(-\xi)d\xi$$

together with the fact (see Section 25.14) that for an invertible symmetric matrix A,

$$(e^{ix^T Ax/(2\hbar)})\hat{\,}(\xi) = \hbar^{d/2}|\det A|^{-1/2}e^{i\pi \mathrm{sgn}(A)/4}e^{-i\hbar\xi^T A^{-1}\xi/2},$$

to obtain

$$I(\hbar) = \hbar^{d/2}|\det S''(\bar{\phi})|^{-1/2}e^{i\pi \mathrm{sgn}(S''(\bar{\phi})/4}e^{iS(\bar{\phi})/\hbar} \int_{\mathbb{R}^d} e^{-i\hbar\xi^T (S''(\bar{\phi}))^{-1}\xi/2}\hat{b}(-\xi)d\xi$$

$$= \hbar^{d/2}|\det S''(\bar{\phi})|^{-1/2}e^{i\pi \mathrm{sgn}(S''(\bar{\phi})/4}e^{iS(\bar{\phi})/\hbar} \int_{\mathbb{R}^d} (1 + O(\hbar|\xi|^2))\hat{b}(-\xi)d\xi,$$

where $b(x) := f(\Phi(\bar{\phi} + x))|\det D\Phi(\bar{\phi} + x)|\chi_{|x|\leq\delta}$. Finally, observing that $\int_{\mathbb{R}^d} \hat{b}(-\xi)d\xi = (2\pi)^{d/2}b(0) = (2\pi)^{d/2}f(\bar{\phi})$, we arrive at the second statement of Theorem 16.1.

Now we would like to formally apply the method outlined above to the infinite-dimensional integral (15.3). To this end, we simply plug the path integral expression (15.3) into the stationary phase expansion formula (16.2). The result is

$$e^{-iHt/\hbar}(y,x) = \int_{P_{x,y,t}} e^{iS(\phi)/\hbar} D\phi$$

$$= \sum_{\bar{\phi} \text{ cp of } S} M_{\bar{\phi}} (\det S''(\bar{\phi}))^{-1/2} e^{iS(\bar{\phi})/\hbar} (1 + O(\sqrt{\hbar})) \qquad (16.5)$$

where $M_{\bar{\phi}}$ is a normalization constant, and the sum is taken over all critical points, $\bar{\phi}$, of the action $S(\phi)$, going from x to y in time t. Critical points and Hessians of functionals are discussed in the mathematical supplement, Chapter 26. Note that the Hessian $S''(\bar{\phi})$ is a differential operator. The problems of how to define and compute determinants of Hessians are discussed in the mathematical supplement Section 25.12.

We will determine $M := M_{\bar{\phi}}$, assuming it is independent of $\bar{\phi}$ and V. For $V = 0$, we know the kernel of the propagator explicitly (see (2.23)):

$$e^{-iH_0 t/\hbar}(y,x) = (2\pi i\hbar t/m)^{-d/2} e^{im|x-y|^2/2\hbar t}.$$

So in particular, $e^{-iH_0 t/\hbar}(x,x) = (2\pi i\hbar t/m)^{-d/2}$. Now the right-hand side of the expression (16.5) for $e^{-itH_0/\hbar}(x,y)$ is (to leading order in $\hbar$)

$$M(\det S_0''(\phi_0))^{-1/2} e^{iS_0(\phi_0)/\hbar}$$

where the unique critical point is $\phi_0(s) = x + (y-x)s/t$. Thus $S_0(\phi_0) = m|y-x|^2/2t$, and $S_0''(\phi_0) = -m\partial_s^2$, an operator acting on functions satisfying Dirichlet boundary conditions.

Comparison thus gives us

$$M = (\det(-m\partial_s^2))^{1/2} (2\pi i\hbar t/m)^{-d/2}$$

and therefore

$$e^{-iHt/\hbar}(y,x) = \sum_{\bar{\phi} \text{ cp } S} \left(\frac{2\pi it\hbar}{m} \right)^{-d/2} \left(\frac{\det(-m\partial_s^2)}{\det S''(\bar{\phi})} \right)^{1/2} e^{iS(\bar{\phi})/\hbar} (1 + O(\sqrt{\hbar}))$$

$$(16.6)$$

as $\hbar \to 0$. This is precisely the semi-classical expression we were looking for.

We now give a "semi-rigorous" derivation of this expression. We assume for simplicity that S has only one critical point, $\bar{\phi}$, going from x to y in time t. Let $\{\xi_j\}_{j=1}^{\infty}$ be an orthonormal basis of eigenfunctions of $S''(\bar{\phi})$ acting on $L^2([0,t])$ with zero boundary conditions (the eigenfunctions of such an operator are complete – see the remark in Section 25.11). So $S''(\bar{\phi})\xi_j = \mu_j\xi_j$ for eigenvalues μ_j. For the n-th order finite dimensional approximation to the space of paths in the path integral, we take the n-dimensional space of functions of the form

$$\phi^{(n)} = \bar{\phi} + \sum_{j=1}^{n} a_j \xi_j$$

with $a_j \in \mathbb{R}$. Expanding $S(\phi^{(n)})$ around $\bar{\phi}$ gives

$$S(\phi^{(n)}) = S(\bar\phi) + \frac{1}{2}\langle \xi, S''(\bar\phi)\xi\rangle + O(\|\xi\|^3)$$

where

$$\xi := \phi^{(n)} - \bar\phi = \sum_{j=1}^{n} a_j \xi_j.$$

We also have

$$D\phi^{(n)} = C_n \prod_{j=1}^{n} da_j$$

(C_n some constant). Now using the fact that

$$\langle \xi, S''(\bar\phi)\xi\rangle = \sum_{i,j=1}^{n} a_i a_j \langle \xi_i, S''(\bar\phi)\xi_j\rangle = \sum_{j=1}^{n} \mu_j a_j^2,$$

we have

$$\int_{\phi^{(n)}:x\to y} e^{iS(\phi^{(n)})/\hbar} D\phi^{(n)} = e^{iS(\bar\phi)/\hbar} \int e^{i\sum \mu_j a_j^2/2\hbar}(1 + O(a^3/\hbar))C_n d^n a$$

(as in Section 16.1, the integrals here are not absolutely convergent). Setting $b_j := a_j/\sqrt{\hbar}$ this becomes

$$\hbar^{n/2} C_n e^{iS(\bar\phi)/\hbar} \int e^{i\sum \mu_j b_j^2/2}(1 + O(b^3\sqrt{\hbar}))d^n b$$

which is (see (16.4))

$$C_n(2\pi i\hbar)^{n/2}(\det(S''(\bar\phi)|_{F_n}))^{-1/2} e^{iS(\bar\phi)/\hbar}(1 + O(\sqrt{\hbar}))$$

where $F_n := \{\sum_1^n a_j \xi_j\}$ so that $\det(S''(\bar\phi)|_{F_n}) = \prod_1^n \mu_j$. To avoid determining the constants C_n arising in the "measure" $D\phi$, we compare again with the free ($V = 0$) propagator. Taking a ratio gives us

$$\frac{e^{-iHt/\hbar}(y,x)}{(\frac{2\pi i\hbar t}{m})^{-d/2}} = \lim_{n\to\infty} \frac{C_n(2\pi i\hbar)^{n/2}(\det(S''(\bar\phi)|_{F_n}))^{-1/2} e^{iS(\bar\phi)/\hbar}}{C_n(2\pi i\hbar)^{n/2}(\det(-m\partial_s^2|_{F_n}))^{-1/2}}(1 + O(\sqrt{\hbar}))$$

which reproduces (16.6).

16.2 Semi-classical Asymptotics of Green's Function

Definition 16.2 *Green's function* $G_A(x,y,z)$ *of an operator* A *is* $(A - z)^{-1}(y,x)$, *the integral kernel of the resolvent* $(A - z)^{-1}$.

For A self-adjoint,

$$(A-z)^{-1} = \frac{i}{\hbar}\int_0^\infty e^{-iAt/\hbar+izt/\hbar}dt$$

converges if $Im(z) > 0$. Taking $z = E + i\epsilon$ (E real, $\epsilon > 0$ small), and letting $\epsilon \to 0$, we define

$$(A-E-i0)^{-1}(y,x) = \frac{i}{\hbar}\int_0^\infty e^{-iAt/\hbar}(y,x)e^{iEt/\hbar}dt.$$

Note that the $-i0$ prescription is essential only for $E \in \sigma_{ess}(A)$, while for $E \in \mathbb{R}\backslash\sigma_{ess}(A)$, it gives the same result as the $+i0$ prescription. Here we are interested in the second case, and so we drop the $-i0$ from the notation.

The above formula, together with our semi-classical expression (16.6) for the propagator $e^{-iHt/\hbar}$, yields in the leading order as $\hbar \to 0$

$$(H-E)^{-1}(y,x) = \frac{i}{\hbar}\int_0^\infty \sum_{\bar\phi \text{ cp } S} K_{\bar\phi}e^{i(S(\bar\phi)+Et)/\hbar}dt$$

where the sum is taken over the critical paths, $\bar\phi$, from x to y in time t, and

$$K_{\bar\phi} := \left(\frac{m}{2\pi it\hbar}\right)^{d/2}\left(\frac{\det(-m\partial_s^2)}{\det S''(\bar\phi)}\right)^{1/2}.$$

We would like to use the stationary phase approximation again, but this time in the variable t. Denote by $\bar t = \bar t(x,y,E)$ the critical points of the phase $S(\bar\phi) + Et$. They satisfy the equation

$$\partial S(\bar\phi)/\partial t = -E.$$

The path $\omega_E := \bar\phi|_{t=\bar t}$ is a classical path at energy E (see Lemma 16.13 of Section 16.5). Introduce the notation $S_0(x,y,t) := S(\bar\phi)$ for a classical path going from x to y in time t. Then the stationary phase formula gives (in the leading order as $\hbar \to 0$)

$$(H-E)^{-1}(y,x) = \frac{i}{\hbar}\sum_{\omega_E} D_{\omega_E}^{1/2}e^{iW_{\omega_E}/\hbar}, \tag{16.7}$$

where the sum is taken over classical paths going from x to y at energy E. Here we have used the notation $D_{\omega_E}^{1/2} := K_{\omega_E}(2\pi i\hbar)^{1/2}(\partial^2 S_0/\partial t^2)^{-1/2}|_{t=\bar t}$ and we have defined $W_{\omega_E}(x,y,E) := (S_0(x,y,t)+Et)|_{t=\bar t}$ (so W_{ω_E} is the Legendre transform of S_0 in the variable t).

Lemma 16.3

$$D_{\omega_E} = \left(\frac{1}{2\pi i\hbar}\frac{\partial^2 S_0}{\partial t^2}\right)^{d-1}\det\left[\frac{\partial^2 W_{\omega_E}}{\partial x\partial y}\frac{\partial^2 W_{\omega_E}}{\partial E^2} - \frac{\partial^2 W_{\omega_E}}{\partial x\partial E}\frac{\partial^2 W_{\omega_E}}{\partial y\partial E}\right]. \tag{16.8}$$

Proof. We just sketch the proof. The first step is to establish

$$\frac{\det(-m\partial_s^2)}{\det(S''(\omega))} = \left(-\frac{m}{t}\right)^{-d} \det\left(\frac{\partial^2 S_0}{\partial x \partial y}\right) \tag{16.9}$$

(we drop the subscript E for ease of notation). To see this, we use the fact that if for an operator A we denote by J_A the $d \times d$ matrix solving $AJ = 0$ (the Jacobi equation) with $J(0) = 0$ and $\dot{J}(0) = 1$, then (see (25.54))

$$\frac{\det(-m\partial_s^2)}{\det(S''(\omega))} = \frac{\det J_{-m\partial_s^2}(t)}{\det J_{S''(\omega)}(t)}.$$

Next, we use

$$-\frac{1}{m} J_{S''(\omega)}(t) = \left(\frac{\partial^2 S_0}{\partial x \partial y}\right)^{-1}$$

(Equation (16.13) of Section 16.5), and for the free classical path $\phi_0 = x + (y-x)(s/t)$,

$$-\frac{1}{m} J_{-m\partial_s^2}(t) = \left(\frac{\partial^2 S(\phi_0)}{\partial x \partial y}\right)^{-1} = -\left(\frac{m}{t}\right)^{-1} 1$$

to arrive at (16.9).

We can then show that

$$\det\left(\frac{\partial^2 S_0}{\partial x \partial y}\right) \left(\frac{-\partial^2 S_0}{\partial t^2}\right)^{-d}\Bigg|_{t=\bar{t}}$$

equals the determinant on the right hand side in equation (16.8) (see the appendix to this section for details). $\square$

We will show in Section 16.5 (see Lemma 16.14) that the function $W_{\omega E}$ (the action at energy E) satisfies the Hamilton-Jacobi equation

$$h(x, -\frac{\partial W_{\omega E}}{\partial x}) = E.$$

Differentiating this equation with respect to y gives

$$\frac{\partial h}{\partial k} \frac{\partial^2 W_{\omega E}}{\partial x \partial y} = 0,$$

and we see that the matrix $(\partial^2 W_{\omega E}/\partial x \partial y)$ has a zero-eigenvalue. Thus its determinant is zero. So if $d = 1$, (16.8) yields

$$D_{\omega E} = -\frac{\partial^2 W_{\omega E}}{\partial x \partial E} \frac{\partial^2 W_{\omega E}}{\partial y \partial E} \qquad (d = 1). \tag{16.10}$$

Formula (16.7), together with (16.8) or (16.10), is our desired semi-classical expression for Green's function $(H - E)^{-1}(y, x)$.

16.2.1 Appendix

Proposition 16.4 At $t = \bar{t}$,

$$\det \left(\frac{\partial^2 S_0}{\partial x \partial y} \right) = \left(\frac{-\partial^2 S_0}{\partial t^2} \right)^d \det \left[\frac{\partial^2 W_{\omega E}}{\partial x \partial y} \frac{\partial^2 W_{\omega E}}{\partial E^2} - \frac{\partial^2 W_{\omega E}}{\partial x \partial E} \frac{\partial^2 W_{\omega E}}{\partial y \partial E} \right].$$

Proof. We drop the subscripts from S_0 and $W_{\omega E}$ to simplify the notation. Differentiating $W = S + Et|_{t=\bar{t}}$ with respect to x, we obtain

$$\frac{\partial W}{\partial x} = \frac{\partial S}{\partial x} + \frac{\partial S}{\partial t} \frac{\partial t}{\partial x} + E \frac{\partial t}{\partial x},$$

which due to the relation $\partial S / \partial t = -E$ gives

$$\frac{\partial W}{\partial x} = \frac{\partial S}{\partial x}.$$

Similarly,

$$\frac{\partial W}{\partial y} = \frac{\partial S}{\partial y} \quad \text{and} \quad \frac{\partial W}{\partial E} = t.$$

This last equation, together with $\partial S / \partial t = -E$ yields

$$\frac{\partial^2 W}{\partial E^2} = \frac{\partial t}{\partial E} = - \left(\frac{\partial^2 S}{\partial t^2} \right)^{-1}.$$

Furthermore,

$$\frac{\partial^2 S}{\partial x \partial y} = \frac{\partial^2 W}{\partial x \partial y} + \frac{\partial^2 W}{\partial x \partial E} \frac{\partial E}{\partial t} \frac{\partial t}{\partial y}$$

$$= \frac{\partial^2 S}{\partial t^2} \left[- \frac{\partial^2 W}{\partial x \partial y} \frac{\partial^2 W}{\partial E^2} + \frac{\partial^2 W}{\partial x \partial E} \frac{\partial^2 W}{\partial E \partial y} \right]$$

and the result follows. $\square$

16.3 Bohr-Sommerfeld Semi-classical Quantization

In this section we derive a semi-classical expression for the eigenvalues (energy levels) of the Schrödinger operator $H = -\frac{\hbar^2}{2m} \Delta + V$. We use the Green's function expansion (16.7) from the last chapter. For simplicity, we will assume $d = 1$.

Application of the expression (16.7) requires a study of the classical paths at fixed energy. Consider the trajectories from x to y at energy E. We can write them (using informal notation) as

$$\phi_n = \phi_{xy} \pm n\alpha$$

where α is a periodic trajectory (from y to y) of minimal period, at the energy E, while ϕ_{xy} is one of the four "primitive" paths from x to y at energy E, sketched in Fig. 16.1.

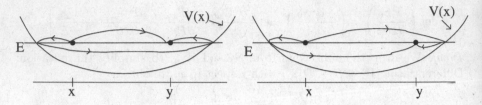

Fig. 16.1. Primitive paths at energy E.

All these paths are treated in the same way, so we consider only one, say the shortest one. The space time picture of ϕ_n in this case is sketched in Fig. 16.2.

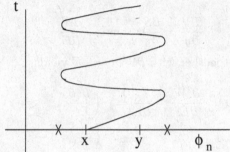

Fig. 16.2. Turning points of ϕ_n.

For this path we compute

$$W_{\phi_n} = W_\phi + nI$$

where $\phi = \phi_{xy}$ and

$$I = \int_0^t L(\alpha, \dot{\alpha})ds + Et.$$

But α is a critical path so its energy is conserved (see Lemma 4.7)

$$\frac{m}{2}\dot{\alpha}^2 + V(\alpha) = E$$

and so

$$I = \int_0^t \{m\dot{\alpha}^2 - E\} + Et = \int_\alpha k \cdot dx,$$

where $k(s) = m\dot{\alpha}(s)$ and $dx = \dot{\alpha}(s)ds$.

Let ω be a classical path at energy E, and let us determine D_ω. We will show later (see (16.18) and Lemma 16.9) that

$$\frac{\partial W_\omega}{\partial x} = \frac{\partial S_0}{\partial x}|_{t=\bar{t}} = -k(0)|_{t=\bar{t}} = \mp\sqrt{2m(E - V(x))}$$

and

$$\frac{\partial W_\omega}{\partial y} = \frac{\partial S_0}{\partial y}|_{t=\bar{t}} = k(t)|_{t=\bar{t}} = \pm\sqrt{2m(E - V(y))}.$$

Differentiating these relations with respect to E and using (16.10), we obtain

$$D_\omega = -\frac{m^2}{k(x)k(y)}.$$

At a turning point x_0, $k(x_0) = 0$ and k changes sign (we think about $k(x)$ as a multi-valued function, or a function on the Riemann surface of $\sqrt{z}$, so at a turning point $\sqrt{k(x)}$ crosses to a different sheet of the Riemann surface).

Because k changes sign at each of the two turning points of the periodic trajectory, we conclude that

$$D_{\phi_n}^{1/2} = D_\phi^{1/2}(-1)^n.$$

So our semi-classical expression (16.7) for Green's function $G_E(y, x)$ is

$$G_E(y, x) = \sum_{n=0}^{\infty} N \exp[i(W_\phi/\hbar + n\left[\frac{1}{\hbar}\int_\alpha k \cdot dx - \pi\right])]$$

$$= Ne^{iW_\phi/\hbar}\frac{1}{1 - e^{i(\int_\alpha k \cdot dx/\hbar - \pi)}}.$$

(N is a constant). We conclude that as $\hbar \to 0$, $G_E(y, x)$ has poles (and hence H has eigenvalues) when

$$\boxed{\int_\alpha k \cdot dx = 2\pi\hbar(j + 1/2)}$$

for an integer j. This is the *Bohr-Sommerfeld semi-classical quantization condition* (for $d = 1$). It is an expression for the quantum energy levels (the energy E appears in the left hand side through the periodic path α at energy E), which uses purely classical data!

Problem 16.5 Show that for the harmonic oscillator potential, the Bohr-Sommerfeld condition gives all of the energy levels exactly.

16.4 Semi-classical Asymptotics for the Ground State Energy

Here we derive a semi-classical expression for the ground state energy (the lowest eigenvalue) of the Schrödinger operator $H = -\frac{\hbar^2}{2m}\Delta + V$ when $V(x) \to$

∞ as $|x| \to \infty$ (as fast as some power of $|x|$, say); i.e., $V(x)$ is a confining potential.

We first define a couple of quantities which are familiar from statistical mechanics (see Chapter 18 for details and discussions).

Definition 16.6 The *partition function*, $Z(\beta)$, at inverse temperature $\beta > 0$ is

$$Z(\beta) := \text{tr } e^{-\beta H}$$

(the trace is well-defined since $\sigma(H) = \{E_n\}_0^\infty$ with $E_n \to \infty$ sufficiently fast).

Definition 16.7 The *free energy*, F, is

$$F(\beta) := -\frac{1}{\beta} \ln Z(\beta).$$

The free energy is a useful quantity for us here because of the following connection with the ground state energy of the Schrödinger operator H:

$$\lim_{\beta \to \infty} F(\beta) = E_0.$$

This is the Feynman-Kac theorem of Section 18.3.

Our goal is to find the semi-classical asymptotics for E_0 by deriving an asymptotic expression for $Z(\beta)$ from a path integral.

As we have seen (Problem 15.1), the path integral expression for $Z(\beta)$ is

$$Z(\tau/\hbar) = \int_{\phi \text{ a path of period } \tau} e^{-S_e(\phi)/\hbar} D\phi \qquad (16.11)$$

where $S_e(\phi) = \int_0^\beta \{\frac{m}{2}|\dot{\phi}|^2 + V(\phi)\}$ (note that this is not the usual action - the potential enters with the opposite sign).

Remark 16.8 The path integral appearing in (16.11) can be rigorously defined. We refer the reader to [272, 135, 245] for details.

Mimicking the procedure we used for the Schrödinger propagator (i.e., the stationary phase method, which in the present context is called the *Laplace method*), we see that the semi-classical expression for $Z(\tau/\hbar)$ is

$$Z(\tau/\hbar) \approx \sum_{\text{minimal paths } \omega} N B_\omega^{1/2} e^{-S_e(\omega)/\hbar} \qquad (16.12)$$

(N a constant) where by a minimal path, we mean a path minimizing S_e, and where
$$B_\omega := \frac{\det S_0''(\omega_0)}{\det S_e''(\omega)}$$

with $S_0(\phi) = \int_0^\tau (m/2)|\dot{\phi}|^2$. A critical path for S_e is a classical path for the inverted potential $-V$. We specialize to $d = 1$ for simplicity, and we assume V

has only one minimum, at x_0. Then the minimal path is $\omega(s) \equiv x_0$ (a constant path), and

$$S_e''(\omega) = -m\partial_s^2 + V''(x_0).$$

Because x_0 minimizes V, $V''(x_0) > 0$ and we write $V''(x_0) = m\omega^2$. Then using the method (25.54) of computing ratios of determinants (see Section 25.12 below), we easily obtain

$$B_\omega = \frac{2\omega\tau}{e^{\omega\tau} - e^{-\omega\tau}}.$$

Also, $S_e(\omega) = \tau V(x_0)$. In this way, we arrive at the leading-order expression

$$F(\tau/\hbar) \approx V(x_0) + \hbar\omega/2 + O(1/\tau)$$

as $\hbar \to 0$. Letting $\beta = \tau/\hbar \to \infty$ and using the Feynman-Kac formula, we obtain

$$\boxed{E_0 \approx V(x_0) + \frac{1}{2}\hbar\omega}$$

which is the desired asymptotic (as $\hbar \to 0$) expression for the ground state energy. It is equal to the classical ground state energy, $V(x_0)$, plus the ground state of the harmonic oscillator with frequency $\sqrt{V''(x_0)/m}$. This suggests that the low energy excitation spectrum of a particle in the potential $V(x)$ is the low energy spectrum of this harmonic oscillator.

16.5 Mathematical Supplement: The Action of the Critical Path

In this section we consider the situation of Example 26.1 no. 6, and its special case, Example 26.1 no. 5. Thus we set

$$X = \{\phi \in C^1([0,t]; \mathbb{R}^m) \mid \phi(0) = x, \phi(t) = y\},$$

and

$$S(\phi) = \int_0^t L(\phi(s), \dot\phi(s))ds.$$

Suppose $\bar\phi$ is a critical path for S with $\bar\phi(0) = x$ and $\bar\phi(t) = y$. We will denote the action of $\bar\phi$ by $S_0(x, y, t) := S(\bar\phi)$ (the action from x to y in time t).

Recall that the *momentum* at time s, is $k(s) := (\partial L/\partial\dot\phi)(\bar\phi(s), \dot{\bar\phi}(s))$.

Lemma 16.9 We have

$$\frac{\partial S_0}{\partial x} = -k(0) \qquad \text{and} \qquad \frac{\partial S_0}{\partial y} = k(t).$$

Proof. Again, we specialize to $L = m|\dot{\phi}|^2/2 - V(\phi)$. Using the chain rule and integration by parts, we find

$$\partial S(\bar{\phi})/\partial x = \int_0^t \{m\dot{\bar{\phi}} \cdot \partial\dot{\bar{\phi}}/\partial x - \nabla V(\bar{\phi}) \cdot \partial\bar{\phi}/\partial x\}ds$$

$$= \int_0^t \{(-m\ddot{\bar{\phi}} - \nabla V(\bar{\phi})) \cdot \partial\bar{\phi}/\partial x\} + m\dot{\bar{\phi}} \cdot \partial\bar{\phi}/\partial x|_0^t$$

which, since $\bar{\phi}$ is a critical point, $(\partial\bar{\phi}/\partial x)(t) = 0$, and $(\partial\bar{\phi}/\partial x)(0) = 1$, is just $-m\dot{\bar{\phi}}(0) = -k(0)$, as claimed. The corresponding statement for $\partial S_0/\partial y$ is proved in the same way, using $(\partial\bar{\phi}/\partial y)(0) = 0$, and $(\partial\bar{\phi}/\partial y)(t) = 1$. $\sqcup$

This lemma implies $\partial k(0)/\partial y = -\partial^2 S_0(x,y,t)/\partial x\partial y$. On the other hand, for $L = \frac{m}{2}|\dot{\phi}|^2 - V(\phi)$, $\partial k(0)/\partial y = (\partial y/\partial k(0))^{-1} = mJ^{-1}(t)$ (as $m(\partial y/\partial k(0))$ is the derivative of the classical path $\bar{\phi}$ at t with respect to the initial velocity $k(0)/m = \dot{\bar{\phi}}(0)$). This gives

$$\frac{\partial^2 S_0(x,y,t)}{\partial x\partial y} = -mJ^{-1}(t) \tag{16.13}$$

which establishes the following result:

Proposition 16.10 If y is a conjugate point to x then $\det(\frac{\partial^2 S_0(x,y,t)}{\partial x\partial y}) = \infty$.

The following exercise illustrates this result for the example of the classical harmonic oscillator.

Problem 16.11 Consider the one-dimensional harmonic oscillator, whose Lagrangian is $L = \frac{m}{2}\dot{\phi}^2 - \frac{m\omega^2}{2}\phi^2$. Compute

$$S_0(x,y,t) = \frac{\omega}{2\sin(\omega t)}[(x^2 + y^2)\cos(\omega t) - 2xy]$$

and so compute

$$\partial^2 S_0(x,y,t)/\partial x\partial y = -\frac{\omega}{\sin\omega t}.$$

Note that this is infinite for $t = n\pi/\omega$ for all integers n. Thus the points $\phi(n\pi/\omega)$ are conjugate to $\phi(0)$.

Lemma 16.12 (Hamilton-Jacobi equation) The action $S_0(x,y,t)$ satisfies the Hamilton-Jacobi equation

$$\partial S_0/\partial t = -h(y, \partial S_0/\partial y) \tag{16.14}$$

where h is the classical Hamiltonian function associated with L.

Proof. The integrands below depend on s (as well as the parameters x,y, and t), and $\dot{\bar{\phi}}$ denotes $\partial\bar{\phi}/\partial s$. Since $S_0 = S(\bar{\phi}) = \int_0^t L(\bar{\phi}, \dot{\bar{\phi}})ds$, we have

$$\partial S(\bar{\phi})/\partial t = L(\bar{\phi}, \dot{\bar{\phi}})|_{s=t} + \int_0^t (\partial L/\partial \bar{\phi} \cdot \partial \bar{\phi}/\partial t + \partial L/\partial \dot{\bar{\phi}} \cdot \partial \dot{\bar{\phi}}/\partial t) ds$$

$$= L(\bar{\phi}, \dot{\bar{\phi}})|_{s=t} + \partial L/\partial \dot{\bar{\phi}} \cdot \partial \bar{\phi}/\partial t|_{s=0}^{s=t}$$

$$+ \int_0^t (\partial L/\partial \bar{\phi} - d/ds(\partial L/\partial \dot{\bar{\phi}})) \cdot \partial \bar{\phi}/\partial t.$$

Since $\bar{\phi}(s) = y + \dot{\bar{\phi}}(t)(s-t) + O((s-t)^2)$ (here we used $\bar{\phi}(t) = y$), we have

$$\frac{\partial \bar{\phi}}{\partial t}|_{s=t} = -\dot{\bar{\phi}}(t).$$

Using this, the fact that $\bar{\phi}$ is a critical point of S, and $\partial \bar{\phi}/\partial t|_{s=0} = 0$, we find

$$\partial S(\bar{\phi})/\partial t = -((\partial L/\partial \dot{\bar{\phi}}) \cdot \dot{\bar{\phi}} - L(\bar{\phi}, \dot{\bar{\phi}}))|_{s=t}$$

$$= -h(\bar{\phi}, \partial L/\partial \dot{\bar{\phi}})|_{s=t}.$$

Since $\bar{\phi}|_{s=t} = y$ and, by Lemma 16.9, $(\partial L/\partial \dot{\bar{\phi}})|_{s=t} = \partial S_0/\partial y$, the result follows. □

We want to pass from a time-dependent to a time-independent picture of classical motion. We perform a Legendre transform on the function $S_0(x, y, t)$ to obtain the function $W_{\bar{\phi}}(x, y, E)$ via

$$W_{\bar{\phi}}(x, y, E) = (S_0(x, y, t) + Et)|_{t:\partial S_0/\partial t = -E}. \qquad (16.15)$$

We denote by $\bar{t} = \bar{t}(x, y, E)$, solutions of

$$\partial S_0/\partial t|_{t=\bar{t}} = -E. \qquad (16.16)$$

There may be many such solutions, so in the notation $W_{\bar{\phi}}$, we record the classical path $\bar{\phi}$ we are concerned with (for which $\bar{\phi}(0) = x$, $\bar{\phi}(\bar{t}) = y$).

Note that by the definition of the Hamilton function, $\frac{\partial L}{\partial \dot{\phi}} \cdot \dot{\phi} - L = h(\phi, \frac{\partial L}{\partial \dot{\phi}})$, and from the energy conservation law (see Lemma 4.7 of Section 4.7) $\left(\frac{\partial L}{\partial \dot{\phi}} \cdot \dot{\phi} - L\right)|_{\phi=\bar{\phi}}$ is constant, and therefore

$$\left(\frac{\partial L}{\partial \dot{\phi}} \cdot \dot{\phi} - L\right)|_{\phi=\bar{\phi}} = E.$$

By (16.15), $W_{\bar{\phi}}(x, y, E) = \int_0^{\bar{t}} (L(\bar{\phi}(s), \dot{\bar{\phi}}(s)) + E) ds$ and therefore we have

$$W_{\bar{\phi}}(x, y, E) = \int_0^t \frac{\partial L}{\partial \dot{\bar{\phi}}} \cdot \dot{\bar{\phi}} ds = \int_{\bar{\phi}} k \cdot dx,$$

where $\int_{\bar{\phi}} k \cdot dx := \int_0^t k(s) \cdot \dot{\bar{\phi}}(s) ds$ and, recall, $k(s) = \partial L/\partial \dot{\phi}|_{\phi=\bar{\phi}}$.

Lemma 16.13 $\bar{\phi}|_{t=\bar{t}}$ is a classical path at energy E.

Proof. By the Hamilton-Jacobi equation (16.14) and the conservation of energy (Lemma 4.7), $\bar{\phi}|_{t=\bar{t}}$ is a classical path with energy $-\partial S_0(x, y, t)/\partial t|_{t=\bar{t}}$, which, by (16.16), is just E. $\square$

Lemma 16.14 $W_{\bar{\phi}}$ satisfies the Hamilton-Jacobi equation

$$h(x, -\partial W_{\bar{\phi}}/\partial x) = E. \tag{16.17}$$

Proof. Using (16.16) and Lemma 16.9, we compute

$$\frac{\partial W_{\bar{\phi}}}{\partial x} = \frac{\partial S_0}{\partial x}|_{t=\bar{t}} + (\frac{\partial S_0}{\partial t} + E)|_{t=\bar{t}}\frac{\partial \bar{t}}{\partial x} = \frac{\partial S_0}{\partial x}|_{t=\bar{t}} = -k(0). \tag{16.18}$$

So by conservation of energy, $E = h(x, k(0)) = h(x, -\partial W_{\bar{\phi}}/\partial x)$. $\square$

16.6 Appendix: Connection to Geodesics

The next theorem gives a geometric reinterpretation of classical motion.

We consider a classical particle in $\mathbb{R}^d$ with a potential $V(x)$. Recall the notation $f(x)_+ := \max(f(x), 0)$.

Theorem 16.15 (Jacobi theorem) The classical trajectory of a particle at an energy E is a geodesic in the Riemannian metric

$$\langle u, v \rangle_x = 2(E - V(x))_+ u \cdot v$$

(where $u \cdot v$ is the inner product in $\mathbb{R}^n$) on the set $\{x \in \mathbb{R}^n | V(x) \leq E\}$ (the classically allowed region).

Proof. By the conservation of energy (Lemma 4.7), a classical path $\bar{\phi}$ has a fixed energy $E = m|\dot{\bar{\phi}}|^2/2 + V(\bar{\phi})$. Hence $\bar{\phi}$ is a critical point of the action $S(\phi) = \int(m|\dot{\phi}|^2/2 - V(\phi))ds$ among paths in $M := \{\phi \mid m|\dot{\phi}|^2/2 + V(\phi) \equiv E\}$. Using the relation $m|\dot{\phi}|^2/2 + V(\phi) = E$, we can write $m|\dot{\phi}|^2/2 - V(\phi)$ as

$$m|\dot{\phi}|^2/2 - V(\phi) = m|\dot{\phi}|^2 = m\sqrt{2(E - V(\phi))/m}|\dot{\phi}|.$$

Hence $\bar{\phi}$ is a critical point of the functional

$$L(\phi) := \int m|\dot{\phi}|\sqrt{2(E - V(\phi))/m}ds$$

on M. This functional gives the length of the path in the metric above.

On the other hand, we can re-parameterize any path with $V(\phi) < E$ so that it satisfies $m|\dot{\phi}|^2/2 + V(\phi) = E$. Indeed, replacing $\phi(s)$ with $\phi(\lambda(s))$, we note $|\dot{\phi}(\lambda)|^2 = (\dot{\lambda}(s))^2|\dot{\phi}(\lambda(s))|^2$. We must solve

$$(\dot{\lambda}(s))^2 = \frac{2}{m}\frac{E - V(\phi(\lambda(s)))}{|\dot{\phi}(\lambda(s))|^2}$$

which we can re-write as

$$\int_0^s \frac{|\dot{\phi}(\lambda(s'))|\dot{\lambda}(s')}{\sqrt{E - V(\phi(\lambda(s')))}}ds' = \sqrt{\frac{2}{m}}s$$

which can be solved. Since the functional $L(\phi)$ is invariant under reparame-terizations (if $\lambda = \alpha(s)$, $\alpha' > 0$, then $|\dot{\phi}(s)|ds = |\dot{\phi}(\lambda)|\frac{\partial\lambda}{\partial s}\frac{\partial s}{\partial\lambda}d\lambda = |\dot{\phi}(\lambda)|d\lambda$), if $\tilde{\phi}$ is a critical point of $L(\phi)$, then so are different reparameterizations of $\tilde{\phi}$, and in particular the one, $\bar{\phi}$, with the energy E. This $\bar{\phi}$ is also a critical point of $L(\phi)$ on M, and, by the above, a critical point of $S(\phi)$ at energy E. Thus classical paths are geodesics up to re-parameterization (so they coincide as curves). □

Problem 16.16 Check that the Euler-Lagrange equation for critical points of $L(\phi)$ on M yields Newton's equation.

Resonances

The notion of a resonance is a key notion in quantum physics. It refers to a metastable state – i.e., to a state which behaves like a stationary (bound) state for a long time interval, but which eventually breaks up. In other words, the resonances are states of the essential spectrum (i.e. scattering states), which for a long time behave as if they were bound states. In fact, the notion of a bound state is an idealization: most of the states which are (taken to be) bound states in certain models, turn out to be resonance states in a more realistic description of the system.

In this chapter, we sketch briefly the mathematical theory of resonance states and apply it to the analysis of the important physical phenomenon of tunneling, on which we illustrate some of the mathematics and physics involved. In Chapter 23, we apply the resonance theory to the problem of radiation.

17.1 Complex Deformation and Resonances

In this section we introduce the powerful tool of complex deformations, which allows for an efficient way to define resonances. We begin with a definition.

Definition 17.1 A family of operators, $H(\theta)$, for θ in a complex disk $\{|\theta| < \epsilon\}$, will be called a *complex deformation* of H if $H(0) = H$, $H(\theta)$ is analytic in $\{|\theta| < \epsilon\}$, and $H(\theta)$ is an analytic continuation of the family

$$H(\theta) = U(\theta)^{-1} H U(\theta) \tag{17.1}$$

for $\theta \in \mathbb{R}$, where $U(\theta)$, $\theta \in \mathbb{R}$, is a one-parameter unitary group, leaving the domain of H invariant.

(We say that the family $H(\theta)$ of unbounded operators is *analytic* iff all $H(\theta)$ have the same domain, D, for every $\psi \in D$, the family of vectors $H(\theta)\psi$ is analytic and the family $(H(\theta) - z)^{-1}$ of bounded operators is analytic, as long as the spectrum of $H(\theta)$ stays away from z.)

© Springer-Verlag GmbH Germany, part of Springer Nature 2020
S. J. Gustafson and I. M. Sigal, *Mathematical Concepts of Quantum Mechanics*, Universitext, https://doi.org/10.1007/978-3-030-59562-3_17

For Schrödinger operators, $H := -\frac{\hbar^2}{2m}\Delta + V(x)$, acting on $L^2(\mathbb{R}^d)$, the choice of the family $H(\theta)$ depends on analytic properties of the potentials $V(x)$. The simplest and most important choice is provided by the notion of dilatation analyticity. Let $U(\theta)$ be the one-parameter *group of dilatations*:

$$U(\theta) : \psi(x) \mapsto e^{d\theta/2}\psi(e^\theta x)$$

for $\psi \in L^2(\mathbb{R}^d)$. This is a unitary implementation of the rescaling $x \mapsto e^\theta x$ for $\theta \in \mathbb{R}$ and is the key example of a one-parameter unitary group $U(\theta)$ used for complex deformation. We compute

$$U(\theta)^{-1}HU(\theta) = e^{-2\theta}(-\frac{\hbar^2}{2m}\Delta) + V(e^\theta x). \qquad (17.2)$$

Assume the family $\theta \to V(e^\theta x)$ has an analytic continuation into a complex disk $\{|\theta| < \epsilon\}$ as operators from the Sobolev space $H^2(\mathbb{R}^d)$ to $L^2(\mathbb{R}^d)$ (the corresponding potentials are called *dilatation analytic*). Then the family on the r.h.s. of (17.2) is analytic in $\{|\theta| < \epsilon\}$ and therefore is a complex deformation, $H(\theta)$, of H.

As an example of the above procedure we consider the complex deformation of the hydrogen atom Hamiltonian $H_{\text{hydr}} := -\frac{1}{2m}\Delta - \frac{\alpha}{|x|}$:

$$H_{\text{hydr}\theta} = e^{-2\theta}(-\frac{\hbar^2}{2m}\Delta) - e^{-\theta}\frac{\alpha}{|x|}.$$

Let e_j^{hydr} be the eigenvalues of the hydrogen atom. Then the spectrum of this deformation is

$$\sigma(H_{\text{hydr}\theta}) = \{e_j^{\text{hydr}}\} \cup e^{-2\operatorname{Im}\theta}[0, \infty).$$

In general, one can **show that**:

1) The real eigenvalues of H_θ, $\operatorname{Im}\theta > 0$, coincide with eigenvalues of H and complex eigenvalues of H_θ, $\operatorname{Im}\theta > 0$, lie in the complex half-plane $\mathbb{C}^-$;

2) The complex eigenvalues of H_θ, $\operatorname{Im}\theta > 0$, are locally independent of θ. The typical spectrum of H_θ, $\operatorname{Im}\theta > 0$, is shown in Fig. 17.1

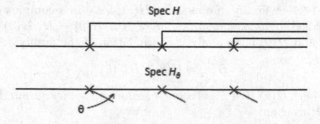

Fig. 17.1. Typical spectrum of H_θ.

We call complex eigenvalues of H_θ, with $\operatorname{Im}\theta > 0$, the *resonances* of H.

Often resonances arise as a result of perturbation of eigenvalues embedded into the essential (continuous) spectrum; that is, when the operator H is of the form $H = H_0 + \kappa W$, where H_0 is a self-adjoint operator with an eigenvalue λ_0 embedded in its essential spectrum, W is a symmetric operator and κ is a small real parameter (coupling constant). This happens, for example, for an atom in a constant electric field (the *Stark effect*).

Another example is the problem of time-periodic perturbations, we considered in Section 11.3. There the isolated eigenvalues E_m of the unperturbed operator H_0 lead to the eigenvalues, E_{mn}, of the unperturbed Bloch-Floquet Hamiltonian K_0 embedded into the essential spectrum of K_0. The computations of Section 11.3 suggest that they turn into resonance eigenvalues, $E_{\kappa mn}$, of the perturbed operator, K_κ. (In Section 11.3 we considered only the case of $n = 0$.) This can be proved rigorously by using the complex deformation theory above, which allows also to derive the expansions (11.24) and (11.27). We explain this briefly. Apply the complex deformation with the dilatation group to the Bloch-Floquet Hamiltonian K_κ to obtain, as in (17.2),

$$U(\theta)^{-1} K_\kappa U(\theta) := H_{\kappa\theta}(t) - i\hbar \frac{\partial}{\partial t} \quad \text{on} \quad L^2(\mathbb{R}^n \times S_T).$$

Here, for $\theta \in \mathbb{R}$, $H_{\kappa\theta} := H_{0\theta} + \kappa W_\theta(t)$, $H_{0\theta} = -e^{-2\theta} \frac{\hbar^2}{2m} \Delta + V(e^\theta x)$, $W_\theta(t) := W(e^\theta x, t)$. If the potentials $V(x)$ and $W(x,t)$ are dilatation analytic in the sense of the definition above, then the family on the r.h.s. which we denote $K_{\kappa\theta}$ can be continued analytically into a neighbourhood $\{|\theta| < \epsilon\}$. Again, the spectrum of $K_{0\theta}$ can be easily computed:

$$\sigma_{ess}(K_{0\theta}) = \bigcup_{n=-\infty}^{\infty} \hbar\omega n + e^{-2\operatorname{Im}\theta}[0, \infty)$$

plus the collection of the eigenvalues $E_{mn} = E_m + \hbar\omega n$, $n \in \mathbb{Z}$, $m = 0, 1, \ldots$, where, recall, $\omega = 2\pi/T$, T is the time-period of W, and E_m are the eigenvalues of H_0. Now, barring an accidental degeneracy (i.e. $E_{mn} = \hbar\omega n'$ for some m, n, n'), the eigenvalues $E_{mn} = E_m + \hbar\omega n$ are *isolated* and have finite multiplicity. The application of the Feshbach-Schur map becomes a standard affair and gives, for κ sufficiently small, the eigenvalues, $E_{\kappa mn}$ of $K_{\kappa\theta}$ emerging from E_{mn}. (Here we assumed for simplicity that the eigenvalues E_m are of multiplicity one.) These eigenvalues are, in general, complex, and, as we saw above, are independent of θ and in general have negative imaginary parts, $\operatorname{Im} E_{\kappa,j}$. They can be computed by the perturbation expansion (11.16) - (11.17) to give (11.24) and (11.27).

Resonances as poles. We know from Section 25.9 of the mathematical supplement that the resolvent $(H - z)^{-1}$ of the Hamiltonian H is analytic away from its spectrum. One can show that H has an isolated eigenvalue at a point z_0 iff matrix elements of the resolvent $(H - z)^{-1}$ have a pole at z_0. Similar to eigenvalues, we would like to characterize the resonances in terms of poles of matrix elements of the resolvent $(H - z)^{-1}$. To this end we have to go beyond the spectral analysis of H. Let $\Psi_\theta = U_\theta \Psi$, etc., for $\theta \in \mathbb{R}$ and $z \in \mathbb{C}^+$. Use the unitarity of U_θ for real θ, to obtain

$$\langle \Psi, (H - z)^{-1} \Phi \rangle = \langle \Psi_{\bar\theta}, (H_\theta - z)^{-1} \Phi_\theta \rangle. \tag{17.3}$$

Assume now that for a dense set of Ψ's and Φ's (say, $\mathcal{D}$, defined below), Ψ_θ and Φ_θ have analytic continuations into a complex neighbourhood of $\theta = 0$, and continue the r.h.s of (17.3) analytically, first in θ into the upper half-plane, and then in z across the continuous spectrum (the Combes argument). This meromorphic continuation has the following properties:

- The real eigenvalues of H_θ give real poles of the r.h.s. of (17.3) and therefore they are the eigenvalues of H.
- The complex eigenvalues of H_θ are poles of the meromorphic continuation of the l.h.s. of (17.3) across the spectrum of H onto the second Riemann sheet.

The complex poles manifest themselves physically as bumps in the scattering cross-section or poles in the scattering matrix.

An example of the dense set $\mathcal{D}$ mentioned above is given by

$$\mathcal{D} := \bigcup_{a>0} \mathrm{Ran}\big(\chi_{|T| \le a}\big). \tag{17.4}$$

Here T is the self-adjoint generator of the one-parameter group U_θ, $\theta \in \mathbb{R}$. (It is not hard to show that it is dense: $\forall \psi \in \mathcal{H}$, $\chi_{|T| \le a} \psi \to \psi$, as $a \to \infty$.)

Resonance states as metastable states. While bound states are stationary solutions of the Schrödinger equation, one expects that resonance eigenvalues lead to almost stationary, long-living solutions. This is proven, so far, only for resonances arising from a perturbation of bound states with eigenvalues embedded into the essential (continuous) spectrum. In this case, for initial condition ψ_0 localized in a small energy interval around the unperturbed eigenvalue, λ_*, or a small perturbation of the corresponding eigenfunction ϕ_*, one shows that the solutions, $\psi = e^{-iHt/\hbar}\psi_0$, of the time-dependent Schrödinger equation, $i\hbar \partial_t \psi = H\psi$, are of the form

$$\psi = e^{-iz_* t/\hbar} \phi_* + O_{\mathrm{loc}}(t^{-\alpha}) + O_{\mathrm{res}}(\kappa^\beta), \tag{17.5}$$

for some $z_* \in \mathbb{C}^-$, $z_* = \lambda_* + O(\kappa)$, and $\alpha > 0$ (depending on ψ_0). These are metastable states with resonance eigenvalue $z_* \in \mathbb{C}^-$. Here κ is the perturbation parameter (the coupling constant) and the error term $O_{\mathrm{loc}}(t^{-\alpha})$ satisfies, for some $\nu > 0$,

$$\|(1 + |x|)^{-\nu} O_{\mathrm{loc}}(t^{-\alpha})\| \le C t^{-\alpha}.$$

Eqn (17.5) implies that the negative of the imaginary part of the resonance eigenvalue, $-\mathrm{Im} z_*$, called the resonance width, gives the decay probability per unit time, and $(-\mathrm{Im} z_*)^{-1}$, can be interpreted as the life-time of the resonance.

17.2 Tunneling and Resonances

Consider a particle in a potential $V(x)$, of the form shown in Fig. 17.2; i.e., $V(x)$ has a local minimum at some point x_0, and $V(x_0) > \limsup_{x\to\infty} V(x)$, for x in some cone, say.

If $V(x) \to -\infty$ as $x \to \infty$ (in some cone of directions), then the corresponding Schrödinger operator, H, is not bounded from below.

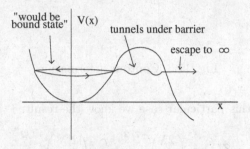

Fig. 17.2. Unstable potential.

If the barrier is very thick, then a particle initially located in the well spends lots of time there, and behaves as if it were a bound state. However, it eventually tunnels through the barrier (*quantum tunneling*) and escapes to infinity. Thus the state of the particle is a scattering state. It is intuitively reasonable that

1. the energy of the resonance $\approx$ the energy of a bound state in the well
2. the resonance lifetime is determined by the barrier thickness and height, and $\hbar$.

Since the resonances are very close to bound states if the barrier is large or $\hbar$ is small (there is no tunneling in classical mechanics), we try to mimic our semi-classical treatment of the ground state (Section 16.4). But right away we run into a problem: if $V(x) \not\to \infty$ as $x \to \infty$ in some directions, then

$$Z(\beta) = \mathrm{Tr} e^{-\beta H} = \infty.$$

The paradigm for this problem is the divergence of the integral

$$Z(\lambda) = \int_0^\infty e^{-\lambda a^2/2} da$$

for $\lambda \le 0$. However, we can define this integral by an analytic continuation. $Z(\lambda)$ is well-defined for $Re(\lambda) > 0$, and it can be continued analytically into $\lambda \in \mathbb{R}^-$ as follows. Move λ from $Re(\lambda) > 0$ into $Re(\lambda) \le 0$, at the same time deforming the contour of integration in such a way that $Re(\lambda a^2) > 0$ (see Fig. 17.3).

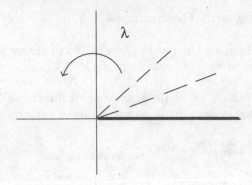

Fig. 17.3. Contour deformation.

Of course, in this particular case we know the result:

$$Z(\lambda) = \left(\frac{2\lambda}{\pi}\right)^{-1/2} = -i\left(\frac{2|\lambda|}{\pi}\right)^{-1/2}$$

for $\lambda < 0$ (which is purely imaginary!).

There is a powerful method of rotating the contour which is applicable much beyond the simple integral we consider. It goes as follows. For $\theta \in \mathbb{R}$, we change variables via $a = e^{-\theta}b$. This gives

$$Z(\lambda) = e^{-\theta} \int_0^{\infty} e^{-\lambda e^{-2\theta}b^2/2}db. \tag{17.6}$$

The integral here is convergent and analytic in θ as long as

$$\text{Re}(\lambda e^{-2\theta}) > 0. \tag{17.7}$$

We continue it analytically in θ and λ, preserving this condition. In particular, for $\lambda \in \mathbb{R}^-$, we should have $\pi/4 < Im(\theta) < 3\pi/4$.

Now observe that the right hand side of (17.6) is independent of θ. Indeed, it is analytic in θ as long as (17.7) holds, and is independent of $Re(\theta)$ since the latter can be changed without changing the integral, by changing the variable of integration ($b \mapsto e^{-\theta'}b$, $\theta' \in \mathbb{R}$). Thus we have constructed an analytic continuation of $Z(\lambda)$ with $\text{Re}(\lambda) > 0$ into a region with $\text{Re}(\lambda) < 0$. In fact, we have continued this function onto the second Riemann sheet!

Finally, we define $Z(\lambda)$, for $\lambda < 0$, by (17.6) with θ obeying (17.7).

17.3 The Free Resonance Energy

With a bit of wisdom gained, we return to the problem of defining the partition function $Z(\beta)$ and free energy $F(\beta)$ (see Definitions 16.6) in the case when

$V(x) \not\to \infty$, as $x \to \infty$, in some directions (or more precisely, $\sup_{x \in \Gamma} V(x) < \infty$ for some cone Γ).

Assume that we can construct a complex deformation, $H(\theta)$, of H, such that

$$Z(\beta) = \mathrm{Tre}^{-\beta H(\theta)} < \infty \qquad (17.8)$$

for $\mathrm{Im}(\theta) > 0$, or more generally for $|\theta| \leq \epsilon$, $\mathrm{Im}(\theta) > 0$. (Let, for example, $V(x) = -Cx^3$ for $x \geq 0$ and $= 0$ for $x \leq 0$. Then $V(e^\theta x) = -Ce^{3\theta}x^3$ for $x \geq 0$ and $= 0$ for $x \leq 0$. Take $\theta = -i\pi/3$. Then $V(e^\theta x) = Cx^3$ is positive for $x \geq 0$. In fact, it is not a simple matter to define the exponential $e^{-\beta H(\theta)}$ rigorously – see [269]. Below we will deal formally with $e^{-\beta H(\theta)}$, assuming it has all the properties which can be derived from the power series expression for the exponential.)

Proposition 17.2 If $\mathrm{Tre}^{-\beta H(\theta)} < \infty$ for $\theta \in \Omega \subset \{|\theta| < \epsilon\}$, then $\mathrm{Tre}^{-\beta H(\theta)}$ is independent of θ.

Proof. $e^{-\beta H(\theta)}$ is analytic in $\{|\theta| \leq \epsilon\}$, and satisfies

$$e^{-\beta H(\theta+s)} = U(s)^{-1}e^{-\beta H(\theta)}U(s)$$

for $s \in \mathbb{R}$. This last relation can be derived using the expression $H(\theta + s) = U(s)^{-1}H(\theta)U(s)$ (which follows from (17.1)) and a power series expansion of the exponential (or Equation (25.43)). By cyclicity of the trace ($\mathrm{Tr}(AB) = \mathrm{Tr}(BA)$),

$$\mathrm{Tre}^{-\beta H(\theta+s)} = \mathrm{Tre}^{-\beta H(\theta)}.$$

Hence $\mathrm{Tre}^{-\beta H(\theta)}$ is independent of $\mathrm{Re}(\theta)$, and so is independent of θ. $\square$

If there is a complex deformation, $H(\theta)$, of H, such that (17.8) holds, we call $Z(\beta) = \mathrm{tre}^{-\beta H(\theta)}$ an *adiabatic partition function* for H, and $F(\beta) = -(1/\beta)\ln Z(\beta)$ the *resonance free energy* for H. We interpret

$$E(\beta) := \mathrm{Re}\, F(\beta)$$

as the *resonance energy* at the temperature $1/\beta$,

$$\Gamma(\beta) := -\,\mathrm{Im}\, F(\beta)$$

as the *resonance decay probability per unit time* (or *resonance width*) at the temperature $1/\beta$, and

$$T(\beta) := \frac{1}{\Gamma(\beta)}$$

as the *resonance lifetime* at the temperature $1/\beta$. The *resonance eigenvalue* for zero temperature is given by

$$z_r = E_r - i\Gamma_r := \lim_{\beta \to \infty} F(\beta).$$

Usually, $|\operatorname{Im} Z(\beta)| \ll |\operatorname{Re} Z(\beta)|$. Hence,

$$E(\beta) = \operatorname{Re} F(\beta) \approx -\frac{1}{\beta} \ln(\operatorname{Re} Z(\beta))$$

and

$$\Gamma(\beta) = -\operatorname{Im} F(\beta) = \frac{1}{\beta} \operatorname{Im} \ln(1 + i\frac{\operatorname{Im} Z(\beta)}{\operatorname{Re} Z(\beta)}) \approx \frac{1}{\beta}\frac{\operatorname{Im} Z(\beta)}{\operatorname{Re} Z(\beta)}.$$

In fact, one can show that for $\frac{1}{\Delta E} \ll t \ll \Gamma^{-1}$, where ΔE is the average gap between eigenvalues of $H(\theta)$ ($\frac{1}{\Delta E}$ gives a time scale for H),

$$e^{-iHt/\hbar}\psi_0 = e^{-iz_r t/\hbar}\psi_0 + \text{small}$$

if ψ_0 lies near E_r in the spectral decomposition of H (see [269, 125]). Note that

$$e^{-iz_r t/\hbar} = e^{-\Gamma_r t/\hbar}e^{-iE_r t/\hbar}$$

exhibits exponential decay at the (slow) rate Γ_r. This is consistent with our picture of a resonance as a metastable state.

Remark 17.3 The example given after Definition 17.1 does not lead to a unique self-adjoint Schrödinger operator $H = -\frac{\hbar^2}{2m}\Delta + V(x)$. Presumably, $F(\beta)$ is independent of the self-adjoint extension chosen. For a large class of self-adjoint Schrödinger operators, Condition (17.8) does not hold, and the trace has to be regularized (see [269]). In such a case, the potential can be modified at infinity in such a way that for the modified potential, Condition (17.8) holds. We expect that such a modification can be chosen so that it leads to a sufficiently small error in the tunneling probabilities. In any case, the results we discuss below (which are obtained by applying another non-rigorous technique – path integrals) coincide with those given by more involved rigorous analysis, wherever the latter is possible.

17.4 Instantons

To compute $Z(\beta)$ for the potential sketched in Fig. 17.2, we proceed as in the ground state problem; we represent $Z(\beta)$ formally as a path integral, and then derive the formal semi-classical expansion (see (16.12)):

$$Z(\tau/\hbar) = \sum_\omega NB_\omega^{1/2}e^{-S_e(\omega)/\hbar}$$

(as usual, we ignore the factor $(1 + O(\sqrt{\hbar}))$). The sum is taken over critical points ω of the "action" $S_e(\phi)$, of period τ. N is a normalization factor independent of ω and H, and

$$B_\omega = \frac{\det S_0''(\omega_0)}{\det S_e''(\omega)}$$

where the operators $S_e''(\omega)$ and $S_0''(\omega_0)$ are defined on $L^2([0,\tau])$ with zero boundary conditions. Now ω is a periodic classical path in imaginary time (or in the inverted potential $-V(x)$), with period τ, which we take large (see Fig. 17.4).

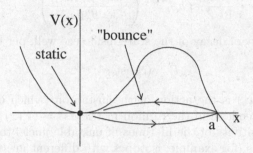

Fig. 17.4. Paths in inverted potential.

Two periodic solutions of arbitrarily large period are

$$\omega_s(s) \equiv 0$$

(the subscript "s" for "static") and

$$\omega_b(s) : 0 \mapsto a \mapsto 0$$

("b" for "bounce"). The solution ω_b is called an *instanton* or "bounce". Since ω_s is a minimum of V, $V''(\omega_s) > 0$, and so

$$S_e''(\omega_s) = -\partial_s^2 + \Omega^2$$

where $\Omega^2 = V''(0)$. We computed earlier

$$B_{\omega_s} = \frac{\Omega\tau}{\sinh(\Omega\tau)} \approx \frac{2\Omega\tau}{e^{\Omega\tau}}$$

for τ large. Moreover, $S_e(\omega_s) = 0$. We will show later (Section 17.6) that

$$B_{\omega_b}^{1/2} = -i\tau S_b^{-1/2} \left(\frac{|\det^\perp S_e''(\omega_b)|}{\det S_0''(\omega_0)} \right)^{-1/2} \tag{17.9}$$

where

$$S_b := S_e(\omega_b) = \int_{\omega_b} k \cdot dx$$

is the action of the "bounce", and

$$\det{}^\perp A := \det(A|_{(\text{Null } A)^\perp}) \tag{17.10}$$

is the determinant of A restricted to the orthogonal complement of the null space, Null$A$, of A. Collecting these results, we have (for large τ)

$$E \approx -\frac{\hbar}{\tau} \ln(\operatorname{Re} Z) \approx \frac{\hbar\Omega}{2}$$

and

$$\Gamma \approx -\frac{\hbar}{\tau}\frac{\operatorname{Im} Z}{\operatorname{Re} Z} \approx \hbar S_b^{-1/2}\left(\frac{|\det^{\perp} S_e''(\omega_b)|}{\det S_e''(\omega_s)}\right)^{-1/2} e^{-S_b/\hbar}.$$

So the probability of decay of the state inside the well, per unit time, is

$$\Gamma = (\text{const})e^{-S_b/\hbar}$$

where $S_b = S_e(\omega_b)$ is the action of the instanton (which equals the length of the minimal geodesic in the Agmon metric $ds^2 = (V(x) - E)_+ dx^2$). This explains the sensitivity of the lifetimes of unstable nuclei to small variations of the parameters (for example, isotopes with different masses can have very different lifetimes).

Finally, we note that

$$\det S_e''(\omega_s) \approx \frac{e^{\Omega\tau}}{2\Omega}.$$

17.5 Positive Temperatures

Here we consider quantum tunneling at positive temperatures ($T = \beta^{-1} > 0$). We use the same approach as above, but let the parameter β be any positive number. Now we have to consider all three critical paths of period $\tau = \hbar\beta$ (see Fig. 17.5): $\omega_1 = \omega_s \equiv x_{min}$, ω_2, and $\omega_3 \equiv x_{max}$, where ω_2 is a classical periodic trajectory in the potential $-V(x)$ of period $\tau = \hbar\beta$.

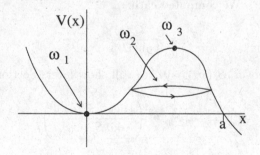

Fig. 17.5. Paths of period τ.

Since $V''(x_{min}) > 0$, ω_1 is a minimal trajectory. As we will see, ω_2 is a saddle point of Morse index 1 (see Section 26.4). Finally, $V''(x_{max}) < 0$, and so ω_3 is also a saddle point. For $k = 1, 3$,

$$S_e(\omega_k) = V(\omega_k)\tau$$

and for $k = 2, 3$,
$$S_e(\omega_1) < S_e(\omega_k).$$
The semi-classical expression for the decay probability works out to be

$$\Gamma = -\frac{1}{\tau B_{\omega_1}}(\operatorname{Im} B_{\omega_2} e^{-S(\omega_2)/\hbar} + \operatorname{Im} B_{\omega_3} e^{-S(\omega_3)/\hbar}).$$

Through which trajectory, ω_2 or ω_3, does the tunneling take place? ω_3 corresponds to a thermally driven escape (due to thermal fluctuations), and ω_2 corresponds to a quantum tunneling escape. If τ is very small (large temperature), the transition occurs through ω_3, as only ω_3 can have arbitrarily small period. On the other hand, if τ is very large (small temperature), ω_2 sits close to the bottom of the well, and one can show that $S_e(\omega_2) < S_e(\omega_3)$. In this case, the transition occurs through ω_2.

There is a critical value of τ, $\tau_c \approx 2\pi/\Omega_{max}$ where $\Omega_{max}^2 = -V''(x_{max})$, at which a transition occurs; the transition is between the situations in which decay is due to tunneling, and in which it is due to thermal fluctuations. (Note that for $\tau < \tau_c$, the decay rate differs from Γ by the factor $\frac{\Omega_{max}\tau}{2\pi}$ (see [8])). This transition can take place either continuously or discontinuously, depending on whether the energy of the periodic classical trajectory in the inverted potential $-V(x)$ depends on its period continuously or discontinuously. In the first case, as temperature decreases below $1/\tau_c$ (i.e. τ increases above τ_c) the tunneling trajectory bifurcates from ω_3 and slips down the barrier (see Fig. 17.6). For $\tau < \tau_c$ tunneling takes place through ω_3.

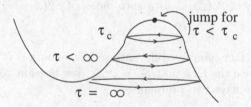

Fig. 17.6. Continuous transition.

In the second case (see Fig. 17.7), there are no closed trajectories with period $> \tau_c$, so the transition is discontinuous: decay jumps from ω_3 to a trajectory at the bottom of the barrier.

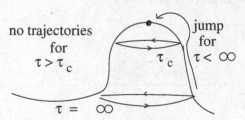

Fig. 17.7. Discontinuous transition.

Thus for intermediate temperatures, the nature of decay depends radically on the geometry of the barrier.

The results above support the following physical picture of the tunneling process. With the Boltzmann probability $(\text{const})e^{-E/T}$, the particle is at an energy level E. The probability of tunneling from an energy level E is $(\text{const})e^{-S_E/\hbar}$ where S_E is the action of the minimal path at energy E. The probability of this process is $(\text{const})e^{-E/T-S_E/\hbar}$. Thus the total probability of tunneling is

$$(\text{const})\int e^{-E/T-S_E/\hbar} \approx (\text{const})e^{\,E_0/T-S_{E_0}/\hbar}$$

where E_0 solves the stationary point equation

$$\frac{\partial}{\partial E}\left(\frac{E}{T}+\frac{S_E}{\hbar}\right) = \frac{1}{T}+\frac{1}{\hbar}\frac{\partial S_E}{\partial E} = 0.$$

Here $-\partial S_E/\partial E$ is the period of the trajectory under the barrier at the energy level $-E$, and $S_{E_0} + \hbar E_0/T$ is the action of a particle (in imaginary time) at energy E_0 corresponding to the period $\hbar/T = \tau$.

17.6 Pre-exponential Factor for the Bounce

The bounce solution, ω_b, presents some subtleties. Since ω_b breaks the translational symmetry of $S_e(\phi)$, $\dot{\omega}_b$ is a zero-mode of $S_e''(\omega_b)$:

$$S_e''(\omega_b)\dot{\omega}_b = 0.$$

To establish this fact, simply differentiate the equation $\partial S_e(\omega_b) = 0$ with respect to s and use the fact that $S_e'' = \partial^2 S_e$ (see Section 26.3).

As a result, we have two problems:

1. $S_e''(\omega_b)$ has a zero eigenvalue, so formally

$$[\det S_e''(\omega_b)]^{-1/2} = (\text{const})\int e^{-\langle \xi, S_e''(\omega_b)\xi\rangle/2\hbar} D\xi = \infty \qquad (17.11)$$

2. $\dot{\omega}_b$ has one zero (see Fig. 17.8), and so the Sturm-Liouville theory (from the study of ordinary differential equations) tells us that, in fact, $S_e''(\omega_b)$ has exactly one negative eigenvalue.

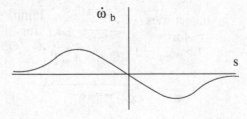

Fig. 17.8.

This gives a second reason for the integral (17.11) to diverge.

To illustrate these divergences, we change variables. Let $\{\xi_k\}$ be an orthonormal basis of eigenfunctions of $S_e''(\omega_b)$ with eigenvalues λ_k, in increasing order. For ϕ near ω_b, write $\phi = \omega_b + \xi$ with

$$\xi = \sum_{k=0}^{\infty} a_k \xi_k.$$

Then

$$S_e(\phi) \approx S_e(\omega_b) + \sum_{0}^{\infty} \lambda_k a_k^2.$$

But $\lambda_0 < 0$ and $\lambda_1 = 0$, hence we have two divergent integrals:

$$\int_{-\infty}^{\infty} e^{-\lambda_j a_j^2/2\hbar} da_j = \infty$$

for $j = 0, 1$. We already know that we can define the first integral by an analytic continuation to be

$$\int_{-\infty}^{\infty} e^{-\lambda_0 a_0^2/2\hbar} da_0 = \left(\frac{2\lambda_0}{\pi\hbar}\right)^{-1/2} = -i \left|\frac{2\lambda_0}{\pi\hbar}\right|^{-1/2}.$$

The second integral, correctly treated, is shown to contribute (see the following section)

$$S_b^{-1/2} \tau \sqrt{2\pi\hbar} \tag{17.12}$$

where S_b is the action of the "bounce", $S_e(\omega_b)$. Hence

$$\int_{\text{near } \omega_b} e^{-S_e(\phi)/\hbar} \approx B_{\omega_b} e^{-S_b/\hbar}$$

where

$$B_{\omega_b}^{1/2} = -i\tau S_b^{-1/2} \left(\frac{|\det^\perp S_e''(\omega_b)|}{\det S_0''(\omega_0)}\right)^{-1/2}$$

(this is (17.9)) and $\det^\perp$ is defined in (17.10).

17.7 Contribution of the Zero-mode

The virial theorem of classical mechanics gives

$$S_b = S_e(\omega_b) = \int \dot{\omega}_b{}^2.$$

Define the normalized zero eigenfunction

$$\xi_1 = S_b^{-1/2}\dot{\omega}_b.$$

Then

$$(\omega_b + c_1\xi_1)(s) = \omega_b(s) + c_1 S_b^{-1/2}\dot{\omega}_b(s) \approx \omega_b(s + c_1 S_b^{-1/2}).$$

Hence

$$\phi \approx \omega_b(s + c_1 S_b^{-1/2}) + \sum_{n \neq 1} c_n\xi_n$$

and therefore

$$\prod_n dc_n = S_b^{-1/2} ds \prod_{n \neq 1} dc_n.$$

Integrating in s from 0 to t gives (17.12).

17.8 Bohr-Sommerfeld Quantization for Resonances

The goal of this section is to derive a semi-classical formula for the resonance eigenvalues of a Schrödinger operator with a tunneling potential. We proceed by analogy with the treatment of a confining potential in Section 16.3 which led to the Bohr-Sommerfeld quantization rule.

As in the rest of this chapter, we consider a tunneling potential of the form sketched in Fig. 17.9.

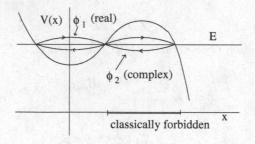

Fig. 17.9. Resonance potential.

The path-integral expression for Green's function of H is, as in Section 16.2,

$$G_H(E, y, x) = \frac{i}{\hbar} \int_0^\infty \int_{P_{x,y,t}} e^{i(S(\phi,t)+Et)/\hbar} D\phi\, dt. \qquad (17.13)$$

We seek critical points (because, as always, we wish to apply the method of stationary phase) which are closed trajectories ($x = y$) at the fixed energy E. The trajectories in phase space are shown in Fig. 17.10.

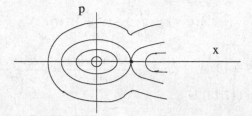

Fig. 17.10. Phase portrait.

At energy E, phase space is partitioned into classically allowed, and classically forbidden regions. Classical trajectories at energy E are shown in Fig. 17.11.

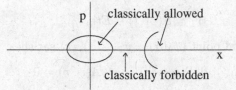

Fig. 17.11. Phase portrait at fixed energy.

If we complexify the phase space

$$\mathbb{R} \times \mathbb{R} \mapsto \mathbb{C} \times \mathbb{C}$$

then the phase space at a fixed energy E becomes connected as shown in Fig. 17.12.

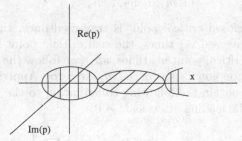

Fig. 17.12. Complexified phase space at fixed energy.

Thus, in addition to real paths, $\phi(s)$, we consider complex paths of the form $\alpha(\sigma) = \psi(-i\sigma)$. Setting $t = -i\tau$, the action for such a path is

$$S(\alpha, -i\tau) = \int_0^\tau (-\frac{m}{2}\dot{\psi}^2 - V(\psi))(-i)d\sigma = iA(\psi, \tau),$$

where

$$A(\psi, \tau) = \int_0^\tau \frac{m}{2}\dot\psi^2 + V(\psi),$$

and so

$$\partial_\alpha S = i\partial_\psi A, \quad \text{and} \quad \frac{\partial S(\alpha, -i\tau)}{\partial \tau} = i\frac{\partial A(\psi, \tau)}{\partial \tau}.$$

Thus the phase in (17.13) is

$$S(\alpha, -i\tau) + E(-i\tau) = i(A(\psi, \tau) - E\tau).$$

Now, the real critical point $(\phi_1(s), t)$ satisfies

$$\partial_\phi S - 0, \quad \frac{\partial S(\phi_1, t)}{\partial t} = -E,$$

so ϕ_1 has period t, and $m\ddot\phi_1 = -\nabla V(\phi_1)$ (as in Fig. 17.9). This has a phase

$$W_1 = S(\phi_1, t) - Et|_{\partial S/\partial t = -E}.$$

The complex critical point $(\phi_2(\sigma) = \psi_2(-i\sigma), i\tau)$ satisfies

$$\partial_\phi S(\alpha, -i\tau) = i\partial_\psi A(\psi, \tau) = 0$$

and

$$\frac{\partial S(\phi_2, -i\tau)}{\partial \tau} = i\frac{\partial A(\psi_2, \tau)}{\partial \tau} = iE,$$

so ψ_2 has period τ, and $m\ddot\psi_2 = \nabla V(\psi_2)$ (as in Fig. 17.9). Hence the phase is

$$iW_2 = i(A(\psi_2, \tau) - E\tau)|_{\frac{\partial A(\psi_2, \tau)}{\partial \tau} = E}$$

We can characterize a general closed critical orbit by the list

$$(1, m_1, 1, m_2, 1, m_3, \ldots),$$

meaning the real closed critical point is traversed once, the complex closed critical point is traversed m_1 times, the real critical point is followed again, then the complex critical point m_2 times, etc. (we follow the real critical point several times in succession if some of the m_i are zero). Applying the stationary phase method, we obtain the following contribution to the path integral (up to a constant, in the leading order as $\hbar \to 0$):

$$\sum_{n=0}^\infty \sum_{m_1 \ldots m_n} e^{i(1+n)W_1/\hbar - (m_1 + \cdots + m_n)W_2/\hbar} = e^{iW_1/\hbar} \sum_{n \ge 0} \left(e^{iW_1/\hbar} \sum_{m=0}^\infty e^{-mW_2/\hbar} \right)^n$$

$$= e^{iW_1/\hbar} \sum_{n=0}^\infty \left(e^{iW_1/\hbar} \frac{1}{1 - e^{-W_2/\hbar}} \right)^n$$

$$= e^{iW_1/\hbar} \frac{1}{1 - e^{iW_1/\hbar} \frac{1}{1 - e^{-W_2/\hbar}}}$$

$$= \frac{e^{iW_1/\hbar}(1 - e^{-W_2/\hbar})}{1 - e^{iW_1/\hbar} - e^{-W_2/\hbar}}.$$

We want to identify values of E for which Green's function has a singularity, with resonance eigenvalues. Writing the lowest resonance eigenvalue as $E_0 - i\Delta E$ and expanding $e^{iW_1(E)/\hbar}$ to first order around E_0, and $e^{-W_2(E)/\hbar}$ to zeroth order, gives the equation

$$e^{iW_1(E_0)/\hbar} = 1, \qquad \text{or} \qquad W_1(E_0) = 2\pi\hbar n, \quad n = 0, \pm 1, \ldots$$

for E_0 (i.e. E_0 is the ground state energy, as before), and the expression

$$\Delta E = \hbar \left(\frac{\partial W_1(E_0)}{\partial E} \right)^{-1} e^{-W_2(E_0)/\hbar}$$

for ΔE. The last two equations represent the Bohr-Sommerfeld quantization for resonances.

Quantum Statistics

In this chapter, we extend quantum mechanics to the situation when only partial information about the system of interest is available. Here the notion of wave function (i.e. a square integrable function of the particle coordinates – an element of $L^2(\mathbb{R}^3)$) is replaced with the notion of *density matrix* (or *density operator*), a positive, trace class operator on the L^2 state space. This topic is closely related to quantum statistical mechanics.

The notions of *trace* and *trace class operators*, which are extensively used in this section, are defined in Mathematical Supplement, Section 25.11.

18.1 Density Matrices

Consider a physical system described by a quantum Hamiltonian H acting on a Hilbert space $\mathcal{H}$ (say, $H = -\frac{\hbar^2}{2m}\Delta + V(x)$ on $L^2(\mathbb{R}^3)$). Let $\{\psi_j\}$ be an orthonormal system (possibly, a basis) in $\mathcal{H}$ and let

$$\psi = \sum a_j \psi_j.$$

Given an arbitrary observable A (say, position or a characteristic function of position), its average in the state ψ is given by

$$\langle A \rangle_\psi := \langle \psi, A\psi \rangle = \sum_{m,n} \bar{a}_m a_n \langle \psi_m, A\psi_n \rangle . \tag{18.1}$$

Now suppose that we know only that, for each n, the system is in the state ψ_n with a probability p_n. We thus have much less information than before. Now the average, $\langle A \rangle$, of an observable A is given by the expression

$$\langle A \rangle = \sum_n p_n \langle \psi_n, A\psi_n \rangle . \tag{18.2}$$

© Springer-Verlag GmbH Germany, part of Springer Nature 2020
S. J. Gustafson and I. M. Sigal, *Mathematical Concepts of Quantum Mechanics*, Universitext, https://doi.org/10.1007/978-3-030-59562-3_18

This corresponds to the situation when the parameters a_n in (18.1) are independent random variables with zero mean and variance $E(|a_n|^2) = p_n$. Observe that (18.2) can be written as

$$\langle A \rangle = \text{Tr}(A\rho), \tag{18.3}$$

where Tr is the trace, $\rho = \sum_n p_n P_{\psi_n}$. Here P_ψ stands for the rank-one orthogonal projection onto the vector ψ, i. e. $P_\psi f = \langle \psi, f \rangle \psi$, or $P_\psi = |\psi\rangle\langle\psi|$ in Dirac's notation (see Sections 25.7 and 25.11 for the definition and discussion of projections and trace).

Problem 18.1 Show that

$$\text{Tr}(AP_\psi) = \langle \psi, A\psi \rangle. \tag{18.4}$$

Note that ρ is a trace class, positive (since $p_n \geq 0$ and $P_{\psi_n} \geq 0$) operator, with trace 1: $\text{Tr}\rho = \sum_n p_n = 1$. We extrapolate from this the assumption that generalized states are given by positive, trace class operators ρ on $\mathcal{H}$, normalized so that $\text{Tr}\rho = 1$. Such operators are called *density matrices* or *density operators*.

If vectors ψ_n evolve according to the Schrödinger equation,

$$i\hbar \frac{\partial \psi}{\partial t} = H\psi, \tag{18.5}$$

then the equation governing the state given by the expression $\rho = \sum p_n P_{\psi_n}$ is:

$$i \frac{\partial \rho}{\partial t} = \frac{1}{\hbar}[H, \rho]. \tag{18.6}$$

One takes this equation to be the *basic dynamical equation* of Quantum Statistical Mechanics, or Quantum Statistics. We call it the *von Neumann-Landau equation*, or the *quantum Liouville equation*.

Problem 18.2 Derive equation (18.6) for $\rho = \sum p_n P_{\psi_n}$ with ψ_n as above.

Problem 18.3 Show that a solution of the equation (18.6) with the initial condition $\rho|_{t=0} =: \rho_0$ is given by

$$\rho = e^{-\frac{iHt}{\hbar}} \rho_0 e^{\frac{iHt}{\hbar}}.$$

Hint: Use the result of Problem 2.21.

Problem 18.4 Show that if the initial condition $\rho|_{t=0}$ is a positive operator, then so is the solution of the equation (18.6), and $\text{Tr}\rho = \text{Tr}\rho|_{t=0}$. (Hint: use the result of the previous problem.)

Problem 18.5 Show that if ψ_i are bound states of H (i.e. $H\psi_i = \lambda_i\psi_i$), then $\rho = \sum_i p_i P_{\psi_i}$, for any $p_i \geq 0$, $\sum p_i = 1$, is a static solutions of the equation $i\frac{\partial\rho}{\partial t} = \frac{1}{\hbar}[H, \rho]$.

An important example of density matrices are rank-one orthogonal projections $\rho = P_\psi$. Indeed, we have

$$P_\psi \geq 0 \quad \text{and} \quad \operatorname{Tr} P_\psi = \|\psi\|^2 = 1$$

Problem 18.6 Show these relations.

There is one-to-one correspondence between one-dimensional subspaces and rank-one orthogonal projections: $\{e^{i\alpha}\psi\} \rightarrow P_\psi$ and any rank-one orthogonal projection P can be written as P_ψ for any normalized function $\psi \in \operatorname{Ran} P$. Wave functions, ψ, or rank-one projections, P_ψ, are called *pure states*. Density operators, ρ, such $\rho \neq P_\psi$ for any ψ, are called *mixed states*. Thus $p_1 P_{\psi_1} + p_2 P_{\psi_2}$ is a mixed state.

To summarize: if only partial information about a quantum system is available – namely, we know only that the system occupies certain states with certain probabilities – we can describe states of such a system by positive trace-class operators $\rho \geq 0$ (normalized by $\operatorname{Tr}\rho = 1$), called density operators, or density matrices, with the equation of motion given by (18.6), and averages of observables computed according to the prescription (18.3).

Denote the spaces of bounded observables and of trace class operators on $\mathcal{H}$ as $L^\infty(\mathcal{H})$ and $L^1(\mathcal{H})$, respectively. There is a duality between density matrices and observables

$$\langle \rho, A \rangle = \operatorname{Tr}(A\rho) \tag{18.7}$$

for $A \in L^\infty(\mathcal{H})$ and $\rho \in L^1(\mathcal{H})$.

Consider the evolution of density matrices, $i\frac{\partial \rho}{\partial t} = \frac{1}{\hbar}[H, \rho]$ (the von Neumann equation) and of observables, $i\frac{\partial A}{\partial t} = -\frac{1}{\hbar}[H, A]$ (the Heisenberg equation). Denote by α_t and α_t^* the corresponding flows, $\alpha_t : \rho \rightarrow \rho_t$ and $\alpha_t^* : A \rightarrow A_t$. We have shown above that they have the following explicit representations

$$\alpha_t(\rho) := e^{-\frac{iHt}{\hbar}} \rho\, e^{\frac{iHt}{\hbar}} \quad \text{and} \quad \alpha_t^*(A) = e^{\frac{iHt}{\hbar}} A\, e^{-\frac{iHt}{\hbar}}.$$

In the sense of the duality (18.7), the evolution of density matrices, α_t, and of observables, α_t^*, are dual:

$$\langle \rho, \alpha_t^*(A) \rangle \;=\; \langle \alpha_t(\rho), A \rangle,$$

where the coupling is defined by the trace $\langle \rho, A \rangle = \operatorname{Tr}(A\rho)$.

18.2 Quantum Statistics: General Framework

We formalize the theory above by making the following postulates:

- States: positive trace-class operators on $\mathcal{H}$ (as usual, up to normalization);
- Evolution equation : $i\hbar\frac{\partial\rho}{\partial t} = [H,\rho]$;
- Observables : self-adjoint operators on $\mathcal{H}$;
- Averages : $\langle A\rangle_\rho := \mathrm{Tr}(A\rho)$.

We call the theory described above *Quantum Statistics*. The last two items lead to the following expressions for the probability densities for the coordinates and momenta:

- $\rho(x; x)$ - probability density for coordinate x;
- $\hat{\rho}(k; k)$ - probability density for momentum p.

Above, $\hat{\rho}(k; k')$ is the integral kernel of the operator $\hat{\rho} := \mathcal{F}\rho\mathcal{F}^{-1}$, i.e.

$$\hat{\rho}(k; k') = (2\pi\hbar)^{-3} \int\int e^{-\frac{ik\cdot x}{\hbar}} e^{\frac{ik'\cdot x'}{\hbar}} \rho(x; x')\, dx\, dx'. \qquad (18.8)$$

Problem 18.7 Show that the integral kernel of the operator $\hat{\rho} := \mathcal{F}\rho\mathcal{F}^{-1}$ is given by (18.8).

In particular, if $\rho = P_\psi$, then

$$\rho(x; x) = |\psi(x)|^2$$
$$\hat{\rho}(k; k) = |\hat{\psi}(k)|^2$$

as should be the case according to our interpretation.

Note that the state space here is not a linear space but a positive cone in a linear space. It can be identified with the space of all positive (normalized) linear functionals $A \to \omega(A) := \mathrm{Tr}(A\rho)$ on the space of bounded observables.

Quantum mechanics is a special case of quantum statistics, and is obtained by restricting the density operators to be rank-one orthogonal projections:

$$\{e^{i\alpha}\psi,\ \forall\alpha\in\mathbb{R}\} \quad\Leftrightarrow\quad \rho = P_\psi$$
$$\langle\psi, A\psi\rangle \quad=\quad \mathrm{Tr}(AP_\psi)$$
$$i\hbar\frac{\partial}{\partial t}(e^{\frac{i\mu t}{\hbar}}\psi) = H(e^{\frac{i\mu t}{\hbar}}\psi),\ \text{for some } \mu\in\mathbb{R}, \quad\Leftrightarrow\quad i\frac{\partial\rho}{\partial t} = \frac{1}{\hbar}[H,\rho]\ \text{for } \rho = P_\psi\,.$$

Problem 18.8 *Show these properties.*

We show the $\Leftarrow$ direction in the last statement. Using Dirac's notation we have that $i\frac{\partial\rho}{\partial t} = \frac{1}{\hbar}[H,\rho]$ and $\rho = P_\psi$ imply $|\chi\rangle\langle\psi| - |\psi\rangle\langle\chi| = 0$ where $\chi = i\hbar\frac{\partial\psi}{\partial t} - H\psi$, which yields that $\chi = \mu\psi$ for some real μ. This implies that the family of vectors $\tilde{\psi} := e^{\frac{i\mu t}{\hbar}}\psi$ satisfies $i\hbar\frac{\partial\tilde{\psi}}{\partial t} = H\tilde{\psi}$. $\square$

Thus quantum statistics applied to the rank-one projections is equivalent to quantum mechanics.

Quantum Statistics and Probability Theory. We have shown that quantum mechanics is a special case of quantum statistics (with $\rho \to P_\psi$). Now we show that another special case of quantum statistics is probability. Recall that the average of A in a state ρ is $\langle A \rangle_\rho = \text{Tr}(A\rho)$. We introduce the following interpretation:

- A *quantum random variable* is an observable A.
- A *quantum event* is an orthogonal projection operator P ($\leftrightarrow$ subspace $\text{Ran}P$).
- The probability of event P in state ρ is $\text{Prob}_\rho(P) = \text{Tr}(P\rho)$.

Consider observables and orthogonal projections which are multiplication operators by measurable functions. For a projection P this means that it is the multiplication operator by a characteristic function, χ_Q, of a measurable set $Q \subseteq \mathbb{R}^3$ (i.e. $\chi_Q(x) = \begin{cases} 1 & x \in Q \\ 0 & x \notin Q \end{cases}$). For an observable, A, which is a multiplication operator by measurable function $\xi : \mathbb{R}^3 \to \mathbb{R}$, the average is

$$\langle A \rangle_\rho = \text{Tr}(A\rho) = \int_{\mathbb{R}^3} \xi(x)\rho(x,x)\,dx,$$

while the probability of the event $P = \chi_Q$ is

$$\text{Prob}_\rho(P) = \text{Tr}(P\rho) = \int_{\mathbb{R}^3} \chi_Q(x)\rho(x,x)\,dx = \int_Q \rho(x,x)\,dx. \qquad (18.9)$$

Thus, if we restrict ourselves to observables and projections, both of which are multiplication operators (or, more generally, are elements of a commutative subalgebra of the algebra of all observables), we obtain a standard probabilistic theory:

- The probability space $(\mathbb{R}^3, \mathbb{P})$ where $d\mathbb{P}(x) = \rho(x,x)dx$;
- Random variables which are measurable functions $\xi : \mathbb{R}^3 \to \mathbb{R}$;
- Events are measurable subsets $Q \subseteq \mathbb{R}^3 \leftrightarrow$ characteristic functions, χ_Q.

18.3 Stationary States

Stationary, i.e. time-independent, solutions of equation (18.6) are given by various functions, $f(H)$ (defined, say, by the formula (2.20)), of the quantum Hamiltonian H. Indeed, such operators, if well-defined, commute with H. However, they represent density matrices (up to normalization) only if they are positive and trace-class. The latter holds if and only if

- the functions $f(\lambda)$ are supported on the discrete spectrum of H,
- if $\sigma(H)$ extends to infinity, the $f(\lambda)$ decay at infinity sufficiently fast.

As an example, consider an operator H whose spectrum consists of isolated eigenvalues (of finite multiplicity) converging to ∞. (For example, H is a Schrödinger operator with the potential $|x|^4$.) Then an operator $f(H)$ is trace class for any function f vanishing at ∞ sufficiently fast (see Section 25.11). Thus, for any such positive function, $f(H)$ is a density matrix (up to normalization).

To summarize: if the operator H has purely discrete spectrum, then Eqn (18.6) has an infinite-dimensional space of time-independent solutions – stationary states – which are density operators. These operators are of the form $f(H)$, where f is a positive function, decaying sufficiently fast at infinity.

However, the states above are not seen in nature if the number of particles is very large. What is seen in this case, are the (thermal) *equilibrium states*. These can be isolated as follows. Assume we have only one conserved quantity – the energy. Then, following the *second law of thermodynamics*, we can characterize the equilibrium states in a finite volume as states ρ which maximize the *von Neumann entropy*,

$$S(\rho) := -\mathrm{Tr}(\rho \ln \rho),$$

given the *internal energy* $E(\rho) := \mathrm{Tr}(H\rho)$:

$$\rho \text{ maximizes } S(\rho), \text{ provided } E(\rho) \text{ is fixed } (E(\rho) = E, \text{ say}). \tag{18.10}$$

The criterion above is called the *principle of maximum entropy*. This principle can be extended in an appropriate form to infinite systems.

In (18.10), ρ varies over the convex set $\mathcal{S} := \{\rho \in \text{ the (Banach) space of trace-class operators, } \rho \geq 0, \mathrm{Tr}\rho = 1\}$. Assuming that the maximum in (18.10) is achieved in the interior of $\mathcal{S}$, one can show, using standard techniques of variational calculus that the maximizer satisfies the Euler-Lagrange equation

$$E'(\rho) - TS'(\rho) - \mu\mathbf{1} = 0, \quad \text{or, explicitly, } H + T(\ln\rho + 1) - \mu = 0, \tag{18.11}$$

where T and μ are Lagrange multipliers corresponding to the constraints $E(\rho) = E$ and $\mathrm{Tr}\rho = 1$ respectively (see the Mathematical Supplement, Chapter 26, on variational calculus). The latter equation can be easily solved to give the following one-parameter family of positive operators

$$\rho_T = e^{-H/T}/Z(T), \text{ where } Z(T) := \mathrm{Tr}\,e^{-H/T}, \tag{18.12}$$

as equilibrium states (for a definition of the operator $e^{-H/T}$, see Section 2.3). These states are called the *Gibbs states* and the Lagrange multiplier T is called the *temperature* (the Lagrange multiplier μ is called the *chemical potential*). The quantity $Z(T) = \mathrm{Tr}\,e^{-H/T}$ (or $Z(\beta) = \mathrm{Tr}\,e^{-\beta H}$ for $\beta = 1/T$) is called the *partition function* (at temperature T, of the system described by the Hamiltonian H).

It is conjectured that in the absence of conserved quantities other than the energy, all equilibrium states of infinite systems of infinite degrees of freedom can be obtained as (weak) limits of Gibbs states.

The Lagrange multiplier theorem of variational calculus (see Section 26.5) implies that an equilibrium state minimizes the *Helmholtz free energy*

$$F_T(\rho) := E(\rho) - TS(\rho) \tag{18.13}$$

where T, or $\beta = T^{-1}$, is the Lagrange multiplier to be found from the relation $\mathrm{Tr}(H\rho_T) = E$.

Problem 18.9 Show this.

By a straightforward computation, the equilibrium free energy, $F(T) := F_T(\rho_T)$, is given by

$$F(T) = -T \ln Z(T) .$$

The next result connects Gibbs states and the free energy, to ground states and the ground state energy. Let ψ_0 be the (unique) ground state of the Hamiltonian H, and E_0 the corresponding ground state energy. Let P_ψ denote the rank-one projection onto the vector ψ. We have

Theorem 18.10 (Feynman-Kac Theorem) As $T \to 0$,

$$\rho_T \to P_{\psi_0} \quad \text{and} \quad F(T) \to E_0 .$$

Proof. Let $E_0 < E_1 \le E_2 \le \cdots$ be the eigenvalues of H (our standing assumption is that H has purely discrete spectrum, running off to ∞), and let $\psi_0, \psi_1, \ldots$ be corresponding orthonormal eigenstates. Then by completeness of the eigenstates, and the spectral mapping theorem (see Section 25.11), $\rho_T = \sum_{n=0}^{\infty} p_n P_{\psi_n}$ where $p_n = e^{-E_n/T}/Z(T)$. We can rewrite p_n as

$$p_n = e^{-(E_n - E_0)/T} / \sum_{n'=0}^{\infty} e^{-(E_{n'} - E_0)/T} .$$

We see that $p_n \le 1$ and as $T \to 0$

$$p_n \to \begin{cases} 1 & n = 0 \\ 0 & n \ge 1 \end{cases}$$

It follows easily that $\|\rho_T - P_{\psi_0}\| \to 0$ as $T \to 0$. Furthermore, since

$$F(T) = -T \ln \left(\sum_{n=0}^{\infty} e^{-E_n/T} \right) = E_0 - T \ln \left(1 + \sum_{n=1}^{\infty} e^{-(E_n - E_0)/T} \right),$$

we see that $F(T) \to E_0$ as $T \to 0$. $\square$

18.4 Hilbert Space Approach

Quantum statistical dynamics can be put into a Hilbert space framework as follows. Consider the space $\mathcal{H}_{HS}$ of Hilbert-Schmidt operators acting on the Hilbert space $\mathcal{H}$. These are the bounded operators, K, such that K^*K is trace-class (see Section 25.11). There is an inner-product on $\mathcal{H}_{HS}$, defined by

$$\langle F, K \rangle := \operatorname{Tr}(F^*K). \tag{18.14}$$

Problem 18.11 Show that (18.14) defines an inner-product.

This inner-product makes $\mathcal{H}_{HS}$ into a Hilbert space (see [45, 244]). On the space $\mathcal{H}_{HS}$, we define an operator L via

$$LK = \frac{1}{\hbar}[H, K],$$

where H is the Schrödinger operator of interest. The operator L is symmetric. Indeed,

$$\hbar\langle F, LK \rangle = \operatorname{Tr}(F^*[H, K]).$$

Using the cyclicity of the trace, the right hand side can be written as

$$\operatorname{Tr}(F^*HK - F^*KH) = \operatorname{Tr}(F^*HK - HF^*K) = \operatorname{Tr}([F^*, H]K)$$
$$= \operatorname{Tr}([H, F]^*K) = \hbar\langle LF, K \rangle$$

and so $\langle F, LK \rangle = \langle LF, K \rangle$ as claimed. In fact, for self-adjoint Schrödinger operators, H, of interest, L is also self-adjoint.

Now consider the Landau-von Neumann equation

$$i\frac{\partial k}{\partial t} = Lk \tag{18.15}$$

where $k = k(t) \in \mathcal{H}_{HS}$. Since $k(t)$ is a family of Hilbert-Schmidt operators, the operators $\rho(t) = k^*(t)k(t)$ are trace-class, positive operators. Because $k(t)$ satisfies (18.15), the operators $\rho(t)$ obey the equation

$$i\frac{\partial \rho}{\partial t} = L\rho = \frac{1}{\hbar}[H, \rho]. \tag{18.16}$$

If ρ is normalized – i.e., $\operatorname{Tr}\rho = 1$ – then ρ is a density matrix satisfying the Landau-von Neumann equation (18.16). The stationary solutions to (18.15) are just eigenvectors of the operator L with eigenvalue zero.

To conclude, we have shown that instead of density matrices, we can consider Hilbert-Schmidt operators, which belong to a Hilbert space, and dynamical equations which are of the same form as for density matrices. Moreover, these equations can be written in the Schrödinger-type form (18.15), with self-adjoint operator L, sometimes called the *Liouville operator*.

18.5 Semi-classical Limit

Unlike the Schrödinger equation and the wave function, the von Neumann equation, $\frac{\partial \rho}{\partial t} = -\frac{i}{\hbar}[H, \rho]$, as well as the density matrix, has a well-defined semi-classical limit, i.e. the limit as $\hbar/(\text{typical classical action}) \to 0$. In this section we explore this limit.

We pass to physical units in which a typical classical action in our system is 1, so that $\hbar$ is now the ratio of the Planck constant to the classical action (cf. Chapter 16). Let the operators $T_{y,k}$ and I be defined as

$$T_{y,k} = e^{-i(kx+yp)/\hbar} \qquad \text{and} \qquad I : \psi(x) \to \psi(-x). \tag{18.17}$$

We introduce the following transformation of density operators $\rho \to W_\rho(y, k)$, where

$$W_\rho(y, k) := (2\pi\hbar)^{-\frac{d}{2}} \operatorname{Tr}(T_{2y,k}I\rho), \tag{18.18}$$

called the Wigner transform. It maps density matrices (quantum statistical states) into functions of the classical phase space which look like classical statistical states.

To formulate properties of the Wigner transform we recall a few definitions, beginning with that of the Fourier transform $\hat{\rho}(k, k')$ of $\rho(x, x')$, as given by (18.8). Let A be the Weyl quantization of the classical observable (symbol) $a(y, k)$ (cf. Section 4.1):

$$A = (2\pi\hbar)^{-d} \int\int \hat{a}(\xi, \eta) e^{i(\xi x + \eta p)/\hbar} d\xi d\eta, \tag{18.19}$$

where

$$a(y, k) = (2\pi\hbar)^{-d} \int\int \hat{a}(\xi, \eta) e^{i(\xi x + \eta k)/\hbar} d\xi d\eta. \tag{18.20}$$

Theorem 18.12 (Properties of Wigner transformation) We have
1) $\rho = \rho^* \implies W_\rho$ is real;
2) $\int dk W_\rho(y, k) = \rho(y, y)$ (probability density in y);
3) $\int dy W_\rho(y, k) = \hat{\rho}(k, k)$ probability distribution in k;
4) Assuming that our system consists of two subsystems labeled as 1 and 2, $\rho_1 = \operatorname{Tr}_2(\rho) \implies W_{\rho_1}(y_1, k_1) = \int dy_2 dk_2 W_\rho(y_1, y_2, k_1, k_2)$.
5) $\operatorname{Tr}(A\rho) = \int\int a W_\rho dy dk$.

Discussion. 2) and 3) imply that $\int\int W_\rho dy dk = \operatorname{Tr}\rho$ and $\operatorname{Prob}_\rho(x \in \Omega) = \int_\Omega \int_{\mathbb{R}^3} W_\rho dy dk$, $\operatorname{Prob}_\rho(k \in \Omega^*) = \int_{\mathbb{R}^3} \int_{\Omega^*} W_\rho dy dk$ where $\operatorname{Prob}_\rho(x \in \Omega) = \int_\Omega \rho(x, x) dx$, and similarly, for $\operatorname{Prob}_\rho(k \in \Omega^*)$. Note that $W_\rho(y, k)$ does not have to be positive and therefore cannot be interpreted as a probability distribution in the phase space. (However, it becomes positive as $\hbar \to 0$.) This interpretation is confirmed by the equation in 5): the r.h.s of this equation is like the classical average of the classical observable a in the "probability distribution" W_ρ. The above property implies that $W_\rho(y, k)$ is an approximate probability distribution in the phase space:

$$\text{Prob}(x \in \Omega, p \in \Omega') \approx \int_{\Omega} \int_{\Omega'} W_{\rho} dy dk. \qquad (18.21)$$

Before proving these statements we find a convenient representation of W_{ρ}. To this end we use the Baker-Campbell-Hausdorff formula:

$$e^{\frac{-ikx}{\hbar}} e^{\frac{-iyp}{\hbar}} = e^{-i\frac{(kx+yp)}{\hbar}} e^{-i\frac{yk}{2\hbar}} \qquad (18.22)$$

(see (4.8)). Using (18.22) in (18.18), we obtain that $W_{\rho}(y, k) = (2\pi\hbar)^{-\frac{d}{2}} e^{\frac{iyk}{\hbar}}$ $\times \text{Tr } B.$ where $B = e^{-\frac{ikx}{\hbar}} e^{-i\frac{2yp}{\hbar}} I\rho$. Compute the integral kernel $B(x, x')$ of B: $B(x, x') = e^{-\frac{ikx}{\hbar}} \rho(-x + 2y, x')$, where $\rho(x, x')$ is the integral kernel of ρ. This gives

$$W_{\rho}(y, k) = (2\pi\hbar)^{-\frac{d}{2}} \int e^{-\frac{ik(x-y)}{\hbar}} \rho(2y - x, x) dx. \qquad (18.23)$$

Changing the variable of integration as $x \to x' = x - y$, we obtain

$$W_{\rho}(y, k) = (2\pi\hbar)^{-\frac{d}{2}} \int e^{-\frac{ikx}{\hbar}} \rho(y - x, y + x) dx. \qquad (18.24)$$

Problem 18.13 Prove 1), 3), 4), 5). Hint for 1): use that $\rho(x, x') = \rho(x', x)$. Hint for 3): use that $x = \frac{1}{2}(x + y) + \frac{1}{2}(x - y)$. Hint for 4): 4) follows from 3) and $y \cdot k = y_1 \cdot k_1 + y_2 \cdot k_2$, etc.

We show 2). We use that for functions f, with integrable Fourier transforms, $(2\pi\hbar)^{-\frac{d}{2}} \int dk \hat{f}(k) = f(0)$, which follows by setting $x = 0$ in $(2\pi\hbar)^{-\frac{d}{2}} \int dk e^{\frac{ixk}{\hbar}}$ $\times \hat{f}(k) = f(x)$. Using this relation we obtain $\int dk W_{\rho}(y, k) = \rho(y - x, x + y)|_{x=0} = \rho(y, y)$, which is 2).

To formulate the main result of this section we recall the definition of the Poisson bracket of classical mechanics:

$$\{a, b\} = \sum_{j=1}^{n} (\partial_{y_j} a \partial_{k_j} b - \partial_{y_j} b \partial_{k_j} a). \qquad (18.25)$$

Theorem 18.14 (Semi-classical limit) If $h(y, k) = \frac{1}{2m} |k|^2 + V(y)$ and $H = -\frac{\hbar^2}{2m} \Delta + V(x)$, then

$$W_{\frac{i}{\hbar}[H, \rho]} = \{h, W_{\rho}\} + O(\hbar \nabla_k^2 W_{\rho})$$

and therefore

$$\frac{\partial \rho}{\partial t} = -\frac{i}{\hbar}[H, \rho] \implies \partial_t W_{\rho} = -\{h, W_{\rho}\} + O(\hbar \nabla_k^2 W_{\rho}).$$

Discussion.

a) This theorem implies that in the semi-classical limit the Landau-von Neumann equation $\frac{\partial \rho}{\partial t} = -\frac{i}{\hbar}[H, \rho]$ becomes the classical Liouville equation of statistical physics,

$$\partial_t w = -\{h, w\}, \tag{18.26}$$

for the semi-classical limit w of the Wigner transform W_ρ of the density operator ρ. Thus, in the limit $\hbar \to 0$, quantum statistics becomes classical statistics.

b) In classical mechanics the equation (18.26) is obtained as follows. Let $w(y, k)$ be the particle density in the classical phase space. Assume now that y and k satisfy the Hamilton equations and we want to see how $w(y, k)$ moves with the flow. Differentiate $w(y, k)$ w.r.to time to obtain

$$\partial_t w(y, k) = \partial_y w \dot{y} + \partial_k w \dot{k} = \partial_y w \partial_k h - \partial_k w \partial_y h, \tag{18.27}$$

which is (18.26).

c) Taking a rank-one, orthogonal projection as the initial condition for the Landau-von Neumann equation, we relate Schrödinger's equation to Newton's. *Proof of Theorem 18.14.* Integrating by parts several times, we obtain

$$
\begin{aligned}
W_{\frac{i}{\hbar}[H_0, \rho]}(y, k) &= \frac{i}{\hbar} \frac{-\hbar^2}{2m} (2\pi\hbar)^{-\frac{d}{2}} \int (\partial_y^2 - \partial_{y'}^2)\rho(y - x, y' + x')|_{y'=y, x'=x} e^{\frac{ixk}{\hbar}} dx \\
&= -\frac{i\hbar}{2m} \frac{1}{\hbar^{\frac{d}{2}}} \int (-\partial_y \partial_x - \partial_{y'} \partial_{x'})\rho(y - x, y' + x')|_{x=x', y=y'} e^{\frac{ixk}{\hbar}} dx \\
&= -\frac{i\hbar}{2m} \frac{1}{\hbar^{\frac{d}{2}}} \int [(\partial_y \partial_{x'} + \partial_{y'} \partial_x)\rho + \frac{ik}{\hbar}(\partial_y + \partial_{y'})\rho]|_{x'=x, y'=y} e^{\frac{ixk}{\hbar}} dx \\
&= -\frac{i\hbar}{2m} \frac{1}{\hbar^{\frac{d}{2}}} \int [(\partial_y \partial_{y'} - \partial_{y'} \partial_y)\rho|_{x'=x, y'=y} + \frac{ik}{\hbar} \partial_y \rho] e^{ixk} \hbar dx \\
&= \frac{k}{m} \partial_y \frac{1}{\hbar^{\frac{d}{2}}} \int \rho e^{ixk} \hbar dx = \frac{k}{m} \nabla_y W_\rho
\end{aligned}
$$

where we used that $(\partial_y + \partial_{y'})\rho|_{x'=x, y'=y} = \partial_y(\rho|_{x'=x, y'=y})$. Furthermore, we have

$$
\begin{aligned}
W_{\frac{i}{\hbar}[V, \rho]}(y, k) &= \frac{i}{\hbar^2} \int (V(y - \frac{1}{2}x) - V(y + \frac{1}{2}x))\rho e^{\frac{ixk}{\hbar}} dx \\
&= \frac{i}{\hbar^2} \int [-\nabla V(y)x + O(x^2)]\rho e^{\frac{ixk}{\hbar}} dx \\
&= \frac{i}{\hbar^2} \int [-\nabla V(y)(-i\hbar\nabla_k) + O((\hbar\nabla)^2)]\rho e^{\frac{ixk}{\hbar}} dx \\
&= (-\nabla V(y)\nabla_k + O(\hbar\nabla_k^2))W_\rho.
\end{aligned}
$$

Recall the definitions of the classical Hamiltonian, $h(y, k) = \frac{1}{2m}|k|^2 + V(y)$, and of the Poisson bracket, (18.25), which give $(\frac{k}{m}\nabla_y - V(y)\nabla_k)W_\rho = \{h, W_\rho\}$. Hence the sum of the last two equations gives the desired equation. $\square$

18.6 Generalized Hartree-Fock and Kohn-Sham Equations

In this section we describe the natural extension of the Hartree-Fock equation (HFE), (14.13), to density operators, i.e. non-negative, trace-class operators, and its natural and powerful modification - the Kohn-Sham equation.

Generalized Hartree-Fock equation. Our starting point is equation (14.13), which we reproduce here

$$i\frac{\partial \gamma}{\partial t} = [h(\gamma), \gamma], \tag{18.28}$$

where $h(\gamma) := h + v * n_\gamma - ex(\gamma)$, with $h := -\Delta + V$, $n_\gamma(x,t) := \gamma(x,x,t)$ and $ex(\gamma) \equiv -v^\sharp \gamma$ where $v^\sharp \gamma$ is the operator with integral kernel $ex(\gamma)(x,y,t) := v(x-y)\gamma(x,y,t)$. Here and in what follows, $A(x,y)$ stands for the integral kernel of an operator A.

Eq. (18.28) was derived for finite-rank, orthogonal projections, which as we learned in this chapter is a special case of density operators. Now, we extend it to general density operators.

Remark on notation: in keeping with the standard notation in this subfield, we use here a different notation, γ, for the density operator, rather than the ρ used in the rest of this chapter and in the next one.[1] Moreover, in this section, we denote the particle density (which is a non-negative function of x and t) by $n(x,t)$, rather than $\rho(x,t)$, as we did in Section 14.1.

We may consider this equation for fermions and bosons. For instance, (18.28) with $ex(\gamma) = 0$ generalizes also the Hartree equation (14.10). For fermions (the HFE), the density operators satisfy

$$0 \leq \gamma = \gamma^* \leq 1, \tag{18.29}$$

while for bosons (the Hartree equation), they are not restricted from above and satisfy only $\gamma \geq 0$. As the eigenvalues of γ give the number of particles in the correspond states, the upper bound in (18.29) expresses the Pauli principle of having at most one fermion per state.

Recall that for a density operator γ, the function $n_\gamma(x) := \gamma(x,x)$ (omitting the time argument) is interpreted as the one-particle density, so that $\text{Tr}\gamma = \int \gamma(x,x)dx = \int n_\gamma(x)dx$ is the total number of particles.

Now, we allow a more general class of terms $ex(\gamma)$ and call the resulting equation the *generalized Hartree-Fock equation*. As special cases, we can consider (18.28) with

- $ex(\gamma) := 0$ for the Hartree (or reduced Hartree-Fock, $\gamma \leq 1$) equation;

---

[1] The origins of the operators γ and ρ are somewhat different: γ comes as a one-particle reduction from a many-particle theory, i.e. from integrating out the coordinates of the remaining particles, while ρ comes from integrating out the degrees of freedom of an environment.

- $ex(\gamma) := -v^\sharp \gamma$ for the Hartree-Fock case;
- $ex(\gamma) :=$ an appropriate real-valued function of $n_\gamma(x) := \gamma(x,x)$, for density functional theory.

In the third case, we arrive at the time-dependent extension of the celebrated *Kohn-Sham equation*, which lies at the foundation of density functional theory.

We observe that (18.28) has the following properties (cf. Theorem 19.13 below; we write γ_t for a solution to (18.28) at time t):

1) (18.28) is positivity preserving: if the initial condition $\gamma_{t=0}$ is positive, then so is the solution γ_t;

2) (18.28) preserves the trace: $\mathrm{Tr}\gamma_t = \mathrm{Tr}\gamma_{t=0}$;

3) (18.28) conserves the energy and the number of particles

$$E(\gamma) := \mathrm{Tr}((h + \frac{1}{2}v * n_\gamma)\gamma) - Ex(\gamma), \qquad (18.30)$$

$$N(\gamma) := \mathrm{Tr}\gamma = \int n_\gamma, \qquad (18.31)$$

where $Ex(\gamma) := -\frac{1}{2}\mathrm{Tr}(\gamma v^\sharp \gamma)$. Note that

$$\mathrm{Tr}((v * n_\gamma)\gamma) = \int n_\gamma v * n_\gamma dx = \int\int n_\gamma(x)v(x-y)n_\gamma(y)dxdy,$$

$$\mathrm{Tr}(\gamma v^\sharp \gamma) = \int\int v(x-y)|\gamma(x,y)|^2 dxdy.$$

To prove the first two statements, we note that for γ's from a reasonable class, $h(\gamma)$ is a self-adjoint operator and let $U_\gamma(t,s)$ be the evolution generated by the operator $h(\gamma)$ (see Theorem 25.32 of Appendix 25.6) and $U_\gamma(t) \equiv U_\gamma(t,0)$. Then the solution γ_t can be written as

$$\gamma_t = U_\gamma(t)\gamma_{t=0}U_\gamma(t)^{-1} \qquad (18.32)$$

(cf. Problem 18.3). To show the first property, we use that the evolution $U_\gamma(t)$ is unitary, $U_\gamma(t)^* = U_\gamma(t)^{-1}$, to find

$$\langle \psi, \gamma_t\psi \rangle = \langle U_\gamma(t)^*\psi, \gamma_{t=0}U_\gamma(t)^*\psi \rangle \geq 0.$$

For the second property, we use the cyclic property of the trace, $\mathrm{Tr}(AB) = \mathrm{Tr}(BA)$, to find $\mathrm{Tr}\gamma_t = \mathrm{Tr}(U_\gamma(t)^{-1}U_\gamma(t)\gamma_{t=0}) = \mathrm{Tr}\gamma_{t=0}$.

Problem 18.15 Complete the proof of the properties above.

Stationary generalized Hartree-Fock equation. Clearly, γ is a static solution to (18.28) if and only if γ is time-independent and solves the equation

$$[h(\gamma), \gamma] = 0. \qquad (18.33)$$

For any reasonable function f and for numbers $T > 0$ and $\mu \in \mathbb{R}$, solutions of the equation

$$\gamma = f(\frac{1}{T}(h(\gamma) - \mu)), \tag{18.34}$$

solve (18.33). Under some conditions, the converse is also true. (The parameters $T > 0$ and μ, the temperature and chemical potential, are introduced here for future reference.) The chemical potential μ is determined by the condition that $\mathrm{Tr}\,\gamma = $ some constant, say ν.

The function f is selected on physical grounds by either a thermodynamic limit (Gibbs states) or by a contact with a reservoir (or imposing the maximum entropy principle). It is given by the Fermi-Dirac distribution

$$f(h) = \frac{1}{1 + e^h}, \tag{18.35}$$

as we are dealing with fermions. For bosons, f_{FD} would be replaced by the Bose-Einstein distribution

$$f_{BE}(h) = \frac{1}{e^h - 1}. \tag{18.36}$$

Inverting the function f and letting $f^{-1} =: g'$, we rewrite the stationary generalized Hartree-Fock equation as

$$h_{\gamma,\mu} - Tg'(\gamma) = 0, \tag{18.37}$$

Here $h_{\gamma,\mu} := h(\gamma) - \mu = -\Delta + V + v * n_\gamma - ex(\gamma) - \mu$, $T \geq 0$ (temperature), and $\mu \geq 0$ (chemical potential). It follows from the equations $g' = f^{-1}$ and (18.35), the function g is given by

$$g(\lambda) = -(\lambda \ln \lambda + (1 - \lambda) \ln(1 - \lambda)), \tag{18.38}$$

so that

$$g'(\lambda) = -\ln \frac{\lambda}{1 - \lambda}. \tag{18.39}$$

Kohn-Sham equation. According to the Hohenberg-Kohn, Levy and Lieb results ([163, 200, 196, 197]), the ground state energy functional for a quantum many-body system of identical particles (fermions) can be written as a functional of the one-particle density of the many-body ground state.

However, this result says nothing about the form of this functional. The latter is suggested by replacing in the Hartree-Fock equation the HF exchange term, $-ex(\gamma)$, which is an operator, by a one-particle density dependent function (or a multiplication operator) $-xc(n_\gamma(x))$, called the *exchange-correlation* self-interaction, which is much handier for computations. One expects $-xc(n_\gamma)$ is small in an appropriate sense. As a result we arrive at the equation

$$i\frac{\partial \gamma}{\partial t} = [h_{n_\gamma}, \gamma] \tag{18.40}$$

where, recall, $n_\gamma(x) := \gamma(x, x)$ and, with $h := -\Delta + V$,

$$h_n := h + v * n - \dot{x}c(n). \tag{18.41}$$

The function $xc(\lambda)$ is given empirically, with the expression $xc(\lambda) = -c\lambda^{4/3}$, going back to Dirac, used most often.[2] (18.40) is the *time-dependent Kohn-Sham equation.*

The most studied is the static case. In this case, Eq (18.40) becomes

$$\gamma = f(\frac{1}{T}(h_{n_\gamma} - \mu)), \tag{18.42}$$

where h_{no} is given in (18.41). To this we add the condition $\int n_\gamma = \text{const}$, which determines the chemical potential μ.

Introducing the map den from operators, A, to functions, $\text{den}A(x) := A(x, x)$, we can rewrite (18.42) as an equation for just the one-particle density n:

$$n = \text{den } f(\frac{1}{T}(h_n - \mu)), \tag{18.43}$$

where f is given by (18.35) and h_n by (18.41).

Eqs (18.42), or (18.43), is the *Kohn-Sham* equation, the main equation of density functional theory.

Note that because of the minimal coupling, there is *no (pure) density functional theory* when the system in question is coupled to a magnetic field.

---

[2] $xc(\lambda) = -c\lambda^{4/3}$ is the leading term in the semi-classical expansion for the Hartree-Fock term $ex(\gamma) := -v^\sharp \gamma$.

Open Quantum Systems

In this chapter, we introduce and study the notion of open quantum system, generalizing that of the quantum system we have dealt with so far. This notion lies at the foundation of quantum information theory.

19.1 Information Reduction

Consider a large system (total system T) containing a subsystem we are interested in. We call the latter a *system of interest*, or just a system, and denote it symbolically as S. The rest of the total system will be called the *environment* and denoted E. We write $T = S + E$. Let x and y be the coordinates of the system and environment, respectively. The state spaces of S, E and T are the Hilbert spaces $\mathcal{H}_S = L^2(dx)$, $\mathcal{H}_E = L^2(dy)$ and $\mathcal{H}_T = L^2(dxdy)$.

We will be interested in system observables, i.e. operators acting on the space $\mathcal{H}_S = L^2(dx)$. We will associate with them operators (denoted by the same letters) on $\mathcal{H}_T = L^2(dxdy)$ acting on the variable x.

Assume we are interested in measuring only properties of the system S, e.g. averages for its various observables. Assume the total system, $T = S + E$, is described by a wave function, say $\psi(x, y)$. The question we would like to address is: is there a wave function $\varphi(x)$ for S, such that measuring the average of any observable A associated with S, in the state $\varphi(x)$, gives the same result as measuring it in the state $\psi(x, y)$; i.e. is, for any $A = A_x$,

$$\langle \psi, A\psi \rangle_T = \langle \varphi, A\varphi \rangle_S \ ?$$

Here the inner-products on the l.h.s. and r.h.s. are in the spaces $\mathcal{H}_T$ and $\mathcal{H}_S$, respectively. The answer is that this holds if and only if the subsystem and environment are not correlated, i.e. $\psi(x, y) = \varphi(x)\eta(y)$ for some η. If we take, for example, $\psi(x, y) = \alpha_1\varphi_1(x)\eta_1(y) + \alpha_2\varphi_2(x)\eta_2(y)$, then $\langle \psi, A\psi \rangle \neq \langle \varphi, A\varphi \rangle$ for any $\varphi(x)$.

So what does it take to describe a state of S in this case without referring to the environment? The answer is: use density operators. Namely, with any

© Springer-Verlag GmbH Germany, part of Springer Nature 2020
S. J. Gustafson and I. M. Sigal, *Mathematical Concepts of Quantum Mechanics*, Universitext, https://doi.org/10.1007/978-3-030-59562-3_19

total wave function, $\psi(x,y)$, we can associate a density operator, $\rho = \rho_\psi$, for our subsystem, so that

$$\langle \psi, A\psi \rangle = \mathrm{Tr}_\mathrm{S}(A\rho_\psi) \tag{19.1}$$

for any system observable A. Here Tr_S is trace of the system's degrees of freedom. Indeed, the operator ρ_ψ is defined by its integral kernel

$$\rho_\psi(x,x') = \int \overline{\psi(x,y)}\psi(x',y)dy . \tag{19.2}$$

Problem 19.1 Check that (19.1) holds for any operator A acting on the variable x, provided ρ_ψ is given by (19.2).

To develop an abstract theory, we define the important notion of the partial trace. For any trace-class operator, R, on $\mathcal{H}_T$, we define the operator $\mathrm{Tr}_\mathrm{E}R$ on $\mathcal{H}_S$ by

$$\langle \phi, (\mathrm{Tr}\,R)\psi \rangle = \sum_i \langle \phi\chi_i, R\psi\chi_i \rangle, \tag{19.3}$$

for any $\phi, \psi \in \mathcal{H}_\mathrm{S}$ and for any orthonormal basis $\{\chi_j\}$ in $\mathcal{H}_\mathrm{E}$. Here $(\psi\chi)(x,y) := \psi(x)\chi(y)$.

Problem 19.2 Show that the r.h.s. of (19.3) is independent of the choice of the orthonormal basis $\{\chi_j\}$.

Problem 19.3 Show that the integral kernel of $\mathrm{Tr}_\mathrm{E}\,P_\psi$ is given by (19.2).

We may consider Tr_E as a map from (linear) operators on $\mathcal{H}_T$ to operators on $\mathcal{H}_\mathrm{S}$. We record its properties. In what follows, Tr_S and Tr_T denote the standard traces on $\mathcal{H}_\mathrm{S}$ and $\mathcal{H}_T$ and we omit the subindex T in Tr_T so that $\mathrm{Tr} \equiv \mathrm{Tr}_T$. We have
  1) Tr_E is a linear map;
  2) Tr_E is positive (or positivity preserving);
  3) Tr_E is trace preserving: $\mathrm{Tr}_\mathrm{S} \circ \mathrm{Tr}_\mathrm{E} = \mathrm{Tr}$.
In addition, Tr_E commutes with multiplication by system observables, i.e., for any system observable A,

$$\mathrm{Tr}_\mathrm{E}(AR) = A\,\mathrm{Tr}_\mathrm{E}(R). \tag{19.4}$$

Here and below, we consider operators acting on $\mathcal{H}_\mathrm{S} = L^2(dx)$ also as operators on $\mathcal{H}_T = L^2(dxdy)$ acting just on the variable x (see Remark 19.6 for a discussion).

The properties of Tr_E listed above imply the following properties of the operator $\mathrm{Tr}_\mathrm{E}\,R$:
  a) $\mathrm{Tr}_\mathrm{E}\,R$ is a linear operator;
  b) $\mathrm{Tr}_\mathrm{E}\,R$ is positive if R is positive;
  c) $\mathrm{Tr}_\mathrm{S}\,\mathrm{Tr}_\mathrm{E}\,R = \mathrm{Tr}\,R$.

Problem 19.4 Show all the properties listed above.

Next, we have the following key result:

Theorem 19.5 *For any system observable A we have*

$$\mathrm{Tr}(AR) = \mathrm{Tr_S}(A\rho), \quad where \quad \rho = \mathrm{Tr_E}\, R. \tag{19.5}$$

Proof. By property (c) above, we have $\mathrm{Tr}(AR) = \mathrm{Tr_S}\left(\mathrm{Tr_E}(AR)\right)$ and by (19.4), we obtain furthermore $\mathrm{Tr_E}(AR) = A\,\mathrm{Tr_E}(R)$, which, together with the previous relation, implies (19.5), since $\rho = \mathrm{Tr_E}\, R$. □

Thus if R is a density operator of the total system T=S+E, then $\rho := \mathrm{Tr_E}\, R$ is a density operator of the system S. It is called the *reduced density operator*.

Now, as was mentioned above, the wave function, ψ, of the total system can be associated with a density operator acting on the total system coordinates, namely the rank-one projection $R := P_\psi$, so that

$$\langle \psi, A\psi \rangle = \mathrm{Tr}(AP_\psi) \tag{19.6}$$

for any operator (observable) A. With this, if A is a system observable, i.e., an operator which acts only on the variables x, then (19.1), with $\rho_\psi := \mathrm{Tr_E}\, P_\psi$, follows from Theorem 19.5.

To summarize: if the total system is described by a density operator R, then the results of observations on the system S are given in terms the system's density operator $\rho = \mathrm{Tr_E}\, R$, without further reference to R.

A system under consideration, whose interaction with the environment cannot be neglected but which is described in terms of its own degrees of freedom, is called an *open system*.

In fact, every system, unless it is the entire universe, can be considered as a subsystem of a larger system. Hence, in reality every quantum system is an open system.

Remark 19.6 In tensor product notation we write $\psi \otimes \chi$, and $A \otimes \mathbf{1_E}$, where $\mathbf{1_E}$ is the identity operator on $\mathcal{H_E}$, for $\psi(x)\chi(y)$ and an observable A acting only on x, respectively. Moreover, using the notion of tensor product (see Appendix 25.13), we can pass to a general framework of abstract Hilbert spaces, $\mathcal{H}_T = \mathcal{H_S} \otimes \mathcal{H_E}$, without referring to the variables describing their degrees of freedom.

One can also define the partial trace, $\rho = \mathrm{Tr_E}\, R$, in terms of the integral kernels $R(x, y, x', y')$ and $\rho(x, x')$ of R and ρ, by generalizing the formula (19.2) as

$$\rho(x, x') := \int R(x, y, x', y)dy. \tag{19.7}$$

Problem 19.7 Check that the definitions of partial trace given by (19.7) and (19.3) are equivalent.

The results of the next problem are useful in applications.

Problem 19.8 Given a density matrix ρ_e of E, introduce the map Tr'_E from operators on $\mathcal{H}_T$ to operators on $\mathcal{H}_T$ by

$$\mathrm{Tr}'_E R := (\mathrm{Tr}_E R) \otimes \rho_e.$$

Show that (i) Tr'_E is a projection, (ii) its dual (on the space of bounded operators) is given by

$$\mathrm{Tr}^*_E A := (\mathrm{Tr}_E A\rho_e) \otimes \mathbf{1}.$$

Consider also the symmetric version

$$\mathrm{Tr}''_E R := (\mathrm{Tr}_E R\sqrt{\rho_e}) \otimes \sqrt{\rho_e}$$

and show that the dual to Tr''_E is of the same form.

Conditional Expectation. In Section 18.2, we explained that quantum statistics is a non-commutative extension of classical probability theory. We further elaborate this thesis by introducing and discussing the quantum (non-commutative) extension of the key probabilistic notion of conditional expectation.

Given a density matrix R of T, we define, for any observable, A, the system observable (SO)

$$Ex^S_R(A) := \mathrm{Tr}_E(AR)/\mathrm{Tr}_E(R). \tag{19.8}$$

This is a map from total observables to system ones. It extends the notion of *conditional expectation* from classical probability to the non-commutative setting, as confirmed by the following properties:

(i) Ex^S_R is a linear map;
(ii) $Ex^S_R(\mathbf{1}) = \mathbf{1}$;
(iii) $Ex^S_R(BAB') = BEx^S_R(A)B'$ for any SOs B and B';
(iv) $Ex^S_R(A^*) = Ex^S_R(A)^*$;
(v) $Ex^S_R(A) \geq 0$, if $A \geq 0$.

Problem 19.9 Prove properties (i) -(iii) above.

Properties (ii)-(iii) imply that $Ex^S_R(B) = B$ for any SO B (take $A = \mathbf{1}$ and $B' = \mathbf{1}$ in (iii)), that

$$Ex^S_R(Ex^S_R(A)) = Ex^S_R(A)$$

(follows from the previous equation by taking $B = Ex^S_R(A)$), and that

$$\mathrm{Tr}\left(Ex^S_R(A)R\right) = \mathrm{Tr}(AR). \tag{19.9}$$

Problem 19.10 Prove (19.9).

Properties (iv) and (v) above are more subtle and we prove them here. For brevity, we write $Ex_R^S(A) \equiv Ex(A)$ and

$$\omega(A) := \mathrm{Tr}(AR).$$

Using properties (iii) and (19.9) and letting B and C be any SOs, we obtain $\omega(BEx(A^*)C) = \omega(Ex(BA^*C)) = \omega(BA^*C)$. Since $BA^*C = (C^*AB^*)^*$ and $\omega(A^*) = \overline{\omega(A)}$, this yields

$$\omega(BEx(A^*)C) = \overline{\omega(C^*AB^*)}.$$

Now, moving in the reverse direction, we find $\omega(C^*AB^*) = \omega(F)$, where $F := Ex(C^*AB^*)$, which, together with $\overline{\omega(F)} = \omega(F^*)$, $F^* = B(Ex(A))^*C$ and (iii), gives $\overline{\omega(C^*AB^*)} = \omega(F^*) = \omega(B(Ex(A))^*C)$. This and the previous conclusion imply

$$\omega(BEx(A^*)C) = \omega(B(Ex(A))^*C).$$

Since this is true for every SOs B and C, we have $Ex(A^*) = (Ex(A))^*$. Indeed, since $Ex(A)$ is a SO, this follows from the following

Lemma 19.11 *The relation $\omega(BEC) = 0$, for a given SO E and for arbitrary SOs B and C, implies that $E = 0$.*

Proof. We use (19.5) to write $\omega(BEC) = \mathrm{Tr}_S(BEC\rho)$, where $\rho = \mathrm{Tr}_E R$, and then take $B = |\psi\rangle\langle\chi|$ and $C = |\varphi\rangle\langle\psi|$, with ψ such that $\langle\psi, \rho\psi\rangle \neq 0$, to obtain

$$\omega(BEC) = \langle\chi, E\varphi\rangle\langle\psi, \rho\psi\rangle.$$

This implies that $\langle\chi, E\varphi\rangle = 0$, for every χ and φ, which yields that $E = 0$. □

This concludes the proof of (iv).

For property (v), proceeding as in the first part of the proof of (iv), we obtain
$$\omega(BEx(A)B^*) = \omega(BAB^*) \geq 0$$
if $A \geq 0$. Since $Ex(A)$ is a SO and since the above inequality is true for every SO B, we have that $Ex(A) \geq 0$. Indeed, this follows from the following

Lemma 19.12 *If $\omega(BEB^*) \geq 0$, for a given SO E and for every SO B, then $E \geq 0$.*

Proof. Proceeding as above, we take $B = |\psi\rangle\langle\varphi|$, with ψ such that $\langle\psi, \rho\psi\rangle \neq 0$, where $\rho = \mathrm{Tr}_E R$. Then

$$0 \leq \omega(BEB^*) = \langle\varphi, E\varphi\rangle\langle\psi, \rho\psi\rangle,$$

which implies that $\langle\varphi, E\varphi\rangle \geq 0$ for every φ, which yields that $E \geq 0$. □

Thus, we conclude that property (v) holds.

19.2 Reduced dynamics

Consider a system of interest (S) interacting with an environment (E). We found in Section 19.1, that if the total system is described by a density operator R, acting on the total system state space $\mathcal{H}_T = L^2(dxdy)$ (or $\mathcal{H}_T = \mathcal{H}_S \times \mathcal{H}_E$), then the density operator $\rho = Tr_E R$ of S gives the same results for averages of the system's observables as R. Here, recall, Tr_E is the partial trace over the environment degrees of freedom defined in (19.3). Then, for any observable $A = A_x$, associated with the system

$$\text{Tr}(AR) = \text{Tr}_S(A\rho), \quad \rho = \text{Tr}_E R. \qquad (19.10)$$

(Recall that $\text{Tr} \equiv \text{Tr}_T$.) Recall that the operator $\rho = \text{Tr}_E R$, called the reduced density operator, acts on the system state space and has the following properties:

1) ρ is a linear operator on the state space of the system;
2) ρ is positive if R is positive;
3) $\text{Tr}_S \rho = \text{Tr} R$.

Assume now our total system is described by the Schrödinger operator, H_T, acting on the state space $\mathcal{H}_T = L^2(dxdy)$. A model to keep in mind is the one given by

$$H_T = H_S \otimes \mathbf{1}_E + \mathbf{1}_S \otimes H_E + \lambda v, \qquad (19.11)$$

where H_S and H_E are Schrödinger operators of this system and environment acting on $\mathcal{H}_S$ and $\mathcal{H}_E$, respectively, λ is a real parameter, called the coupling constant, and v is an operator acting on $\mathcal{H}_T = \mathcal{H}_S \otimes \mathcal{H}_E$. Here λv describes the interaction of the system and the environment.

The evolution of the total system is given by the von Neumann-Landau equation

$$i\frac{\partial R_t}{\partial t} = \frac{1}{\hbar}[H_T, R_t], \quad R_{t=0} = R_0. \qquad (19.12)$$

We write the solution of this initial value problem as $R_t = \alpha_t(R_0)$. Let $U_t := e^{-\frac{iHt}{\hbar}}$. It is shown in Homework 18.3 that

$$\alpha_t(R) = U_t R U_t^*. \qquad (19.13)$$

A map α on a space, $\mathcal{B}$, of operators is called *positive* iff $\alpha(R) \geq 0$, whenever $R \geq 0$. With this definition, we have

Theorem 19.13 For each t, the map α_t has the following properties
1) α_t is linear;
2) α_t is positive;
3) α_t is trace preserving: $\text{Tr}\alpha_t(R) = \text{Tr}R$;
4) $\alpha_t(R^*) = \alpha_t(R)^*$;
5) α_t is invertible.

Problem 19.14 Prove this theorem. (Hint: To prove the trace preservation property, one can use either the definition of the trace or its cyclic property, $\text{Tr}(AB) = \text{Tr}(BA)$, and properties of $U_t := e^{-\frac{iH_T t}{\hbar}}$, see Homework 2.21.)

Here we prove the second statement. We omit the subindex tot at H_T. Using the relations (19.13) and $\langle \psi, U_t^* \phi \rangle = \langle U_t \psi, \phi \rangle$ (the latter shown in Homework 2.21), we find

$$\langle \psi, \alpha_t(R)\psi \rangle = \langle \psi, U_t R U_t^* \psi \rangle = \langle U_t^* \psi, R U_t^* \psi \rangle \geq 0.$$

Now, the reduced density operator of the system S at time t is given by

$$\rho_t := \text{Tr}_E R_t.$$

We consider the particular class of R_0 of the form $R_0 = \rho_0 \otimes \rho_e$ for some fixed density operator ρ_e for E and define

$$\beta_t(\rho_0) = \text{Tr}_E \alpha_t(\rho_0 \otimes \rho_e). \tag{19.14}$$

The family β_t is called the *reduced evolution*. We observe

1) The evolution β_t depends on ρ_e.

2) Starting with a pure state, the evolution β_t produces mixed states, unless S and E do not interact.

3) If S and E do not interact then $\beta_t = \alpha_t^S$, where $\alpha_t^S(\rho) := U_t \rho U_t^*$, the evolution of S. (Indeed, in this case, $\alpha_t := \alpha_t^S \otimes \alpha_t^E$ and the result follows from the fact that α^S and Tr_E commute.)

4) In general, $\beta_t(\rho_0) = \text{Tr}_E \alpha_t(R_0)$, where R_0 is s.t. $\text{Tr}_E R_0 = \rho_0$, is not even a linear map for fixed t.

What can we say about the reduced evolution β_t? As we deal with a fixed t, we can concentrate on a single map α. For a fixed map α acting on density operators on the total state space $\mathcal{H}_T = L^2(dxdy)$ and an environment's density operator ρ_e, we define the reduced map β acting on system density operators on the system space $\mathcal{H}_S = L^2(dx)$ as

$$\beta(\rho) = \text{Tr}_E \alpha(\rho \otimes \rho_e). \tag{19.15}$$

Moreover, we assume that α is of the form $\alpha(R) = URU^*$ where U is a unitary (or more generally, invertible) operator on $\mathcal{H}_T$. Then we have

Theorem 19.15 The map (19.15) has the following properties
  (a) β is linear;
  (b) β is positive;
  (c) β is trace preserving: $\text{Tr}\beta(\rho) = \text{Tr}\rho$;
  (d) $\beta(\rho^*) = \beta(\rho)^*$.

Proof. We show only property (b), leaving the other properties as the exercise. This property follows immediately as β is a composition of two positivity preserving maps, Tr_E and α.

Problem 19.16 Show (a) - (d). (Hint: To prove (c) show that $\mathrm{Tr}_S \mathrm{Tr}_E = \mathrm{Tr}_T$ and use the cyclic property of the trace, $\mathrm{Tr}_E(AB) = \mathrm{Tr}_E(BA)$.)

Theorem 19.17 The map β is of the form

$$\beta(\rho) = \sum_n V_n \rho V_n^* \tag{19.16}$$

where V_n are bounded operators satisfying (strong convexity)

$$\sum_n V_n^* V_n = 1. \tag{19.17}$$

We prove this theorem at the end of this subsection.

Problem 19.18 Show that (19.16) implies Theorem 19.13.

Is the converse true, i.e. do the properties stated in Theorem 19.13 imply (19.16)? The answer is no. However, the following strengthening of the positivity property does the job.

Definition 19.19 A map β on a space, $\mathcal{B}$, of operators is called *completely positive* if and only if $\beta \otimes 1$ is positive on $\mathcal{B} \otimes M_k$ for any $k \geq 1$. Here M_k is the space of $k \times k$ matrices.

Problem 19.20 Show directly that maps of the form (19.16) are completely positive.

The next theorem shows that the reverse, which is more subtle, is also true.

Theorem 19.21 (Kraus) (i) Linear, completely positive maps β are of the form (19.16). If, in addition, β are trace preserving, then (19.17) holds also. (ii) Representation (19.16) is unique up to a unitary transformation, i.e. if $\sum_n V_n \rho V_n^* = \sum_n W_n \rho W_n^*$, for every density matrix ρ, then there is a unitary operator U s.t. $W_n = U V_n$, for every n.

The operators V_n in (19.16) are called the *Kraus operators*. We prove this result below. This theorem illustrates the importance of the notion of complete positivity. This notion also differentiates between quantum and classical situations, see Remark 19.27.

Definition 19.22 Linear, completely positive, trace preserving (TPCP) maps are called *quantum (dynamical) maps* or *quantum (communication) channels*.

In the opposite direction, a quantum map can be lifted to a unitary evolution of a larger (total) system. To describe this, it is convenient to pass from quantum maps of density operators to the *dual maps* of observables. The map (automorphism) β', acting on (an algebra of) bounded of operators on the system Hilbert space, is said to be dual to a map (automorphism) β on density operators if it satisfies

$$(\rho, \beta'(A)) = (\beta(\rho), A), \tag{19.18}$$

where, recall, (ρ, A) is the coupling between density operators and observables:

$$(\rho, A) = \mathrm{Tr}_S(A\rho). \tag{19.19}$$

A map β' on (an algebra of) bounded of operators on the system Hilbert space, is said to be *positive* if and only if $\beta'(A) \geq 0$, whenever $A \geq 0$. Similarly, one defines the complete positivity. β' is called *unital* if and only if $\beta'(1) = 1$.

A quantum map β satisfies the conclusions of Theorem 19.17 if and only if its dual, β', satisfies
1) β' is linear;
2) β' is completely positive;
3) β' is unital: $\beta'(1) = 1$;
4) $\beta'(A^*) = \beta'(A)^*$
5) $\beta'(A) = \sum_n V_n^* A V_n$,
where V_n are bounded operators satisfying $\sum_n V_n^* V_n = 1$ (strong convexity).

By Kraus' theorem 19.21, properties 2) and 3) are equivalent to property 5). Furthermore, β' is unital if and only if β is the trace preserving. Indeed, $\mathrm{Tr}\beta(\rho) = (\beta(\rho), 1) = (\rho, \beta'(1))$ and $\mathrm{Tr}\rho = (\rho, 1)$ imply that $\mathrm{Tr}\beta(\rho) = \mathrm{Tr}\rho$ if and only if $\beta'(1) = 1$.

Finally, property 3) shows that β' has the eigenvalue 1 (with the eigenvector 1). This suggests that β also has the eigenvalue 1. The corresponding eigenvector would be an equilibrium state.

Problem 19.23 Show that (1) the composition of two quantum maps is again a quantum map and (2) the composition of two quantum maps with Kraus operators $\{V_n\}$ and $\{W_n\}$, has the Kraus operators $\{V_m W_n\}$.

In the direction opposite to the one of Theorem 19:17, we have

Theorem 19.24 (Stinespring) $\beta' : \mathcal{B}(\mathcal{H}) \to \mathcal{B}(\mathcal{H})$ *is an adjoint quantum map if and only if it is of the form*

$$\beta'(A) = W^*(A \otimes 1_{\mathcal{K}})W, \tag{19.20}$$

where $\mathcal{K}$ is a Hilbert space and $W : \mathcal{H} \to \mathcal{H} \otimes \mathcal{K}$, a bounded operator.

The artificial environment space $\mathcal{K}$ is called the *ancilla* space[1].

Derivation of Theorem 19.24 from Theorem 19.21 (finite-dimensional $\mathcal{H}$). Let $d := dim\mathcal{H}$. Define $W : \mathcal{H} \otimes \mathbb{C}^d \to \mathcal{H}$ by $W : f \to \oplus_i V_i f$. Then $W^* : \mathcal{H} \to \mathcal{H} \otimes \mathbb{C}^d$ is given by $W^* : \oplus_i f_i \to \sum_i V_i^* f_i$ and (19.20), with $\mathcal{K} := \mathbb{C}^d$, follows. $\square$

---

[1] In quantum information theory, the ancilla is an auxiliary quantum system. One of its common uses is for indirect measurements, in which a measurement is done on the ancilla coupled to S. After the measurement is finished, the ancilla is discarded.

For other proofs of this result see [93, 177]. ([93], Theorem 4.6, shows a more general result implying Theorem 19.24.)

Problem 19.25 Show that the operators W and W^* is the proof above (a) are mutually adjoint and (b) satisfy $W^*(A \otimes 1_\mathcal{K})W = \sum_i V_i^* A V_i$.

In the opposite direction, Stinespring's theorem implies Kraus' Theorem 19.21, as we show in Appendix 19.3 below.

The first proof of Theorem 19.21 (finite-dimensional $\mathcal{H}$). Let $\mathcal{K}$ be an abstract Hilbert space of the same dimension as $\mathcal{H}$. We construct the function

$$\Psi := \sum_i \theta_i \otimes \chi_i \in \mathcal{H} \otimes \mathcal{K},$$

where $\{\theta_i\}$ and $\{\chi_i\}$ are some orthonormal bases in $\mathcal{H}$ and $\mathcal{K}$, respectively. The function Ψ provides the map from $\mathcal{K}$ to $\mathcal{H}$ as

$$K : \phi \in \mathcal{K} \to \psi := \langle \bar{\Psi}, \phi \rangle_\mathcal{K} \in \mathcal{H},$$

whose inverse is given by

$$K^{-1} : \psi \in \mathcal{H} \to \phi := \langle \Psi, \psi \rangle_\mathcal{H} \in \mathcal{K}.$$

(This demonstrates that any two Hilbert spaces of the same dimension are isometrically isomorphic.)

Problem 19.26 Show that the two expressions above define operators inverse of each other. (Hint: use the expansion in the bases $\{\theta_i\}$ and $\{\chi_i\}$.)

It is convenient to introduce the maps $T_\phi : \mathcal{H} \otimes \mathcal{K} \to \mathcal{H}$ and $T_\phi^* : \mathcal{H} \to \mathcal{H} \otimes \mathcal{K}$, defined by $T_\phi \xi := \langle \phi, \xi \rangle_\mathcal{K}$ and $T_\phi^* \eta := \eta \otimes \phi$ (the Dirac bra and ket operators). Then $K\phi = T_\phi \Psi$.

Let $\psi = K\phi = T_\phi \Psi$. Then $P_\psi = T_\phi P_\Psi T_\phi^*$, where $\phi = K^{-1}\psi$, and therefore

$$\beta(P_\psi) = T_\phi(\beta \otimes 1)(P_\Psi)T_\phi^*. \tag{19.21}$$

Since the map $\beta \otimes 1$ is completely positive, the operator $(\beta \otimes 1)(P_\Psi)$ is positive and therefore can be written as

$$(\beta \otimes 1)(P_\Psi) = \sum_k |\Phi_k\rangle\langle\Phi_k|, \tag{19.22}$$

where Φ_k are non-normalized, but mutually orthogonal vectors from $\mathcal{H} \otimes \mathcal{K}$ (Φ_k are eigenvectors of the operator $(\beta \otimes 1)(P_\Psi)$ with the norms equal to the square roots of the corresponding eigenvalues). Using relations (19.21) and (19.22), we find

$$\beta(P_\psi) = \sum_k |T_\phi \Phi_k\rangle\langle T_\phi \Phi_k|, \tag{19.23}$$

where, recall, $\phi = K^{-1}\psi$.

Now we have $T_\phi \xi := \langle \bar{\phi}, \xi \rangle_\mathcal{K} = \langle \bar{\xi}, \phi \rangle_\mathcal{K}$. For $\phi = K^{-1}\psi$, this gives $T_\phi \xi = \langle \bar{\xi}, K^{-1}\psi \rangle_\mathcal{K} = \langle (K^*)^{-1}\bar{\xi}, \psi \rangle_\mathcal{H}$ and therefore

$$T_\phi \xi = \tilde{T}_{(K^*)^{-1}\bar{\xi}}\psi, \tag{19.24}$$

where $\tilde{T}_\psi : f \to \langle \psi, f \rangle_\mathcal{H}$ mapping $\mathcal{H} \otimes \mathcal{K}$ to $\mathcal{K}$. The last two relations imply

$$\beta(P_\psi) = \sum_k |V_k\psi\rangle\langle V_k\psi|,$$

with the operators V_k on $\mathcal{H}$ given by $V_k := \tilde{T}_{(K^*)^{-1}\bar{\Phi}_k}$. Since, in the Dirac bra-ket notation, $P_\psi = |\psi\rangle\langle\psi|$, the latter equation can be rewritten as (show this)

$$\beta(P_\psi) = \sum_k V_k P_\psi V_k^*.$$

Extending this relation, by linearity, to arbitrary density matrices gives (19.16).   □

Remark 19.27 One can show (see [93], Theorem 4.3) that a positive linear map between two algebras of observables[2] one of which is abelian is completely positive.

19.3 Some Proofs

First, we give a lengthier but instructive direct proof of Theorem 19.13(b) which we will use below. Let $\{\chi_i\}$ be an orthonormal basis in the environment space $L^2(dy)$. By the definition of β, (19.15), and the property $\alpha(R) = URU^*$, we have $\forall \phi, \psi \in L^2(dx)$

$$\langle \phi, \beta(\rho)\psi \rangle = \sum_i \langle \phi\chi_i, \alpha(\rho \otimes \rho_e)\psi\chi_i \rangle$$
$$= \sum_i \langle U^*\phi\chi_i, \rho \otimes \rho_e U^*\psi\chi_i \rangle. \tag{19.25}$$

Taking here $\phi = \psi$, we see the r.h.s is non-negative, provided ρ is non-negative. This gives Theorem 19.13(b).   □

Proof of Theorem 19.17. To prove (19.16), we write the inner product in $L^2(dxdy)$ as $\langle \cdot \rangle = \langle\langle \cdot \rangle_E\rangle_S$, i.e. first as the inner product in $L^2(dy)$ and then in

[2] Technically our algebras of observables are C^*-algebras. See [45] for the definition of the C^*-algebra. A good example to keep in mind is a subset of the space of bounded operators (on some Hilbert space) close under algebraic operations of addition, multiplication and taking adjoints.

$\mathcal{H}_S = L^2(dx)$. Let χ_i be an orthonormal basis in $\mathcal{H}_E = L^2(dy)$ of eigenfunctions of ρ_e with eigenvalues λ_j. Then $\rho_e = \sum \lambda_j P_{\chi_j} = \sum \lambda_j |\chi_j\rangle\langle\chi_j|$, so that, using (19.25), we obtain, for any $\phi, \psi \in \mathcal{H}_S$,

$$\langle\phi, \beta(\rho)\psi\rangle = \sum_{i,j}\langle\sqrt{\lambda_j}\langle\chi_j, U^*\phi\chi_i\rangle_E, \rho\sqrt{\lambda_j}\langle\chi_j, U^*\psi\chi_i\rangle_E\rangle_S$$

$$= \sum_{i,j}\langle V_{ij}^*\phi, \rho_0 V_{ij}^*\psi\rangle_S, \qquad (19.26)$$

where $V_{ij}^*\phi := \sqrt{\lambda_j}\langle\chi_j, U^*\phi\chi_i\rangle_E$, and therefore $\langle\phi, \beta(\rho)\psi\rangle = \langle\phi, \sum_{i,j} V_{ij}\rho V_{ij}^*\psi\rangle$. Now

$$\langle V_{ij}\phi, \psi\rangle_S = \langle\phi, V_{ij}^*\psi\rangle_S = \sqrt{\lambda_j}\langle\phi, \langle\chi_j, U^*\psi\chi_i\rangle_E\rangle_S$$

$$= \sqrt{\lambda_j}\langle U\phi\chi_j, \psi\chi_i\rangle = \langle\sqrt{\lambda_j}\langle U\phi\chi_j, \chi_i\rangle_E, \psi\rangle_S.$$

This implies

$$V_{ij}\phi = \sqrt{\lambda_j}\langle\chi_i, U\phi\chi_j\rangle_E,$$

which, in turn, gives

$$\sum_{i,j} V_{ij}^* V_{ij}\phi = \sum_{i,j} V_{ij}^*\sqrt{\lambda_j}\langle\chi_i, U\phi\chi_j\rangle_E$$

$$= \sum_{ij}\lambda_j\langle\chi_j, U^*\langle\chi_i, U\phi\chi_j\rangle_E\chi_i\rangle_E$$

$$= \sum_j \lambda_j\langle\chi_j, U^*\sum_i\langle\chi_i, U\phi\chi_j\rangle_E\chi_i\rangle_E.$$

Since $\sum_i\langle\chi_i, U\phi\chi_j\rangle_E\chi_i = U\phi\chi_j$, this gives

$$\sum_{i,j} V_{ij}^* V_{ij}\phi = \sum_j \lambda_j\langle\chi_j, U^*U\phi\chi_j\rangle_E = \sum\lambda_j\langle\chi_j, \phi\chi_j\rangle_E. \qquad (19.27)$$

Since $\langle\chi_j, \phi\chi_j\rangle_E = \phi$ and $\sum\lambda_j = \mathrm{Tr}\,\rho_e = 1$, we have $\sum_{i,j} V_{ij}^* V_{ij}\phi = \phi$. $\square$

The second proof of Theorem 19.21. (Derivation from Theorem 19.24 for arbitrary $\mathcal{H}$.) For $\phi, \psi \in \mathcal{H}$, we use Stinespring's Theorem 19.24 to write $\langle\phi, \beta^*(A)\psi\rangle = \langle\phi, W^*(A\otimes 1_\mathcal{K})W\psi\rangle = \langle W\phi, (A\otimes 1_\mathcal{K})W\psi\rangle$ for some ancilla space $\mathcal{K}$. Let $\{\chi_i\}$ be an orthonormal basis in $\mathcal{K}$. Inserting the partition of unity $\sum_i P_{\chi_i} = 1$, where recall $P_{\chi_i} \equiv |\chi_i\rangle\langle\chi_i|$ is the rank one projection on χ_i, and writing the inner product in $\mathcal{H}\otimes\mathcal{K}$ as $\langle\cdot\rangle_{\mathcal{H}\otimes\mathcal{K}} = \langle\langle\cdot\rangle_\mathcal{K}\rangle_\mathcal{H}$, i.e. first as the inner product in $\mathcal{K}$ and then in $\mathcal{H}$, we find

$$\langle\phi, \beta^*(A)\psi\rangle_\mathcal{H} = \sum_i\langle W\phi, (A\otimes P_{\chi_i})W\psi\rangle_{\mathcal{H}\otimes\mathcal{K}}$$

$$= \sum_i\langle\langle\chi_i, W\phi\rangle_\mathcal{K}, A\langle\chi_i, W\psi\rangle_\mathcal{K}\rangle_\mathcal{H}. \qquad (19.28)$$

Now, define the operators V_i on $\mathcal{H}$ by $V_i \phi := \langle \chi_i, W\phi \rangle_{\mathcal{K}}$, so that the last equation reads $\langle \phi, \beta^*(A)\psi \rangle_{\mathcal{H}} = \sum_i \langle V_i \phi, AV_i \psi \rangle_{\mathcal{H}}$, which is equivalent to

$$\beta^*(A) = \sum_i V_i^* AV_i. \qquad (19.29)$$

Since W is bounded, the operators V_i are bounded as well. By definition (19.18), Eq. (19.29) implies (19.16).

Next, if β is trace preserving, then definition (19.18) implies that β^* is unital (i.e. $\beta^*(\mathbf{1}) = \mathbf{1}$) and (19.29), with $A = \mathbf{1}$, becomes (19.17). This proves the theorem.   □

19.4 Communication Channels

The quantum communication channel is one of the key notions in quantum information theory. In this section we compare classical and quantum channels and describe various ways to process information.

We begin by explaining the notion of information. Classical information consists of words constructed from a finite collection of symbols, called an alphabet, with the simplest one being the binary alphabet consisting of two symbols, called bits, say $\{0, 1\}$. A more complicated example consists of a written language alphabet. More generally, classical information could be modelled by a probability distribution, $p(x)$, on the set X of words or letters ($p : X \to \mathbb{R}$, s.t. $p(x) \geq 0$ and $\sum_x p(x) = 1$). (We use the symbol $\sum$ in both the discrete and continuous cases. In the latter case, $\sum$ stands for an integral, $\int$.)

A classical (communication) channel is a map β_{cl} from probability distributions into probability distributions, satisfying

1) β_{cl} is linear;
2) β_{cl} is positive (i.e. positivity preserving: $p \geq 0 \implies \beta_{cl}(p) \geq 0$);
3) β_{cl} preserves the total probability, $\sum_x \beta_{cl}(p)(x) = \sum_x p(x)$.

A map β_{cl} having the above properties is called a *stochastic map*. Note that we can write 3) as $\langle \beta_{cl}(p), \mathbf{1} \rangle = \langle p, \mathbf{1} \rangle$ $\forall p$ or $\beta_{cl}^*(\mathbf{1}) = \mathbf{1}$, where $\langle p, f \rangle = \sum_x p(x) f(x)$, and β_{cl}^*, defined on bounded functions, is the map dual to β_{cl} defined by

$$\langle \beta_{cl}(p), f \rangle = \langle p, \beta_{cl}^*(f) \rangle.$$

Quantum information is represented by density operators ρ ($\rho \geq 0$ and $\mathrm{Tr}\,\rho = 1$). The simplest physical system – the one with state space $\mathbb{C}^2$ – is called the *qubit* (quantum bit). One can realize a qubit as the *lowest two states* of an atom or of a Josephson junction between superconductors (with the rest of the spectrum considered as an environment).

While the classical binary digit (also called the bit) takes only two values, 0 or 1, the qubit (in a pure state) takes values on the 3 dimensional sphere $\{\psi \in \mathbb{C}^2 : \|\psi\| = 1\}$. If we take the standard basis, $|0\rangle := \begin{pmatrix} 1 \\ 0 \end{pmatrix}$ and $|1\rangle := \begin{pmatrix} 0 \\ 1 \end{pmatrix}$, in

$\mathbb{C}^2$ (playing the role of the classical information bits), then a qubit can be in any superposition of the states $|0\rangle$ and $|1\rangle$: $\alpha|0\rangle + \beta|1\rangle$ with $|\alpha|^2 + |\beta|^2 = 1$.

Clearly, a quantum communication channel should map density matrices into density matrices. Since a transmission of information could be viewed as an evolution process, and since evolution of an open quantum system is given by quantum (dynamical) maps, the quantum channel is, by the definition, a quantum (dynamical) map.

Recall that a quantum (dynamical) map is an automorphism on density operators (i.e. a map of density operators into density operators), which is linear, completely positive and trace preserving.

Thus depending on the application, the same object is called either a quantum (dynamical) map or a *quantum communication channel*, or simply *quantum channel*.

One introduces the following definitions:

Preparation: The preparation is a linear, bounded, positive map $\Phi : p \to \rho$ from probability densities, $p := \{p(x)\}$, to density operators, ρ (i.e. from classical to quantum information). Given a basis, $\{\rho_x\}$, of density operators ρ_x (DMB), Φ can be constructed as

$$\Phi(p) = \sum_x p(x)\rho_x.$$

Reception: The reception is a linear, bounded, positive map $R : \rho \to p$ from density matrices to probability densities. Given a collection $\{M_x\}$ of positive operators satisfying $\sum_x M_x = 1$ (i.e. $\{M_x\}$ is a positive operator-valued resolution of the identity), called *positive operator-valued measure or POVM*, a reception map can be given by

$$R(\rho)(x) := \mathrm{Tr}(M_x\rho).$$

Starting with von Neumann's axiomatization of the measurement process, one can show that a measurement is a reception map (see [242], Section 3.1).

Note that the preparation and reception maps allow us to embed classical channels, β_{cl}, into quantum ones as

$$\beta_{\mathrm{cl}} \to \Phi \circ \beta_{\mathrm{cl}} \circ R.$$

Assume our transmission (computation) is quantum, but the input and/or extraction of information could be either quantum or classical. Further, assume that the originator possesses a DMB, $\{\rho_x\}$, and the receiver a POVM, $\{M_x\}$. Thus we can introduce the following refinements:

quantum-quantum channel $\beta_{\mathrm{qq}} : \rho \to \rho'$
classical-quantum channel $\beta_{\mathrm{clq}} : p \to \rho = \sum p_x\rho_x \to \rho'$.
quantum-classical channel $\beta_{\mathrm{qcl}} : \rho \to \rho' \to p_x = Tr(M_x\rho')$.
classical-classical channel $\beta_{\mathrm{clcl}} : p \to \rho = \sum p_x\rho_x \to \rho' \to p_x = \mathrm{Tr}(M_x\rho')$.

The last case is the one one expects in most quantum computations. One starts with classical information, converts it into a quantum one, which is manipulated by a quantum computer or quantum channel, and at the end one converts the quantum output into a classical one.

Remark 19.28 (Measuring classical and quantum information) Let a source be modelled as a probability distribution $\{p_x\}$ (or a random variable X with probability distribution, $p_x = P(X = x)$). To measure classical information contained in a probability distribution $p(x)$, Shannon defined the quantity

$$H(p) = -\sum_x p(x) \log p(x), \tag{19.30}$$

which is now called *Shannon's information entropy* or just Shannon's information. It is analogous to the Boltzmann entropy of Statistical Mechanics. The quantum analogue of Shannon's information is von Neumann's entropy, which has already appeared in Section 18.3 and will be discussed in Section 19.6.

19.5 Quantum Dynamical Semigroups

Now, we address *dynamical* properties of the reduced evolution (see (19.14)) and *restore t* in the notation.

Definition 19.29 A family β_t, $\forall t \geq 0$, of quantum maps is called an *open quantum evolution*. (Thus a reduced evolution is an open quantum evolution.)

Note that a single quantum map, β, generates the discrete time flow, $\beta^k, k = 0, 1, \ldots$.

Now we may introduce the notion of an an *abstract open quantum system* as a pair: a space of density operators and an open quantum evolution on it.

We further refine the notion of open quantum evolution to the following:

Definition 19.30 A *quantum dynamical semigroup (qds)* is an open quantum evolution which has the Markov property, i.e. it is a semigroup:

$$\beta_t \circ \beta_s = \beta_{t+s} \; \forall t, s \geq 0. \tag{19.31}$$

For a quantum dynamical semigroup β_t, we define the generator by

$$\mathcal{L}(\rho) := \partial_t \beta_t(\rho)|_{t=0}, \tag{19.32}$$

for those density operators ρ for which $\beta_t(\rho)$ is differentiable at $t = 0$. Then $\beta_t(\rho)$ satisfies

$$\partial_t \beta_t(\rho) = \mathcal{L}(\beta_t(\rho)), \; \beta_{t=0}(\rho) = \rho. \tag{19.33}$$

Problem 19.31 Show that (19.31) and (19.32) imply (19.33).

Problem 19.32 Show that α_t given in (19.13) is a qds and find its generator.

Theorem 19.33 Under certain technical continuity conditions, the generators of Markov evolutions are of the form

$$\mathcal{L}(\rho) = -\frac{i}{\hbar}[H, \rho] + \sum_{j=0}^{\infty}(W_j\rho W_j^* - \frac{1}{2}\{W_j^*W_j, \rho\}), \tag{19.34}$$

where H is a self-adjoint operator (system Hamiltonian), $\{A, B\} := AB + BA$ and W_j are bounded operators s.t. $\sum W_j^*W_j$ converges strongly.

For a proof of this theorem see [93, 184] and also [7]. Eq (19.33) with (19.34) is called the *Lindblad equation*.

If an open quantum evolution β_t has the Markov property, then so has its dual, β_t'. The generator, $\mathcal{L}'$, of β_t' is the operator dual to $\mathcal{L}$ (in the coupling between density operators and observables given by (19.19)). Hence β_t' satisfies the differential equation $\frac{\partial}{\partial_t}\beta_t' = \mathcal{L}'\beta_t'$. Furthermore, for $\mathcal{L}$ of the form (19.34), $\mathcal{L}'$ is given by

$$\mathcal{L}'(A) = \frac{i}{\hbar}[H, A] + \sum_i(W_j^*AW_j - \frac{1}{2}\{W_j^*W_j, A\}). \tag{19.35}$$

Write $\mathcal{L} = L_0 + G$, where $L_0\rho := -\frac{i}{\hbar}[H, \rho]$ and $G := L - L_0$. Then, as follows from (19.34), the operator L_0 is formally anti-symmetric, while the operator G is formally symmetric (for self-adjoint W_j's), in the sense that

$$\mathrm{Tr}(AL_0\rho) = -\mathrm{Tr}((L_0A)\rho), \qquad \mathrm{Tr}(AG\rho) = \mathrm{Tr}((GA)\rho),$$

or (abusing notation as primed and not primed operators act on different spaces) $L_0' = -L_0$ and $G' = G$. In addition, G satisfies

$$\mathrm{Tr}(\rho^*G\rho) \leq 0 \tag{19.36}$$

(and therefore is dissipative in the sense of the following definition: An operator G on density operators is said to be *dissipative* if for each density operator ρ there exists an observable A, such that $\mathrm{Re}\,\mathrm{Tr}(AG\rho) \leq 0$.)

Problem 19.34 Show that $\mathrm{Tr}(\rho^*G\rho) \leq 0$ for self-adjoint W_j's. Hint: for self-adjoint W_j's,

$$\mathrm{Tr}(\rho^*G\rho) = -\frac{1}{2}\sum_{j=0}^{\infty}\mathrm{Tr}(([W_j, \rho])^*([W_j, \rho])) \leq 0. \tag{19.37}$$

The fact that β_t' has the eigenvalue 1 (with the eigenvector **1**) is now translated into the property that $\mathcal{L}'$ has the eigenvalue 0 (with the eigenvector

1), which is apparent from expression (19.35). As above, this suggests that $\mathcal{L}$ also has the eigenvalue 0, with the eigenvector being an equilibrium state.

In fact, for the Lindblad equation (19.33)-(19.34) derived, non-rigorously, by the 2nd order and Markov approximations, from reduced dynamics (19.14), with the initial state ρ_e of the environment being the Gibbs state

$$\rho_{Te} = e^{-H_e/T}/Z_{Te}, \tag{19.38}$$

where $Z_{Te} := \mathrm{Tr} e^{-H_e/T}$, at temperature T, see (18.12), we have the following property:

- the operator (19.34) satisfies $L(A\rho_T) = (L'A)\rho_T$,

where $\rho_T \equiv \rho_{Ts}$ is the Gibbs state of the system, $\rho_T = e^{-H/T}/Z_T$, with $Z(T) := \mathrm{Tr} e^{-H/T}$. This property is called the *detailed balance condition*. Taking here $A = 1$ and using $L'1 = 0$, we see that it implies $L(\rho_T) = 0$, i.e.

- the Gibbs state ρ_T is a stationary state of (19.33)-(19.34).

To understand where the latter property comes from, let the Lindblad equation (19.33)-(19.34) be derived as describeded above. Let ρ_{Ts} and ρ_{Te} be the Gibbs states for the system and environment at temperature T. Take $\rho_{Ts} \otimes \rho_{Te}$ as the initial condition in (19.14). Of course, unless the system and environment are decoupled, $\rho_{Ts} \otimes \rho_{Te}$ is not a static solution for the total system (19.11)-(19.12). However, it is a static solution in the leading order in the coupling constant, so that ρ_{Ts} gives a static solution for the approximate reduced dynamics (19.14) generated by $\mathcal{L}$.

Given the properties above, we see that in the inner product $\langle A, B \rangle_T := \mathrm{Tr}(A^* B \rho_T)$ on the algebra of observables, we have that L_0 is anti-self-adjoint, while G is self-adjoint,

$$L_0^* = -L_0, \quad G^* = G^*. \tag{19.39}$$

Problem 19.35 Show the relations (19.39).

Note that if $G = 0$, i.e. for the flow generated by L_0, the energy and entropy, as defined in Section 18.3, are conserved.

Problem 19.36 Prove this statement.

For a general evolution (19.33)-(19.34), this is not true anymore. Instead, as we will show in Section 19.6, if ρ_T is a static solution, then the free energy and relative entropy are non-increasing functions of time (see Proposition 19.40 and Theorem 19.38).

Finally, recall that quantum maps are characterized by either complete positivity, or the explicit form (19.16), which are equivalent by Theorem 19.21 and Homework 19.20. These characterizations extend to quantum dynamical semigroups, since the latter are families of quantum maps. So we expect the generators of these semigroups to be characterized correspondingly. So far, we characterized their generators ($\mathcal{L}$) by the explicit form (19.34) corresponding to (19.16). Thus a natural question is which property of $\mathcal{L}$ corresponds to the complete positivity of β_t? The answer is the *complete dissipativity*, which is defined by lifting the property

$$D_{\mathcal{L}}(A, A) \geq 0, \tag{19.40}$$

where $D_{\mathcal{L}}(A, B)$ is the dissipation function defined by

$$D_{\mathcal{L}}(A, B) := \mathcal{L}'(A^*B) - \mathcal{L}'(A^*)B - A^*\mathcal{L}'(B). \tag{19.41}$$

to $\mathcal{A} \otimes M_k$, for any $k \geq 1$, where $\mathcal{A}$ is an algebra of observables and M_k is the space of $k \times k$ matrices. The form $D_{\mathcal{L}}$ has some remarkable properties:

- $D_{\mathcal{L}}(A, A) = 0$ if and only if $\mathcal{L}'A = \frac{i}{\hbar}[H, A]$ for some self-adjoint operator H; [3]
- $D_{\mathcal{L}}(c\mathbf{1}, c\mathbf{1}) = 0$;
- if $D_{\mathcal{L}}(A, A) = 0$ implies $A = c\mathbf{1}$, then there is a unique stationary state ρ_* and for every ρ, we have $\beta_t(\rho) \to \rho_*$, as $t \to \infty$.

(If ρ_* is an equilibrium state, then the last property is called *return to equilibrium*.)

A linear map $\mathcal{L}'$ of a C^*-algebra satisfying (19.40) and $\mathcal{L}'A^* = (\mathcal{L}'A)^*$ is called a *dissipation*. It is shown in [178] that dissipations are dissipative in the sense of the definition given before (19.36).

19.6 Irreversibility

A quantum (dynamical) map β is said to be *irreversible* iff β is not invertible (within the class of quantum maps) and *reversible*, otherwise. An example of a reversible map is

$$\beta(\rho) = V\rho V^*, \tag{19.42}$$

where V is an invertible operator. In fact, we have (cf. [242], Section 3.2)

Proposition 19.37 Any invertible quantum map β is of form (19.42).

Proof. Assume β is invertible. Then β^{-1} exists and is a quantum map. By Theorem 19.17 (or 19.21), both, β and β^{-1}, are of the form (19.16) - (19.17). Now, denote the V_n-operators for β^{-1} by W_n and rewrite the equation $\beta^{-1} \circ \beta = 1$ as

$$\beta^{-1}(\beta(\rho)) = \sum_{m,n} W_m V_n \rho V_n^* W_m^* = \rho, \tag{19.43}$$

with the bounded operators V_n and W_m satisfying $\sum_n V_n^* V_n = \mathbf{1}$ and $\sum_m W_m^* W_m = \mathbf{1}$. Taking here $\rho = P_\psi$, for an arbitrary normalized vector

---

[3] In the terminology of C^*-algebras, $\mathcal{L}'$ is a *derivation*. In general, a *derivation* is a linear map δ of a C^*-algebra, satisfying $\delta(AB) = \delta(A)B + A\delta(B)$, for all A, B in the domain of δ.

ψ, we see that (19.43) can be rewritten as $\sum_{m,n} P_{W_m V_n \psi} = P_\psi$. Testing this relation on various functions, we see that it holds if and only if every $W_m V_n \psi$ is proportional to ψ:

$$W_m V_n \psi = a_{mn} \psi,$$

for some numbers a_{mn} and for each pair m, n. On the other hand, by the completeness relations above, we have $V_n^* V_k = \sum_m V_n^* W_m^* W_m V_k$. Since $V_n^* W_m^* = (W_m V_n)^*$, this gives

$$V_n^* V_k = \sum_m \bar{a}_{mn} a_{mk} \mathbf{1}.$$

Now, using the polar decomposition $V_k = U_k |V_k|$, where U_k are some unitary operators and $|A| := \sqrt{A^* A}$, the notation $b_{nk} := \sum_m \bar{a}_{mn} a_{mk}$, the observation that $b_{kk} \geq 0, \forall k$ and the relation $|V_k| = \sqrt{b_{kk}}$, we find

$$\sum_m \bar{b}_{nk} \mathbf{1} = \sqrt{b_{nn} b_{kk}} U_n^* U_k$$

and therefore $U_n = \frac{b_{nk}}{\sqrt{b_{nn} b_{kk}}} U_k$ for each n and k. Hence $V_n = b_{nn} U_n = c_{nk} U_k$, where $c_{nk} := b_{nk} \sqrt{\frac{b_{nn}}{b_{kk}}}$, for every n and some fixed k. Remembering (19.16), this gives (19.42) with $V := \sqrt{\sum_n |c_{nk}|^2} V_k$. Since β is invertible, V should be invertible too. $\qquad\square$

How to quantify the notion of irreversibility? In classical mechanics we encounter irreversibility when we pass from Newton's equation to the Boltzmann equation. While Newton's equation is reversible $((x(t), p(t)) \rightarrow (x(-t), -p(-t))$ is a symmetry of Newton's equation), the Boltzmann equation is irreversible: the Boltzmann entropy

$$H(f) = -\int f \log f, \tag{19.44}$$

for the particle densities, $f(x, v, t)$, which solve the Boltzmann equation, increases along the evolution (the celebrated Boltzmann H-theorem).

In quantum mechanics, while the Schrödinger equation is reversible, the reduced evolution, say, $\rho_t = \beta_t(\rho_0)$ (or $\frac{\partial}{\partial t} \rho_t = \mathcal{K} \rho_t$, in the Markov case) it leads to when some information is 'integrated out', is expected to be irreversible. To prove this, we look for an analogue of the Boltzmann entropy, and it is natural to define

$$S(\rho) = -\mathrm{Tr}(\rho \log \rho), \tag{19.45}$$

which is nothing but the von Neumann entropy, defined by von Neumann 20 years earlier, and which was already used in Section 18.3, where the definition of the operator $\log \rho$ was discussed. We list properties of $S(\rho)$:

1) $S(\rho) = 0$ if and only $\rho = P_\psi$ is a pure state;

2) $S(U\rho U^*) = S(\rho)$, for unitaries U;

3) $S(\sum \lambda_j \rho_j) \geq \sum \lambda_j S(\rho_j)$, for $\lambda_j \geq 0, \sum \lambda_j = 1$;

4) For any density operator, ρ_{AB}, of a composed system $A + B$,

$$S(\rho_{AB}) \leq S(\rho_A) + S(\rho_B) \tag{19.46}$$

where ρ_A and ρ_B are the marginals of ρ_{AB}:

$$\mathrm{Tr}_B \rho_{AB} = \rho_A \text{ and } \mathrm{Tr}_A \rho_{AB} = \rho_B.$$

To prove Property 1), we let λ_j and ϕ_j be eigenfunctions and eigenvalues of ρ so that $\rho = -\sum \lambda_j P_{\phi_j}$ and use that due to the relation if, then

$$S(\rho) = -\sum \lambda_j \log \lambda_j = H(\{\lambda_j\}) \tag{19.47}$$

(which can be also used to define $S(\rho)$; remember $0 \leq \rho \leq 1$ and therefore $0 \leq \lambda_j \leq 1$ and $S(\rho) \geq 0$). Then the equation $S(\rho) = 0$ can be rewritten as $\sum \lambda_j \log \lambda_j = 0$. Since every term in the sum on the l.h.s. is non-positive, it should vanish separately. So λ_j's are either 1 or 0. Next, 2) follows from the cyclicity of the trace and 3), from concavity of the function $x \log x$. Finally, the subadditivity of the entropy, (19.46), was proven in [75, 187].

However, there is no H-theorem for $S(\rho)$, i.e, in general, $S(\rho)$ does not decrease (or increase) under the evolution. We look for a more general object which has monotonicity properties. Such a candidate is the relative entropy defined as

$$S(\rho_1, \rho_2) := \mathrm{Tr}(\rho_1(\log \rho_1 - \log \rho_2)), \tag{19.48}$$

if $\overline{\mathrm{Ran}\rho_1} = \overline{\mathrm{Ran}\rho_2}$ and ∞, otherwise. We have the following result whose proof can be found in [212] and [292], Statement 3.1.12:

Theorem 19.38 (Generalized H-theorem (Lindblad)) If β is a quantum map, then

$$S(\beta(\rho_1), \beta(\rho_2)) \leq S(\rho_1, \rho_2). \tag{19.49}$$

Note that, if β is reversible, then (19.49) holds with equality.

This shows that $S(\beta_t(\rho_1), \beta_t(\rho_2))$ is monotonically non-increasing, as t increases, and is constant if β_t is reversible.

It is shown in [229] that *for finite-dimensional spaces*, any transformation β, which satisfies the equality in (19.49) – i.e., β preserves the relative entropy – is reversible:

$$S(\beta(\rho_1), \beta(\rho_2)) = S(\rho_1, \rho_2) \quad \Longleftrightarrow \quad \beta \text{ is reversible.} \tag{19.50}$$

Thus, by this theorem, the strict decrease of the relative entropy implies that the map is irreversible.

We say a quantum (dynamical) map β is *unitary* if and only if it is of the form (19.42) with V unitary. [4]

If ρ_* is a stationary state, then Theorem 19.38 implies that for a quantum dynamical semigroup, β_t, the relative entropy $S(\beta_t(\rho), \rho_*)$ is monotonically non-increasing, while (19.50) says it is constant if and only if β_t is unitary. This can be strengthened to the following result ([282]):

Theorem 19.39 Let β_t be a quantum dynamical semigroup on a finite-dimensional space. Assume β_t has a stationary state ρ_*. Then $S(\beta_t(\rho), \rho_*)$ is differentiable in t and the entropy production, defined by the relation

$$\sigma(\rho) := -\partial_t S(\beta_t(\rho), \rho_*)|_{t=0}, \tag{19.51}$$

is non-negative and convex. Moreover, it is equal to 0 if and only if β_t is unitary (and consequently, its generator $\mathcal{L}$ is a commutation, $\mathcal{L}\rho = i[H', \rho]$ for some H').

Recall from Section 18.3 the definition of for the free energy (see (18.13))

$$F_T(\rho) := E(\rho) - TS(\rho). \tag{19.52}$$

Proposition 19.40 Let β_t be a quantum dynamical semigroup on a finite-dimensional space. Assume ρ_T is a static state of β_t. Then the free energy F_T is non-increasing under β_t, and is constant if and only if β_t is unitary.

Proof. By definitions (19.38) and (19.48), the free energy (19.52) can be expressed in terms of the relative entropy for ρ and ρ_T as

$$F_T(\rho) = TS(\rho, \rho_T) + T \ln Z_T. \tag{19.53}$$

This relation, together with (19.51) (for $\rho_* = \rho_T$) and the semi-group property of β_t, implies

$$\partial_t F_T(\rho_t) = -T\sigma(\rho_t), \tag{19.54}$$

where $\rho_t := \beta_t(\rho)$, which, by Theorem 19.39, yields the result.

---

[4] If V is unitary, then the adjoint quantum map β^* is a homomorphism, i.e. it satisfies $\beta^*(AB) = \beta^*(A)\beta^*(B)$.

19.7 Decoherence and Thermalization

Consider an open quantum system with a state space $\mathcal{H}_S$ and an evolution β_t. Let $\{\varphi_j\}_{j\geq 1}$ be a *preferred* orthonormal basis in $\mathcal{H}_S$, which will be specified below. The system is said to exhibit (full) *decoherence (in the preferred basis)* if for any initial condition ρ_0, the off-diagonal matrix elements $\langle\varphi_m, \beta_t(\rho_0)\varphi_n\rangle$ of its evolution $\beta_t(\rho_0)$ vanish in the limit of large time:

$$\lim_{t\to\infty} \langle\varphi_m, \beta_t(\rho_0)\varphi_n\rangle = 0, \qquad (19.55)$$

whenever $m \neq n$. Denote $\bar{\rho}_t := \beta_t(\rho_0)$ and $[\bar{\rho}_t]_{m,n} := \langle\varphi_m\bar{\rho}_t\varphi_n\rangle$. We claim then that decoherence implies that after some time

$$\mathrm{Tr}(A\bar{\rho}_t) \sim \sum_m p_m\langle\varphi_m, A\varphi_m\rangle, \qquad (19.56)$$

where $p_m := [\bar{\rho}_t]_{m,m}$, which is a statistical sum. Indeed, inserting the completeness relation, $\sum_m |\varphi_m\rangle\langle\varphi_m| = \mathbf{1}$, for the basis $\{\varphi_m\}$ on both sides of A, we find

$$\mathrm{Tr}(A\bar{\rho}_t) = \sum_{m,n}[\bar{\rho}_t]_{m,n}\langle\varphi_n, A\varphi_m\rangle. \qquad (19.57)$$

Eqs (19.55) and (19.57) imply the desired relation (19.56).

Now assume our open system originates by reduction from a total system which includes an environment. If $H \equiv H_S$ is the original hamiltonian of the system of interest (S), then the preferred basis is the one of (generalized) eigenvectors of H_S, called the *energy* basis, and the quantum evolution β_t is the reduced one, $\beta_t(\rho_0) = \mathrm{Tr}_E \alpha_t(\rho_0 \otimes \rho_e)$.

The environment plays a crucial role in the phenomenon of decoherence. Indeed, if the system and environment do not interact, then there is no decoherence.

To gain more insight into the role of the environment, let the evolution of the total system be given by $\Psi = \Psi_t$. Expanding this in a basis $\{\varphi_m\}$ in $L^2(dx)$, we find $\Psi = \sum_m \varphi_m \otimes \alpha_m$, for some $\alpha_m \in L^2(dy)$. We compute the reduced density operator for this state. Let $\{\chi_i\}$ be an orthonormal basis in $\mathcal{H}_E$. Then $\mathrm{Tr}_E P_\Psi = \sum_i\langle\chi_i, \Psi\rangle_E\langle\Psi, \chi_i\rangle_E = \sum_{m,n,i}\langle\chi_i, \alpha_m\rangle_E|\varphi_m\rangle\langle\alpha_n, \chi_i\rangle_E\langle\varphi_n|$. Since $\{\chi_i\}$ is an orthonormal basis, we have $\sum_i\langle\alpha_n, \chi_i\rangle_E\langle\chi_i, \alpha_m\rangle_E = \langle\alpha_n, \alpha_m\rangle_E$, which together with the previous relation gives

$$\bar{\rho}_t = \mathrm{Tr}_E P_\Psi = \sum_{m,n}[\bar{\rho}_t]_{m,n}|\varphi_m\rangle\langle\varphi_n|,$$

where $[\bar{\rho}_t]_{m,n} := \langle\varphi_m, \bar{\rho}_t\varphi_n\rangle_S = \langle\alpha_m, \alpha_n\rangle_E$. Hence decoherence means that the environmental dynamics forces $\langle\alpha_m, \alpha_n\rangle_E \to 0$, whenever $m \neq n$.

According to Subsection 18.3, the equilibrium states of the total system, and environment, at temperature $T = 1/\beta$ are given by $R_\beta := e^{-\beta H_T}/Z(\beta)$, $Z(\beta) := \mathrm{Tr}e^{-\beta H_T}$, and $\rho_{E,\beta} := e^{-\beta H_E}/Z_E(\beta)$, $Z_E(\beta) := \mathrm{Tr}e^{-\beta H_E}$. (The subindex T for "total" should not be confused with the temperature $T = 1/\beta$.) We say that the total system $T = S + E$ has the property of *return to equilibrium* iff

$$\lim_{t\to\infty} \mathrm{Tr}_{S+E}(\alpha_t(A)(\rho_0 \otimes \rho_{E,\beta})) = \mathrm{Tr}_{S+E}(AR_\beta),$$

for all observables (of the joint system) A, and for all initial density matrices ρ_0 on $\mathcal{H}_S$. If this happens, then the large time limit of the reduced density matrix of such a system is given by

$$\overline{\rho}_\infty := \lim_{t\to\infty} \overline{\rho}_t = \mathrm{Tr}_E(R_\beta) = \rho_{S,\beta} + O(\lambda),$$

where λ is the coupling constant between in the system S and environment (see (19.11)). In the second relation, we used that, by perturbation theory of equilibrium states, $R_\beta = \rho_{S,\beta} \otimes \rho_{E,\beta} + O(\lambda)$. The leading term of $\overline{\rho}_\infty$ (for small coupling constant λ) is just the Gibbs state of the system (see e.g. [45, 224]). In this sense, the system undergoes the process of *thermalization*.

The Second Quantization

In this chapter we describe a powerful technique used in the analysis of quantum many-body systems – the method of second quantization. In rough terms, it allows one, instead of working with a fixed number of particles, to let the number of particles fluctuate, while keeping the average number fixed. We apply this method to derive a useful mean-field limit for many-body dynamics. The method also provides a natural language for quantum field theory (see Chapters 21 and 22) and can be used to preview, in a much simpler setting, some of the issues arising there.

20.1 Fock Space and Creation and Annihilation Operators

Consider a system of n identical particles moving in an external potential $W(x)$ and interacting via pair potentials $v(x_i - x_j)$. Its Schrödinger operator is given by

$$H_n = \sum_{i=1}^{n} (-\frac{\hbar^2}{2m}\Delta_{x_i} + W(x_i)) + \frac{1}{2}\sum_{i \neq j} v(x_i - x_j). \qquad (20.1)$$

We assume the particles are bosons. In this case H_n acts on the space $L^2_{sym}(\mathbb{R}^{3n})$ of L^2 functions on $\mathbb{R}^{3n}$ which are symmetric w.r.t permutations of the particle coordinates $x_1, ..., x_n \in \mathbb{R}^3$. We define the new Hilbert space

$$\mathcal{F}_{bos} := \oplus_{n=0}^{\infty}\mathcal{F}_n, \qquad (20.2)$$

of sequences $\Phi = (\Phi_0, \Phi_1, \dots) \equiv \oplus_{n=0}^{\infty}\Phi_n$, $\Phi_n \in \mathcal{F}_n$, where $\mathcal{F}_0 = \mathbb{C}$, and $\mathcal{F}_n = L^2_{sym}(\mathbb{R}^{3n}), n \geq 1$, equipped with the inner product

$$\langle \Psi, \Phi \rangle = \sum_{n=0}^{\infty} \int \overline{\Psi_n(x_1, ..., x_n)}\Phi_n(x_1, ..., x_n)d^n x, \qquad (20.3)$$

© Springer-Verlag GmbH Germany, part of Springer Nature 2020
S. J. Gustafson and I. M. Sigal, *Mathematical Concepts of Quantum Mechanics*, Universitext, https://doi.org/10.1007/978-3-030-59562-3_20

where Ψ_n and Φ_n are the n–th components of Ψ and Φ. The space $\mathcal{F}_{bos}$ is called the *bosonic Fock space*. On $\mathcal{F}_{bos}$ we define the operator

$$H = \oplus_{n=0}^{\infty} H_n, \tag{20.4}$$

where $H_0 = 0$, $H_1 = -\frac{\hbar^2}{2m}\Delta_x + W(x)$ and H_n, $n \geq 2$, are as above, so that $H\Psi = \oplus_{n=0}^{\infty} H_n \Psi_n$. The operator H is called the 2nd quantized Schrödinger operator. The reason for this name will become clear later. The vector $\Omega := (1, 0, 0, ...) \in \mathcal{F}_{bos}$ is called the *vacuum vector* in $\mathcal{F}_{bos}$. Note that, by construction, $H\Omega = 0$.

Problem 20.1 Assuming the pair potentials, $v(x_i - x_j)$, are real and bounded, show that H is self-adjoint.

One of the advantages of the 2nd quantization is the representation of operators on the Fock space $\mathcal{F}_{bos}$ in terms of annihilation and creation operators (raising and lowering the number of particles). These are the operator-valued distributions $f \to a(f)$ and $f \to a^*(f)$, where $f \in L^2(\mathbb{R}^3)$ and $a(f)$ and $a^*(f)$ are operators on $\mathcal{F}_{bos}$, defined as

$$(a(f)\Psi)_n = \sqrt{n+1} \int \overline{f(x)}\Psi_{n+1}(x, x_1, ..., x_n)dx, \tag{20.5}$$

and $a(f)\Omega = 0$, and

$$(a^*(f)\Phi)_{n+1} = \sqrt{n+1}(f \circledS \Phi_n) \tag{20.6}$$

for $n \geq 0$. Here $f \circledS \Phi_n = P_{n+1}^S(f \otimes \Phi_n)$, with P_n^S the orthogonal projection from $L^2(\mathbb{R}^{3n})$ to $L^2_{sym}(\mathbb{R}^{3n})$:

$$(P_n^S f)(x_1, ..., x_n) := \frac{1}{n!} \sum_{\pi \in S_n} f(x_{\pi(1)}, ..., x_{\pi(n)}),$$

where, recall, S_n is the symmetric group of permutations of n indices. The operators $a(f)$ and $a^*(f)$ are unbounded and satisfy

$$\langle a^*(f)\Phi, \Psi \rangle = \langle \Phi, a(f)\Psi \rangle. \tag{20.7}$$

Indeed, we compute, using (20.5), on vectors Φ, Ψ, with finite numbers of components,

$$\langle \Phi, a(f)\Psi \rangle = \sum_n \langle \Phi_n, (a(f)\Psi)_n \rangle$$

$$= \sum_n \sqrt{n+1} \int dx_1 ... dx_{n+1} \overline{\Phi_n(x_1, ...x_n)f(x_{n+1})}\Psi_{n+1}(x_1, ..., x_{n+1}).$$

It is easy to see by relabeling the variables of integration that $\Phi_n(x_1, ...x_n)f(x_{n+1})$ can be replaced by

$$P_{n+1}^S(f(x_{n+1})\Phi_n(x_1,...,x_n))$$

$$= \frac{1}{n+1} \sum_j f(x_j)\Phi_n(x_1,\ldots,x_{j-1},x_{j+1},\ldots,x_{n+1})$$

to obtain

$$\langle \Phi, a(f)\Psi \rangle$$

$$= \sum_n \sqrt{n+1} \int d^{n+1}x P_{n+1}^S(\overline{\Phi_n(x_1,...,x_n)}f(x_{n+1}))\Psi_{n+1}(x_1,...,x_{n+1}),$$

which, together with (20.6), implies (20.7). Moreover, $a^*(f)$ is adjoint to the operator $a(f) : a^*(f) = a(f)^*$.

Problem 20.2 Show that

$$[a(f), a^*(g)] = \langle f, g \rangle, \qquad [a(f), a(g)] = [a^*(f), a^*(g)] = 0. \tag{20.8}$$

The operators $a(f), a^*(f)$ are operator-valued distributions and it is convenient to introduce the formal notation $a^\#(x) = a^\#(\delta_x)$, so that, formally,

$$a(f) = \int \overline{f(x)}a(x)dx, \qquad a^*(f) = \int f(x)a^*(x)dx. \tag{20.9}$$

We consider $a(x), a^*(x)$ as formal symbols satisfying

$$[a(x), a^*(y)] = \delta(x-y), \quad [a^\#(x), a^\#(y)] = 0. \tag{20.10}$$

Representation of vectors in terms of creation operators.

Proposition 1. *Any* $\Phi = \oplus \Phi_n$ *can be written as*

$$\Phi = \sum_n \frac{1}{\sqrt{n!}} \int \Phi_n(x_1,...,x_n) \prod_{j=1}^n a^*(x_j)\Omega d^n x. \tag{20.11}$$

Proof. Using the definition of $a^*(x)$, we compute

$$(a^*(y_n)...a^*(y_1)\Omega)_m = \sqrt{n!}P_n^S(\prod_{j=1}^n \delta(x_i - y_i))\delta_{n,m}. \tag{20.12}$$

Problem 20.3 Show (20.12). Hint: Use induction starting with

$$(a^*(y_1)\Omega)_n(x_1) = \sqrt{1}P_1^S(\delta(x_1 - y_1))\delta_{n,1}$$

and

$$(a^*(y_2)a^*(y_1)\Omega)_n(x_1, x_2) = \sqrt{2}P_2^S(\delta(x_1 - y_1)\delta(x_2 - y_2))\delta_{n,2}.$$

The equation (20.12) implies that

$$\frac{1}{\sqrt{n!}} \int \Phi_n(y_1,...,y_n) \prod a^*(y_j)\Omega d^n y = P_n^S \Phi_n(x_1,...,x_n), \tag{20.13}$$

which implies the statement of the proposition. $\square$

20.2 Many-body Hamiltonian

One-particle operators. Let b be an operator on $L^2(\mathbb{R}^3)$. We write it as an integral operator $b_x f(x) = \int b(x, y) f(y) dy$, where $b(x, y)$ is the integral kernel of b. We think of $b_x a(x)$ as the operator b acting on the parameter x, $b_x a(x) = \int b(x, y) a(y) dy$. We define an operator B on $\mathcal{F}_{bos}$ by the formula

$$B = \int a^*(x) b_x a(x) dx. \qquad (20.14)$$

We call operators of this type one-particle operators. The equation (20.14) is an integral of the product of operator-valued distributions and it is not clear whether it is well-defined to begin with. However, if one thinks about (20.14) as a formal expression and uses formally the definitions of $a(x)$ and $a^*(x)$, one finds that the operator B defined by (20.14) is equal to

$$B = \oplus_{n=0}^{\infty} B_n, \ B_0 = 0, \ B_n := \sum_{i=1}^{n} b_i, \ n \geq 1, \qquad (20.15)$$

where $b_i \equiv b_{x_i}$ stands for the operator b acting on the variable x_i. The latter expression makes a perfect sense.

One can demonstrate (20.15) formally as follows. Using the definitions of $a(x) = a(\delta_x)$ and $a^*(x) = a^*(\delta_x)$ and (20.5) and (20.6)

$$
\begin{aligned}
(B\Phi)_n &= \sqrt{n} \int dx P_n^S [\delta(x_1 - x)(b_x a(x)\Phi)_{n-1}(x_2, ..., x_n)] \\
&= \sqrt{n} \int dx P_n^S [\delta(x_1 - x) b_x \sqrt{n} \Phi_n(x, x_2, ..., x_n)] \\
&= n P_n^S [b_{x_1} \Phi_n(x_1, ..., x_n)] \\
&= \sum_{i=1}^{n} b_{x_i} \Phi_n(x_1, ..., x_n),
\end{aligned}
$$

as claimed.

There is a different representation of (20.14) which is well defined directly. Let $\{f_i\}$ be an orthonormal basis in $L^2(\mathbb{R}^3)$ and define the operators $a_i := a(f_i)$ and $a_i^* := a^*(f_i)$. Assume $f_i \in D(b)$ and let $b_{ij} := \langle f_i, b f_j \rangle$ be the matrix associated with the operator b and the basis $\{f_i\}$. Consider the expression

$$B = \sum_{ij} a_i^* b_{ij} a_i, \qquad (20.16)$$

defined on vectors for which the sum on the r.h.s. converges (say on vectors with finite number of components).

Problem 20.4 Show that the operator $\sum_{ij} a_i^* b_{ij} a_i$ acts as (20.15).

Formally, the expression (20.16) is obtained from (20.14) one by inserting the partition of unity $\sum_n |f_n\rangle\langle f_n| = \mathbf{1}$ into the latter expression. Though (20.14) needs an additional interpretation, it is much more convenient to work with and is used commonly.

Consider a few examples. The free Schrödinger operator:

$$H_0 = \int a^*(x)(-\frac{\hbar^2}{2m}\Delta_x)a(x)dx. \tag{20.17}$$

The equation (20.15) shows that $H_0 = \oplus_{n=0}^\infty H_{0n}$, where $H_{0n} = \sum_{i=1}^n(-\frac{\hbar^2}{2m}\Delta_{x_i})$ is the free n-particle Schrödinger operator ($V = 0$).

We define the *number operator* $N = \int a^*(x)a(x)dx$ and *momentum operator* $P = \int a^*(x)(-i\hbar\nabla_x)a(x)dx$. Then $N = \oplus_{n=0}^\infty n\mathbf{1}$ and $P = \oplus_{n=0}^\infty P_n$ where $P_n = \sum_{j=1}^n p_j$, respectively. We have

$$[H_0, N] = 0, \quad [H_0, P] = 0. \tag{20.18}$$

These equations imply the conservation of the particle number and total momentum. We prove $[H_0, P] = 0$:

$$[H_0, P] = [\int dx a^*(x)(-\frac{\hbar^2}{2m}\Delta_x)a(x), \int dy a^*(y)(-i\hbar\nabla_y)a(y)]$$

$$= \int\int dx dy a^*(x)(-\frac{\hbar^2}{2}\Delta_x)[a(x), a^*(y)](-i\hbar\nabla_y)a(y)$$

$$+ \int\int a^*(y)(-i\hbar\nabla_y)[a(x), a^*(y)](-i\hbar\Delta_x)a(x)$$

$$= \int dx a^*(x)[-\frac{\hbar^2}{2m}\Delta, -i\hbar\nabla]a(x) = 0,$$

where in the last step we used that $[a(x), a^*(y)] = \delta(x - y)$.

Problem 20.5 Show that $[a^*(x)\Delta_x a(x), a(y)] = [a^*(x), a(y)]\Delta_x a(x)$.

Problem 20.6 Show (20.18).

In mathematics, one denotes the operator in (20.15) as $d\Gamma(b)$.

Problem 20.7 Show formally that $[d\Gamma(b_1), d\Gamma(b_2)] = d\Gamma([b_1, b_2])$.

We can generalize (20.17) to a particle moving in an external field W:

$$H_W = \int a^*(x)(-\frac{\hbar^2}{2m}\Delta_x + W(x))a(x)dx. \tag{20.19}$$

Two-particle operators. Let

$$V = \frac{1}{2}\int a^*(x)a^*(y)v(x - y)a(x)a(y)dxdy. \tag{20.20}$$

(The integrand on the r.h.s. annihilates particles at x and y, acts with $v(x-y)$, and then creates particles at x and y.) We claim that

$$(V\Phi)_n = \frac{1}{2} \sum_{i \neq j} v(x_i - x_j)\Phi_n(x_1, ..., x_n). \qquad (20.21)$$

Indeed, consider $a(x)a(y)\prod_1^n a^*(x_j)\Omega$. Using this expression and (20.20) and using $a(y)a^*(z) = a^*(z)a(y) + \delta(y-z)$, we pull $a(y)$ through $\prod_{j=1}^n a^*(x_k)$ and use $a(y)\Omega = 0$, to obtain

$$a(x)a(y)\prod_{i=1}^n a^*(x_j)\Omega = a(x)\sum_{i=1}^n \delta(y-x_i)\prod_{k \neq i} a^*(x_i)\Omega. \qquad (20.22)$$

Then we pull similarly $a(x)$ trough $\prod_{k \neq i} a^*(x_k)$ and use $a(x)\Omega = 0$ to find

$$a(x)a(y)\prod_{i=1}^n a^*(x_j)\Omega = \sum_{i=1}^n \delta(y-x_i)\sum_{j \neq i}\delta(x-x_j)\prod_{k \neq i,j}^n a^*(x_k)\Omega. \qquad (20.23)$$

Let $\Phi_n = \frac{1}{\sqrt{n!}} \int \Phi_n(x_1, ..., x_n)\prod_1^n a^*(x_j)\Omega d^n x$. Using (20.23) and (20.20) and integrating the delta-functions, we arrive at

$$V\Phi_n = \frac{1}{\sqrt{n!}} \sum_{i=1}^n \sum_{j \neq i} v(x_i - x_j)\Phi_n(x_1, ..., x_n)\prod_{i=1}^n a^*(x_k)\Omega, \qquad (20.24)$$

which gives (20.21).   □

The equations (20.19) and (20.21) show that the Hamiltonian (20.4) introduced at the beginning of this chapter can be written as

$$H = \int a^*(x)(-\frac{\hbar^2}{2m}\Delta_x + W(x))a(x)dx$$
$$+ \frac{1}{2}\int a^*(x)a^*(y)v(x-y)a(x)a(y)dxdy. \qquad (20.25)$$

Problem 20.8 Show that $[V, N] = 0$, $[V, P] = 0$, $[H, N] = 0$, $[H_{W=0}, P] = 0$.

20.3 Evolution of Quantum Fields

Consider the Heisenberg evolution of $a(x)$: $a(x, t) = e^{\frac{iHt}{\hbar}} a(x)e^{-\frac{iHt}{\hbar}}$. Then $a(x, t)$ satisfies the Heisenberg equation,

$$i\hbar\frac{\partial}{\partial t}a(x, t) = -[H, a(x, t)]. \qquad (20.26)$$

Using that $[H, a(x, t)] = e^{\frac{iHt}{\hbar}}[H, a(x)]e^{-\frac{iHt}{\hbar}}$, and computing the commutator $[H, a(x)]$, one can show that $a(x, t)$ satisfies the differential equation

$$i\hbar\frac{\partial}{\partial t}a(x,t) = (-\frac{\hbar^2}{2m}\Delta_x + W(x))a(x,t)$$
$$+ \int dy v(x-y)a^*(y,t)a(y,t)a(x,t).$$
(20.27)

Problem 20.9 Show (20.26) and (20.27).

Note that if $v = 0$, then (20.27) is the Schrödinger equation but for the operator valued function $a(x,t)$, called a *quantum field*. This is the origin of the term the second quantization.

20.4 Relation to Quantum Harmonic Oscillator

Assume the potential W is positive and confining, so that the operator $-\frac{\hbar^2}{2m}\Delta_x + W(x)$ has positive, purely discrete spectrum $\{\lambda_j\}$ accumulating to infinity. Let $\{f_j\}$ be an orthonormal basis of its eigenfunctions. Then applying to the operator H_W the formula (20.16), with this basis and using that the matrix of the operator $-\frac{\hbar^2}{2m}\Delta_x + W(x)$ in this basis is diagonal with the diagonal elements $\{\lambda_j\}$, we obtain

$$H_W = \sum_j \lambda_j a_j^* a_j.$$
(20.28)

On the other hand, consider the r-dimensional quantum harmonic oscillator, described by the Schrödinger operator

$$H_{ho} = -\frac{\hbar^2}{2m}\Delta + \frac{1}{2}\sum_{i=1}^{r} m\omega_i^2 x_i^2$$
(20.29)

on $L^2(\mathbb{R}^r)$. As we know from Section 7.4, it can be rewritten in a form similar to (20.28):

$$H_{ho} = \sum_{i=1}^{r} \hbar\omega_i \left(a_i^* a_i + \frac{1}{2}\right),$$
(20.30)

where, recall, a_i and a_i^* are the harmonic oscillator annihilation and creation operators,

$$a_j := \frac{1}{\sqrt{2m\hbar\omega_j}}(m\omega_j x + ip_j) \text{ and } a_j^* := \frac{1}{\sqrt{2m\hbar\omega_j}}(m\omega_j x_j - ip_j).$$
(20.31)

(These operators satisfy the commutation relation $[a_i, a_j^*] = \delta_{ij}$.) We see that, modulo the additive constant, $\frac{1}{2}\sum_{i=1}^{r}\hbar\omega_i$, (20.30), with $\hbar\omega_j = \lambda_j$, is a finite-dimensional approximation to (20.28).

20.5 Scalar Fermions

As before, to simplify notation slightly we consider scalar (or spinless) fermions. (For fermions with spin we would have to introduce one more variable - spin- for each particle and require that functions below are antisymmetric with respect to permutations of particle coordinates and spins.) We define the Fock space for scalar fermions as

$$\mathcal{F}_{fermi} := \oplus_{n=0}^{\infty} \mathcal{F}_n, \tag{20.32}$$

where $\mathcal{F}_0 = \mathbb{C}$, and $\mathcal{F}_n = L^2_{asym}(\mathbb{R}^{3n}), n \geq 1$, with $L^2_{asym}(\mathbb{R}^{3n})$ the space of functions, $\Psi \in L^2(\mathbb{R}^{3n})$, anti-symmetric w.r.t permutations of coordinates $x_1, ..., x_n \in \mathbb{R}^3$:

$$\Psi(x_{\pi(1)}, ..., x_{\pi(n)}) = (-1)^{\#(\pi)}\Psi(x_1, ..., x_n), \ \pi \in S_n,$$

where, recall S_n is the symmetric group of permutations of n indices and $\#(\pi)$ is the number of transpositions in the permutation π. We equip this space with the inner product (20.3). Again, the vector $\Omega := (1, 0, 0, ...) \in \mathcal{F}_{fermi}$, is called the *vacuum vector* in $\mathcal{F}_{fermi}$. The second quantized hamiltonian H is again defined by (20.4) and $H\Omega = 0$.

The creation and annihilation operators are defined now as

$$(a(f)\Psi)_n = \sqrt{n+1} \int \overline{f(x)}\Psi_{n+1}(x, x_1, ..., x_n)dx, \tag{20.33}$$

for $n \geq 0$, and $a(f)\Omega = 0$, and

$$(a^*(f)\Phi)_{n+1} = \sqrt{n+1}P^A_{n+1}(f(x_1)\Phi_n(x_2, ..., x_{n+1})), \tag{20.34}$$

where P^A_n is the orthogonal projection of $L^2(\mathbb{R}^{3n})$ to $L^2_{asym}(\mathbb{R}^{3n})$. As before, $a^*(f)$ is adjoint to the operator $a(f) : a^*(f) = a(f)^*$. However, unlike the bosonic annihilation and creation operators, the fermionic ones anti-commute

$$[a(f), a^*(g)]_+ = \langle f, g \rangle, \qquad [a(f), a(g)]_+ = [a^*(f), a(g)]_+ = 0, \tag{20.35}$$

where $[A, B]_+ := AB + BA$. A remarkable fact is that the fermionic annihilation and creation operators are bounded.

Problem 20.10 Show this. Hint: Use the anti-commutation relation $a(f)a^*(f) + a^*(f)a(f) = \|f\|^2$.

The second-quantized operator (20.4) is still expressed in the form (20.25) and the derivations above concerning (20.25) remain true.

20.6 Mean Field Regime

We introduce the coupling constant g into our Hamiltonian (20.25) by replacing v by gv and consider regime where $g \ll 1$ and the number of particles $n \gg 1$ but $ng = O(1)$ so that the kinetic and potential energy terms in (20.1) are of the same order. We rescale our Hamiltonian by defining:

$$a(f) = \frac{1}{\sqrt{g}}\psi_g(f) \text{ and } a^*(f) = \frac{1}{\sqrt{g}}\psi_g^*(f).$$

The rescaled creation and annihilation operators are operator-valued distributions obeying commutation relations

$$[\psi_g(x),\, \psi_g^*(y)] \;=\; g\,\delta^3(x-y), \text{ and rest} = 0.$$

The Hamiltonian H can be written in terms of $\psi_g(x)$ and $\psi_g^*(x)$ as $H = g^{-1}\hat{H}$ with

$$
\begin{aligned}
\hat{H} \;=\; & \int \psi_g^*(x)\left(-\frac{\hbar^2}{2m}\Delta + W(x)\right)\psi_g(x)d^3x \\
& + \int \psi_g^*(x_1)\,\psi_g^*(x_2)\,v(x_1-x_2)\,\psi_g(x_1)\psi_g(x_2)d^3x_1 dx_2.
\end{aligned}
\tag{20.36}
$$

In the mean-field limit, as $g \to 0$, the rescaled creation and annihilation operators, $\psi_g(x)$ and $\psi_g^*(x)$, commute, and our quantum theory converges to the classical one, which is a Hamiltonian theory with the phase space given by a space of differentiable functionals (*classical field observables*) $A(\psi, \overline{\psi})$, with the *Poisson bracket* defined (for two functionals $A(\psi,\overline{\psi})$ and $B(\psi,\overline{\psi})$) as

$$\{A, B\}(\psi,\overline{\psi}) = \frac{i}{\hbar}\int (\partial_{\psi(x)}A\partial_{\overline{\psi}(x)}B - \partial_{\overline{\psi}(x)}A\partial_{\psi(x)}B)dx, \tag{20.37}$$

and the *Hamiltonian functional* given by

$$
\begin{aligned}
H(\psi,\overline{\psi}) := & \int \overline{\psi}(x)\,h_x\,\psi(x)d^3x \\
& + \int \overline{\psi}(x_1)\,\overline{\psi}(x_2)\,v(x_1-x_2)\,\psi(x_1)\psi(x_2)d^3x_1 d^3x_2,
\end{aligned}
\tag{20.38}
$$

with $h_x := -\frac{\hbar^2}{2m}\Delta + W(x)$. With ψ considered as an evaluation functional $\psi \to \psi(x)$, the Hamilton equation is given by

$$\partial_t\psi = \{H(\psi,\overline{\psi}), \psi\}, \tag{20.39}$$

which has the form similar to (20.26). (For details and discussions see mathematical supplement, Chapter 26, and especially Section 26.6.) Using (20.38) and (20.37) this equation can be written explicitly as

$$i\hbar\partial_t\psi = (-\frac{\hbar^2}{2m}\Delta + W(x) + v * |\psi|^2)\psi, \qquad (20.40)$$

which is the *Hartree equation* introduced in Section 14.1. Thus we arrived at the Hartree Hamiltonian system with the Hamiltonian functional $H(\psi,\overline{\psi})$ and the Poisson bracket $\{A,B\}(\psi,\overline{\psi})$ and with dynamics given by the Hartree equation (20.40). For more details see Sections 4.7, 26.6 and 26.7.

To formalize this heuristic analysis we consider analytic functionals, $A(\psi,\overline{\psi})$, i.e. functionals of the form

$$A(\psi,\overline{\psi}) := \sum_{pq} \int \prod_1^p \psi(x_i)a_{pq}(x_1,\ldots,x_p;y_1,\ldots,y_q)\prod_1^q \psi(y_i)dxdy, \quad (20.41)$$

where $a_{pq}(x_1,...,x_p;y_1,...,y_q)$ are the integral kernels of bounded operators a_{pq}, with series converging in an appropriate topology.

Now, consider the Wick quantization, i.e. association with classical observables $A = A(\psi,\overline{\psi})$ their quantum counterparts:

$$A = A(\psi,\overline{\psi}) \longrightarrow \hat{A} = A(\psi_g,\psi_g^*),$$

according to the rule

$$\hat{A} := \sum_{pq} \int\int \prod_{i=1}^p \psi_g^*(x_i)a_{pq}(x_1,\ldots,x_p;y_1,\ldots,y_q)\prod_{i=1}^q \psi_g(y_i)d^pxd^qy. \quad (20.42)$$

The operator $\hat{A}$ is said to be the *Wick quantization* of the classical field observable A and A is said to be the *Wick symbol* of the operator $\hat{A}$.

Let Φ_t be the flow generated by the Hartree equation (20.39), or (20.40), i.e. for any reasonable ψ_0, $\Phi_t(\psi_0)$ is the solution of (20.40) with the initial condition ψ_0. It defined the classical evolution of classical observables according to $\alpha_t^{cl}(A) := A \circ \Phi_t$ where $(A \circ \Phi_t)(\psi,\overline{\psi}) = A(\Phi_t(\psi),\overline{\Phi_t(\psi)})$. Furthermore, let $\alpha_t(\hat{A}) := e^{i\hat{H}t/g}\hat{A}e^{-i\hat{H}t/g}$ be the Heisenberg dynamics of quantum observables. Finally, let $N := g^{-1}\int \psi_g^*(x)\,\psi_g(x)d^3x$ be the particle number operator and $\tau := \frac{\hbar}{8\|v\|_\infty}$. One can show (see [124, 16]) that for a certain class of classical observables A and on states, Ψ, satisfying $\|N\Psi\| \leq c/g$, we have for $t \leq \tau$, as $g \to 0$,

$$\alpha_t(\hat{A}) = \widehat{\alpha_t^{cl}(A)} + O(g). \qquad (20.43)$$

We sketch a proof of (20.43). Using the Duhamel principle (i.e. writing $\alpha_{-t}(\widehat{\alpha_t^{cl}(A)})-\hat{A}$ as the integral of derivative $\partial_r\alpha_{-r}(\widehat{\alpha_r^{cl}(A)}) = \alpha_{-r}([\hat{H},\widehat{\alpha_r^{cl}(A)}]-\{H,\widehat{\alpha_r^{cl}(A)}\}))$, we obtain

$$\alpha_t(\hat{A}) - \widehat{\alpha_t^{cl}(A)} = \int_0^t ds\alpha_{t-s}(R(\alpha_s^{cl}(A))), \qquad (20.44)$$

where $R(A) := \frac{i}{g}[\hat{H},\hat{A}] - \widehat{\{H,A\}}$, which gives

$$\|\alpha_t(\hat{A}) - \widehat{\alpha_t^{cl}(A)}\|_{\mathcal{B}(\mathcal{F}_{N\leq n})} \leq \int_0^t ds \|R(A \circ \Phi_s)\|_{\mathcal{B}(\mathcal{F}_{N\leq n})}. \qquad (20.45)$$

Note that the full evolution α_t drops out of the estimate. Now, we have to obtain appropriate estimates of the remainder $R(A)$ and the classical observable $A \circ \Phi_s$. First, we find a convenient expression for the remainder $R(A) := \frac{i}{g}[\hat{H}, \hat{A}] - \widehat{\{H, A\}}$. Let the operators $\hat{A}$ and $\hat{B}$ have the symbols, $A(\psi, \overline{\psi})$ and $B(\psi, \overline{\psi})$ and assume $B(\psi, \overline{\psi})$ is a monomial of degree m in ψ and $\overline{\psi}$, separately. Then the commutator operator $\hat{C} = [\hat{A}, \hat{B}]$, has the symbol $C(\psi, \overline{\psi})$ given by (see e.g. [13])

$$C(\overline{\psi}, \psi) = \sum_{k=1}^m \frac{g^k}{k!} \{A, B\}_k(\overline{\psi}, \psi), \qquad (20.46)$$

where

$$\{A, B\}_k := \int \int (\partial_\psi^k A, \partial_{\overline{\psi}}^k B - \partial_\psi^k B, \partial_{\overline{\psi}}^k A). \qquad (20.47)$$

The definition of the Hamiltonian $H(\overline{\psi}, \psi)$ and Eqns (20.46) and (20.47) imply

$$i[\hat{H}, \hat{A}] = g\widehat{\{H, A\}} + \frac{g^2}{2}\widehat{\{V, A\}_2}. \qquad (20.48)$$

It is not hard to show that $\|\widehat{\{V, A\}_2}\| \leq C\|A\|_*$ with $C < \infty$ and $\|A\|_*$ an appropriate norm of A. This implies (cf. (4.11))

$$i[\hat{H}, \hat{A}] = g\widehat{\{H, A\}} + O(g^2). \qquad (20.49)$$

Comparing the relation (20.48) with the expression for the remainder $R(A)$ we see that $R(A) = \frac{g}{2}\widehat{\{V, A\}_2} = O(g)$ which together with (20.44) implies, in turn, (20.43). Finally, one has to estimate the norm $\|A \circ \Phi_s\|_*$ of the classical evolution and this is the place where our restrictions enter and which we skip here. This completes our sketch.

Thus the mean-field regime is nothing else but the quasiclassical regime of the quantum many-body field theory with the reciprocal, $\frac{1}{n} = g$, of the number of particles playing role of the quasiclassical parameter. As the number of particle increases a quantum system starts behaving classically. In the opposite direction it is shown in ([124]) that the many-body theory described by the quantum Hamiltonian (20.25) can be obtained by quantizing the classical field theory described by the classical Hamiltonian (20.38) and Poisson brackets (20.37).

20.7 Appendix: the Ideal Bose Gas

As an example of an application of the second quantization, we compute the partition function, pressure, and the equation of state for an ideal bose gas (i.e.

a gas with no interparticle interactions), placed in the box $\Lambda \equiv [-\frac{L}{2}, \frac{L}{2}]^d \subset \mathbb{R}^d$. The Schrödinger operator for such a system is

$$H_\Lambda = \int_\Lambda a^*(x)(-\frac{\hbar^2}{2m}\Delta_\Lambda)a(x)dx, \tag{20.50}$$

where Δ_Λ is the Laplacian on $L^2(\Lambda)$ with periodic boundary conditions. The Gibbs equilibrium state with a fixed average number of particles and for the inverse temperature β (*grand canonical ensemble*) is

$$\rho_{\beta,\mu} := e^{-\beta(H_\Lambda - \mu N_\Lambda)}/Z_\Lambda(\beta, \mu),$$

where $N_\Lambda = \int_\Lambda a^*(x)a(x)dx$ is the number of particles operator in the volume $\Lambda \subset \mathbb{R}^3$ and $Z_\Lambda(\beta, \mu) = \operatorname{Tr} e^{-\beta(H_\Lambda - \mu N_\Lambda)}$ is the partition function. Here μ is the *chemical potential* entering as a Lagrange multiplier due to fixing the average number of particles. This state is obtained by maximizing the entropy, while leaving the average energy and average number of particles fixed (hence two Lagrange multipliers, β and μ, appear, see Section 18.3).

Note that in our case the one particle configuration space is the flat torus which is $\Lambda \equiv [-\frac{L}{2}, \frac{L}{2}]^d$ with opposite sides identified (i.e. $\mathbb{R}^d/\frac{2\pi}{L}\mathbb{Z}^d$) and the corresponding momentum space is the lattice $\frac{2\pi}{L}\mathbb{Z}^d$. Using separation of variables, and the result of Section 7.1 for a single particle in a box with periodic boundary conditions, or verifying directly that $\prod_{k \in \frac{2\pi}{L}\mathbb{Z}^d}(a^*(k))^{n_k}\Omega$ are eigenvectors of H_Λ with the eigenvalues $\sum_{k \in \frac{2\pi}{L}\mathbb{Z}^d}\varepsilon_k n_k$, we conclude that the spectrum of the operator H_Λ is

$$\sigma(H_\Lambda) = \{ \sum_{k \in \frac{2\pi}{L}\mathbb{Z}^d} \varepsilon_k n_k \mid n_k = 0, 1, \ldots \forall k\}.$$

Here $\varepsilon_k = \frac{1}{2m}|k|^2$ and each eigenvalue is of the multiplicity 1. Using equation (25.50) from the Mathematical Supplement, which expresses the trace of an operator in terms of its eigenvalues, we obtain the following expression for $Z_\Lambda(\beta, \mu)$:

$$Z_\Lambda(\beta, \mu) = \sum_{n_k} z^{\sum n_k} e^{-\beta \sum \varepsilon_k n_k}, \tag{20.51}$$

where $z = e^{\beta\mu}$, is called the *fugacity*. The latter expression can be transformed as

$$Z_\Lambda(\beta, \mu) = \prod_k \left(\sum_{n=0}^\infty z^n e^{-\beta \varepsilon_k n} \right) = \prod_k (1 - z e^{-\beta \varepsilon_k})^{-1}. \tag{20.52}$$

Note in passing that if for a self-adjoint operator A with a purely discrete spectrum $\{\lambda_j\}$ accumulating at 1, we define $\det A := \prod_{j=1}^\infty \lambda_j$ whenever this is finite, then we can rewrite (20.52) as

$$Z_\Lambda(\beta, \mu) = \det (1 - z e^{-\beta H_{1,\Lambda}})^{-1}$$

where $H_{1,\Lambda} = -\frac{1}{2m}\,\Delta_\Lambda$ is the one particle Hamiltonian acting on $L^2(\Lambda)$ with periodic boundary conditions.

Next, we consider the quantity $P_V(\beta,\mu) := \frac{1}{V\beta}\,\ln Z_\Lambda(\beta,\mu)$, where $V :=$ $vol(\Lambda)$, called the *pressure*. We have

$$
\begin{aligned}
P_V(\beta,\mu) &= -\frac{1}{V\beta}\sum_k \ln\left(1 - z\,e^{-\beta\,\varepsilon_k}\right) \\
&= -\frac{1}{V\beta}\,\mathrm{Tr}\,\ln(1 - z\,e^{-\beta\,H_{1,\Lambda}}).
\end{aligned}
\tag{20.53}
$$

Using that, as $V \to \infty$,

$$
\frac{(2\pi)^d}{V}\sum_{k \in \frac{2\pi}{L}\,\mathbb{Z}^d} f(k) \to \int f(k)\,dk\,,
$$

we see that formally $P_V(\beta,\mu)$ converges, as $V \to \infty$, to

$$
\begin{aligned}
P(\beta,\mu) &= -\frac{1}{(2\pi)^d\,\beta}\int \ln\left(1 - z\,e^{-\beta\,\varepsilon_k}\right)dk \\
&= -\frac{1}{(2\pi)^d\,\beta^{\frac{d+2}{2}}}\int \ln\left(1 - z\,e^{-\varepsilon_k}\right)dk.
\end{aligned}
\tag{20.54}
$$

To get the last integral, we changed variables $k \to k/\sqrt{\beta}$.

We compute average number of particles, $\bar{n} = \mathrm{Tr}(N_\Lambda\rho_{\beta,\mu})$. This definition implies the relation $\bar{n} = z\frac{\partial}{\partial z}\ln Z_\Lambda(\beta,\mu)$, from which we obtain the expression

$$
\bar{n}/V = \frac{1}{V}\sum_{k \in \frac{2\pi}{L}\,\mathbb{Z}^d} \frac{z\,e^{-\beta\,\varepsilon_k}}{1 - z\,e^{-\beta\,\varepsilon_k}}.
\tag{20.55}
$$

To have $\bar{n} \geq 0$, we should take $0 \leq z \leq 1$. The terms in the sum on the right hand side are, as can be easily checked, the average numbers of particles having momenta k,

$$
\bar{n}_k = \mathrm{Tr}\left(N_k\,\rho_{\beta,\mu}\right)
$$

where $N_k := a^*(k)a(k)$. To show this, one uses that, by (20.51), $\bar{n}_k = -\frac{1}{\beta}\,\partial_{\varepsilon_k} Z_\Lambda(\beta,\mu)$, where $Z_\Lambda(\beta,\mu)$ is considered as a functional of $\varepsilon = \varepsilon_k$, to obtain

$$
\bar{n}_k = \frac{z\,e^{-\beta\,\varepsilon_k}}{1 - z\,e^{-\beta\,\varepsilon_k}}.
$$

Equations (20.54) and (20.55) constitute the equation of state of the ideal Bose gas (parameterized by z). More precisely, solving Equation (20.55) for z as a function of the density $\rho = \bar{n}/V$, and temperature $T = 1/\beta$, and substituting the result into Equation (20.54), we find the pressure P as a function of ρ and T.

However, if we seek the equation of state as a relation between P, E and V (which is, of course, equivalent to the expression involving P, ρ and T), the answer is much simpler. Indeed, using the definition of E

$$E = E(\beta, \mu, V) := \mathrm{Tr}\,(H_\Lambda\, \rho_{\beta,\mu,\Lambda})$$

and considering the partition function, $Z_\Lambda(\beta, z)$, as a function of β and z (and V) rather than of β and μ (and V), and similarly for the pressure, we find

$$E = -\frac{\partial}{\partial\beta}\,\ln Z_\Lambda(\beta, z) = -\frac{\partial}{\partial\beta}\,(V\beta\, P(\beta, z))\,. \tag{20.56}$$

Taking into account (20.54) and setting $d = 3$ we find

$$E = \frac{3}{2}\,PV\,. \tag{20.57}$$

This is the equation of state of the ideal Bose gas.

Problem 20.11 It is an instructive exercise to re-derive the results of this section for the ideal Fermi gas.

Now, we consider the ideal Bose gas in the domain Λ with a fixed number, n, of particles. Its Hamiltonian is

$$H_{\Lambda,n} = \sum_{i=1}^{n} -\frac{1}{2m}\,\Delta_{x_i}$$

acting on the space $\circledS_1^n L^2(\Lambda) := L^2_{\text{sym}}(\Lambda^n)$ with periodic boundary conditions. Here $\circledS$ is the symmetric tensor product, and $L^2_{\text{Sym}}(\Lambda^n)$ is the L^2 space of functions symmetric with respect to permutations of variables belonging to different factors in the product Λ^n. Assume we want to compute the canonical partition function,

$$Z_{\Lambda,n}(\beta) = \mathrm{Tr}\,e^{-\beta H_{\Lambda,n}}\,. \tag{20.58}$$

This is not a simple matter (try it!). We show how to derive it from the grand canonical one, $Z_\Lambda(\beta, \mu)$. The considerations below are heuristic, but can be made rigorous.

Using expression (20.52) one can show that as a function of $z = e^{\beta\mu}$, $Z_\Lambda(\beta, \mu)$ is analytic in the disk $\{|z| \le \varepsilon\}$ for some $\varepsilon > 0$. Next, by the definition of $Z_\Lambda(\beta, \mu)$, we have $Z_\Lambda(\beta, \mu) = \sum_{n=0}^{\infty} Z_{\Lambda,n}(\beta)z^n$. Hence $Z_{\Lambda,n}$ can be computed by the Cauchy formula

$$Z_{\Lambda,n}(\beta) = \frac{1}{2\pi i}\oint_{|z|=\varepsilon} Z_\Lambda(\beta, \mu)\frac{dz}{z^{n+1}}\,. \tag{20.59}$$

By the definition of $P_\Lambda(\beta, \mu)$, we can write $Z_\Lambda(\beta, \mu) = e^{V\beta P_\Lambda(\beta,\mu)}$. Writing also $z^{-n} = e^{-n\ln z} = e^{-Vv\ln z}$, where $v = n/V$, (20.59) becomes

$$Z_{\Lambda,n}(\beta) = \frac{1}{2\pi i} \oint_{|z|=\varepsilon} e^{V(\beta P_\Lambda(\beta,\mu) - \nu \ln z)} \frac{dz}{z}.$$

Now we take $V = |\Lambda|$ and n large, while $\nu = n/V$ remains fixed. Taking into account the fact that $P_\Lambda(\beta,\mu)$ has a limit as $V \to \infty$, and applying (formally) the method of steepest descent to the integral above, we find

$$Z_{\Lambda,n} \approx c e^{-\beta\mu n} Z_\Lambda(\beta,\mu), \qquad (20.60)$$

where $|\ln c|$ is uniformly bounded in V, and μ solves the stationary phase equation

$$\mu : \beta z \partial_z P_\Lambda(\beta,\mu) = n/V,$$

or equivalently (passing from z to μ) $\mu : \partial_\mu P_\Lambda(\beta,\mu) = n/V$. Define $P_{\Lambda,n}(\beta) := \frac{1}{\beta V} \ln Z_{\Lambda,n}(\beta)$. Then relation (20.60) can be rewritten as

$$P_{\Lambda,n}(\beta) \approx (-\beta\mu n + P_\Lambda(\beta,\mu))|_{\partial_\mu P_\Lambda(\beta,\mu) = n/V}.$$

That is, $P_{\Lambda,n}(\beta)$, as a function of n, is (in the leading order as $V \to \infty$) the Legendre transform of $P_\Lambda(\beta,\mu)$, considered as a function of μ.

Similarly, we can pass from $Z_\Lambda(\beta,\mu)$ to $Z_{\Lambda,n}(\beta)$ by taking the Legendre transform in the variable $\nu = n/V$.

20.7.1 Bose-Einstein Condensation

We analyze formula (20.55) for the average number of particles. From now on we set $d = 3$. We would like to pass to the thermodynamic limit, $V \to \infty$. The point is that Equation (20.55) is the relation between the average number of particles $\bar{n}$ (or the average density $\rho = \bar{n}/V$), the temperature $T = 1/\beta$, and the chemical potential μ (or fugacity $z = e^{\beta\mu}$). Recall also that $0 \le z \le 1$. As long as ρ and β are such that $z < 1$, the right hand side of (20.55) converges to the integral

$$\frac{1}{(2\pi)^3} \int \frac{z e^{-\beta\varepsilon_k}}{1 - z e^{-\beta\varepsilon_k}}\, dk$$

as $V \to \infty$. However, if the solution of Equation (20.55) for z yields, e.g., $z = 1 - O(V^{-1})$, then we have to consider the $k = 0$ term in the sum on the right hand side of (20.55) separately. In this case we rewrite (20.55) approximately as

$$\rho = \frac{\bar{n}_0}{V} + \int \frac{z e^{-\beta\varepsilon_k}}{1 - z e^{-\beta\varepsilon_k}}\, dk, \qquad (20.61)$$

where we put $\bar{n}_0 = \frac{z}{1-z}$. Now using $\varepsilon_k = \frac{|k|^2}{2m}$, changing the variable of integration as $k' = \sqrt{\frac{\beta}{2m}}\, k$, and passing to spherical coordinates, we obtain

$$\int \frac{z e^{-\beta\varepsilon_k}}{1 - z e^{-\beta\varepsilon_k}}\, dk = \lambda^{-3} g_{3/2}(z),$$

where $\lambda = \sqrt{2\pi \beta/m}$ (the *thermal wave length*), and

$$g_{3/2}(z) = \int_0^\infty \frac{z\,e^{-x^2}}{1 - z\,e^{-x^2}}\, x^2\, dx\,.$$

Thus Equation (20.61) can be rewritten as

$$\rho = \frac{\overline{n}_0}{V} + \lambda^{-3}\, g_{3/2}(z)\,. \tag{20.62}$$

Recall that $\overline{n}_0 = \frac{z}{1-z}$, and that this equation connects the density ρ, the thermal wave length λ (or temperature $T = 2\pi/m\,\lambda^2$), and the fugacity z (or chemical potential $\mu = \frac{1}{\beta}\ln z$), and is supposed to be valid in the entire range of values, $0 < z < 1$, of z.

Can z really become very close to 1 (within $O(1/V)$), or is the precaution we took in the derivation of this equation by isolating the term $\overline{n}_0/V$ spurious? To answer this question we have to know the behaviour of the function $g_{3/2}(z)$ for $z \in (0,1)$. One can see immediately that

$$g_{3/2}(0) = 0\,,\ \ g'_{3/2}(z) > 0\ \text{ and }\ g'_{3/2}(1) = \infty\,.$$

The function $g_{3/2}(z)$ is sketched below.

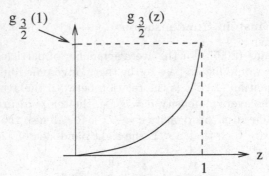

Fig. 20.1. Sketch of $g_{3/2}(z)$.

We see now that if $\rho\lambda^3 < g_{3/2}(1)$ (with $g_{3/2}(1) - \rho\lambda^3 \geq$ a positive number, independent of V), then the equation

$$\rho\lambda^3 = g_{3/2}(z)\,, \tag{20.63}$$

which is obtained from (20.62) by omitting the V-dependent term $\frac{\overline{n}_0}{V}$, has a unique solution for z which is less than 1, and is independent of V. Consequently, taking into account the term $\overline{n}_0/V = \frac{z}{V(1-z)}$ would lead to an adjustment of this solution by a term of order $O(1/V)$, which disappears in the thermodynamic limit $V \to \infty$.

However, for $\rho\lambda^3 = g_{3/2}(1)$, the solution of this equation is, obviously, $z = 1$, and for $\rho\lambda^3 > g_{3/2}(1)$, the above equation has no solutions at all. Thus for $\rho \lambda^3 \geq g_{3/2}(1)$ we do have to keep the term $\overline{n}_0/V$. Moreover, we have an estimate

$$\frac{\overline{n}_0}{V} = \rho \lambda^3 - g_{3/2}(z_*) \geq \rho \lambda^3 - g_{3/2}(1)$$

(here z_* is the solution to Equation (20.62)) which shows that in the case $\rho \lambda^3 > g_{3/2}(1)$, a macroscopically significant (i.e. proportional to the volume or the total number of particles) fraction of the particles is in the single, zero momentum – or *condensed* – state. This phenomenon is called *Bose-Einstein condensation*. The critical temperature, T_c, at which this phenomenon takes place can be found by solving the equation

$$\rho \lambda^3 = g_{3/2}(1)$$

describing the borderline case for λ, and remembering that $T = \frac{2\pi}{m \lambda^2}$. As a result we have

$$T_c = \frac{2\pi}{m} \left(\frac{\rho}{g_{3/2}(1)} \right)^{2/3}.$$

From Equation (20.62) we can also find the fraction of particles, $\overline{n}_0/V$, in the zero momentum (condensed) state as a function of temperature. This dependence is shown in the diagram below.

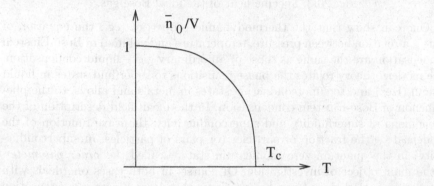

Fig. 20.2.

In this elementary situation, we have stumbled upon one of the central phenomena in macrosystems – the phenomenon of phase transition. Indeed, the states for which all the particles are in the single quantum state corresponding to zero momentum, and those for which the macroscopic fraction of the particles in the quantum state of zero momentum (and consequently in every single quantum state) is zero, can be considered two distinct pure phases of ideal Bose matter (gas). The first pure phase – called the *condensate* – occurs at $T = 0$, while the second pure phase takes place for $T \geq T_c$. In the interval $0 < T < T_c$ of temperatures, the Bose matter is in a mixed state in which both phases coexist.

Bose-Einstein condensation exhibits a typical property of phase transitions of the second kind: though all the thermodynamic functions and their first derivatives are continuous at the phase transition, some of the second derivatives are not. Typically one looks at the specific heat

$$C_V := \frac{\partial E(T, \mu, V)}{\partial T},$$

the change of heat or energy per unit of temperature. Using Equations (20.54) and (20.56), one can show that while C_V is continuous at $T = T_c$, its derivative with respect to T is not. C_V as a function of T is plotted below (see [166], Sect. 12.3]):

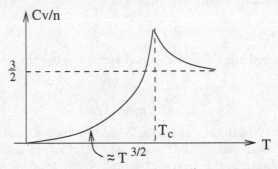

Fig. 20.3. Specific heat of the ideal Bose gas.

One can show that the thermodynamic properties (eg., the equation of state – a relation between pressure, temperature and volume) of Bose-Einstein condensation are the same as those of an ordinary gas – liquid condensation. The modern theory relates the phase transitions to superfluid states in liquid helium (He^2) and to superconducting states in metals and alloys, to the phenomenon of Bose-Einstein condensation. In the mean field description of the phenomena of superfluidity and superconductivity, the wave function of the condensate – the fraction of particles (or pairs of particles, in superconductivity) in the quantum zero momentum state – called the *order parameter*, is the main object of investigation. Of course, in both cases one deals with interacting particles, and one has to argue that Bose-Einstein condensation persists, at least for weakly interacting Bose matter.

Problem 20.12 Extend the above analysis to an arbitrary dimension d.

Quantum Electro-Magnetic Field - Photons

To have a theory of emission and absorption of electromagnetic radiation by quantum systems, not only should the particle system be quantized, but the electro-magnetic field as well. Hence we have to quantize Maxwell's equations. We do this by analogy with the quantization of classical mechanics as we have done this in Section 4.1. This suggests we have to put the classical electro-magnetic field theory, which is originally given in terms of Maxwell's PDEs, into Hamiltonian form. As before we do this in two steps: by introducing the action principle, and performing a Legendre transform. Then we define the quantization map by associating with canonically conjugate classical fields the corresponding operators, and quantizing observables correspondingly. Since Maxwell's equations are wave equations for vector fields with constraints, to provide the reader with a simpler guide, we first quantize the scalar Klein-Gordon equation, which gives the wave equation in the limit of vanishing mass. The reader familiar with the quantization of the Klein-Gordon equation can proceed directly to the next section on quantization of the Maxwell equations. In what follows, we work in physical units in which the Planck constant and speed of light are equal to 1: $\hbar = 1$, $c = 1$.

21.1 Klein-Gordon Classical Field Theory

21.1.1 Principle of minimum action

We construct the Hamiltonian formulation of the Klein-Gordon equation. We consider a scalar (real or complex) field $\phi(x, t)$ on $\mathbb{R}^d$ satisfying the evolution equation

$$(\Box + m^2)\phi = 0, \qquad (21.1)$$

where, recall, $\Box := \partial_t^2 - \Delta$ is the *D'Alembertian* operator and the parameter $m \geq 0$ is interpreted as mass. For $m > 0$ this is the Klein-Gordon equation, once proposed to describe relativistic particles, and for $m = 0$ this is the wave

© Springer-Verlag GmbH Germany, part of Springer Nature 2020
S. J. Gustafson and I. M. Sigal, *Mathematical Concepts of Quantum Mechanics*, Universitext, https://doi.org/10.1007/978-3-030-59562-3_21

equation (assumed to describe massless particles). The corresponding theory is called the Klein-Gordon classical field theory.

We write equation (21.1) as a Hamiltonian system. This is done in two steps: introducing the action principle, and performing a Legendre transform. Then we quantize the resulting infinite dimensional Hamiltonian system. To fix ideas, we consider from now on only real fields. We remark on complex fields at the end.

As for classical mechanics, we begin with the principle of minimal action (properly, of "stationary" action). Recall that it states that an evolution equation for physical states is an Euler-Lagrange equation for a certain functional called the *action*.

More precisely, one considers a space of functions ϕ, defined on space-time, called the *fields*. The equation of motion for ϕ is given by $S'(\phi) = 0$, where S is an action functional on the space of fields. This functional is of the form

$$S(\phi) = \int_0^T \int_{\mathbb{R}^d} \mathcal{L}(\phi(x,t), \nabla_x \phi(x,t), \dot{\phi}(x,t)) dx dt \qquad (21.2)$$

for $\phi : \mathbb{R}^d_x \times \mathbb{R}_t \to \mathbb{R}$ (for the moment we consider only real fields). Here, $\mathcal{L} : \mathbb{R} \times \mathbb{R}^d \times \mathbb{R} \to \mathbb{R}$ is the *Lagrangian density*. The space integral of the Lagrangian density,

$$L(\phi, \dot{\phi}) := \int_{\mathbb{R}^d} \mathcal{L}(\phi(x,t), \nabla_x \phi(x,t), \dot{\phi}(x,t)) dx, \qquad (21.3)$$

is called the *Lagrangian functional*. Recalling from in Section 4.7 the definitions of critical points and the derivation of the Euler-Lagrange equations (see also Section 26.2 of Mathematical supplement), it is easy to show that critical points of satisfy the Euler-Lagrange equation

$$- \partial_t (\partial_{\dot{\phi}} \mathcal{L}(\phi, \dot{\phi})) + \partial_\phi \mathcal{L}(\phi, \dot{\phi}) = 0. \qquad (21.4)$$

Now, we turn to the Klein-Gordon equation. Let $f : \mathbb{R} \to \mathbb{R}$ be a differentiable function. Consider the Lagrangian functional

$$L(\phi, \dot{\phi}) = \int_{\mathbb{R}^d} \left\{ \frac{1}{2} |\dot{\phi}|^2 - \frac{1}{2} |\nabla_x \phi|^2 - f(\phi) \right\} dx \qquad (21.5)$$

defined on some subspace of $H^1(\mathbb{R}^d) \times L^2(\mathbb{R}^d)$ s.t. $f(\phi(x))$ is integrable. The corresponding Lagrangian density is $\mathcal{L}(\phi, \dot{\phi}) = \frac{1}{2} |\dot{\phi}|^2 - \frac{1}{2} |\nabla_x \phi|^2 - f(\phi)$. The critical point equation for the corresponding action functional $S(\phi) = \int_0^T L(\phi(t), \dot{\phi}(t)) d^d x dt$ is the (nonlinear) *Klein-Gordon equation*:

$$\Box \phi + f'(\phi) = 0. \qquad (21.6)$$

One can generalize the above construction by considering the action $S(\phi) = \int_0^T L(\phi(t), \dot{\phi}(t)) dt$, defined on the space of paths

$$\mathcal{P}_{\phi_0, \phi_T} = \{\phi \in C^1([0,T]; X) \mid \phi(0) = \phi_0, \phi(T) = \phi_T\},$$

for some $\phi_0, \phi_T \in X$, where the Lagrangian functional $L(\phi, \eta)$ is defined on $X \times V$. Here X is an open subset of a normed space V (or manifold). X is called the *configuration space* of the physical system, and its elements are called fields. In an examples above, X is some functional space, say $X = H^1(\mathbb{R}^d, \mathbb{R})$. (If X is a non-linear space, then the Lagrangian functional would be defined on (a subset of) TX, the tangent bundle of X.) The Euler-Lagrange equation in this case is

$$- \partial_t(\partial_{\dot{\phi}} L(\phi, \dot{\phi})) + \partial_\phi L(\phi, \dot{\phi}) = 0 \qquad (21.7)$$

where $\partial_\phi L$ and $\partial_{\dot{\phi}} L$ are variational or Gâteaux derivative of L with respect ϕ and $\dot{\phi}$, respectively. (See Mathematical Supplement, Section 26.2 for the definition.)

21.1.2 Hamiltonians

We generalize the construction we used for classical mechanics. Suppose the dynamics of a system are determined by the minimum action principle with a Lagrangian functional $L : X \times V \to \mathbb{R}$, which is differentiable. Here V and X are a Banach space and an open subset of V. We pass to the new variables $(\phi, \dot{\phi}) \to (\phi, \pi)$, where $\pi \in V^*$, as a function of ϕ and $\dot{\phi}$, is given by

$$\pi = \partial_{\dot{\phi}} L(\phi, \dot{\phi}). \qquad (21.8)$$

(Recall that V^* is the space dual to V, see Mathematical Supplement, Section 25.1.) We assume that the equation (4.41) has a unique solution for $\dot{\phi}$. (Typically, L is convex in the second variable.) With this in mind, we define the *Hamiltonian functional*, $H : X \times V^* \to \mathbb{R}$, as

$$H(\phi, \pi) = \left(\langle \pi, \dot{\phi} \rangle - L(\phi, \dot{\phi}) \right)\big|_{\dot{\phi} : \partial_{\dot{\phi}} L(\phi, \dot{\phi}) = \pi}. \qquad (21.9)$$

As in Classical Mechanics, the space $Z := X \times V^*$ is called a *phase space* of the system.

Theorem 21.1 If $L(\phi, \dot{\phi})$ and $H(\phi, \pi)$ are related by (21.8)-(21.9), then the Euler-Lagrange equation (21.4) for the action (21.2) is equivalent to the Hamilton equations

$$\dot{\phi} = \partial_\pi H(\phi, \pi), \quad \dot{\pi} = -\partial_\phi H(\phi, \pi). \qquad (21.10)$$

The proof of this theorem is a straightforward generalization of the proof of Theorem 4.8.

Problem 21.2 Prove this theorem.

This gives a hamiltonian formulation of CFT.

For the Klein-Gordon classical field theory, the Lagrange functional is $L(\phi, \chi) = \int \left\{ \frac{1}{2}(|\chi|^2 - |\nabla\phi|^2) - f(\phi) \right\} dx$, and, consequently, the Klein-Gordon Hamiltonian is

$$H(\phi, \pi) = \int \left\{ \frac{1}{2}(|\pi|^2 + |\nabla \phi|^2) + f(\phi) \right\} dx. \qquad (21.11)$$

Problem 21.3 Show (21.11).

21.1.3 Hamiltonian System

Suppose that $Z = X \times V^*$ is a space of functions $\Phi(x) = (\phi(x), \pi(x))$ on $\mathbb{R}^d$. The functional on X which maps $X \ni \phi \mapsto \phi(x)$ is called the *evaluation functional* (at x), which we denote (with some abuse of notation) as $\phi(x)$, and similarly for V.

Problem 21.4 Show that $\partial_\phi \phi(y) = \delta_y$ and $\partial_\pi \pi(y) = \delta_y$.

Now, we recognize that the Hamilton equations equations (21.10) can be written, for all x, as

$$\dot{\Phi}_t(x) = \{\Phi(x), H\}(\Phi_t) \qquad (21.12)$$

where $\{F, G\}$ is the Poisson bracket on Z defined for any pair differentiable functionals F, G on Z as

$$\begin{aligned}
\{F, G\} &= \langle \partial_\phi F, \partial_\pi G \rangle - \langle \partial_\pi F, \partial_\phi G \rangle \\
&= \int \{\partial_\pi F \partial_\phi G - \partial_\phi F \partial_\pi G\} dx.
\end{aligned} \qquad (21.13)$$

(See Section 4.7 for the notion and another example of the Poisson brackets.)

Problem 21.5 Prove this.

One can show, formally, that with the Poisson bracket given in (21.13),

$$\{\pi(x), \phi(y)\} = \delta(x - y), \quad \{\phi(x), \phi(y)\} = 0, \quad \{\pi(x), \pi(y)\} = 0. \qquad (21.14)$$

Problem 21.6 Show (21.14).

Equation (21.14) says that the evaluation functionals, π, and ϕ, are *canonical coordinates*. To have a rigorous interpretation of the first equation in (21.14), we introduce, for $f \in C_0^\infty(\mathbb{R}^d)$, the functionals

$$\phi(f) : \phi \mapsto \langle f, \phi \rangle \qquad \text{and} \qquad \pi(f) : \pi \mapsto \langle f, \pi \rangle.$$

Then (21.14) means that $\{\pi(f), \phi(g)\} = \langle f, g \rangle$, etc, for all $f, g \in C_0^\infty(\mathbb{R}^d)$. Note that a path Φ_t in Z solves (21.12) iff, for all functionals F,

$$\frac{d}{dt} F(\Phi_t) = \{F(\Phi_t), H\}. \qquad (21.15)$$

Definition 21.7 A *Hamiltonian system* is a triple, $(Z, \{\cdot, \cdot\}, H)$, a Poisson space $(Z, \{\cdot, \cdot\})$ (a Banach space Z, with a Poisson bracket) together with a Hamiltonian functional H defined on that space.

For the Klein-Gordon classical field theory the phase, or state, space is $Z = H^1(\mathbb{R}^d) \times H^1(\mathbb{R}^d)$, the Hamiltonian is given by (21.11), and the Poisson brackets are defined on it as

$$\{F, G\} = \int \{\partial_\pi F \partial_\phi G - \partial_\phi F \partial_\pi G\}. \tag{21.16}$$

(In principle we could consider a larger phase space, $Z = H^1(\mathbb{R}^d) \times L^2(\mathbb{R}^d)$, but the space above is more convenient.)

For a more general situation see Mathematical Supplement, Section 26.6.

Remark 21.8 Our definition of a Hamiltonian system differs from the standard one in using the Poisson bracket instead of a symplectic form. The reason for using the Poisson bracket is its direct relation to the commutator.

21.1.4 Complexification of the Klein-Gordon Equation

With the view to quantization, it is convenient to pass from the real phase space $Z = H^1(\mathbb{R}^d) \times H^1(\mathbb{R}^d) \equiv H^1(\mathbb{R}^d, \mathbb{R}^2)$, with the Poisson bracket (21.16), and with the canonical real fields, $\phi(f)$ and $\pi(f)$, to the complex one, $Z^c = H^1(\mathbb{R}^d, \mathbb{C})$, with the Poisson bracket

$$\{F, G\} = i \int \{\partial_{\alpha(x)} F \partial_{\bar{\alpha}(x)} G - \partial_{\bar{\alpha}(x)} F \partial_{\alpha(x)} G\} dx, \tag{21.17}$$

where $\partial_{\alpha(x)} := \partial_{\operatorname{Re}\alpha(x)} - i\partial_{\operatorname{Im}\alpha(x)}$ and $\partial_{\bar{\alpha}(x)} := \partial_{\operatorname{Re}\alpha(x)} - i\partial_{\operatorname{Im}\alpha(x)}$, and with the canonical complex field $\alpha(f)$, and its complex conjugate $\bar{\alpha}(f)$, defined by

$$\alpha(f) = \frac{1}{\sqrt{2}} \left(\phi(C^{-1/2}f) + i\pi(C^{1/2}f) \right)$$
$$\bar{\alpha}(f) = \frac{1}{\sqrt{2}} \left(\phi(C^{-1/2}f) - i\pi(C^{1/2}f) \right), \tag{21.18}$$

where C is an operator on $L^2(\mathbb{R}^d)$. The Poisson brackets (21.16) and (21.17), imply, for $f, g \in C_0^\infty$, the Poisson brackets

$$\{\alpha(f), \bar{\alpha}(g)\} = \langle \bar{f}, g \rangle, \qquad \{\alpha(f), \alpha(g)\} = \{\bar{\alpha}(f), \bar{\alpha}(g)\} = 0. \tag{21.19}$$

Problem 21.9 Prove the above statement.

Thus (21.18) is a canonical transformation. Consider the free classical field theory, i.e. the Hamiltonian (21.11) with

$$F(\phi) = \frac{1}{2}m^2|\phi|^2, \tag{21.20}$$

with $m > 0$. Since $-\Delta + m^2 > 0$, we can take

$$C = (-\Delta + m^2)^{-1/2}. \tag{21.21}$$

Then the free Klein-Gordon Hamiltonian is expressed in terms of the fields (21.18) as

$$H(\alpha, \bar{\alpha}) = \int \bar{\alpha} C^{-1} \alpha dx, \tag{21.22}$$

21.2 Quantization of the Klein-Gordon Equation

We are now ready to attempt to quantize the free Klein-Gordon theory, i.e. (21.11) - (21.16), with (21.20). The Klein-Gordon equation (21.1), or the corresponding Hamiltonian system, (21.11) - (21.16), with (21.20), look like a continuum of classical harmonic oscillators. The link becomes especially explicit if pass in the classical fields $\phi(x,t)$ and $\pi(x,t)$ to the Fourier transform in the x variable: $\hat{\phi}(k,t)$ and $\hat{\pi}(k,t)$. The Fourier transformed fields satisfy the equation $(\partial_t^2 + |k|^2)\hat{\phi} = 0$, and have the Hamiltonian

$$H(\hat{\phi}, \hat{\pi}) = \frac{1}{2} \int \{|\hat{\pi}|^2 + \omega(k)^2|\hat{\phi}|^2\} d^d k, \qquad \omega(k) := \sqrt{|k|^2 + m^2}. \tag{21.23}$$

Thus we want to quantize the free Klein-Gordon classical field as a continuum of harmonic oscillators. In finite dimensional case we can quantize the corresponding hamiltonian system in any set of canonically conjugate variables. In the infinite dimensional case, the choice of such variables makes a difference between a meaningful and meaningless quantum theory. Recall, that we obtained a representation (7.22) of the harmonic oscillator, which can be rewritten, with $m = 1$ and $\hbar = 1$, as $H = \sum_{j=1}^r \omega_j a_j^* a_j$ on the space

$$L^2(\mathbb{R}^r, d\mu_c), \; d\mu_c(x) := [\det(2\pi c)]^{-1/2} e^{-\langle x, c^{-1}x \rangle} d^r x,$$

where $c := \text{diag}(\hbar/\omega_i)$, so that $\langle x, c^{-1}x \rangle = \sum_{i=1}^r \omega_i x_i^2$, $d^r x = \prod_{i=1}^r dx_i$ is the Lebesgue measure on $\mathbb{R}^r$, and the operators a_j and their adjoints a_j^*, are defined on the space $L^2(\mathbb{R}^r, d\mu_c)$ by

$$a_j := \sqrt{\frac{1}{2\omega_j}} \partial_{x_j}, \; a_j^* = -\sqrt{\frac{1}{2\omega_j}} \partial_{x_j} + \sqrt{2\omega_j} x_j. \tag{21.24}$$

$d\mu_c$ is Gaussian measure of mean zero and covariance c. The point here is that unlike the standard representation, where the harmonic oscillator Hamiltonian acts on the space $L^2(\mathbb{R}^r, d^r x)$, which has no $r \to \infty$ limit, the new representation is on the space $L^2(\mathbb{R}^r, d\mu_c)$, which can be considered as a finite-dimensional approximation of the Gaussian space $L^2(Q_s, d\mu_C)$. Here

$$Q_s = \{f \mid (-\Delta + |x|^2 + 1)^{-s/2} f \in L^2(\mathbb{R}^d)\},$$

for s sufficiently large, and $d\mu_C$ is the Gaussian measure on Q_s of mean 0, and covariance operator C which acts on $L^2(\mathbb{R}^d)$ and has matrix c as a finite-dimensional approximation. Elements of the space $L^2(Q_s, d\mu_C)$ are functionals $F(\phi)$ on Q_s such that

$$\int_{Q_s} |F(\phi)|^2 d\mu_C(\phi) < \infty.$$

We explain this in more detail later, in Subsection 21.3 below. The representation (21.22) can be thought of as coming from quantization of the classical hamiltonian system with the hamiltonian $h(\alpha, \bar{\alpha}) := \langle \alpha, c^{-1}\alpha \rangle$, where $\alpha := (\alpha_j)$.

Motivated by this representation, we associate with the complex fields $\alpha(f)$, and $\bar{\alpha}(f)$, the *annihilation operator* $a(f)$, and its adjoint, the *creation operator* $a^*(f)$, acting on the space $L^2(Q_s, d\mu_C)$, with some covariance operator C, for s sufficiently large, according to the relations

$$a(f) = \langle C^{1/2}f, \partial_\phi \rangle, \quad a^*(f) = -\langle C^{1/2}f \partial_\phi \rangle + \langle C^{-1/2}f \phi \rangle. \tag{21.25}$$

(More precisely, $a(f)F = \langle C^{1/2}f, \partial_\phi F \rangle = \partial_\lambda|_{\lambda=0} F(\phi + \lambda C^{1/2}f)$, etc.) These operators satisfy the commutation relations

$$[a(f), a^*(g)] = \langle \bar{f}, g \rangle,$$

$$[a(f), a(g)] = [a^*(f), a^*(g)] = 0,$$

where $f, g \in C_0^\infty$. (For representation of canonical commutation relation, see [45].) It is convenient to introduce the operator-valued distributions, $a(x)$ and $a^*(x)$ by $a(f) = \int a(x)\bar{f}(x)dx$ and $a^*(f) = \int a^*(x)f(x)dx$. Then

$$a(x) = C^{1/2}\partial_{\phi(x)}, \quad a^*(x) = -C^{1/2}\partial_{\phi(x)} + C^{-1/2}\phi(x). \tag{21.26}$$

With this correspondence, the Hamiltonian functional $H(\alpha, \bar{\alpha}) = \int \bar{\alpha} C^{-1}\alpha$ is mapped into the quantum Hamiltonian operator

$$H_f = \int a(x)^* C^{-1} a(x) dx, \tag{21.27}$$

acting on the space $L^2(Q_s, d\mu_C)$. (The subindex f stands for the "field".) Here we put the creation operator $a(x)^*$ on the left of the annihilation operator $a(x)$ for a reason, to be explained later. Recall, that for the Klein-Gordon CFT, the covariance operator is given by (21.21). In this case it is convenient to pass to the Fourier representation, $a(k) = (2\pi)^{-d/2} \int e^{-ik\cdot x} a(x)dx$, and use the Plancherel theorem, to obtain

$$H_f = \int \omega(k)a^*(k)a(k)dk, \tag{21.28}$$

where $\omega(k) = \sqrt{|k|^2 + m^2}$. (The meaning of the integral on the r.h.s. is explained in Section 20.2.)

Define the particle number and the momentum operators (cf. Section 26.8 of Mathematical Supplements and Problem 21.13)

$$N = \int a^*(x)a(x)dx \quad \text{and} \quad P_f = \int a^*(x)(-i\nabla_x)a(x)dx. \tag{21.29}$$

Problem 21.10 Show (formally) that the operators P, H, and N commute.

21.3 The Gaussian Spaces

We sketch a definition of the Gaussian measure, $d\mu_C(\phi)$, on Q_s, of mean 0 and covariance operator C, acting on $L^2(\mathbb{R}^d)$. (For detailed exposition see [135, 270, 271].) One way to describe $d\mu_C$ is through finite-dimensional approximations, X, of $Q_s^* = Q_{-s}$ given in terms of cylinder sets. For each $f \in Q_s^*$, we define the linear functional on Q_s,

$$f(\phi) := \int f\phi dx$$

where the integral on the r.h.s. is understood as $\int (-\Delta + |x|^2 + 1)^{s/2} f \cdot (-\Delta + |x|^2 + 1)^{-s/2} \phi d^3x$. Let $n = \dim X$. Given n linearly independent vectors $f_1, \ldots, f_n$ from X, we associate with each Borel set B in $\mathbb{R}^n$, the $X-$ cylinder set

$$C_B := \{\phi \mid (f_1(\phi), \ldots, f_k(\phi)) \in B\}.$$

With every measurable function g on $\mathbb{R}^n$ we associate the functional $G(\phi) = g(\phi(f_1), \ldots, \phi(f_n))$ on Q_s^*, called a $X-$cyliner function. We define $d\mu_C(\phi)$ by giving its integrals of $X-$cyliner functions over $X-$cylinder sets

$$\int_{C_B} G(\phi)d\mu_C(\phi) = \int_B g(x_1, \ldots, x_n)d\mu_C^X(x) \tag{21.30}$$

for every B and g and for increasing sequence of finite-dimensional spaces X, whose limit is Q_s^*. Here $d\mu_C^X$ is the finite-dimensional Gaussian measure on $\mathbb{R}^n$,

$$d\mu_C^X(\phi) = (\det 2\pi C_X)^{-1/2} e^{-\langle x, C_X^{-1} x\rangle/2} d^n x$$

where $d^n x$ is the usual Lebesgue measure on $\mathbb{R}^n$ (here we display the dimension of the measure) and C_X is the matrix $(\langle f_i, Cf_j\rangle)$. $d\mu_C(\phi)$ can be extended to a measure on the σ−algebra generated by cylinder sets which leads to the notion of the integral of measurable w.r.to this algebra complex functionals.

Next, as usual we define $L^2(Q_s, d\mu_C)$ to be the space of measurable complex functionals $F(\phi)$ on Q_s such that $\int_{Q_s} |F(\phi)|^2 d\mu_C(\phi) < \infty$. The basic examples of square integrable functionals are

$$ F : \phi \in Q_s \mapsto p(\phi(f_1), \ldots, \phi(f_n)) $$

where $p(t_1, \ldots, t_n)$ are polynomials in $t_1, \ldots, t_n$ and $f_1, \ldots, f_n \in Q_s^*$. In fact, one can show that, by our construction,

the span of vectors of the form $\prod_1^n \phi(f_j)\Omega$, $n \geq 1$, is dense in $L^2(Q_s, d\mu_C)$.

$$ (21.31) $$

Proving this latter fact requires some general considerations from the theory of functional spaces. (See however the corresponding proof for the harmonic oscillator.)

We introduce the *expectation* of a functional F with respect to $d\mu_C$:

$$ E(F) := \int F(\phi) d\mu_C(\phi). $$

The terminology "mean 0" and "covariance C" corresponds to the properties

$$ E(\phi(f)) = 0 \quad \text{and} \quad E(\phi(f)\phi(g)) = \langle f, Cg \rangle, $$

or, formally,

$$ E(\phi(x)) = 0 \quad \text{and} \quad E(\phi(x)\phi(y)) = C(x, y) $$

where $C(x, y)$ is the integral kernel of C. These formulae can be easily proven by using finite dimensional approximations (21.30).

What should s be? Compute the expectation of the square of the norm functional $\phi \to \|\phi\|_{Q_s}^2 := \langle \phi, (-\Delta + |x|^2 + 1)^{-s}\phi \rangle = \int \phi(x)K_s(x, y)\phi(y)dxdy$, where $K_s(x, y)$ is the integral kernel of the operator $(-\Delta + |x|^2 + 1)^{-s}$. We have formally

$$ E(\|\phi\|_{Q_s}^2) = \int\int \delta(x - y)K_s(x, y)E(\phi(x)\phi(y))dxdy $$

$$ = \int\int \delta(x - y)K_s(x, y)C(x, y)dxdy. $$

Since $C(x, y) = C(y, x)$, this gives $E(\|\phi\|_{Q_s}^2) = \text{Tr}((-\Delta_y + |x|^2 + 1)^{-s}C)$. Now, the operator C is bounded and the operator $(-\Delta_y + |x|^2 + 1)^{-s}$ is trace class if $s > d = 3$. So in this case $E(\|\phi\|_{Q_s}^2) < \infty$. If we know more about C, we can relax the condition $s > d = 3$. For example, for the Klein-Gordon theory, $C = (-\Delta + m^2)^{-1/2}$ and we have $s > d - 1$ for dimension $d \geq 2$.

The integration by parts formula in our space is

$$\int_{Q_s} \bar{F}(-i\partial_\phi G)d\mu_C(\phi) = \int_{Q_s} \overline{(-i\partial_\phi + iC^{-1}\phi)F}Gd\mu_C(\phi). \qquad (21.32)$$

To prove (21.32) we use

$$\int_{Q_s} \bar{F}(-i\partial_\phi G)d\mu_C(\phi) = \int_{Q_s} \overline{(-i\partial_\phi F)}Gd\mu_C(\phi) + i\int_{Q_s} \bar{F}G\partial_\phi d\mu_C(\phi)$$

and the relation

$$\partial_\phi d\mu_C(\phi) = -C^{-1}\phi d\mu_C(\phi).$$

To see the latter relation, think formally about the Gaussian measure as being

$$d\mu_C(\phi) = \text{const } e^{-\langle\phi, C^{-1}\phi\rangle/2}D\phi.$$

The integration by parts formula implies that $-i\partial_\phi$ is not symmetric on $L^2(Q_s, d\mu_C)$:

$$(-i\partial_\phi)^* = -i\partial_\phi + iC^{-1}\phi.$$

However, the above implies that the operator $\pi := -i\partial_\phi + \frac{1}{2}iC^{-1}\phi$ is symmetric. In fact, it is self-adjoint: $\pi^* = \pi$.

21.4 Wick Quantization

We now describe a systematic way to quantize observables, called *Wick quantization*. To see why we need a special procedure let us quantize naively the free Klein-Gordon Hamiltonian (21.11), with (21.20), which we rewrite as $H(\phi^{cl}, \pi^{cl}) = \frac{1}{2}\int\{(\pi^{cl})^2 + (C^{-1}\phi^{cl})^2\}dx$, where C is given in (21.21) and where we denote for the moment the classical coordinate and momentum field by ϕ^{cl} and π^{cl}, respectively. First we observe that the quantum fields corresponding to the classical fields ϕ^{cl} and π^{cl} are

$$\phi(x) = \frac{1}{\sqrt{2}}C^{1/2}(a(x) + a^*(x)) \text{ and } \pi(x) = \frac{1}{i\sqrt{2}}C^{-1/2}(a(x) - a^*(x)). \quad (21.33)$$

Here, remember, $\phi(x)$ is considered as an operator of multiplication by the evaluation functional $\phi(x)$. The second equation gives

$$\pi := -i\partial_\phi + \frac{1}{2}iC^{-1}\phi.$$

We have shown in Subsection 21.3 that this operator is symmetric. Using this expression, one can show that

$$i[\pi(x), \phi(y)] = \delta(x - y) \qquad (21.34)$$

$$i[\pi(x), \pi(y)] = i[\phi(x), \phi(y)] = 0.$$

Problem 21.11 Derive formally the commutation relations above.

Now in the expression (21.11), with (21.20), we replace the classical fields by the quantum ones above to obtain

$$H(\phi, \pi) = \frac{1}{2} \int \{\pi^2 + (C^{-1}\phi)^2\} dx$$

$$= \frac{1}{2} \int \{[\frac{i}{2}C^{-1/2}(a^* - a)]^2 + [\frac{1}{2}C^{-1/2}(a^* + a)]^2\} dx$$

$$= \frac{1}{4} \int \{C^{-1/2}a^* C^{-1/2}a + C^{-1/2}aC^{-1/2}a^*\} dx$$

$$= \frac{1}{2} \int a^* C^{-1}a + \frac{1}{4} \int (C^{-1}\delta_x) dx$$

where we have used the commutation relation for a and a^*, and the self-adjointness of $C^{-1/2}$. The first term on the right hand side is non-negative, and the second is infinite (recall (21.21), which gives $C^{-1} = (-\Delta + m^2)^{1/2}$), and therefore the r.h.s. is infinite.

We can make this argument rigorous as follows. First, we move to momentum space via the Fourier transform: $a(x) \mapsto a(k)$, $a^*(x) \mapsto a^*(k)$ (we omit the 'hats' over $a(k)$ and $a^*(k)$). We wish to show that $\int \omega(k)a(k)a^*(k)dk = \infty$, where $\omega(k) = (|k|^2 + m^2)^{1/2}$ (the *dispersion law*). Let $\mathbb{R}^d = \cup_{\alpha \in \mathbb{Z}^d} B_\alpha$ (a disjoint union), where B_α is the unit box with center at a lattice point in $\mathbb{Z}^d$, $a_\alpha = \int_{B_\alpha} a$, and $\omega_\alpha = \min_{k \in B_\alpha} \omega(k)$. Then

$$[a_\alpha, a_\beta^*] = \delta_{\alpha,\beta}$$

so that

$$\int \omega aa^* \geq \sum_{\alpha \in \mathbb{Z}^3} \omega_\alpha a_\alpha a_\alpha^*.$$

On the other hand,

$$\sum \omega_\alpha a_\alpha a_\alpha^* = \sum \omega_\alpha a_\alpha^* a_\alpha + \sum \omega_\alpha = \infty.$$

Now we describe the quantization procedure which avoids the above problem. Fix the covariance C. Recall that we denote for the moment the classical coordinate and momentum fields by ϕ^{cl} and π^{cl}. Classical observables are functionals $A(\phi^{cl}, \pi^{cl})$ of ϕ^{cl} and π^{cl}. We express $A(\phi^{cl}, \pi^{cl})$ in terms of the functions α and $\bar{\alpha}$, using

$$\phi^{cl} = \frac{1}{\sqrt{2}}C^{1/2}(\alpha + \bar{\alpha}) \text{ and } \pi^{cl} = \frac{1}{i\sqrt{2}}C^{-1/2}(\alpha - \bar{\alpha})$$

(see (21.18)): $A(\phi^{cl}, \pi^{cl}) = B(\alpha, \bar{\alpha})$. We consider functionals $B(\alpha, \bar{\alpha}) : L^2(\mathbb{R}^d) \times L^2(\mathbb{R}^d) \to \mathbb{C}$, analytic in α and $\bar{\alpha}$, of the sense of the convergent power series,

$$B(\alpha, \bar{\alpha}) = \sum_{m,n} \int B_{m,n}(x_1, \ldots, x_{m+n}) \prod_{i=1}^{m} \bar{\alpha}(x_i) \prod_{i=m+1}^{m+n} \alpha(x_i) d^{(n+m)}x.$$

We associate with classical observables, $A(\phi^{cl}, \pi^{cl})$, quantum observables, i.e. operators A on the state space $L^2(Q_s, \mu_C)$, according to the rule

$$A(\phi^{cl}, \pi^{cl}) \mapsto \quad A :=: A(\phi, \pi) : \equiv : B(a^*, a) : \tag{21.35}$$

where $: B(a^*, a) :$ is a *Wick ordered* operator defined by

$: B(a^*, a) :$

$$= \sum_{m,n} \int B_{mn}(x_1, \ldots, x_{m+n}) \prod_{i=1}^{m} a^*(x_i) \prod_{i=m+1}^{m+n} a(x_i) d^{(n+m)}x. \tag{21.36}$$

Here are some examples of Wick ordering:

Example 21.12

$$\begin{aligned} : \phi^2 : \; &= \; : [C^{-1/2}(a + a^*)]^2 : \\ &= \; : (C^{-1/2}a)^2 + (C^{-1/2}a^*)^2 + C^{-1/2}a^*C^{-1/2}a + C^{-1/2}aC^{-1/2}a^* : \\ &= (C^{-1/2}a)^2 + (C^{-1/2}a^*)^2 + 2C^{-1/2}a^*C^{-1/2}a. \end{aligned}$$

Using above, we compute

$$H_f = \frac{1}{2} \int : \{\pi^2 + |\nabla\phi|^2 + m^2\phi^2\} : dx = \frac{1}{2} \int a^*C^{-1}a dx. \tag{21.37}$$

Problem 21.13 Show that the momentum operator $P_f = \int a^*(x)(-i\nabla_x) \times a(x)dx$ can be written as (cf. (26.26))

$$P_f = \int : \pi(x)\nabla_x\phi(x) : dx.$$

Problem 21.14 (see [135]) Let $c = \langle f, Cf \rangle = E(\phi(f)^2)$. Show

1. $: \phi(f)^n := c^{n/2}P_n(c^{-1/2}\phi(f))$ where P_n is the nth Hermite polynomial
2. $: e^{\phi(f)} := e^{\phi(f) - c/2}$.

21.5 Fock Space

Using the definitions (21.26) of the annihilation operator $a(x)$, we see that the only solution to the equation $a\Omega = 0$ is $\Omega = $ const. We thus set $\Omega \equiv 1$, and call it the *vacuum*. Now we show that any element of $L^2(Q, d\mu_C)$ can be obtained by applying creation operators to the vacuum. From now on, we omit the subindex s in Q_s. Let ⓢ denote the symmetrized tensor product and $L^2_{sym}(\mathbb{R}^{nd})$, the subspace of $L^2(\mathbb{R}^{nd})$ consisting of functions which are symmetric with respect to permutations of the n variables $x_j \in \mathbb{R}^d$.

Theorem 21.15 Any vector $F \in L^2(Q, d\mu_C)$ can be written uniquely as

$$F = \sum_{n=0}^{\infty} \frac{1}{\sqrt{n!}} \int F_n(x_1, \ldots, x_n) a^*(x_1) \cdots a^*(x_n) \Omega dx_1 \cdots dx_n \qquad (21.38)$$

where $F_n \in L^2_{sym}(\mathbb{R}^{nd}) = \circledS_1^n L^2(\mathbb{R}^d)$.

Proof. For simplicity, we will denote the right hand side in (21.38) by $\sum_n \frac{1}{\sqrt{n!}} \int F_n(a^*)^n \Omega$. We first remark that for any $G_n \in L^2(\mathbb{R}^{nd})$,

$$\int G_n(a^*)^n \Omega = \int \mathcal{G}_n^{sym}(a^*)^n \Omega \qquad (21.39)$$

where

$$G^{sym}(x_1, \ldots, x_n) = \frac{1}{n!} \sum_{\pi \in S_n} G(x_{\pi(1)}, \ldots, x_{\pi(n)}) \in L^2_{sym}(\mathbb{R}^{nd})$$

is the *symmetrization* of ϕ. Here S_n is the group of permutations of the n variables, and for $\pi \in S_n$, $\pi(1, \ldots, n) = (\pi(1), \ldots, \pi(n))$.

Next, we have by straightforward computation

$$\langle \int F_n(a^*)^n \Omega, \int G_m(a^*)^m \Omega \rangle = \begin{cases} 0 & n \neq m \\ n! \langle F_n, G_n \rangle & n = m. \end{cases} \qquad (21.40)$$

Problem 21.16 Show (21.40).

Thus, for two vectors, F and G, of the form (21.38), we have

$$\langle F, G \rangle := \sum_{n=1}^{\infty} \langle F_n, G_n \rangle. \qquad (21.41)$$

Now we use the fact that the span of vectors of the form $\prod_1^n \phi(f_j) \Omega, n \geq 1$, is dense in L^2 (see (21.31)). By the relation (21.33), we have

$$\prod_1^n \phi(f_j) = \prod_1^n \frac{1}{\sqrt{2}} (a(C^{1/2} f_j) + a^*(C^{1/2} f_j)).$$

Using the commutation relation, $a(f)a(f)^* = a(f)^* a(f) + \langle f, g \rangle$, to pull a's to the right of a^*'s, we transform the product $\prod(a + a^*)$ to the normal form $\prod_1^n (a + a^*) = \sum_{k+l \leq n} \int A_{kl} (a^*)^k a^l$. But $(a^*)^k a^l \Omega = 0$ unless $l = 0$, so $\prod_1^n \phi(f_j) \Omega = \sum_{k \leq n} \int A_{k0} (a^*)^k \Omega$. By (21.39), this can be rewritten as

$$\prod_1^n \phi(f_j) \Omega = \sum_{k \leq n} \int \frac{1}{\sqrt{n!}} F_k(a^*)^k \Omega$$

where $F_k \in L^2_{sym}(\mathbb{R}^{kd}) = \bigotimes^k_{1sym} L^2(\mathbb{R}^d)$. Thus vectors of the form

$$\sum \frac{1}{n!} \int F_n(a^*)^n \Omega$$

are dense in L^2. On the other hand, by (21.41), the subspace

$$\left\{ F = \sum_n \frac{1}{\sqrt{n!}} \int F_n(a^*)^n \Omega \mid F_n \in L^2_{sym}(\mathbb{R}^{nd}) \right\} \qquad (21.42)$$

is closed. Since by above, it contains a dense set, it is the whole L^2 space.

Definition 21.17 The (bosonic) Fock space is

$$\mathcal{F} := \bigoplus_{n=0}^{\infty} [\bigotimes_1^n L^2(\mathbb{R}^d)] \qquad \text{with} \qquad \langle F, G \rangle := \sum_{n=1}^{\infty} \langle F_n, G_n \rangle.$$

We call $\mathcal{F}_n := \bigotimes_1^n L^2(\mathbb{R}^d)$ the *n-particle sector*. By convention, $\mathcal{F}_0 = \mathbb{C}$.

The previous theorem provides a unitary isomorphism

$$L^2(Q, d\mu_C) \simeq \mathcal{F}$$

given by

$$\sum \frac{1}{\sqrt{n!}} \int \phi_n(a^*)^n \Omega \leftrightarrow \bigoplus_{n=0}^{\infty} \phi_n \equiv (\phi_0, \phi_1, \phi_2, \dots).$$

Moreover it is easily checked that

$$E(F) := \int_Q F d\mu_C = \langle \Omega, F\Omega \rangle$$

and on $\mathcal{F}$,

$$a(f) : \phi_n \in \mathcal{F}_n \mapsto \sqrt{n}\langle f, \phi_n \rangle \in \mathcal{F}_{n-1}$$

and

$$a^*(f) : \phi_n \in \mathcal{F}_n \mapsto \sqrt{n+1} f \bigotimes \phi_n \in \mathcal{F}_{n+1}.$$

Proposition 2. In the Fock space representation, the Hamiltonian, $H_f = \int a^* C^{-1} a$, the momentum operator $P_f = \int a^*(-i\nabla)a$ and *particle number*, $N = \int a^* a$, operators (see (21.27) and (21.29)) are of the form

$$H_f \phi \leftrightarrow \bigoplus_{n=0}^{\infty} \left(\sum_1^n C_{x_j}^{-1} \phi_n \right), \qquad P_f \phi \leftrightarrow \bigoplus_{n=0}^{\infty} \left(\sum_1^n p_{x_j} \phi_n \right),$$

$$N\phi \leftrightarrow \bigoplus_{n=0}^{\infty} (n\phi_n),$$

where the subscript x_j in C_{x_j} indicates that this operator acts on the variable x_j.

This is simply a matter of using the commutation relations. We leave the proof as an exercise.

Thus we have obtained a very simple realization of our state space $L^2(Q, d\mu_C)$ which is independent of C, and in which the Klein-Gordon Hamiltonian acts as a direct sum of simple operators in a finite but increasing number of variables:

$$H_f = \oplus_{n=0}^{\infty} \sum_1^n \sqrt{-\Delta_{x_i} + m^2}.$$

In particular, the spectrum of H_f is

$$\sigma(H_f) = \{0\} \cup \{\cup_{n \geq 1} [nm, \infty)\}$$

where the zero-eigenfunction is the vacuum, Ω. Physically, this theory describes non-interacting particles (bosons) of mass m.

Problem 21.18 Find the spectrum of the momentum operator

$$P_f = \int a^*(x)(-i\nabla_x)a(x)dx.$$

Remark 21.19 For connections with stochastic fields and with infinite dimensional pseudodifferential calculus, see [304] and Berezin [38], respectively. There is also a relation to the Wiener chaos expansion used in stochastic differential equations.

21.6 Quantization of Maxwell's Equations

Maxwell's equations as a hamiltonian system. As a prelude to quantizing Maxwell's equations we have first to write them in Hamiltonian form. Recall that we work in physical units in which the Planck constant and speed of light are equal to 1: $\hbar = 1$, $c = 1$. The Maxwell equations for vector fields $E : \mathbb{R}^{3+1} \to \mathbb{R}^3$ (the electric field) and $B : \mathbb{R}^{3+1} \to \mathbb{R}^3$ (the magnetic field) in vacuum are

$$\nabla \cdot E = 0 \qquad \nabla \times B = \frac{\partial E}{\partial t} \tag{21.43}$$

$$\nabla \times E = -\frac{\partial B}{\partial t} \qquad \nabla \cdot B = 0. \tag{21.44}$$

The third and fourth equations are actually constraints. They can be used to reduce number of unknowns by introducing the potentials $U : \mathbb{R}^{3+1} \to \mathbb{R}$ and $A : \mathbb{R}^{3+1} \to \mathbb{R}^3$ such that

$$B = \nabla \times A, \qquad E = -\partial A/\partial t - \nabla U. \tag{21.45}$$

Then the last the equations (21.44) are satisfied automatically. There is still a redundancy in the choice of A and U. Specifically, any *gauge transformation*

$$A \mapsto A + \nabla \chi, \qquad U \mapsto U - \partial \chi / \partial t \qquad (21.46)$$

for $\chi : \mathbb{R}^{3+1} \mapsto \mathbb{R}$, results in new potentials A and U which yield the same fields E and B. By appropriate choice of χ, we may take

$$\nabla \cdot A = 0$$

which is called the *Coulomb gauge*. From now on, we work in this gauge. The equation for U is $\Delta U = 0$, so we can take $U = 0$, which gives

$$E = -\partial A / \partial t, \qquad B = \nabla \times A.$$

Now, the first, third and fourth equations in (21.43) - (21.44) are automatically satisfied while the second equation results in

$$\Box A = 0, \qquad \nabla \cdot A = 0. \qquad (21.47)$$

Thus the vector potential A is a transverse vector field satisfying a wave equation. (A vector field $f : \mathbb{R}^3 \to \mathbb{R}^3$ is called *transverse* (or *incompressible* or *divergence free*) if $\nabla \cdot f = 0$.)

Similarly as for the Klein-Gordon equation, one shows that Equation (21.47) is the Euler-Lagrange equation for the action

$$S(A) = \frac{1}{2} \int_0^T \int_{\mathbb{R}^3} \{|\dot{A}|^2 - |\nabla \times A|^2\} dx dt \qquad (21.48)$$

where the variation is among transverse vector fields. Again repeating the steps we went through in the Klein-Gordon case, we arrive, for Maxwell's equations, at

1. the phase space, which is the direct sum of two Sobolev spaces of transverse vector fields (i.e. the derivatives are considered in distributional sense),
$$Z = H^{1,trans}(\mathbb{R}^3; \mathbb{R}^3) \oplus H^{1,trans}(\mathbb{R}^3; \mathbb{R}^3);$$

2. the Hamiltonian functional (the Legendre transform of the Lagrangian $L(A, \dot{A}) := \int_{\mathbb{R}^3} \{|\dot{A}|^2 - |\nabla \times A|^2\}$),

$$\begin{aligned} H(A, E) &= \frac{1}{2} \int \{|E|^2 + |\nabla \times A|^2\} dx \\ &= \frac{1}{2} \int \{|E|^2 + |B|^2\} dx \end{aligned} \qquad (21.49)$$

where $-E$ is the canonically conjugate field to $A : \partial_{\dot{A}} L(A, \dot{A}) = \dot{A} = -E$, and $\nabla \cdot E = 0$;

3. the Poisson bracket

$$\{F, G\} := \langle \partial_A F, T \partial_E G \rangle - \langle \partial_E F, T \partial_A G \rangle \qquad (21.50)$$

(on Z), where T be the projection operator of vector fields onto transverse vector fields:

$$TF := F - (\Delta)^{-1} \nabla (\nabla \cdot F);$$

4. Canonically conjugate fields are A and E. If $T_{ij}(x - y)$ is the matrix integral kernel of the operator T, then

$$\{E_i(x), A_j(y)\} = T_{ij}(x - y).$$

Note that by the definition, $\partial_A F = T \partial_A F$ and similarly for $\partial_E F$. The Hamilton equations for the above system are

$$\dot{\Phi} = J_T \partial_\Phi H(\Phi), \quad J_T := \begin{pmatrix} 0 & -T \\ T & 0 \end{pmatrix}, \quad \Phi = (A, E). \tag{21.51}$$

Note that the first Hamilton equation yields, as expected,

$$\dot{A} = T \partial_E H(A, E) = -TE = -E.$$

Problem 21.20 Check that $\nabla \cdot (TF) = 0$.

Problem 21.21 Check that Maxwell's equations are equivalent to the Hamilton equations (21.51). (Cf. Problem 21.2.)

Quantization of the EM field. We quantize the Maxwell hamiltonian system in the same way as the Klein-Gordon one, but replacing scalar generalized operator-functions, say $a(x)$, by generalized transverse operator-vector-fields, say $a(x) = (a_1(x), a_2(x), a_3(x))$ with $\nabla \cdot a(x) = 0$, and using the projection T onto the be the transverse vector fields, when needed. Here $\nabla \cdot a(x)$ is understood as a distributional divergence of the operator-valued distribution,

$$\int \nabla \cdot a(x) \bar{f}(x) dx = - \int a(x) \cdot \nabla \bar{f}(x) dx = -a(\nabla f),$$

i.e. a symbolic way to represent the operator-valued functional $-a(\nabla f)$. The resulting theory can be summarized as follows:

1. The quantum state space is $L^2(Q^{trans}, d\mu_C)$, where, for s sufficiently large,

$$Q^{trans} = \{A : \mathbb{R}^3 \to \mathbb{R}^3 \mid (-\Delta + |x|^2 + \mathrm{id})^{-s/2} A \in L^2, \ \mathrm{div} A = 0\}$$

and $d\mu_C$ is the Gaussian measure of Q^{trans} with the covariance operator $C = (-\Delta)^{-1/2}$.

2. The quantized $A(x)$ and $E(x)$ are operator-valued transverse vector fields, acting on the space $L^2(Q^{trans}, d\mu_C)$ as the multiplication operator by $A(x)$ and $E(x) := -i\partial_{A(x)} + \frac{1}{2}iC^{-1}A(x)$, respectively. They give canonically conjugate fields with the non-trivial commutation relation

$$i[E_k(x), A_l(y)] = T_{kl}(x - y)\mathbf{1}$$

where $T(x - y)$ is the integral kernel of the projection operator, T, onto the transverse vector fields (recall our convention $\hbar = 1$).

3. $A(x)$ and $E(x)$ are expressed in terms of the annihilation and creation operators, $a(x)$ and $a^*(x)$, as

$$A(x) = \frac{1}{\sqrt{2}}C^{1/2}(a(x)+a^*(x)), \ E(x) = \frac{1}{i\sqrt{2}}C^{-1/2}(a(x)-a^*(x)). \ (21.52)$$

Here $a(x)$ and $a^*(x)$ are operator-valued transverse vector-fields, $a(x) = (a_1(x), a_2(x), a_3(x))$ with $\nabla \cdot a(x) = 0$, satisfying the commutation relations

$$[a_i(x), a_j^*(y)] = T_{kl}(x - y)\mathbf{1}, \quad [a_i(x), a_j(y)] = 0 = [a_i^*(x), a_j^*(y)]. \quad (21.53)$$

4. The quantum Hamiltonian operator acts on the space $L^2(Q^{trans}, d\mu_C)$ and is given by

$$H_f = \frac{1}{2}\int :|E|^2 + |\nabla \times A|^2 : dx = \int a(x)^* \cdot \sqrt{-\Delta}a(x)dx, \quad (21.54)$$

where $a(x)^* \cdot \sqrt{-\Delta}a(x) = \sum_i a_i(x)^* \cdot \sqrt{-\Delta}a_i(x)$. (Here again the subindex f stands for the "field" and the precise definition of the integral on the r.h.s can be found in Section 21.2.)

Note an essential difference in the EM case: there is no mass ($m = 0$). That is, the covariance operator is $C = (-\Delta)^{-1/2}$.

The equations (21.52) - (21.54) give the full description of the quantized electro-magnetic field in vacuum.

For the Maxwell theory, the Fock space is built on the one-particle space $L^2_{transv}(\mathbb{R}^3, \mathbb{R}^3) := \{f \in L^2(\mathbb{R}^3, \mathbb{R}^3)|\mathrm{div} f = 0\}$, instead of $L^2(\mathbb{R}^d) \equiv L^2(\mathbb{R}^d, \mathbb{C})$. The Maxwell Hamiltonian acts on it as

$$H_f = \oplus_{n=0}^{\infty} \sum_1^n \sqrt{-\Delta_{x_i}}.$$

This representation can be used to obtain the spectrum of H:

$$\sigma(H_f) = \{0\} \cup \{\cup_{n \geq 1}[0, \infty)\}$$

where the zero-eigenfunction is the vacuum, Ω. Physically, this theory describes non-interacting massless particles (bosons), called photons.

The Fourier transforms, $a(k)$ and $a^*(k)$, of the operators $a(x)$ and $a^*(x)$ satisfy $a(k) = (a_1(k), a_2(k), a_3(k))$ with $k \cdot a(k) = 0$. The commutation relations for $a_i(k)$, $a_i^*(k)$ are

$$[a_i(k), a_j^*(k)] = (\delta_{ij} - \frac{k_i k_j}{|k|^2})\delta(k - k'), \quad [a_i(k), a_j(k')] = 0 = [a_i^*(k), a_j^*(k')]. \quad (21.55)$$

In these terms, the Maxwell quantum Hamiltonian is of the form

$$H_f = \int \omega(k) a^*(k) \cdot a(k) dk,$$

with $\omega(k) = |k|$, and the quantized vector potential, $A(x)$, is given by

$$A(x) = \int \{e^{ix\cdot k} a(k) + e^{-ix\cdot k} a^*(k)\} \frac{dk}{\sqrt{\omega(k)}}. \qquad (21.56)$$

In the Fourier representation, we can choose an orthonormal basis $e_\lambda(k) \equiv e(k, \lambda)$, $\lambda \in \{-1, 1\}$ in $k^\perp \subset \mathbb{R}^3$, satisfying $k \cdot e_\lambda(k) = 0$. The vectors $e_\lambda(k) \equiv e(k, \lambda)$, $\lambda \in \{-1, 1\}$ are called polarization vectors. We can write the operator-valued transverse vector fields $a^\#(k)$ as

$$a^\#(k) := \sum_{\lambda \in \{-1, 1\}} e_\lambda(k) a_\lambda^\#(k),$$

where $a_\lambda^\#(k) \equiv a^\#(k, \lambda) := e_\lambda(k) \cdot a^\#(k)$, the components of the creation and annihilation operators $a^\#(k)$ in the direction transverse to k; they satisfy the commutation relations

$$[a_\lambda^\#(k), a_{\lambda'}^\#(k')] = 0, \qquad [a_\lambda(k), a_{\lambda'}^*(k')] = \delta_{\lambda,\lambda'} \delta^3(k - k'). \qquad (21.57)$$

Now the Hamiltonian H_f and the quantized vector potential, $A(x)$, can be written as

$$H_f = \sum_\lambda \int \omega(k) a_\lambda^*(k) a_\lambda(k) dk$$

and

$$A(x) = \sum_\lambda \int \{e^{ix\cdot k} a_\lambda(k) + e^{-ix\cdot k} a_\lambda^*(k)\} e_\lambda(k) \frac{dk}{\sqrt{\omega(k)}}. \qquad (21.58)$$

Elements of $\mathcal{F}_n$ can be written as $\psi_n(k_1, \lambda_1, \ldots, k_n, \lambda_n)$, where $\lambda_j \in \{-1, 1\}$ are the polarization variables. Roughly in the case of photons, compared with scalar bosons, one replaces the variable k by the pair (k, λ) and adds to the integrals in k also the sums over λ.

Standard Model of Non-relativistic Matter and Radiation

In this chapter we introduce and discuss the standard model of non-relativistic quantum electrodynamics (QED). Non-relativistic QED was proposed in the early days of Quantum Mechanics (it was used by Fermi ([98]) in 1932 in his review of theory of radiation). It describes quantum-mechanical particle systems coupled to a quantized electromagnetic field, and appears as a quantization of the system of non-relativistic classical particles interacting with a classical electromagnetic field (coupled Newton's and Maxwell's equations). It is the most general quantum theory obtained by quantizing a classical system.

22.1 Classical Particle System Interacting with an Electro-magnetic Field

We consider a system of n classical particles of masses m_i and electric charges e_i interacting with electro-magnetic field. Recall that we work in physical units in which the speed of light is equal to 1: $c = 1$. The coupled Newton's and Maxwell's equations, describing interacting particles and electromagnetic field are

$$m_i \ddot{x}_i(t) = e_i [E + \dot{x}_i \wedge B](x_i(t), t), \tag{22.1}$$

$$\nabla \cdot E = \rho, \qquad \nabla \times B = \frac{\partial E}{\partial t} + j, \tag{22.2}$$

$$\nabla \times E = -\frac{\partial B}{\partial t}, \qquad \nabla \cdot B = 0. \tag{22.3}$$

Here ρ and j are the charge and current densities: $\rho(x, t) = \sum_i e_i \delta(x - x_i(t))$ and $j(x, t) = \sum_i e_i \dot{x}_i(t) \delta(x - x_i(t))$. The first equation is Newton's equation with the Lorentz force. The last four equations are Maxwell's equations in presence of charges and currents.

To find a hamiltonian formulation of these equation we first present the minimum action principle for this system. As before, we express the electric and magnetic fields, E and B, in terms of the vector potential, $A : \mathbb{R}^{3+1} \mapsto \mathbb{R}^3$,

© Springer-Verlag GmbH Germany, part of Springer Nature 2020
S. J. Gustafson and I. M. Sigal, *Mathematical Concepts of Quantum Mechanics*, Universitext, https://doi.org/10.1007/978-3-030-59562-3_22

and the scalar potential, $U : \mathbb{R}^{3+1} \mapsto \mathbb{R}$ via (21.45). The action functional which gives (22.1) is given by

$$S(\gamma|U, A) = \int_0^T \sum_i \left(\frac{m_i}{2}|\dot{\gamma}_i|^2 - e_i U(\gamma_i) + e_i \dot{\gamma}_i \cdot A(\gamma_i) \right). \qquad (22.4)$$

Indeed, let us compute the Euler-Lagrange equation for this functional (see (4.40)). Using the Lagrangian function $L(\gamma, \dot{\gamma}|U, A) = \sum_i \left(\frac{m_i}{2}|\dot{\gamma}_i|^2 - eU(\gamma_i) - e\dot{\gamma}_i \cdot A(\gamma_i) \right)$, we compute

$$\partial_{\dot{\gamma}} L(\gamma, \dot{\gamma}|U, A) = m\dot{\gamma} + eA(\gamma), \quad \partial_\gamma L(\gamma, \dot{\gamma}|U, A) = e\nabla U(\gamma) + e\nabla\dot{\gamma} \cdot A(\gamma).$$

Plug this into (4.40) and use the relations $\frac{d}{dt} A(\gamma) = \partial_t A(\gamma) + (\dot{\gamma} \cdot \nabla) A(\gamma)$, $\nabla(\dot{\gamma} \cdot A) - (\dot{\gamma} \cdot \nabla) A(\gamma) = \dot{\gamma} \wedge \text{curl} A$, and (21.45) to obtain (22.1).

To (22.4) we add the action of electromagnetic field found in the previous chapter, but with a non-zero scalar potential,

$$S_{em}(U, A) = \frac{1}{2} \int_0^T \int_{\mathbb{R}^3} \left(|\dot{A} + \nabla U|^2 - |\nabla \times A|^2 \right)(x, t) dx dt.$$

(We do not have to assume the Coulomb gauge, $\text{div} A = 0$, here.) Using that $eU(x(t), t) = \int_{\mathbb{R}^3} \rho(x, t) U(x, t) dx$ and $e\dot{x}(t) \cdot A(x(t), t) = \int_{\mathbb{R}^3} j(x, t) \cdot A(x, t) dx$, to rewrite the last two terms in (22.4), we obtain the action functional of the coupled system

$$S(\gamma, U, A) = S_p(\gamma) + S_{em}(U, A) + S_{int}(\gamma, U, A), \qquad (22.5)$$

where $S_p(\gamma) = \int_0^T \sum_i \frac{m_i}{2}|\dot{\gamma}_i|^2 dt$ is the action of the free particle, familiar to us from Section 4.7, $S_{em}(U, A)$ is the action of a free electro-magnetic field given above, and $S_{int}(\phi, U, A)$ is the interaction action, coupling them,

$$S_{int}(\gamma, U, A) = \int_0^T \int_{\mathbb{R}^3} \left(-\rho(x, t) U(x, t) + j(x, t) \cdot A(x, t) \right) dx dt. \qquad (22.6)$$

It is easy to check that the Euler-Lagrange equation for the action (22.5) gives the Newton - Maxwell system (22.1) - (22.3).

Gauge invariance. The fields E and B are not changed under *gauge transformations* (21.46) of A and U. Hence we would like to make sure that the action (22.6) gives the same equations of motion for different A and U, connected by a gauge transformation (21.46). Under (21.46), the action (22.6) changes as $S(\gamma, U, A) \to S(\gamma, U, A) + \Lambda(\gamma, \chi)$, where

$$\Lambda(\gamma, \chi) := \int_0^T \int_{\mathbb{R}^3} \left(\rho \partial_t \chi + j(x, t) \cdot \nabla \chi \right) dx dt \qquad (22.7)$$

$$= \int_0^T \int_{\mathbb{R}^3} (\partial_t(\rho\chi) + \nabla \cdot (j\chi) - (\nabla \cdot j + \partial_t \rho))\,dx\,dt. \qquad (22.8)$$

Hence the gauge transformation leads to an equivalent action/Lagrangian, i.e. to the action which gives the same Euler-Lagrange equation.

Covariant formulation. Geometrically, the electric field is a one-form, while the magnetic field is a two-form. One can form the field tensor $F :=$ $-E \wedge dx^0 + B$, as a two-form on the Minkowskii space M^{3+1}, with coordinates (x^0, x^1, x^2, x^3), where $x^0 = t$, and the metric $g = \mathrm{diag}(1, -1, -1, -1)$. By the last two Maxwell's equations, F is a closed form, $dF = 0$, and therefore it is exact, i.e. there is a one-form $\mathcal{A}$, s.t. $F = d\mathcal{A}$. ($\mathcal{A}$ can be thought of as a connection, and F as a curvature on a $U(1)$−bundle.) Then the action functional can be written as a 'Dirichlet' integral, $S(\mathcal{A}) = -\frac{1}{16\pi} \int_{M^{3+1}} (\|d\mathcal{A}\|^2 + \langle \mathcal{J}, \mathcal{A} \rangle)$, where $\mathcal{J} := (\rho, j)$ on the Minkowskii space M^{3+1}, with the norm and inner product related to the Minkowskii metrics, $g = \mathrm{diag}(1, -1, -1, -1)$. This gives the first two of Maxwell's equations as $d^*\mathcal{F} = \mathcal{J}$, where d^* is the operator adjoint to d.

Elimination of scalar potential. Note that the time derivative of scalar potential U does not enter this Lagrangian. This indicates that U is not a dynamical variable and we can eliminate it from the action. We do this in the Coulomb gauge, $\mathrm{div}\,A = 0$, as follows. Varying the action $S(\gamma, U, A)$ w.r. to U, we obtain the first of Maxwell's equations

$$- \Delta U = \rho. \qquad (22.9)$$

Furthermore, $\mathrm{div}\,A = 0$ implies, after integration by parts, that $\int_{\mathbb{R}^3} |\dot{A} + \nabla U|^2 dx = \int_{\mathbb{R}^3} (|\dot{A}|^2 + |\nabla U|^2)\,dx$. Using this and $\int_{\mathbb{R}^3} |\nabla U|^2 dx = \int_{\mathbb{R}^3} U(-\Delta)U\,dx$ $= \int_{\mathbb{R}^3} U\rho\,dx$, we see that the terms involving U add to $-\frac{1}{2} \int_{\mathbb{R}^3} U\rho\,dx$. Now, solving the Poisson equation (22.9) for U as $U = (-\Delta)^{-1}\rho$, we obtain

$$\int_{\mathbb{R}^3} \rho U\,dx = \int_{\mathbb{R}^3} \rho(-\Delta)^{-1}\rho\,dx = V_{coul}(\underline{x}),$$

where $\underline{x} = (x_1, \ldots, x_n)$, plus the infinite Coulomb self-energy term. Here $V_{coul}(\underline{x}) := \frac{1}{2} \sum_{i \neq j} \frac{e_i e_j}{|x_i - x_j|}$. With this and dropping the Coulomb self-energy term, we can write the action as

$$S^{coul}(\gamma, A) = S_p^{coul}(\gamma) + S_f(A) + S_{int}^{coul}(\gamma, A), \qquad (22.10)$$

where $S_p^{coul}(\gamma) = \int_0^T \left(\sum_i \frac{m_i}{2} |\dot{\gamma}_i|^2 - V_{coul} \right) dt$, $S_f(A) = \frac{1}{2} \int_0^T \int_{\mathbb{R}^3} (|\dot{A}(x,t)|^2 - |\nabla \times A(x,t)|^2)\,dx\,dt$ is the free action of electromagnetic field, encountered in the previous chapter, and

$$S_{int}^{coul}(\phi, A) = \int_0^T \int_{\mathbb{R}^3} j(x,t) \cdot A(x,t)\,dx\,dt. \qquad (22.11)$$

Hamiltonian formulation. Now, we pass to the hamiltonian formulation. To this end we impose the Coulomb gauge, $\mathrm{div}A = 0$ and use the Lagrangian given by the above action. The generalized particle momenta are $k_i = m_i \dot{x}_i + eA(x_i)$, while the field momentum is $-E = \dot{A}$, the same as in the free case. The classical Hamiltonian functional is $H(\underline{x}, \underline{k}, A, E) = \underline{k} \cdot \underline{v} - L(\underline{x}, \underline{v}, A, E)|_{m_i \dot{x}_i = k_i - eA(x_i), \dot{A} = -E}$, which gives

$$H(\underline{x}, \underline{k}, A, E) = \sum_i \frac{1}{2m_i}(k_i - e_i A(x_i))^2 + V_{coul}(\underline{x}) + H_f(A, E)$$

where $\underline{k} = (k_1, \ldots, k_n)$ and $H_f(A, E)$ is the Hamiltonian functional of the free electro-magnetic field. Defining the Poisson bracket as the sum of classical mechanics one, (4.47), and electro-magnetic one, (21.50), we arrive at the hamiltonian formulation for a system of n particles of masses $m_1, \ldots, m_n$ and charges $e_1, \ldots, e_n$ interacting with electromagnetic field, (E, B).

22.2 Quantum Hamiltonian of Non-relativistic QED

According to our general quantization procedure, we replace the classical canonical variables x_i^{cl} and k_i^{cl} and classical fields $A^{cl}(x)$, and $E^{cl}(x)$ by the quantum canonical operators x_i, p_i, $A(x)$, and $E(x)$ (see (4.4) and (21.52)), acting on the state space

$$L^2(\mathbb{R}^{3n}) \otimes L^2(Q^{trans}, d\mu_C) \simeq L^2(\mathbb{R}^{3n}) \otimes \mathcal{F} \equiv \mathcal{H}_{\mathrm{part}} \otimes \mathcal{H}_f. \qquad (22.12)$$

In the units in which the Planck constant divided by 2π and the speed of light are equal to 1, $\hbar = 1$ and $c = 1$, the resulting Schrödinger operator, acting on $\mathcal{H}_{\mathrm{part}} \otimes \mathcal{H}_f$, is

$$H = \sum_{i=1}^{n} \frac{1}{2m_i}(p_i - e_i A(x_i))^2 + V_{coul}(\underline{x}) + H_f, \qquad (22.13)$$

where, recall, m_i and e_i are the mass (in fact, the 'bare' mass, see below) and charge of the i-th particle. Recall the the quantized vector potential $A(x)$ and quantum Hamiltonian H_f are given in terms of the annihilation and creation operators $a(k) = \sum_\lambda e_\lambda(k) a_\lambda(k)$ and $a^*(k) = \sum_\lambda e_\lambda(k) a_\lambda^*(k)$ (obeying canonical commutation relations (21.57)) by

$$A(x) = \int \{e^{ix \cdot k} a(k) + e^{-ix \cdot k} a^*(k)\} \frac{d^3 k}{\sqrt{\omega(k)}} \qquad (22.14)$$

and

$$H_f = \int \omega(k) a(k)^* \cdot a(k) dk. \qquad (22.15)$$

Note that in the units we have chosen, various physical quantities entering the quantum hamiltonian are not dimensionless. We choose these units in order to keep track of the particle mass.

Also note that we omitted the identities 1_{part} and 1_f on $\mathcal{H}_{\text{part}}$ and $\mathcal{H}_f$, respectively, and a careful notation would have $V_{coul}(\underline{x}) \otimes 1_f$ and $1_{\text{part}} \otimes H_f$, instead of $V_{coul}(\underline{x})$ and H_f. In what follows, as a rule, we do not display these identities.

Gauge principle (minimal coupling). We see that the full, interacting Hamiltonian, is obtained by replacing the particle momenta p_j by the covariant momenta $p_{A,j} = p_j - e_j A(x_j)$, and adding to the result the field Hamiltonian, H_f, responsible for the dynamics of $A(x)$. This procedure is called the *minimal coupling* and it is justified by a gauge principle requiring the global gauge symmetry of the particle system also to be the local one.

Ultra-violet cut-off. The quantum Hamiltonian H we obtained above is ill-defined: its domain of definition contains no non-zero vectors. The problem is in the interaction $(p - eA(x))^2 - p^2 = -2eA(x)p + e^2A(x)^2$ (written here in the one-particle case). The $A(x)$ is too rough an operator-valued function. It has the empty domain. E. g.

$$A(x)(f(x) \otimes \Omega) = f(x)A(x)\Omega$$

$$= f(x) \int e^{-ix \cdot k} a^*(k)\Omega \frac{d^3k}{\sqrt{\omega(k)}} \notin L^2(\mathbb{R}^3) \otimes \mathcal{F}.$$

To remedy this we institute an *ultraviolet cut-off*. It consists of replacing $A(x)$ in (22.14) by the operator $A_\chi(x) := \check{\chi} * A(x)$ where $\check{\chi}$ is a smooth well-localized around 0 (i.e. sufficiently fast decaying away from 0) function:

$$A(x) \mapsto A_\chi(x) := \check{\chi} * A(x). \tag{22.16}$$

Recall that $\check{\chi}$ denotes the inverse Fourier transform of a function χ. We choose $\check{\chi}$ to be a positive function whose integral is equal to 1 (a smoothed-out δ-function). In fact, the specific shape of $\check{\chi}$ is not important for us. Now, $A_\chi(x)$ is of the form

$$A_\chi(x) = \int (e^{ik \cdot x}a(k) + h.c.) \frac{\chi(k)}{\sqrt{\omega(k)}} dk. \tag{22.17}$$

Assuming that the ultra-violet cut-off χ decays on the scale κ, we arrive, as a result, at the Hamiltonian of non-relativistic matter interacting with radiation:

$$H_{e\kappa} = \sum_{j=1}^n \frac{1}{2m_j}(p_{A_\chi,j})^2 + e_j^2 V_{coul}(\underline{x}) + H_f \tag{22.18}$$

acting on $\mathcal{H}_{\text{part}} \otimes \mathcal{H}_f$, where $p_{A_\chi,j} = p_j - e_j A_\chi(x_j)$, and $p_j = -i\nabla_{x_j}$ (note that the parameter e enters the definition of the operator $p_{A_\chi,j}$). This is the standard model of non-relativistic QED. We show below that $H_{e\kappa}$ is well-defined and is self-adjoint.

To sum up, we arrived at a physical system with the state space $\mathcal{H} = \mathcal{H}_p \otimes \mathcal{H}_f$ and the quantum Hamiltonian $H_{e\kappa}$. Its dynamics is given by the time-dependent Schrödinger equation with

$$i\partial_t \psi = H_{e\kappa}\psi,$$

where ψ is a differentiable path in the Hilbert space $\mathcal{H} = \mathcal{H}_p \otimes \mathcal{H}_f$. A few comments are now in order.

Choice of κ. We assume that our matter consists of electrons and the nuclei and that the nuclei are infinitely heavy and therefore are manifested through the interactions only). We reintroduce the Planck constant, $\hbar$ and speed of light, c, for a moment. The electron mass is denoted by m_{el}. Since we assume that $\chi(k)$ decays on the scale κ, in order to correctly describe the phenomena of interest, such as emission and absorption of electromagnetic radiation, i.e. for optical and rf modes, we have to assume that the cut-off energy, $\hbar c \kappa$, is much greater than the characteristic energies of the particle motion. The latter motion takes place on the energy scale of the order of the ionization energy, i.e. of the order $\alpha^2 m_{el} c^2$, where $\alpha = \frac{e^2}{4\pi\hbar c} \approx \frac{1}{137}$ is the fine-structure constant. Thus we have to assume $\alpha^2 m_{el} c^2 \ll \hbar c \kappa$.

On the other hand, for energies higher than the rest energy of the the electron ($m_{el}c^2$) the relativistic effects, such as electron-positron pair creation, take place. Thus it makes sense to assume that $\hbar c \kappa \ll m_{el} c^2$. Combining the last two conditions we arrive at the restriction $\alpha^2 m_{el} c^2 \ll \hbar c \kappa \ll m_{el} c^2$ or $\alpha^2 m_{el} c/\hbar \ll \kappa \ll m_{el} c/\hbar$. In our units ($\hbar = 1$, $c = 1$) this becomes

$$\alpha^2 m_{el} \ll \kappa \ll m_{el}.$$

Free parameters. We will see later that the physical mass, m_{el}, is not the same as the parameter $m \equiv m_j$ (the 'bare' mass) entering (22.18), but depends on m and κ. Inverting this relation, we can think of m as a function of m_{el} and κ. If we fix and the particle potential $V(x)$ (e.g. taking it to be the total Coulomb potential), and m_{el} and e, then the Hamiltonian (22.18) depends on one free parameter, the bare electron mass m (or the ultraviolet cut-off scale, κ).

Gauge equivalence. We quantized the system in the Coulomb gauge. Assume we quantized the system in a different gauge, say the Lorentz gauge, how is the latter Hamiltonian related to the former one? As we saw above, classically, different gauges give equivalent descriptions of a classical system. Do they lead to equivalent descriptions of the corresponding quantum system? The answer is yes. One can show that quantum Hamiltonians coming out of different gauges, as well as other observables, are related by unitary (canonical) transformations.

22.2.1 Translation invariance

The system of particles interacting with the quantized electromagnetic fields is invariant under translations of the particle coordinates, $\underline{x} \to \underline{x} + y$, where $\underline{y} = (y, \ldots, y)$ ($n-$ tuple) and the fields, $A(x) \to A(x-y)$, i.e. $H_{e\kappa}$ commutes with the translations

$$T_y : \Psi(\underline{x}, A) \to \Psi(\underline{x} + \underline{y}, t_y A), \qquad (22.19)$$

where $(t_y A)(x) = A(x-y)$. (We say that $H_{e\chi}$ is *translation invariant*.) Indeed, we use (22.19) and the definitions of the operators $A(x)$ and $E(x)$, to obtain $T_y A(x) = (t_y A)(x) T_y$ and $T_y E(x) = (t_y E)(x) T_y$, which, due to the definition of $H_{e\chi}$, give

$$T_y H_{e\kappa} = H_{e\kappa} T_y,$$

In the Fock space representation (22.19) becomes

$$T_y : \oplus_n \Psi_n(\underline{x}, k_1, \ldots, k_n) \to \oplus_n e^{iy \cdot (k_1 + \ldots k_n)} \Psi_n(\underline{x} + \underline{y}, k_1, \ldots, k_n)$$

and therefore can be rewritten as $T_y : \Psi(\underline{x}) \to e^{iy \cdot P_f} \Psi(\underline{x} + \underline{y})$, where P_f is the momentum operator associated to the quantized radiation field,

$$P_f = \sum_\lambda \int dk\, k\, a_\lambda^*(k) a_\lambda(k).$$

As we know from Section 3.4, typically, symmetries of a physical system lead to conservation laws. (In the classical case, this is the content of the Noether theorem.) In our case, of a particular importance is space-time translational symmetry, which leads to conservation laws of the energy, $H_{e\chi}$, and the total momentum. We check this for the spatial translations. It is straightforward to show that T_y are unitary operators and that they satisfy the relations $T_{x+y} = T_x T_y$ (and therefore $y \to T_y$ is a unitary Abelian representation of $\mathbb{R}^3$). Finally, we observe that the group T_y is generated by the total momentum operator, P_{tot}, of the electrons and the photon field: $T_y = e^{iy \cdot P_{tot}}$. Here P_{tot} is the selfadjoint operator on $\mathcal{H}$, given by

$$P_{tot} := \sum_i p_i \otimes 1_f + 1_{\text{part}} \otimes P_f \qquad (22.20)$$

where, as above, $p_j := -i\nabla_{x_j}$, the momentum of the $j-$th electron and P_f is the field momentum given above. Hence $[H_{e\kappa}, P_{tot}] = 0$.

22.2.2 Fiber decomposition with respect to total momentum

Since the Hamiltonian $H_{e\kappa}$ commutes with translations, T_y, it has a fiber decomposition w.r.to the generator P_{tot} of the translations (cf. Section 6.6). We construction this decomposition. Let $\mathcal{H}$ be the direct integral $\mathcal{H} = \int_{\mathbb{R}^3}^{\oplus} \mathcal{H}_P dP$, with the fibers $\mathcal{H}_P := L^2(X) \otimes \mathcal{F}$, where $X := \{x \in \mathbb{R}^{3n} \mid \sum_i m_i x_i = 0\} \simeq \mathbb{R}^{3(n-1)}$, (this means that $\mathcal{H} = L^2(\mathbb{R}^3, dP; L^2(X) \otimes \mathcal{F})$) and define $U : \mathcal{H}_{el} \otimes \mathcal{H}_f \to \mathcal{H}$ on smooth functions, compactly supported in $\underline{x}$, by the formula

$$(U\Psi)_P(\underline{x}') = \int_{\mathbb{R}^3} e^{i(P-P_f)\cdot x_{cm}} \Psi(\underline{x}' + \underline{x}_{cm}) dy, \qquad (22.21)$$

where $\underline{x}'$ are the coordinates of the n particles in the center-of-mass frame and $\underline{x}_{cm} = (x_{cm}, \ldots, x_{cm})$ ($n-$ tuple), with $x_{cm} = \frac{1}{\sum_i m_i} \sum_i m_i x_i$, the center-of-mass coordinate, so that $\underline{x} = \underline{x}' + \underline{x}_{cm}$. Then U extends uniquely to a unitary operator (see below). Its converse is written, for $\Phi_P(\underline{x}') \in L^2(X) \otimes \mathcal{F}$, as

$$(U^{-1}\Phi)(\underline{x}) = \int_{\mathbb{R}^3} e^{-ix_{cm}\cdot(P-P_f)} \Phi_P(\underline{x}') dP. \qquad (22.22)$$

The functions $\Phi_P(\underline{x}') = (U\Psi)_P(\underline{x}')$ are called fibers of Ψ.

Lemma 1. *The operations (23.8) and (22.22) define unitary maps $L^2(\mathbb{R}^{3n}) \otimes \mathcal{F} \to \mathcal{H}$ and $\mathcal{H} \to L^2(\mathbb{R}^{3n}) \otimes \mathcal{F}$, and are mutual inverses.*

Proof. By density, we may assume that Ψ is a C_0^∞ in $\underline{x}$. Then, it follows from standard arguments in Fourier analysis that

$$(U^{-1}U\Psi)(\underline{x}) = \int dP\, e^{-i(P-P_f)x_{cm}} \int dy\, e^{i(P-P_f)\cdot y} \Psi(\underline{x}' + y)$$

$$= \int dy \int dP\, e^{-iP\cdot(x_{cm}-y)} e^{iP_f\cdot(x_{cm}-y)} \Psi(\underline{x}' + y)$$

$$= \int dy\, \delta(x_{cm} - y)\, e^{iP_f\cdot(x_{cm}-y)} \Psi(\underline{x}' + y)$$

$$= \Psi(\underline{x}). \qquad (22.23)$$

On the other hand, for $\Phi \in \mathcal{H}$,

$$(UU^{-1}\Phi)_P(\underline{x}') = \int dy\, e^{i(P-P_f)y} \int dq\, e^{i(q-P_f)y} \Phi_q(\underline{x}')$$

$$= \int dq \int dy\, e^{i(p-q)y} \Phi_q(\underline{x}')$$

$$= \int dq\, \delta(P - q) \Phi_q(\underline{x}')$$

$$= \Phi_P(\underline{x}'). \qquad (22.24)$$

From the density of C_0^∞ in $\underline{x}$ functions, we infer that (22.23) and (22.24) define bounded maps which are mutual inverses. Unitarity can be checked easily. $\square$

Since $H_{e\kappa}$ commutes with P_{tot}, it follows that $H_{e\kappa}$ admits the fiber decomposition

$$U H_{e\kappa} U^{-1} = \int_{\mathbb{R}^3}^{\oplus} H_{e\kappa}(P) dP, \qquad (22.25)$$

where the fiber operators $H_{e\kappa}(P)$, $P \in \mathbb{R}^3$, are self-adjoint operators on the space fibers $\mathcal{H}_P$. The latter means that $U H_{e\kappa} U^{-1} \Phi_P = H_{e\kappa}(P) \Phi_P$ for $\Phi_P(\underline{x}') \in \int_{\mathbb{R}^3}^{\oplus} \mathcal{H}_P dP$. We compute $H_{e\kappa}(P)$. Using $a(k)e^{-iy \cdot P_\mathrm{f}} = e^{-iy \cdot (P_\mathrm{f}+k)} a(k)$ and $a^*(k)e^{-iy \cdot P_\mathrm{f}} = e^{-iy \cdot (P_\mathrm{f}-k)} a^*(k)$, we find $\nabla_y e^{iy \cdot (P-P_\mathrm{f})} A_\chi(x'+y) e^{iy \cdot (P-P_\mathrm{f})} = 0$ and therefore

$$A_\chi(x) e^{iy \cdot (P-P_\mathrm{f})} = e^{iy \cdot (P-P_\mathrm{f})} A_\chi(x-y). \qquad (22.26)$$

Using this and (22.22), we compute

$$H_{e\kappa}(U^{-1}\Phi)(\underline{x}) = \int_{\mathbb{R}^3} e^{ix_{cm} \cdot (P-P_\mathrm{f})} H_{e\kappa}(P) \Phi(P) dP,$$

where the fiber Hamiltonians $H_{e\kappa}(P)$ are given explicitly by

$$H_{e\kappa}(P) = \sum_j \frac{1}{2m_i} \left(P - P_\mathrm{f} - i\nabla_{x'_j} - e_i A_\chi(x'_j) \right)^2 + V_{\mathrm{coul}}(\underline{x}') + H_\mathrm{f} \qquad (22.27)$$

where $x'_i = x_i - x_{cm}$ is the coordinate of the i-th particle in the center-of-mass frame.

22.3 Rescaling and decoupling scalar and vector potentials

In order to simplify the notation and some of the analysis from now on we assume that our matter consists of electrons and the nuclei and that the nuclei are fixed (the Born-Oppenheimer approximation in the case of molecules, see Section 10.1) and therefore are manifested through the interactions only. Recall that we work in physical units in which the Planck constant and speed of light are equal to 1: $\hbar = 1$ and $c = 1$, so that $e^2/4\pi = \alpha \approx \frac{1}{137}$, the fine-structure constant. In the original units, $\alpha = \frac{e^2}{4\pi\hbar c}$.

In the Hamiltonian (22.18) the coupling constant - the electron charge e - enters in two places - into the particle system itself (the Coulomb potential) and into coupling the particle systems to the quantized electro-magnetic field and thus has two different effects onto the total system. To decouple these two effects, we rescale our Hamiltonian as follows:

$$x \to \frac{1}{\alpha} x , \qquad k \to \alpha^2 k ,$$

Under this rescaling, the Hamiltonian $H_{e\kappa}$ is mapped into the Hamiltonian $\alpha^2 H(\varepsilon)$, where

$$H(\varepsilon) = \sum_{j=1}^n \frac{1}{2m} (i\nabla_{x_j} + \varepsilon A^\chi(x_j))^2 + V(\underline{x}) + H_f , \qquad (22.28)$$

where $\varepsilon = 2\sqrt{\pi}\alpha^{3/2}$, $A^\chi(x) = A_\chi(\alpha x)$ with $\chi(k)$ replaced by $\chi(\alpha^2 k)$, and $V(\underline{x}) = V_{\text{coul}}(\underline{x}, \alpha\underline{R})$. We write out the operator $A^\chi(x)$:

$$A^\chi(x) = \int (e^{ik\cdot\alpha x}a(k) + h.c.)\frac{\chi(\alpha^2 k)}{\sqrt{\omega(k)}}d^3k. \tag{22.29}$$

The Hamiltonian $H(\varepsilon)$ is, of course, equivalent to our original Hamiltonian $H_{e\kappa}$.

After the rescaling performed above the new UV cut-off momentum scale, $\kappa' = \alpha^{-2}\kappa$, satisfies

$$m_{\text{el}} \ll \kappa' \ll \alpha^{-2}m_{\text{el}},$$

which is easily accommodated by our estimates (e.g. we can have $\kappa' = O(\alpha^{-1/3}m_{\text{el}})$). From now on we fix κ' and we do not display α in the coupling function in (22.29) (the presence of α multiplying x above only helps our arguments).

At this point we forget about the origin of the potential $V(x)$, but rather assume it to be a general real function satisfying standard assumptions, say

(V) $V \in L^2(\mathbb{R}^{3N}) + L^\infty(\mathbb{R}^{3N})_\epsilon$,

i.e. V can be written as a sum of L^2- and $L^\infty-$functions, where the second component can be taken arbitrary small. It is shown in Section 22.3.1 that a sum of Coulomb potentials satisfies this condition and it is shown in Section 22.3.1 that under conditions (V), the operator

$$H_{\text{part}} := -\sum_{j=1}^{n}\frac{1}{2m}\Delta_{x_j} + V(\underline{x})$$

is self-adjoint on the domain $D(\sum_{j=1}^{n}\frac{1}{2m}\Delta_{x_j})$. The Hamiltonian $H(\varepsilon)$ is a key object of our analysis.

22.3.1 Self-adjointness of $H(\varepsilon)$

The key properties of a quantum Hamiltonian are boundedness below and self-adjointness. One way to approach proving these properties is to use the gaussian space representation (22.12) and the Kato inequality (see [73], cf. Section 10.1). This would give the result for for all coupling constants ε. (See also a proof of the self-adjointness of $H(\varepsilon)$ for an arbitrary ε, using path integrals by [160].) We take a different approach which demonstrates some of estimates we need later, but proves these properties only for ε sufficiently small.

Theorem 22.1 Assume ε is sufficiently small. Then the Hamiltonian $H(\varepsilon)$ is defined on $D(H(0))$ and is self adjoint and bounded below.

Proof. To keep the notation simple we consider one-particle systems ($n = 1$). We decompose the Hamiltonian (22.28) into unperturbed part and perturbation as

$$\cdot H(\varepsilon) \; = \; H(0) + I(\varepsilon). \tag{22.30}$$

In this proof we omit the super-index χ and write $A(x)$ for $A^\chi(x)$ in (22.28). Using (22.28), we find

$$I(\varepsilon) = -\varepsilon p \cdot A(x) + \frac{1}{2}\varepsilon^2 |A(x)|^2. \tag{22.31}$$

Now the self-adjointness of $H(\varepsilon)$ follows from the self-adjointness of $H(0) = H_{\text{part}} \otimes 1_f + 1_{\text{part}} \otimes H_f$, Proposition 22.2 below, the smallness of ε and an abstract result, Theorem 2.11 of Section 2.2 . $\square$

The next statement shows that I_ε is a relatively bounded perturbation in the sense of forms.

Proposition 22.2 There is an absolute constant $c > 0$, s. t.

$$\langle \psi, I(\varepsilon)\psi \rangle \le c\varepsilon\big(\langle \psi, H(0)\psi \rangle + \|\psi\|^2\big). \tag{22.32}$$

Proof. Let $\| \cdot \|_{\mathcal{F}}$ stand for the norm in the Fock space $\mathcal{F}$. First we prove the bounds

$$\|a(f)\psi\|_{\mathcal{F}}^2 \le \int \frac{|f|^2}{\omega}\|H_f^{1/2}\psi\|_{\mathcal{F}}^2 \tag{22.33}$$

and

$$\|a^*(f)\psi\|_{\mathcal{F}}^2 \le \int \frac{|f|^2}{\omega}\|H_f^{1/2}\psi\|_{\mathcal{F}}^2 + \int |f|^2\|\psi\|_{\mathcal{F}}^2. \tag{22.34}$$

The first inequality follows from the relations

$$\begin{aligned}
\|a(f)\psi\|_{\mathcal{F}} &\le \int |f|\|a\psi\|_{\mathcal{F}} \\
&\le \left(\int \frac{|f|^2}{\omega}\right)^{1/2}\left(\int \omega\|a\psi\|_{\mathcal{F}}^2\right)^{1/2}.
\end{aligned} \tag{22.35}$$

(due to the Schwarz inequality) and $\int \omega\|a\psi\|^2 = \langle \psi, \omega a^* a\psi \rangle = \langle \psi, H_f\psi \rangle$. The second inequality follows from the first and the relation

$$\|a(f)^*\psi\|_{\mathcal{F}}^2 = \langle \psi, a(f)a(f)^*\psi \rangle_{\mathcal{F}} = \|a(f)\psi\|_{\mathcal{F}}^2 + \int |f|^2\|\psi\|_{\mathcal{F}}^2. \tag{22.36}$$

The inequalities (22.33) and (22.34) yield the following bound $\|A(x)\psi\|_{\mathcal{F}} \le c(\|H_f^{1/2}\psi\|_{\mathcal{F}} + \|\psi\|_{\mathcal{F}})$, which implies that

$$\|A(x)(H_f + 1)^{-1/2}\| \le c . \tag{22.37}$$

Now, chose c' s.t. $H_{\text{part}} + c'1 \ge 1$. Using the equation $H(0) = H_{\text{part}} \otimes 1_f + 1_{\text{part}} \otimes H_f$ and using repeatedly that for any positive operator B, $\|B^{\frac{1}{2}}\psi\|^2 =$

$\langle \psi, B\psi \rangle$ and that $\langle \psi, H_{\text{part}}\psi \rangle \leq \langle \psi, H(0)\psi \rangle$ (since $H_f \geq 0$) and $\langle \psi, H_f\psi \rangle \leq \langle \psi, (H(0) + c'1)\psi \rangle$, we find that

$$\|(H_f + 1)^{\frac{1}{2}}(H(0) + c'1)^{-\frac{1}{2}}\| \leq 1,$$

$$\||(H_{\text{part}} + c'1)^{\frac{1}{2}}(H(0) + c'1)^{-\frac{1}{2}}\| \leq 1.$$

These inequalities together with the estimates $\|p\psi\| \leq c\|(H_{\text{part}} + c'1)^{\frac{1}{2}}\psi\|$, and (22.37) give,

$$\|(H_{\text{part}} + H_f + c'1)^{-1/2}pA(x)(H_{\text{part}} + H_f + c'1)^{-1/2}\| \leq c. \qquad (22.38)$$

Similarly, (22.37) implies that

$$\|(H_{\text{part}} + H_f + c'1)^{-1/2}A(x)^2(H_{\text{part}} + H_f + c'1)^{-1/2}\| \leq c. \qquad (22.39)$$

The estimates (22.39) and (22.38) together with the relation (22.31) imply (22.32). $\square$

One can also prove a bound on $I(\varepsilon)$ itself, not just on its form $\langle \psi, I(\varepsilon)\psi \rangle$, so that the self-adjointness of $H(\varepsilon)$ would follow from the Kato-Rellich theorem 2.9. This is done in Appendix 23.8.

22.4 Mass Renormalization

In this section we study electron mass renormalization. First we analyze the definition of (inertial) mass in Classical Mechanics. Consider a classical particle with the Hamiltonian $h(x, k) := K(k) + V(x)$, where $K(k)$ is some function describing the kinetic energy of the particle. To find the particle mass in this case we have to determine the relation between the force and acceleration at very low velocities. The Hamilton equations give $\dot{x} = \nabla_k K$ and $\dot{k} = F$, where $F = -\partial_x V$ is the force acting on the particle. Assuming that K has a minimum at $k = 0$, we expand $\nabla_k K(k)$ around 0 to obtain $\dot{x} = K''(0)k$, where $K''(0)$ is the hessian of K at $k = 0$. Differentiating the above relation w.r. to time and using the second Hamilton equation, we obtain $\ddot{x} = K''(0)F(x)$. This suggests to define the mass of the particle as $m = K''(0)^{-1}$, i.e. as the inverse of the Hessian of the energy, in the absence of external forces, as a function of momentum, at 0. ($K(k)$ is called the dispersion relation.) We adopt this as a general definition: *the (effective) mass of a particle interacting with fields is the inverse of the Hessian of the energy of the total system as a function of of the total momentum at 0.*

Now, we consider a single non-relativistic electron coupled to quantized electromagnetic field. Recall that the charge of electron is denoted by $-e$ and its *bare* mass is m. The corresponding Hamiltonian in our units is

$$H_{e\kappa}^{(1)} := \frac{1}{2m}(-i\nabla_x \otimes 1_f + A^\chi(x))^2 + 1_{\text{part}} \otimes H_f, \qquad (22.40)$$

acting on the space $L^2(\mathbb{R}^3) \otimes \mathcal{F} \equiv \mathcal{H}_{\text{part}} \otimes \mathcal{H}_f$. It is the generator for the dynamics of a single non-relativistic electron, and of the electromagnetic radiation field, which interact via minimal coupling. Here recall $A_\chi(x)$ and H_f are the quantized electromagnetic vector potential with ultraviolet cutoff and the field Hamiltonian, defined in (22.17) and (22.15)

The system considered is *translationally invariant* in the sense that $H_{e\kappa}^{(1)}$ commutes with the translations, T_y,

$$T_y H_{e\kappa}^{(1)} = H_{e\kappa}^{(1)} T_y,$$

which in the present case take the form

$$T_y : \Psi(x, A) \to \Psi(x + y, t_y A), \tag{22.41}$$

where $(t_y A)(x, A) = A(x - y)$. This as before leads to $H_{e\kappa}^{(1)}$ commuting with the total momentum operator,

$$P_{tot} := P_{\text{part}} \otimes \mathbf{1}_f + \mathbf{1}_{\text{part}} \otimes P_f, \tag{22.42}$$

of the electron and the photon field: $[H_{e\kappa}^{(1)}, P_{tot}] = 0$. Here $P_{\text{el}} := -i\nabla_x$ and $P_f = \sum_\lambda \int dk\, k\, a_\lambda^*(k) a_\lambda(k)$, the electron and field momenta. Again as in Section 22.2.2, this leads to the fiber decomposition

$$U H_{e\kappa}^{(1)} U^{-1} = \int_{\mathbb{R}^3}^{\oplus} H_{e\kappa}^{(1)}(P)\, dP, \tag{22.43}$$

where the fiber operators $H_{e\kappa}^{(1)}(P)$, $P \in \mathbb{R}^3$, are self-adjoint operators on $\mathcal{F}$. The computation of the operator $H_{e\kappa}^{(1)}(P)$ is the same as of thje corresponding fiber operator in Section 22.2.2. Specifying (22.27) to the present case (i.e. taking $x' = 0$), we obtain

$$H_{e\kappa}^{(1)}(P) = \frac{1}{2m}\left(P - P_f - eA^\chi\right)^2 + H_f \tag{22.44}$$

where $A^\chi := A^\chi(0)$. Explicitly, A^χ is given by

$$A^\chi = \sum_\lambda \int dk\, \frac{\kappa(|k|)}{|k|^{1/2}}\, e_\lambda(k)\, \{\, a_\lambda(k) + a_\lambda^*(k)\, \}. \tag{22.45}$$

Consider the infimum, $E(p) := \inf \sigma(H_{e\kappa}^{(1)}(P))$, of the spectrum of the fiber Hamiltonian $H_{e\kappa}^{(1)}(P)$. Note that for $e = 0$, $E(P)|_{e=0} =: E_0(P)$ is the ground state energy of $H_0(P) := H_{e\kappa}^{(1)}(P)|_{e=0} = \frac{1}{2m}\left(P - P_f\right)^2 + H_f$ with the ground state Ω and is $E_0(P) = \frac{|P|^2}{2m}$. Moreover, it is easy to show that $E(p)$ is spherically symmetric and has a minimum at $P = 0$ and is twice differentiable

$P = 0$. Following the heuristic discussion at the beginning of this section, we define the renormalized electron mass as

$$m_{\text{ren}} := E''(0)^{-1},$$

where, recall, $E''(0)$ is the Hessian of $E(P)$ at the critical point $P = 0$. A straightforward computation gives

$$(\text{Hess}\, E(P))_{ij} = \left(\delta_{ij} - \frac{P_i P_j}{|P|^2}\right)\frac{\partial_{|P|} E(P)}{|P|} + \frac{P_i P_j}{|P|^2}\partial^2_{|P|} E(P) \quad (22.46)$$

Since $E(P)$ is spherically symmetric and C^2 at $P = 0$, and satisfies $\partial_{|P|} E(0) = 0$, we find $\text{Hess}\, E(0))_{ij} = \frac{P_i P_j}{|P|^2}\partial^2_{|P|} E(0)$ and therefore as

$$m_{\text{ren}} := \frac{1}{\partial^2_{|P|} E(0)}.$$

Recalling our discussion at the beginning of this section, the kinematic meaning of this expression is as follows. The ground state energy $E(P)$ can be considered as an effective Hamiltonian of the electron in the ground state. (The propagator $\exp(-itE(P))$ determines the propagation properties of a wave packet formed of dressed one-particle states with a wave function supported near $P = 0$ – which exist as long as there is an infrared regularization.) The first Hamilton equation gives the expression for the electron velocity as

$$v = \partial_P E(P).$$

Expanding the right hand side in P we find $v = \text{Hess}\, E(0)P + O(P^2)$. Since $E(P)$ is spherically symmetric, and C^2 in $|P|$ near $P = 0$, this becomes

$$v = \partial^2_{|P|} E(0, \sigma)\, P + O(P^2).$$

This suggests taking $(\partial^2_{|P|} E(0))^{-1}$ as the renormalized electron mass.

It is shown in [28, 59, 61] that the infimum, $E(P) = \inf \text{spec}(H^{(1)}_{e\kappa}(P))$, of the spectrum of $H^{(1)}_{e\kappa}(P)$ is twice differentiable and $P = 0$ and the expression for m_{ren} is computed to the order e^3.

22.5 Appendix: Relative bound on $I(\varepsilon)$ and Pull-through Formulae

Proposition 22.3 There is an absolute constant $c > 0$, s. t.

$$\|I(\varepsilon)\psi\| \leq c\varepsilon(\|H(0)\psi\| + \|\psi\|). \quad (22.47)$$

Proof. We use the pull-through formula $a(k)f(H_f) = f(H_f+\omega(k))a(k)$, valid for any piecewise continuous, bounded function, f, on $\mathbb{R}$, proven in Appendix 23.8. Applying it to the function $f(\lambda) := (\lambda+1)^{-\frac{1}{2}}$ gives

$$a(k)(H_f+1)^{-\frac{1}{2}} = (H_f+\omega(k)+1)^{-\frac{1}{2}}a(k). \tag{22.48}$$

Using this relation we obtain

$$\|a(g)a(f)\psi\| \leq \int |f|\|a(g)a(k)\psi\|$$
$$= \int |f|\|a(g)(H_f+\omega(k)+1)^{-\frac{1}{2}}a(k)(H_f+1)^{\frac{1}{2}}\psi\|$$
$$\leq \int |f|\|a(g)(H_f+\omega(k)+1)^{-\frac{1}{2}}\|\|a(k)(H_f+1)^{\frac{1}{2}}\psi\|.$$

Applying now the bounds (22.33) and (22.35) and using that $\|H_f^{1/2}(H_f+\omega(k)+1)^{-\frac{1}{2}}\psi\| \leq \|\psi\|$ and $\|H_f^{1/2}(H_f+1)^{\frac{1}{2}}\psi\| \leq \|(H_f+1)\psi\|$, we obtain

$$\|a(g)a(f)\psi\| \leq \left(\int \frac{|f|^2}{\omega}\right)^{1/2}\left(\int \frac{|g|^2}{\omega}\right)^{1/2}\|(H_f+1)\psi\|.$$

Now, using this relation together with (22.33), (22.35) and (22.36), we find

$$\|a^*(g)a(f)\psi\|^2 \leq \int \frac{|f|^2}{\omega}\int \frac{|g|^2}{\omega}\|(H_f+1)\psi\|^2$$
$$+ \int \frac{|f|^2}{\omega}\int |g|^2\|H_f^{1/2}\psi\|^2,$$

and

$$\|a^*(g)a^*(f)\psi\| \leq 2\left(\int \frac{|f|^2}{\omega}\right)^{1/2}\left(\int \frac{|g|^2}{\omega}\right)^{1/2}\|(H_f+1)\psi\|$$
$$+ 3\int \frac{|f|^2}{\omega}\int |g|^2\|H_f^{1/2}\psi\|^2$$
$$+ 3\int |f|^2\int |g|^2\|\psi\|^2.$$

The above estimates imply

$$\|A(x)A(x)\psi\| \leq c(\|H_f\psi\| + \|\psi\|). \tag{22.49}$$

Finally, using the estimates (22.37) and $\|p\psi\| \leq c\|(H_{\text{part}}+c'1)^{\frac{1}{2}}\psi\|$, for c' s.t. $H_{\text{part}}+c'1 \geq 1$, and the fact that H_{part} and $A(x)$ commute, we find

$$\|pA(x)\psi\| \leq c\|A(x)(H_{\text{part}}+c'1)^{\frac{1}{2}}\psi\| \leq c\|(H_f+1)^{\frac{1}{2}}(H_{\text{part}}+c'1)^{\frac{1}{2}}\psi\|. \tag{22.50}$$

Using the equation $H(0) = H_{\text{part}} \otimes 1_f + 1_{\text{part}} \otimes H_f$ and using repeatedly that for any positive operator B, $\|B^{\frac{1}{2}}\psi\|^2 = \langle \psi, B\psi \rangle$ and that $\langle \psi, H_{\text{part}}\psi \rangle \leq \langle \psi, H(0)\psi \rangle$ and $\langle \psi, H_f \psi \rangle \leq \langle \psi, (H(0)+c'1)\psi \rangle$, we find that $\|(H_f+1)^{\frac{1}{2}}(H_{\text{part}}+c'1)^{\frac{1}{2}}\psi\| \leq \|(H(0) + c'1)\psi\|$. This inequality together with (22.50) gives

$$\|pA(x)\psi\| \leq c\|(H(0) + c'1)\psi\|. \tag{22.51}$$

The relation (22.31) and estimates (22.49) and (22.51) imply (22.47).

Now we prove the very useful "pull-through" formulae (see [135, 28]) used above:

$$a(k)f(H_f) = f(H_f + \omega(k))a(k) \tag{22.52}$$

and

$$f(H_f)a^*(k) = a^*(k)f(H_f + \omega(k)), \tag{22.53}$$

valid for any piecewise continuous, bounded function, f, on $\mathbb{R}$.

Problem 22.4 Using the commutation relations above, prove relations (22.52)-(22.53) for $f(H) = (H_f - z)^{-1}$, $z \in \mathbb{C}/\bar{\mathbb{R}}^+$.

Using the Stone-Weierstrass theorem, one can extend (22.52)- (22.53) from functions of the form $f(\lambda) = (\lambda - z)^{-1}$, $z \in \mathbb{C}\backslash\bar{\mathbb{R}}^+$, to the class of functions mentioned above. Alternatively, (22.52)- (22.53) follow from the relation

$$f(H_f)\prod_{j=1}^{N} a^*(k_j)\Omega = f(\sum_{i=1}^{N} \omega(k_i))\prod_{j=1}^{N} a^*(k_j)\Omega.$$

Problem 22.5 Prove this last relation, and derive (22.52)- (22.53) from it.

23

Theory of Radiation

Emission and absorption of electromagnetic radiation by systems of non-relativistic particles such as atoms and molecules is a key physical phenomenon, central to the existence of the world as we know it. Attempts to understand it led, at the beginning of the twentieth century, to the birth of quantum physics. In this chapter we outline the theory of this phenomenon. It addresses the following fundamental physical facts:

(a) a system of matter, say an atom or a molecule, in its lowest energy state is stable and well localized in space, while

(b) the same system placed in the vacuum in an excited state, after awhile, spontaneously emits photons and descends to its lowest energy state.

The starting point of theory of radiation is the Schrödinger equation

$$i\frac{\partial \psi}{\partial t} = H(\varepsilon)\psi,$$

describing quantum particles interacting amongst themselves, and with quantized electro-magnetic field. Here ψ is a path in the state space $\mathcal{H}_{\text{part}} \otimes \mathcal{H}_f$ and the quantum Hamiltonian operator $H(\varepsilon)$ entering it acts on $\mathcal{H}_{\text{part}} \otimes \mathcal{H}_f$ and is given by (see (22.28)):

$$H(\varepsilon) = \sum_{j=1}^{n} \frac{1}{2m}(i\nabla_{x_j} + \varepsilon A^{\chi}(x_j))^2 + V(\underline{x}) + H_f, \qquad (23.1)$$

with the notation explained in the previous chapter. (Recall that we use the units $\hbar = 1$ and $c = 1$ and, as a rule, we do not display these identities $\mathbf{1}_{\text{part}}$ and $\mathbf{1}_f$ on $\mathcal{H}_{\text{part}}$ and $\mathcal{H}_f$, respectively. A careful notation would have $V_{coul}(\underline{x}) \otimes \mathbf{1}_f$ and $\mathbf{1}_{\text{part}} \otimes H_f$, instead of $V_{coul}(\underline{x})$ and H_f.)

The mathematical manifestation of the fact (a) is that $H(\varepsilon)$ has a ground state, which is well localized in the particle coordinates, while the statement (b), rendered in mathematical terms, says that the system in question has no stable states in a neighbourhood of the excited states of the particle system, but 'metastable' ones.

© Springer-Verlag GmbH Germany, part of Springer Nature 2020
S. J. Gustafson and I. M. Sigal, *Mathematical Concepts of Quantum Mechanics*, Universitext, https://doi.org/10.1007/978-3-030-59562-3_23

23.1 Spectrum of the Uncoupled System

To understand the spectral properties of the operator $H(\varepsilon)$ we first examine a system consisting of matter and radiation not coupled to each other. Such a system is described by the Hamiltonian

$$H(0) = H_{\text{part}} \otimes 1_f + 1_{\text{part}} \otimes H_f, \tag{23.2}$$

which is obtained by setting the parameter ε in (22.28) to zero. The corresponding time-dependent Schrödinger equation is

$$i\frac{\partial \psi}{\partial t} = H(0)\psi. \tag{23.3}$$

The dynamics of a system of quantum matter (atom, molecule, etc., with fixed nuclei) is described by the Schrödinger operator

$$H_{\text{part}} = \sum_{j=1}^{n} \frac{1}{2m} p_j^2 + V(\underline{x})$$

acting on $\mathcal{H}_{\text{part}} = L^2(\mathbb{R}^{3n})$ (or a subspace of this space of a definite symmetry type). Recall the spectral structure of the Schrödinger operator H_{part}. By HVZ theorem (see Section 13.4), we have

$$\sigma(H_{\text{part}}) = \{\text{negative EV's, } E_j\} \cup \{\text{ continuum } [0, \infty)\}. \tag{23.4}$$

Here $j = 0, 1, \ldots$ and we assume $E_0 < E_1 \leq \ldots$. The eigenfunction, ψ_0^{part}, corresponding to the smallest eigenvalue, E_0 is called the *ground state*, while the eigenfunctions ψ_j^{part} for the higher eigenvalues E_j with $j \geq 1$, are called the *excited states*. The generalized eigenfunctions of the essential spectrum are identified with the scattering states (see Figure 23.1).

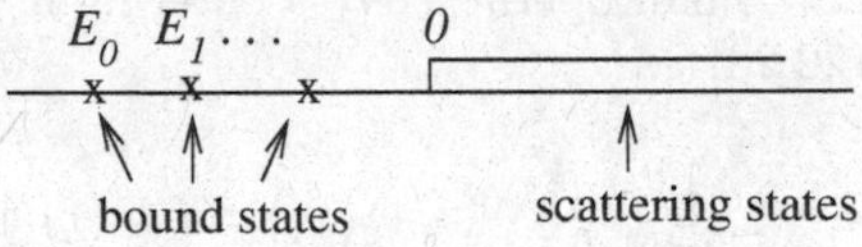

Fig. 23.1. Spectrum of H_{part}

For the field, $\sigma_{ess}(H_f) = [0, \infty)$ and $\sigma_d(H_f) = \{0\}$. The eigenvalue 0 is non-degenerate, and corresponds to the vacuum vector: $H_f \Omega = 0$.

Using separation of variables, we obtain the eigenfunctions and the generalized eigenfunctions,

$$\psi_j^{\text{part}} \otimes \Omega, \quad \text{and} \quad \psi_j^{\text{part}} \otimes \prod_{i=1}^{s} a^*(k_i)\Omega, \tag{23.5}$$

for various $s \geq 1$ and $k_1, \ldots, k_s$ in $\mathbb{R}^3$, of $H(0)$, corresponding to the eigenvalues E_j, $j = 0, 1, \ldots,$ and the spectral points $E_j(k) = E_j + \sum_{i=1}^{s} \omega(k_i) \in [E_j, \infty)$, respectively. This leads to the following stationary solutions

$$e^{-iE_j t}(\psi_j^{\text{part}} \otimes \Omega) \quad \text{and} \quad e^{-iE_j(k)}\Big(\psi_j^{\text{part}} \otimes \prod_{i=1}^{s} a^*(k_i)\Omega\Big) \,,$$

of the time-dependent Schrödinger equation (23.3). The first of these states describes the particle system in the state ψ_j^{part} with no photons around, while the second one corresponds to the system in the state ψ_j^{part} and s photons with momenta $k_1, \ldots, k_s$. Both states are stationary in time. In the absence of coupling between matter and radiation, the system of matter and radiation placed in one of these states remains in the same state forever. Radiation is neither absorbed nor emitted by this system.

Note finally that various eigenfunctions and generalized eigenfunctions above lead to different branches of the spectrum of $H(0)$, which, as a set, is of the form

$$\sigma\big(H(0)\big) = \{\text{EV's } E_j\} \bigcup_{j \geq 0} \{\text{continuum } [E_j, \infty)\} \cup \text{continuum } [0, \infty) \quad (23.6)$$

(see Section 25.13 of the Mathematical Supplements). The spectrum of $H(0)$ is pictured in Figure 23.2.

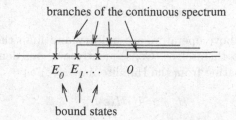

Fig. 23.2. Spectrum of $H(0)$.

The question we want to address is: how does this picture change as an interaction between the matter and radiation is switched on? This is the main problem of the mathematical theory of radiation.

23.2 Complex Deformations and Resonances

In this section we discuss a key notion of the (quantum) *resonance*. It gives a clear-cut mathematical description of processes of emission and absorption of the electro-magnetic radiation. In this description, the process of radiation corresponds to formation of resonances out of the excited states of particle systems. The most effective way to define resonances is to use complex transformation of the Hamiltonian under consideration (see Section 17.1) which we proceed to describe in the present setting.

Define the dilatation transformation, $U_{f\theta}$, on the Fock space, $\mathcal{H}_f \equiv \mathcal{F}$, as the unitary group of operators, given by

$$U_{f\theta}\Omega = \Omega \quad \text{and} \quad U_{f\theta}\prod a^*(f_j)\Omega = \prod a^*(u_\theta f_j)\Omega. \tag{23.8}$$

Here u_θ is the rescaling transformation acting on $L^2(\mathbb{R}^3)$, given by

$$(u_\theta f)(k) = e^{-3\theta/2}f(e^{-\theta}k). \tag{23.9}$$

($U_{f\theta}$ is the second quantization of u_θ.) This definition implies $U_{f\theta} := e^{iT\theta}$ where $T := \int a^*(k)ta(k)dk$ and t is the generator of the group u_θ (see Chapter 20 for the careful definition of the integral $\int a^*(k)ta(k)dk$) and

$$U_{f\theta}a^{\#}(f)U_{f\theta}^{-1} = a^{\#}(u_\theta f) \tag{23.10}$$

where $a^{\#}(f)$ stands for either $a(f)$ or $a^*(f)$. Applying this transformation to the photon Hamiltonian H_f, we find $H_{f\theta} = e^{-\theta}H_f$. Denote by $U_{p\theta}$ the standard dilation group on the particle space: $U_{p\theta} : \psi(x) \to e^{\frac{3n}{2}\theta}\psi(e^\theta x)$ where, recall, n is the number of particles (cf. Section 17.1). We define the dilation transformation on the total space $\mathcal{H} = \mathcal{H}_p \otimes \mathcal{H}_f$ by

$$U_\theta = U_{p\theta} \otimes U_{f\theta}. \tag{23.11}$$

For $\theta \in \mathbb{R}$ the above operators are unitary and map the domains of the operators $H(\varepsilon)$ into themselves. Consequently, we can define the family of Hamiltonians originating from the Hamiltonian $H(\varepsilon)$ as

$$H_{\varepsilon,\theta} := U_\theta H(\varepsilon)U_\theta^{-1} . \tag{23.12}$$

We would like to extend this family analytically into complex θ's. To this end we impose the following condition (in addition to condition (V)):

(A) $V_\theta(x) := V(e^\theta x)$, as a multiplication operator from $D(\Delta_x)$ to $\mathcal{H}_{\text{part}}$, has an analytic continuation, V_θ, in θ from $\mathbb{R}$ into a complex neighbourhood of $\theta = 0$, which is bounded from $D(\Delta_x)$ to $\mathcal{H}_{part}$.

Furthermore, we fix the ultra-violet cut-off $\chi(k)$ from now on so that $\chi_\theta(k) := e^{-3\theta/2}\chi(e^{-\theta}k)$ is an analytic function in a neighbourhood of $\theta = 0$, vanishing sufficiently fast at ∞, e.g. by taking $\chi(k) = e^{-|k|^2}$.

Under Condition (A), there is a family $H(\varepsilon, \theta)$ of operators Type-A analytic ([176, 247, 162]) in the domain $|\text{Im}\theta| < \theta_0$, which is equal to (23.12) for $\theta \in \mathbb{R}$ and s.t. $H^*(\varepsilon, \theta) = H(\varepsilon, \bar{\theta})$ and

$$H(\varepsilon, \theta) = U_{\text{Re}\theta} H(\varepsilon, i\text{Im}\theta)U_{\text{Re}\theta}^{-1}. \tag{23.13}$$

Indeed, using (22.28), we decompose $H(\varepsilon) = H_{\text{part}} + H_f + I(\varepsilon)$, where $I(\varepsilon)$ is defined by this relation: $I(\varepsilon) = \sum_{j=1}^n [i\nabla_{x_j} \cdot \varepsilon A^{\chi}(x_j) + \frac{1}{2}\varepsilon^2 A^{\chi}(x_j)^2]$, where we used that $\text{div}\,A^{\chi} = 0$. Using this decomposition, we write for $\theta \in \mathbb{R}$

$$H(\varepsilon, \theta) = H_p(\theta) \otimes 1_f + e^{-\theta} 1_p \otimes H_f + I(\varepsilon, \theta), \qquad (23.14)$$

Furthermore, using (23.10) and the definitions of the interaction $I(\varepsilon)$, we see that $I(\varepsilon, \theta)$ is obtained from $I(\varepsilon)$ by the replacement $a^{\#}(k) \rightarrow e^{-\frac{3\theta}{2}} a^{\#}(k)$ and, in the coupling functions only,

$$k \rightarrow e^{-\theta} k \qquad \text{and} \qquad x_j \rightarrow e^{\theta} x_j. \qquad (23.15)$$

This gives the required analytic continuation of (23.12). We call $H(\varepsilon, \theta)$ with $\text{Im}\theta > 0$ the complex deformation of $H(\varepsilon)$.

One can show show that:

1) The essential spectrum of moves (e.g. the spectrum of the deformation $H_{f\theta}$ is $\sigma(H_{f\theta}) = \{0\} \cup e^{-\text{Im}\,\theta}[0, \infty))$;

2) The real eigenvalues of $H(\varepsilon, \theta)$, $\text{Im}\theta > 0$, coincide with eigenvalues of $H(\varepsilon)$ and that complex eigenvalues of $H(\varepsilon, \theta)$, $\text{Im}\theta > 0$, lie in the complex half-plane $\mathbb{C}^-$;

3) The complex eigenvalues of $H(\varepsilon, \theta)$, $\text{Im}\theta > 0$, are locally independent of θ.

Let $\Psi_\theta = U_\theta \Psi$, etc., for $\theta \in \mathbb{R}$ and $z \in \mathbb{C}^+$. Use the unitarity of U_θ for real θ, to obtain (the Combes argument)

$$\langle \Psi, (H(\varepsilon) - z)^{-1} \Phi \rangle = \langle \Psi_{\bar\theta}, (H(\varepsilon, \theta) - z)^{-1} \Phi_\theta \rangle. \qquad (23.16)$$

Assume now that for a dense set of Ψ's and Φ's (say, $\mathcal{D}$, defined below), Ψ_θ and Φ_θ have analytic continuations into a complex neighbourhood of $\theta = 0$ and continue the r.h.s analytically first in θ into the upper half-plane and then in z across the continuous spectrum.

• The real eigenvalues of $H(\varepsilon, \theta)$ give real poles of the r.h.s. of (23.16) and therefore they are the eigenvalues of $H(\varepsilon)$.

• The complex eigenvalues of $H(\varepsilon, \theta)$ are poles of the meromorphic continuation of the l.h.s. of (23.16) across the spectrum of H onto the second Riemann sheet.

The poles manifest themselves physically as bumps in the scattering cross-section or poles in the scattering matrix.

An example of the dense set $\mathcal{D}$ for which the r.h.s. of (23.16) has an analytic continuation into a complex neighbourhood of $\theta = 0$, is

$$\mathcal{D} := \bigcup_{n>0, a>0} \text{Ran}\big(\chi_{N \leq n} \chi_{|T| \leq a}\big). \qquad (23.17)$$

Here $N = \int d^3k \, a^*(k) a(k)$ be the photon number operator and T be the self-adjoint generator of the one-parameter group U_θ, $\theta \in \mathbb{R}$. It is easy to show that the set $\mathcal{D}$ is dense.

We define the the *resonances* of $H(\varepsilon)$ as the complex eigenvalues of $H(\varepsilon, \theta)$ with $\mathrm{Im}\,\theta > 0$. Thus to find resonances (and eigenvalues) of $H(\varepsilon)$ we have to locate complex (and real) eigenvalues of $H(\varepsilon, \theta)$ for some θ with $\mathrm{Im}\,\theta > 0$.

23.3 Results

The rigorous answer to the question of how the spectral properties of $H(0)$ change as the interaction is switched on is given in the theorem below. This theorem refers to the notion of resonance described in Section 23.2 (see also Section 17.1). We have

Theorem 23.1 Let $\varepsilon \neq 0$ be sufficiently small. For statements (ii) and (iii), we assume, besides (V), Condition (A). Then

(i) $H(\varepsilon)$ has a unique ground state, ψ_ε. This state converges to $\psi_0^{\mathrm{part}} \otimes \Omega$ as $\varepsilon \to 0$, and is exponentially localized in the particle coordinates: i.e. $\|e^{\delta|x|}\psi_\varepsilon\| < \infty$ for some $\delta > 0$.

(ii) $H(\varepsilon)$ has no other bound states. In particular, the excited states of H_{part} (i.e. $\psi_j^{\mathrm{part}} \otimes \Omega$, $j \geq 1$) are unstable.

(iii) The excited states of H_{part} turn into resonances of $H(\varepsilon)$, $\varepsilon \neq 0$ (see Fig. 23.3).

Statement (ii) uses in addition to (V) and (A) a technical condition called (positivity of) the Fermi Golden Rule, which is satisfied except in a few "degenerate" cases).

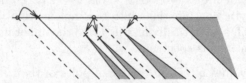

Fig. 23.3. Bifurcation of eigenvalues of $H(0)$(the second Riemann sheet).

This theorem gives mathematical content to the physical picture based on formal calculations performed with the help of perturbation theory. Statement (i) says that a system of matter, say an atom or a molecule, in its lowest energy state is stable and well localized in space, and according to statements (ii) and (iii), the excited states of the particle system disappear and give rise to long-living, metastable states of the total system. The latter are solutions of the time-dependent Schrödinger equation

$$i\frac{\partial \psi}{\partial t} = H(\varepsilon)\psi$$

which are localized for long intervals of time, but eventually disintegrate. (Recall that the metastable state is another term for a resonance.) These metastable states are responsible for the phenomena of emission and absorption of radiation and their life-times tell us how long, on average, we have to wait until a particle system emits (or absorbs) radiation.

The real parts of the resonance eigenvalues – the resonance energies – produce the *Lamb shift*, first experimentally measured by Lamb and Retherford (Lamb was awarded the Nobel prize for this discovery). The imaginary parts of the resonance eigenvalues – the decay probabilities – are given by the *Fermi Golden Rule* (see, eg, [169]).

In the rest of this chapter we outline main steps of the proof of Theorem 23.1, with more machinery given in the next chapter.

23.4 Idea of the proof of Theorem 23.1

A complete proof of statements (i)-(iii) can be found in [265] (see also [29, 27]). We will describe the proof of (i) in this chapter and in Chapter 24. The proofs of (ii) and (iii) are similar. Namely, techniques developed in the proof of (i) are applied to the family $H(\varepsilon, \theta)$, $\operatorname{Im} \theta > 0$, instead of $H(\varepsilon)$.

Since we are dealing only with the ground state, we will not need the analyticity Condition (A) above.

The proof of statement (i) of Theorem 23.1 – existence of the ground state – is done in two steps. On the first step, after performing a convenient canonical transformation, we map the family $H(\varepsilon) - z\mathbf{1}$ into a family $H_0(\varepsilon, z)$, acting on the subspace of the Fock space $\mathcal{H}_f \equiv \mathcal{F}$ corresponding to the photon energies $\leq \rho$ for some $\varepsilon \ll \rho \leq E_1^{\text{part}} - E_0^{\text{part}}$, where E_0^{part} and E_1^{part} are the ground state and the first excited state energies of H_{part}. On the second step, done in Chapter 24, we apply to $H_0(\varepsilon, z)$ a spectral renormalization group which brings it isospectrally into a very simple form which can be analyzed easily.

We explain some intuition underlying the proof. To understand the problem we are facing we look at the spectrum of $H(\varepsilon = 0)$, i.e., when the interaction is turned off (see Figure 23.4). We see that the unperturbed ground state energy is at the bottom of continuous spectrum. This suggests that only the ground state ψ_0^{part} of H_{part} and the low energy states of H_f are essential.

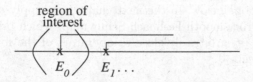

Fig. 23.4. Region of interest w.r.t. spec$H(0)$.

Hence - and this is the key idea - we would like to *project out the inessential parts* of the spectrum without distorting the essential ones. But how do we do this? Let us try the first idea that comes to mind:

$$H \to Q_\rho H Q_\rho \,, \tag{23.18}$$

where P_ρ is the spectral projection for $H(0)$ associated with the interval $[E_0, E_0 + \rho]$. The operator Q_ρ can be written explicitly as

$$Q_\rho = P_0^{\text{part}} \otimes E_{[0,\rho]}(H_f) \,. \tag{23.19}$$

Here P_0^{part} is the orthogonal projection onto ψ_0^{part} and $E_{[0,\rho]}(H_f)$ is the spectral projection of H_f corresponding to the interval $[0,\rho]$, defined as follows. First, recall that $H_f = \bigoplus_{n=0}^\infty H_{f,n}$ on $\mathcal{F} = \bigoplus_{n=0}^\infty \mathcal{F}_n$, where $H_{f,0}$ is the operator of multiplication by 0 on $\mathcal{F}_0 := \mathbb{C}$, and for $n \geq 1$, $H_{f,n} = \sum_{i=1}^n \omega(k_i)$ are multiplication operators on $\mathcal{F}_n := \circledS_1^n L^2(\mathbb{R}^3)$. Next, let χ_Δ be the characteristic function of the set $\Delta \subset \mathbb{R}$. Now we define

$$E_\Delta(H_f) = \bigoplus_{n=0}^\infty E_\Delta(H_{f,n})$$

where for $n \geq 1$, $E_\Delta(H_{f,n})$ is the operator of multiplication by the function $\chi_\Delta(\sum_{k=1}^n \omega(k_i))$, acting on $\mathcal{F}_n$, and $E_\Delta(H_{f,0})$ is the operator of multiplication by $\chi_\Delta(0)$ on $\mathcal{F}_0$. The operator $E_{[0,\rho]}(H_f)$ "cuts-off" the energy states of H_f with energy above ρ. The new operator $Q_\rho H(\varepsilon) Q_\rho$ acts on the subspace $L^2(\mathbb{R}^3) \otimes \mathcal{F}$ which consists of states of the form

$$\psi_0^{\text{part}} \otimes \phi \,, \ \phi \in \text{Ran} E_{[0,\rho]}(H_f)$$

(see Section 12.7). This is exactly what we want. However, the low energy spectrum of the operator $Q_\rho H(\varepsilon) Q_\rho$ is different from that of $H(\varepsilon)$. So we have lost the spectral information we are after. In Section 11.1, we learned how to project operators to smaller subspaces without losing the spectral information of interest. (We will refine this procedure in Sections 23.6 and 24.3.) But, as usual, there is a trade-off involved. While the map (23.18) acting on operators H is linear, the new map we introduce is not. This new map is called the *decimation map* (or *Feshbach-Schur map*). The result of application of this map to the family $H(\varepsilon) - z\mathbf{1}$ is a family $H_z(\varepsilon)$ of operators which act on the subspace $\text{Ran} E_{[0\rho]}(H_f)$ of the Fock space $\mathcal{F}$ and which is a small perturbation of the operator H_f plus a constant. In fact, for technical reasons it is convenient to use not the Feshbach-Schur map, but its extension - the smooth Feshbach-Schur map. (The latter is more difficult to define but much easier to use.) This is done below. In the next chapter we apply a renormalization group, based on the smooth Feshbach-Schur map, which brings isospectrally the operators $H_z(\varepsilon)$ arbitrarily close to operators of the form $E \cdot \mathbf{1} + w H_f$, with $E \in \mathbb{R}$ and $w > 0$.

23.5 Generalized Pauli-Fierz Transformation

In order to improve the infra-red behaviour of the Hamiltonian $H(\varepsilon)$, in this section we apply to it a canonical transformation. To simplify notation we only consider the case $n = 1$ and $m = 1$. (The generalizations to an arbitrary number of particles is straightforward.) In this case we have

$$H(\varepsilon) = \frac{1}{2}(i\nabla_x + \varepsilon A^\chi(x))^2 + V(x) + H_f , \qquad (23.20)$$

where $x \in \mathbb{R}^3$. The vector potential operator $A^\chi(x)$ is still given by (22.29). Now $H_{\text{part}} = -\frac{1}{2}\Delta + V(x)$ acts on $\mathcal{H}_{\text{part}} = L^2(\mathbb{R}^3)$, so that $H(\varepsilon)$ acts on $L^2(\mathbb{R}^3) \otimes \mathcal{H}_f$.

Now, we introduce the generalized Pauli-Fierz transformation

$$H_\varepsilon := e^{-i\varepsilon F(x)} H(\varepsilon) e^{i\varepsilon F(x)}, \qquad (23.21)$$

where $F(x)$ is the self-adjoint operator on the state space $\mathcal{H}$ given by

$$F(x) = \sum_\lambda \int (\bar{f}_{x,\lambda}(k)a_\lambda(k) + f_{x,\lambda}(k)a_\lambda^*(k)) \frac{d^3k}{\sqrt{|k|}}, \qquad (23.22)$$

with the coupling function $f_{x,\lambda}(k)$ chosen as (recall that we do not display α in the coupling function in (22.29))

$$f_{x,\lambda}(k) := e^{-ik\cdot x} \frac{\chi(k)}{\sqrt{|k|}} \varphi(|k|^{\frac{1}{2}} e_\lambda(k) \cdot x). \qquad (23.23)$$

The function φ is assumed to be C^2, bounded, with bounded second derivative, and satisfying $\varphi'(0) = 1$. We assume also that φ has a bounded analytic continuation into the wedge $\{z \in \mathbb{C}| \ |\arg(z)| < \theta_0\}$. We call the resulting Hamiltonian, H_ε, the generalized Pauli-Fierz Hamiltonian. We compute

$$H_\varepsilon = \frac{1}{2}(p - \varepsilon A_1(x))^2 + V_\varepsilon(x) + H_f + \varepsilon G(x) \qquad (23.24)$$

where $A_1(x) = A^\chi(x) - \nabla F(x)$, $V_\varepsilon(x) := V(x) + \frac{\varepsilon^2}{2}\sum_\lambda \int |k||f_{x,\lambda}(k)|^2 d^3k$ and

$$G(x) := -i \sum_\lambda \int |k|(\bar{f}_{x,\lambda}(k)a_\lambda(k) - f_{x,\lambda}(k)a_\lambda^*(k)) \frac{d^3k}{\sqrt{|k|}}. \qquad (23.25)$$

(The terms εG and $V_\varepsilon - V$ come from the commutator expansion $e^{-i\varepsilon F(x)} H_f \times e^{i\varepsilon F(x)} = -i\varepsilon[F, H_f] - \frac{\varepsilon^2}{2}[F, [F, H_f]]$.) Observe that the operator-family $A_1(x)$ is of the form

$$A_1(x) = \sum_\lambda \int (\overline{\chi_{x,\lambda}(k)}a_\lambda(k) + \chi_{x,\lambda}(k)a_\lambda^*(k)) \frac{d^3k}{\sqrt{|k|}}, \qquad (23.26)$$

where the coupling function $\chi_{\lambda,x}(k)$ is defined as follows

$$\chi_{\lambda,x}(k) := e_\lambda(k)e^{-ikx}\chi(k) - \nabla_x f_{x,\lambda}(k).$$

It satisfies the estimates

$$|\chi_{\lambda,x}(k)| \le \mathrm{const}\, \min(1, \sqrt{|k|}\langle x \rangle), \tag{23.27}$$

with $\langle x \rangle := (1 + |x|^2)^{1/2}$, and

$$\int \frac{d^3 k}{|k|} |\chi_{\lambda,x}(k)|^2 < \infty. \tag{23.28}$$

The fact that the operators A_1 and G have better infra-red behavior than the original vector potential A, is used in proving, with a help of a renormalization group, the existence of the ground state and resonances for the Hamiltonian H_ε and therefore for $H(\varepsilon)$.

We note that for the standard Pauli-Fierz transformation, the function $f_{x,\lambda}(k)$ is chosen to be $\chi(k)e_\lambda(k) \cdot x$, which results in the operator G (which in this case is proportional to (the electric field at $x = 0$) $\cdot$ x) growing as x. This transformation can be used if our system is placed in an external confining potential $W(x)$ satisfying the estimate $W(x) \ge c|x|^2$, for $|x| \ge R$, for some $c > 0$ and $R > 0$.

For further reference, we mention that the operator (23.24) can be written as

$$H_\varepsilon = H_{0\varepsilon} + I_\varepsilon, \tag{23.29}$$

where

$$H_{0\varepsilon} = H_0 + \frac{\varepsilon^2}{2} \sum_\lambda \int \left(|f_{x,\lambda}(k)|^2 + \frac{|\chi_{\lambda,x}(k)|^2}{2|k|} \right) d^3 k, \tag{23.30}$$

with $H_0 := H(\varepsilon = 0) = H_{\mathrm{part}} + H_f$ (see (23.20)), and I_ε is defined by this relation. Note that the operator I_ε contains linear and quadratic terms in creation- and annihilation operators, with coupling functions (form-factors) in the linear terms satisfying estimate (23.27) and with coupling functions in the quadratic terms satisfying a similar estimate. Moreover, the operator $H_{0\varepsilon}$ is of the form $H_{0\varepsilon} = H_\varepsilon^{\mathrm{part}} + H_f$ where

$$H_\varepsilon^{\mathrm{part}} := H_{\mathrm{part}} + \frac{\varepsilon^2}{2} \sum_\lambda \int (|f_{x,\lambda}(k)|^2 + |\chi_{\lambda,x}(k)|^2 2|k|) d^3 k, \tag{23.31}$$

where, recall, $H_{\mathrm{part}} = -\frac{1}{2}\Delta + V(x)$. Note that similarly to Proposition 22.3 one can show the following

Proposition 23.2 We have

$$\|G(x)\psi\|, \ \|A(x)\psi\| \le c(\|H_f^{1/2}\psi\| + \|\psi\|). \tag{23.32}$$

This theorem implies that I_ε is a relatively bounded perturbation of $H_{0\varepsilon}$.

23.6 Elimination of Particle and High Photon Energy Degrees of Freedom

Since we are looking at a vicinity of the ground state energy of $H(\varepsilon)$, the degrees of freedom connected to the excited particle states and to high photon energies should not be essential. So we eliminate them isospectrally using the smooth Feshbach-Schur (decimation) map. In this section we construct this map and use it to pass isospectrally from the family $H_\varepsilon - z \cdot \mathbf{1}$ to a family $H_0(\varepsilon, z)$ of operators which act non-trivially only on the subspace $\mathrm{Ran}E_{[0,\rho]}(H_f)$, where recall $E_{[0,\rho]}(\lambda)$ is the characteristic function of the interval $[0, \rho]$, of the Fock space $\mathcal{H}_f \equiv \mathcal{F}$. The advantage of this family is that it is even smaller perturbation of H_f plus a constant and that it can be treated by the renormalization group approach developed in the next chapter. Passing from $H_\varepsilon - z \cdot \mathbf{1}$ to $H_0(\varepsilon, z)$ will be referred to as *elimination of the particle and high photon energy* (actually photon energy $\geq \rho$) *degrees of freedom*.

As was already mentioned, the smooth Feshbach-Schur map is a generalization of the Feshbach-Schur map, discussed in Section 11.1 and it arises when the projections P and $\bar{P} := 1 - P$ are replaced by more general operators P and $\bar{P}$ forming a partition of unity

$$P^2 + \bar{P}^2 = \mathbf{1}. \tag{23.33}$$

Here we give a quick definition of the smooth Feshbach-Schur map. For more details see Section 24.3. Consider operators H on a Hilbert space $\mathcal{H}$ with specified decompositions $H = H_0 + I$. We define

$$H_{\bar{P}} := H_0 + \bar{P}I\bar{P}. \tag{23.34}$$

Assume now that

$$IP \text{ and } PI \text{ extend to bounded operators on } \mathcal{H} \tag{23.35}$$

and that z is s.t.

$$H_{\bar{P}} - z \text{ is (bounded) invertible on } \mathrm{Ran}\,\bar{P}. \tag{23.36}$$

We define *smooth Feshbach-Schur map, F_P^{smooth},* as

$$\begin{aligned} F_P^{\mathrm{smooth}}(H - z\mathbf{1}) := & \ H_0 - z\mathbf{1} + PIP \\ & - PI\bar{P}(H_{\bar{P}} - z)^{-1}\bar{P}IP. \end{aligned} \tag{23.37}$$

Definition 23.3 We say that operators A and B are *isospectral* at a point $z \in \mathbb{C}$ iff

(a) $z \in \sigma(A) \Leftrightarrow z \in \sigma(B)$,
(b) $A\psi = z\psi \Leftrightarrow B\phi = z\phi$

where ψ and ϕ are related by $\phi = P\psi$ and $\psi = Q\phi$ for some bounded operators P and Q.

We have the following theorem :

Theorem 23.4 Assume $H - z\mathbf{1}$ is in the domain of F_P^{smooth}. Then the operators H and $F_P^{\text{smooth}}(H - z\mathbf{1}) + z\mathbf{1}$ are isospectral at z, in the sense of the definition above, with the operator P, the same as in F_P^{smooth} and the operator Q defined by

$$Q_P(H - z\mathbf{1}) := P - \bar{P}(H_{\bar{P}} - z)^{-1}\bar{P}IP. \qquad (23.38)$$

This theorem generalizes Theorem 11.1 of Section 11.1; its proof is similar and is sketched in Section 24.3.

Now we adapt these notions to the families of operators (23.24). In this case, a partition $(P, \bar{P})$ (see (23.33)) is defined as follows. Let $\chi_\rho \equiv \chi_\rho(H_f) := \chi_{H_f/\rho \leq 1}$ and $\bar{\chi}_\rho \equiv \bar{\chi}_\rho(H_f) := \chi_{H_f/\rho \geq 1}$, where $\chi_{\lambda \leq 1}$ and $\chi_{\lambda \geq 1}$ are smooth functions satisfying the relations

$$\chi_{\lambda \leq 1}^2 + \chi_{\lambda \geq 1}^2 = 1, \ 0 \leq \chi_{\lambda \leq 1}, \ \chi_{\lambda \geq 1} \leq 1, \ \chi_{\lambda \leq 1} = \begin{cases} 1 \text{ if } \lambda \leq \frac{4}{5} \\ 0 \text{ if } \lambda > 1 \end{cases}$$

(and the corresponding relation for $\chi_{\lambda \geq 1}$). Thus $\chi_\rho(H_f) = \chi_{H_f \leq \rho}$ is an almost the spectral projection for the operator H_f onto energies $\leq \rho$. With the definition (23.31), let furthermore

$$P_0^{\text{part}} = \text{orthogonal projection onto the ground state eigenspace of } H_\varepsilon^{\text{part}}, \qquad (23.39)$$

and $\bar{P}_0^{\text{part}} := \mathbf{1} - P_0^{\text{part}}$. (Recall that $H_\varepsilon^{\text{part}}$ is defined in (23.31).) Assume the ground state of $H_\varepsilon^{\text{part}}$ is simple and the corresponding eigenspace is spanned by a function ψ_0^{part}. Then, in the Dirac notation, $P_0^{\text{part}} = |\psi_0^{\text{part}}\rangle\langle\psi_0^{\text{part}}|$. We introduce a pair of almost projections

$$P_\rho := P_0^{\text{part}} \otimes \chi_\rho(H_f) \qquad (23.40)$$

and $\bar{P}_\rho := \bar{P}_0^{\text{part}} + P_0^{\text{part}} \otimes \bar{\chi}_\rho(H_f)$, which form a partition of unity $P_\rho^2 + \bar{P}_\rho^2 = \mathbf{1}$. Note that P_ρ and $\bar{P}_\rho$ commute with $H_{0\varepsilon}$. Let E_0 and E_1 be the ground state energy and the first excited state energy of $H_\varepsilon^{\text{part}}$ and set

$$\Omega_\rho := \{z \in \mathbb{C} \mid \text{Re} z \leq E_0 + \frac{1}{4}\rho\}. \qquad (23.41)$$

Now we have

Theorem 23.5 Let $|\varepsilon| \ll \rho \leq E_1 - E_0$. Then $F_{P_\rho}^{\text{smooth}}$ is defined on for the families of operators, $H_\varepsilon - z \cdot \mathbf{1}$, $z \in \Omega_\rho$, where H_ε is given by (23.24), with the decomposition (23.29), and is isospectral (in the sense of Theorem 23.4).

Proof. The first part of the theorem follows from the definition of the domain of the map F_P^{smooth} given above and Propositions 23.9 and 23.10 given in Appendix 23.10 below. The isospectrality follows from the first part of the theorem and Theorem 23.4. $\square$

Let, as above, $E_{[0,\rho]}(\lambda)$ be the characteristic function of the interval $[0,\rho]$. We define the subspace

$$X_\rho := \psi_0^{\text{part}} \otimes E_{[0,\rho]}(H_f)\mathcal{H}_f \approx E_{[0,\rho]}(H_f)\mathcal{H}_f$$

of $\mathcal{H}_{\text{part}} \otimes \mathcal{H}_f$. Here $\approx$ stands for the isomorphic isomorphism. The subspace X_ρ and its orthogonal complement, $X_\rho^\perp$, are invariant under $F_{P_\rho}^{\text{smooth}}(H_\varepsilon - z\mathbf{1})$ and span $\mathcal{H}_{\text{part}} \otimes \mathcal{H}_f$.

Let $\Delta E = E_1 - E_0$. We choose ρ_0 such that $\varepsilon \ll \rho_0 < \min(1, \Delta E)$. We apply $F_{P_{\rho_0}}^{\text{smooth}}$ to the families of operators (23.24) with the decomposition $H_\varepsilon = H_{0\varepsilon} + I_\varepsilon$, as given in (23.29) - (23.30), to obtain the new Hamiltonian

$$H_0(\varepsilon, z) := F_{P_{\rho_0}}^{\text{smooth}}(H_\varepsilon - z\mathbf{1})|_{X_{\rho_0}}. \tag{23.42}$$

We claim that for $\text{Re} z \le E_0 + \frac{1}{4}\rho_0$ and $0 < \rho_0 \le E_1 - E_0$,

$$H_\varepsilon - z\mathbf{1} \quad \text{and} \quad H_0(\varepsilon, z) \quad \text{are isospectral at} \quad 0. \tag{23.43}$$

Indeed, since $P_\rho = 0$ on $X_\rho^\perp$, we have that

$$F_{P_\rho}^{\text{smooth}}(H_\varepsilon - z\mathbf{1})|_{X_\rho^\perp} = (H_{0\varepsilon} - z\mathbf{1})|_{\text{Ran}X_\rho^\perp}. \tag{23.44}$$

Therefore, for $\text{Re} z \le E_0 + \frac{1}{4}\rho$ and $0 < \rho \le E_1 - E_0$,

$$F_{P_\rho}^{\text{smooth}}(H_\varepsilon - z\mathbf{1})|_{X_\rho^\perp} \ge \frac{3}{4}\rho > 0 \tag{23.45}$$

and, in view of Theorem 23.4, (23.43) holds.

Observe that since the projection P_0^{part} has rank 1, the operator $H_0(\varepsilon, z)$ is of the form

$$H_0(\varepsilon, z) = \left[\langle \psi_0^{\text{part}} , F_{P_\rho}^{\text{smooth}}(H_\varepsilon - z\mathbf{1})\psi_0^{\text{part}} \rangle_{\mathcal{H}_{\text{part}}} \right]|_{\text{Ran}E_{[0,\rho_0]}(H_f)}. \tag{23.46}$$

Since we are interested in the part of the spectrum which lies in the set $\{z \in \mathbb{R} \mid \text{Re} z \le E_0 + \frac{1}{4}\rho_0\}$, we can study $H_0(\varepsilon, z)$, which acts on the space $X_{\rho_0} \approx \text{Ran}E_{[0,\rho_0]}(H_f)$, instead of H_ε. Thus we have passed from the operator H_ε acting on $\mathcal{H}_{\text{part}} \otimes \mathcal{H}_f$ to the operator $H_0(\varepsilon, z)$ acting on the subspace, $\text{Ran}E_{[0,\rho]}(H_f)$, of the Fock space $\mathcal{H}_f \equiv \mathcal{F}$, which is isospectral (in the sense of Definition 23.3) to $H_\varepsilon - z\mathbf{1}$ at 0, provided z is in the set (23.41). We eliminated, in an isospectral way, the degrees of freedom corresponding to the particle and to the photon energies $\ge \rho_0$; i.e., we projected out the part $\text{Ran}(\mathbf{1} - P_0^{\text{part}})$ of the particle space $\mathcal{H}_{\text{part}}$, and the part $\text{Ran}(\mathbf{1} - E_{[0,\rho_0]}(H_f))$ of the Fock space $\mathcal{H}_f \equiv \mathcal{F}$. The parameter ρ_0 is called the *photon energy scale*.

23.7 The Hamiltonian $H_0(\varepsilon, z)$

Key properties of $H_0(\varepsilon, z)$ are summarized in the next theorem.

Theorem 23.6 Let $z \in \Omega_{\rho_0}$ and $\varepsilon \ll \rho_0 \leq \Delta E$ (ρ_0 is the scale parameter entering the definition of $H_0(\varepsilon, z)$). The operator $H_0(\varepsilon, z)$ has a generalized normal form,

$$H_0(\varepsilon, z) = H_{0,00} + \sum_{r+s \geq 1} \chi_{\rho_0} H_{0,rs} \chi_{\rho_0}, \quad H_{0,00} := h_{0,00}(H_f)$$

$$H_{0,rs} := \int (\prod_{j=1}^{r} a^*(k_j)) h_{0,rs}(H_f, k_1 \ldots k_{r+s}) \prod_{i=r+1}^{r+s} a(k_i) d^{r+s}k, \tag{23.47}$$

with coupling functions $h_{0,rs}$ which are analytic in $z \in \Omega$, and satisfy the estimates

$$|\partial_\mu^n h_{0,rs}(\mu, \underline{k})| \leq (\text{const})\rho_0 \sum_{i=1}^{r+s} \omega(k_i)^{\frac{1}{2}} \prod_{j=1}^{r+s} (\text{const} \cdot \varepsilon \cdot \rho_0^{-1} \cdot \omega(k_j)^{-\frac{1}{2}}), \tag{23.48}$$

for any i, $1 \leq i \leq r+s$, $\mu \in [0,1]$, and $n = 0, 1$, where the product is absent in the case $r = s = 0$, and

$$|h_{0,00}(0)| \leq (\text{const})|E_0 - z + O(\varepsilon^2)|, \quad |\partial_\mu h_{0,00}(\mu) - 1| \leq (\text{const})\varepsilon^2/\rho_0^2, \tag{23.49}$$

for $\mu \in [0, \infty)$.

We note that the crucial for the future analysis term $\sum_{i=1}^{r+s} \omega(k_i)^{\frac{1}{2}}$ is due to the estimate (23.27) gained in the generalized Pauli-Fierz Hamiltonian (23.24). The proof of this theorem is simple but lengthy. It can be found in [27, 121]. Below we sketch its main ideas.

Sketch of proof of Theorem 23.6. Using the definition of the smooth Feshbach-Schur map $F_{P_{\rho_0}}^{\text{smooth}}(H_\varepsilon - z\mathbb{1})$, we write the operator $H_0(\varepsilon, z)$ in the form

$$H_0(\varepsilon, z) = H_f + \chi_{\rho_0} W \chi_{\rho_0},$$

where with the same definitions as in the proof of Proposition 23.10,

$$W = I_\varepsilon - \langle \psi_0^{\text{part}}, I_\varepsilon \bar{R}_{\rho_0}(z) I_\varepsilon \psi_0^{\text{part}} \rangle_{\mathcal{H}_{\text{part}}},$$

with $\bar{R}_\rho(z) = \bar{P}_\rho(H_{0\varepsilon} + P_{\rho_0} I_\varepsilon P_{\rho_0} - z)^{-1} \bar{P}_\rho$. Now we use estimates obtained in the proof of Proposition 23.10 of Appendix 23.10 below, to expand the resolvent $(H_{0\varepsilon} + P_{\rho_0} I_\varepsilon P_{\rho_0} - z)^{-1}$ in a Neumann series, which with the notation $I = I_\varepsilon$, $P = P_{\rho_0}$, $\bar{P} = \bar{P}_{\rho_0}$, and $R_{0,\bar{P}} = \bar{P}(H_{0\varepsilon} - z)^{-1} \bar{P}$ can be written as

$$W = \sum_{n=0}^{\infty} \langle \psi_0^{\text{part}}, I(-R_{0,\bar{P}} I)^n \psi_0^{\text{part}} \rangle. \tag{23.50}$$

Next we bring each term on the right hand side of (23.50) to the generalized normal form. To this end we observe that the terms on the right hand side of (23.50), with $n \geq 1$, consist of sums of products of five operators: $R_{0,\bar{P}}$,

$\bar{P}$, p, $a(k)$ and $a^*(k)$. We do not touch the operators $R_{0,\bar{P}}, \bar{P}$ and p, while we move the operators $a(k)$ to the extreme right and the operators $a^*(k)$ to the extreme left. In doing this we use the following rules:

1. $a(k)$ is pulled through $a^*(k)$ according to the relation

$$a(k)a^*(k') = a^*(k')a(k) + \delta(k - k')$$

2. $a(k)$ and $a^*(k)$ are pulled through $R_{0,\bar{P}}$ according to the relations

$$a(k)R_{0,\bar{P}} = R_{0,\bar{P}}^{\omega(k)}a(k)$$

$$R_{0,\bar{P}}a^*(k) = a^*(k)R_{0,\bar{P}}^{\omega(k)} \ ,$$

where $R_{0,\bar{P}}^{\omega(k)} = R_{0,\bar{P}}\big|_{H_f \to H_f + \omega(k)}$, and similarly for pulling $a(k)$ and $a^*(k)$ through other functions of H_f, such as $\bar{P}$ (see the equations (22.52) of Appendix 23.8).

The procedure above brings the operator $H_0(\varepsilon, z)$ to generalized normal form (23.47), at least formally. It remains to estimate the coupling functions $h_{0,rs}$ entering (23.47). A direct estimate produces large combinatorial factors which we must avoid. So we use a special technique which amounts to a partial resummation of the series, in order to take advantage of cancelations. This can be found in [27, 121]. $\square$

23.8 Estimates on the operator $H_0(\varepsilon, z)$

Though the operator $H_0(\varepsilon, z)$ looks rather complicated, we show now that the complicated part of it gives a very small contribution. To this end we use Proposition 23.11 of Appendix 23.11 and the estimates (23.48) and

$$\int_{\sum_1^n \omega(k_j) \le \rho} \prod_{j=1}^n \left(\frac{J_j(k_j)^2}{\omega(k_j)} \right) d^{3n}k \le \frac{1}{n!} \prod_{j=1}^n \left(\int \frac{J_j^2}{\omega} \right)$$

(valid for any functions J_i and used for $J_i = \omega^{-1/2}$ or $= 1$), to obtain for $r + s \ge 1$,

$$\|\chi_{\rho_0} H_{0,rs}(\varepsilon, z)\chi_{\rho_0}\| \le (r!s!)^{-1/2}(\sqrt{4\pi}\varepsilon\rho_0^{-1/2})^{r+s}\rho_0^{3/2}. \tag{23.51}$$

With choosing, say, $\rho_0 = \varepsilon^{9/5}$, these estimates and the relation

$$h_{0,00}(H_f) - h_{0,00}(0) - H_f = \int_0^1 (h'_{0,00}(sH_f) - 1)ds \, H_f$$

imply that the operator $H_0(\varepsilon, z)$ is of the form

$$H_0(\varepsilon, z) = (E_0 + \Delta_0 E - z) \cdot \mathbf{1} + w(H_f)H_f + O(\varepsilon^{\frac{9}{5}}) ,$$

where $E_0 + \Delta_0 E - z = h_{0,00}(0)$ and $w(H_f) = 1 + O(\varepsilon^{\frac{1}{5}})$. Here $\Delta_0 E$ is a explicitly computable energy shift of the order $O(\varepsilon^2)$. Since the spectrum of the operator $H_f + E_0 + \Delta_0 E$ fills in the semi-axis $[E_0 + \Delta_0 E, \infty)$, the isospectrality Theorem 23.4 implies that $\operatorname{spec} H(\varepsilon) = \operatorname{spec} H_\varepsilon \subset [E_0 + \Delta_0 E + O(\varepsilon^{\frac{9}{5}}), \infty)$, which implies the following intermediary result:

Theorem 23.7 Assume $\varepsilon \neq 0$ is sufficiently small and that the ground state of the particle Hamiltonian, H_{part}, is non-degenerate. Define $E_0(\varepsilon) = \inf \sigma(H(\varepsilon))$, the ground state energy of $H(\varepsilon)$. Then

$$E_0(\varepsilon) = E_0 + \Delta_0 E + O(\varepsilon^{\frac{9}{5}}),$$

with explicitly computable energy shift $\Delta_0 E$ of the order $O(\varepsilon^2)$.

Applying the renormalization transformation iteratively (see the next chapter) to quantum Hamiltonians of the form (23.47) - (23.49), given in Theorem 23.6, we find energies $E^{(n)} = E_0 + O(\varepsilon^2)$ and numbers $w^{(n)} = 1 + O(\varepsilon^{\frac{1}{5}})$, such that for $\rho = 0(\varepsilon) \ll 1$ and any $n \geq 1$,

$$H_\varepsilon \text{ (or } H(\varepsilon)) \text{ is isospectral to } E^{(n)} + w^{(n)} H_f + O(\varepsilon^2 \rho^n)$$

in the disk $D(E^{(n)}, \rho^n)$. This will give us a much more precise information about the spectral properties of the Hamiltonian $H(\varepsilon)$. The above procedure is at the heart of the renormalization group approach.

Remark 23.8 The property that the interaction vanishes under renormalization transformations, i.e. when we go closer and closer to the ground energy (or farther and farther from the particle system) is called *infrared asymptotic freedom.*

Note that the spectrum of the operator $H_f + E_0 + \Delta_0 E$ contains the eigenvalue $E_0 + \Delta_0 E$, with eigenfunction Ω, and the continuum $[E_0 + \Delta_0 E, \infty)$, with generalized eigenfunctions $\psi_0^{\text{part}} \otimes \Pi a^*(k_j)\Omega$. Hence, extending the results of Theorem 23.4, $Q_{P_{\rho_0}}(H_\varepsilon - z)\Omega$, where $Q_P(H - z\mathbf{1})$ is given by (23.38), gives an approximate eigenfunction of H_ε with approximate eigenvalue, $E_0 + \Delta_0 E$ and similarly for the continuum.

23.9 Ground state of $H(\varepsilon)$

Let $U_{f\theta}$ be a unitary group of operators, given by (23.8). We define the rescaling transformation, S_ρ, on operators on the Fock space $\mathcal{F}$ by

$$S_\rho(H) := U_{f\theta} H U_{f\theta}^{-1}, \qquad \theta = -\ln \rho. \tag{23.52}$$

In particular, we have (see (23.10))

$$S_\rho(a^\#(k)) = \rho^{-3/2}a^\#(\rho^{-1}k), \tag{23.53}$$

so that

$$S_\rho(H_f) = \rho H_f \quad \text{and} \quad S_\rho(\chi_\rho) = \chi_1. \tag{23.54}$$

Define $H^{(0)}(\varepsilon, z) := \rho_0^{-1}S_{\rho_0}(H_0(\varepsilon, z))$, where, recall, $H_0(\varepsilon, z)$ is defined in (23.42). The operator $H^{(0)}(\varepsilon, z)$ is of the form

$$H = H_{00} + \sum_{r+s\geq 1} \chi_1 H_{rs}\chi_1, \quad H_{00} := h_{00}(H_f)$$

$$H_{rs} := \int \prod_{j=1}^{r} a^*(k_j) \, h_{rs}(H_f, k_1\ldots k_{r+s}) \prod_{i=r+1}^{r+s} a(k_i)d^{r+s}k, \tag{23.55}$$

acting on acting on the space $E_{[0,1]}(H_f)\mathcal{F}$, where recall $E_{[0,\rho]}(\lambda)$ denotes the characteristic function of the interval $[0, \rho]$, and with coupling functions h_{rs} satisfying the estimates

$$|\partial_\mu^n h_{rs}(\mu, \underline{k})| \leq (\text{const})\rho_0^n \sum_{i=1}^{r+s} \omega(k_i)^{\frac{1}{2}} \prod_{j=1}^{r+s} (\xi_0 \cdot \varepsilon \cdot \omega(k_j)^{-\frac{1}{2}}), \tag{23.56}$$

for any i, $1 \leq i \leq r+s$, $\mu \in [0,1]$, $\xi_0 > 0$, and $n = 0,1$, where the product is absent in the case $r = s = 0$, and

$$|h_{00}(0)| \leq \frac{\text{const}}{\rho_0}|E_0 - z + O(\varepsilon^2)|, \quad |\partial_\mu h_{00}(\mu) - 1| \leq \frac{\text{const}}{\rho_0^2}\varepsilon^2, \tag{23.57}$$

for $\mu \in [0, \infty)$. To prove this we use (23.53). Then Theorem 23.4 implies that

$$z \in \sigma_\#(H_\varepsilon) \ \leftrightarrow\ 0 \in \sigma_\#(H^{(0)}(\varepsilon, z)) \tag{23.58}$$

as long as z is in the set $\Omega_{\rho_0} := \{z \in \mathbb{R} \mid z \leq E_0 + \frac{1}{4}\rho_0\}$. Next, let $S := \{w \in \mathbb{C} | \text{Re} w \geq 0, |\text{Im} w| \leq \frac{1}{3}\text{Re} w\}$. Theorem 23.6 above implies that for ε sufficiently small, the operators $H^{(0)}(\varepsilon, z)$ (more precisely, $H^{(0)}(\varepsilon, z) - \langle H^{(0)}(\varepsilon, z)\rangle_\Omega$) satisfy the assumptions of Theorem 24.1 of the next chapter, which implies, in particular, that

- $H^{(0)}(\varepsilon, z)$ has a simple eigenvalue $\lambda^{(0)}(\varepsilon, z) \in D(0, c\varepsilon^2)$;
- $\sigma_\#(H^{(0)}(\varepsilon, z)) \subset \lambda^{(0)}(\varepsilon, z) + S$.

With some extra work (in which the analyticity of $\lambda^{(0)}(\varepsilon, z)$ in $z \in \Omega_{\rho_0}$ plays an important role) we show that the equation $\lambda^{(0)}(\varepsilon, z) = 0$ for z has a unique solution. Let $E_0(\varepsilon)$ solve this equation. By (23.58), this gives that $E_0(\varepsilon)$ is a simple eigenvalue of H_ε and $\sigma_\#(H(\varepsilon)) \subset S$. This together with (23.58) shows that $E_0(\varepsilon)$ is the ground state energy of the operator H_ε.

Finally, if ϕ is the eigenfunction of $H^{(0)}(\varepsilon, E_0(\varepsilon))$ corresponding to the eigenvalue 0, then $Q_{P_\rho}(H_\varepsilon - E_0(\varepsilon)\mathbf{1})\phi$, where $Q_P(H - z\mathbf{1})$ is given by (23.38), is an eigenfunction of H_ε corresponding to the eigenvalue $E_0(\varepsilon)$.

Finally, we use the relation $\sigma_\#(H_\varepsilon) = \sigma_\#(H(\varepsilon))$ with a simple relation between the eigenfunctions to transfer the spectral information from H_ε to $H(\varepsilon)$. $\square$

23.10 Appendix: Estimates on I_ε and $H_{\bar{P}_\rho}(\varepsilon)$

Recall that we consider the Hamiltonian H_ε, given in (23.24), whose decomposition into unperturbed part and perturbation is given in (23.29) - (23.30). In this section we omit the subindex 1 in $A_1(x)$, entering (23.24), so that the operator-family $A(x)$ is given by

$$A(x) = \sum_\lambda \int (\overline{\chi_{x,\lambda}(k)} a_\lambda(k) + \chi_{x,\lambda}(k) a_\lambda^*(k)) \frac{d^3 k}{\sqrt{|k|}}, \tag{23.59}$$

with the coupling function $\chi_{\lambda,x}(k)$ satisfying the estimates (23.27) - (23.28). Using (23.29)-(23.30), we find

$$I_\varepsilon = -p \cdot A(x) + \frac{1}{2}\varepsilon |A(x)|^2 + G(x).$$

Proposition 23.9

$$\|I_\varepsilon P_\rho\| \le C|\varepsilon| . \tag{23.60}$$

Proof. We write P, $\bar{P}$ for P_ρ, $\bar{P}_\rho$, respectively. Since p is bounded relative to H_{part}, we have $\|P \cdot p\| \le C$. Finally, since $\rho \le c$, the bound (23.32) implies that

$$\|PG(x)\|, \ \|PA(x)\| \le C.$$

Collecting the last four estimates and using the fact that $\operatorname{Re} z \le E_0 + \rho$, we arrive at (23.60). $\square$

In the present context, the operator $H_{\bar{P}}$, introduced in (23.34), is given by

$$H_{\bar{P}_\rho}(\varepsilon) := H_{0\varepsilon} + \bar{P}_\rho I_\varepsilon \bar{P}_\rho. \tag{23.61}$$

Proposition 23.10 Assume $|\varepsilon| \ll \rho$. Then $\Omega_\rho \subset \rho(H_{\bar{P}_\rho}(\varepsilon))$ and, for $z \in \Omega_\rho$, the inverse of $H_{\bar{P}_\rho}(\varepsilon) - z$ satisfies the estimate

$$\|\bar{P}_\rho (H_{\bar{P}_\rho}(\varepsilon) - z)^{-1}\| \le C\rho^{\frac{1}{2}}. \tag{23.62}$$

Proof. To simplify notation we assume z is real. If we omit the subindexes ρ and ε and denote $P = P_\rho$, $\bar{P} = \bar{P}_\rho$ and $I = I_\varepsilon$ (e.g. $I_{\bar{P}}$ stands for $\bar{P}_\rho I_\varepsilon \bar{P}_\rho$). Recall the definition (23.39) ($P = P_0^{\text{part}} \otimes \chi_\rho(H_f)$) and the definitions before and after this equation. We have

$$\bar{P} = \bar{P}_0^{\text{part}} \otimes 1_f + P_0^{\text{part}} \otimes \bar{\chi}_\rho(H_f) . \tag{23.63}$$

Now using $H_\varepsilon^{\text{part}} \bar{P}_0^{\text{part}} \ge E_1 \bar{P}_0^{\text{part}}$, $H_\varepsilon^{\text{part}} \ge E_0$ and $H_f \ge 0$, $H_f \bar{\chi}_\rho(H_f) \ge \rho \bar{\chi}_\rho(H_f)$, where recall E_0 and E_1 are the ground state energy and the first excited state of $H_\varepsilon^{\text{part}}$, and using that $H_{0\varepsilon} = H_\varepsilon^{\text{part}} \otimes 1_f + 1_{\text{part}} \otimes H_f$, we estimate

$$\bar{P}H_{0\varepsilon}\bar{P} \geq (E_1 + H_f)\bar{P}_0^{\mathrm{part}} \otimes 1_f$$
$$+ (E_0 + \frac{2}{3}\rho + \frac{1}{3}H_f)1_{\mathrm{part}} \otimes \bar{\chi}_\rho(H_f).$$

Setting $\delta := 3\min(E_1 - E_0 - \frac{1}{4}\rho, \frac{1}{3}\rho) \geq \rho$. Since for $z \in \Omega_\rho$, $z \leq E_0 + \frac{1}{4}\rho$, we conclude that

$$\bar{P}(H_{0\varepsilon} - z)\bar{P} \geq \frac{1}{3}(\delta + H_f)\bar{P}^2. \qquad (23.64)$$

Due to (23.64) and the fact that $H_{0\varepsilon}$ commutes with $\bar{P}$, we can define, for any real α, the invertible, positive operator $R_{0,\bar{P}}^\alpha := \bar{P}(H_{0\varepsilon} - z)^{-\alpha} = (H_{0\varepsilon} - z)^{-\alpha}\bar{P}$, satisfying $R_{0,\bar{P}}^\alpha R_{0,\bar{P}}^{-\alpha} = \bar{P}^2$ so that the following identity holds:

$$\bar{P}(H_{\bar{P}}(\varepsilon) - z)\bar{P} = R_{0,\bar{P}}^{-1/2}[1 + K]R_{0,\bar{P}}^{-1/2}, \qquad (23.65)$$

where $K = R_{0,\bar{P}}^{1/2}IR_{0,\bar{P}}^{1/2}$. Next we show that $\|K\| \leq \mathrm{const}\ \varepsilon$. Using the definition of I, we find

$$\|K\| \leq \varepsilon\|pR_{0,\bar{P}}^{1/2}\|\|A(x)R_{0,\bar{P}}^{1/2}\|$$
$$+ \frac{\varepsilon^2}{2}\|R_{0,\bar{P}}^{1/2}A(x)\|\|A(x)R_{0,\bar{P}}^{1/2}\| \qquad (23.66)$$
$$+ \varepsilon^2\|R_{0,\bar{P}}^{1/2}\|\|G(x)R_{0,\bar{P}}^{1/2}\| .$$

The relative bound on $A(x)$ proven above implies that

$$\|A(x)R_{0,\bar{P}}^{1/2}\| \leq c(\|H_f^{1/2}R_{0,\bar{P}}^{1/2}\| + \|R_{0,\bar{P}}^{1/2}\|). \qquad (23.67)$$

Now let $u = R_{0,\bar{P}}^{1/2}v$. Then the estimate (23.64) can be rewritten as

$$\|(\delta + H_f)^{1/2}R_{0,\bar{P}}^{1/2}u\| \leq \sqrt{3}\|\bar{P}u\| .$$

This gives, in particular, that, for $z \in \Omega \cap \mathbb{R}$,

$$\|H_f^{1/2}R_{0,\bar{P}}^{1/2}\| \leq \sqrt{3} \qquad (23.68)$$

and

$$\|R_{0,\bar{P}}^{1/2}\| \leq \sqrt{3}\delta^{-1/2}. \qquad (23.69)$$

Next, using that $\|pv\|^2 = \langle v, p^2 v\rangle \leq \langle v, (H_{0\varepsilon} + C)v\rangle$ and taking $v := R_{0,\bar{P}}^{1/2}u$, we obtain

$$\|pR_{0,\bar{P}}^{1/2}\| \leq C\delta^{-1/2}. \qquad (23.70)$$

Estimates (23.66) - (23.70) imply $\|K\| \leq C\varepsilon\rho^{-1}$, if $z \in \Omega_\rho$. Pick now $|\varepsilon| \ll \delta$ so that $\|K\| \leq 1/2$ and therefore the right hand side of (23.65) is invertible and satisfies (23.62). $\square$

23.11 Appendix: Key Bound

Recall that $E_{[0,\rho]}(\lambda)$ denotes the characteristic function of the interval $[0, \rho]$.

Proposition 23.11 (Key bound) Let H_{rs} be an rs-monomial of the form

$$\int (\prod_{j=1}^{r} a^*(k_j)) h_{rs}(H_f, k_1 \ldots k_{r+s}) \prod_{i=r+1}^{r+s} a(k_i) d^{r+s}k , \qquad (23.71)$$

where $h_{rs}(\mu, \underline{k})$, $\underline{k} = (k_1 \ldots k_{r+s})$, are measurable functions on $[0, 1] \times \mathbb{R}^{3(r+s)}$, called *coupling functions*, and let $\Omega_{rs}(\rho) := \{k \in \mathbb{R}^{3(r+s)} \mid \sum_{j=1}^{r} \omega(k_j) \le \rho, \sum_{j=r+1}^{r+s} \omega(k_j) \le \rho\}$. Then we have the following bound:

$$\|E_{[0,\rho]}(H_f) H_{rs} E_{[0,\rho]}(H_f)\|^2 \le \rho^{r+s} \int_{\Omega_{rs}(\rho)} \frac{\sup_\mu |h_{rs}(\mu, \underline{k})|^2}{\prod_{j=1}^{r+s} \omega(k_j)} d^{r+s}k \qquad (23.72)$$

Proof. In this proof we denote $E_\rho := E_{[0,\rho]}(H_f)$. Using the form (23.71) of H_{rs}, taking the norm under the integral sign, and using the norm inequality for the product of operators, as well as $\|A^*\| = \|A\|$, we have

$$\|E_\rho H_{rs} E_\rho\| \le \int \int \|a^r E_\rho\| \|h_{rs}\| \|a^s E_\rho\| . \qquad (23.73)$$

Here $\|h_{rs}\|$ is the operator norm of $h_{rs}(H_f, \underline{k})$. So $\|h_{rs}\| = \sup_\mu |h_{rs}(\mu, \underline{k})|$. Let f be a positive, continuous function on $\mathbb{R}^{3n}$. We will prove the estimate

$$\int f \|a^n E_\rho\| \le (\rho^n \int_{\sum \omega_j \le \rho} \frac{f^2}{\prod_1^n \omega_j})^{1/2} , \qquad (23.74)$$

where $\omega_j = \omega(k_j)$, which will imply (23.72).

First, we prove the estimate for $n = 1$. By the pull-through formula (22.52), we have

$$a(k) E_\rho = E_\rho (H_f + \omega(k)) a(k) E_\rho.$$

Since $H_f \ge 0$, this implies

$$\|a(k) E_\rho\| \le \chi_{\omega(k) \le \rho} \|a(k) E_\rho\|.$$

Proceeding as in the proof of the relative bound on $a(k)$ (see (22.33)), we obtain

$$\int f\|a(k)E_\rho\| \leq \int_{\omega \leq \rho} f\|a(k)E_\rho\|$$

$$\leq \left(\int_{\omega \leq \rho} \frac{f^2}{\omega}\right)^{1/2} \|H_f^{1/2}E_\rho\| \qquad (23.75)$$

$$\leq \left(\rho \int_{\omega \leq \rho} \frac{f^2}{\omega}\right)^{1/2}$$

which implies (23.74) for $n = 1$. Now we prove (23.74) for arbitrary $n \geq 1$. First of all, applying the pull-through formula (22.52) n times, we find

$$\prod_1^n a(k_j)E_\rho(H_f) = E_\rho(H_f + \sum_1^n \omega(k_j)) \prod_1^n a(k_j)E_\rho.$$

Hence

$$\int f\|a^n E_\rho\| \leq \int_{\sum_1^n \omega_j \leq \rho} f\|a^n E_\rho\|. \qquad (23.76)$$

Secondly, applying the pull-through formula (22.52) n times again, we obtain

$$\prod_1^n a(k_j)E_\rho = \prod_1^n a(k_j)H_f^{-1/2}H_f^{1/2}E_\rho$$

$$= (H_f + \sum_1^n \omega(k_j))^{-1/2} \prod_1^n a(k_j)H_f^{1/2}E_\rho.$$

This formula and inequalities (23.75) and (23.76) give

$$\int f\|a^n \psi\| \leq \int \left(\int_{\sum \omega_j \leq \rho} \frac{f^2}{\omega} dk_n\right)^{1/2} \|a^{n-1}H_f^{1/2}E_\rho\| d^{n-1}k.$$

Proceeding in the same fashion we arrive at (23.74).

Applying the bound (23.74) with $n = r$ to the integral $\int \|a^r E_\rho\|\|h_{rs}\| d^r k$, we bound the r.h.s. of (23.73) by

$$\int \|a^s E_\rho\| \left(\rho^r \int_{\sum_1^r \omega_j \leq \rho} \frac{\|h_{rs}\|^2}{\prod_1^r \omega_j} d^r k\right)^{1/2} d^s k.$$

Applying bound (23.74) with $n = s$ to the outer integral gives (23.72). $\square$

Renormalization Group

In this chapter we investigate spectral properties of quantum Hamiltonians of the form (23.55) - (23.57). To this end we develop a *spectral renormalization group* method. It consists of the following steps (see Sections 24.2-24.5):

- Pass from a single operator H ($= H_\varepsilon$) to a Banach space, $\mathcal{B}$, of Hamiltonian-type operators (of the form (23.55) - (23.57));
- Construct a map, $\mathcal{R}_\rho$, on $\mathcal{B}$, with the following properties:
  (a) $\mathcal{R}_\rho$ is isospectral in the sense of Definition 23.3;
  (b) $\mathcal{R}_\rho$ removes the photon degrees of freedom related to energies $\geq \rho$.
- Relate the dynamics of semi-flow, $\mathcal{R}_\rho^n, n \geq 1$, to spectral properties of individual operators in $\mathcal{B}$.

The map $\mathcal{R}_\rho$ is called the *renormalization map*. It is of the form

$$\mathcal{R}_\rho = \rho^{-1} S_\rho \circ F_\rho$$

where F_ρ is the smooth Feshbach - Schur (decimation) map, and S_ρ is a simple rescaling map (see Sections 23.9 and 24.4). F_ρ maps operators which act non-trivially only on the subspace $\mathrm{Ran} E_{[0,1]}(H_f)$ consisting of states with photon energies ≤ 1, to operators which act non-trivially on states with photon energies $\leq \rho$. The rescaling S_ρ brings us back to the subspace $\mathrm{Ran} E_{[0,1]}(H_f)$. By design, the renormalization map $\mathcal{R}_\rho$ is *isospectral* in the sense that the operators K and $\mathcal{R}_\rho(K)$ have the same spectrum near 0, modulo rescaling.

The renormalization map gives rise to an isospectral (semi-)flow $\mathcal{R}_\rho^n$, $n \geq 1$, (called *renormalization group (RG)*) on the Banach space $\mathcal{B}$. We will see that orbits of this flow, with appropriate initial conditions approach the operators of the form ωH_f (for some $\omega \in \mathbb{C}$) as $n \to \infty$. In fact, $\mathbb{C} H_f$ is a line of fixed points of the flow ($\mathcal{R}_\rho(\omega H_f) = \omega H_f$). By studying the behaviour of the flow near this line of fixed points, we can relate the spectrum of an initial condition, H, near 0 to that of H_f, which we know well. This is the basic idea behind the proof of Theorem 23.1.

© Springer-Verlag GmbH Germany, part of Springer Nature 2020
S. J. Gustafson and I. M. Sigal, *Mathematical Concepts of Quantum Mechanics*, Universitext, https://doi.org/10.1007/978-3-030-59562-3_24

24.1 Main Result

In this section we formulate the main result of this chapter. It concerns the spectra of quantum Hamiltonians of the form (23.55) - (23.57),

$$H = H_{00} + \sum_{r+s\geq 1} \chi_1 H_{rs} \chi_1, \quad H_{00} := h_{00}(H_f), \tag{24.1}$$

$$H_{rs} := \int \prod_{j=1}^{r} a^*(k_j) \, h_{rs}(H_f, k_1 \ldots k_{r+s}) \prod_{i=r+1}^{r+s} a(k_i) d^{r+s}k, \tag{24.2}$$

acting on the space $\mathcal{H}_{\mathrm{red}} := E_{[0,1]}(H_f)\mathcal{F}$, where recall $E_{[0,\rho]}(\lambda)$ denotes the characteristic function of the interval $[0,\rho]$. Let B^r denote the unit ball in $\mathbb{R}^{3r}$ and $I := [0,1]$. Above, $\chi_\rho := \chi_\rho(H_f)$ are the operators defined in Section 23.6 (the shorthand we use from now on) and the coupling functions, $h_{rs} : I \times B^{r+s} \to \mathbb{C}$, satisfy the estimates

$$|\partial_\mu^n h_{rs}(\mu, \underline{k})| \leq \gamma_0 \sum_{i=1}^{r+s} \omega(k_i)^{\frac{1}{2}} \prod_{j=1}^{r+s} \left(\xi_0 \, \omega(k_j)^{-\frac{1}{2}} \right), \tag{24.3}$$

for any $1 \leq i \leq r + s$, $\mu \in [0,1]$, $\xi_0 > 0$, and $n = 0, 1$, where the product is absent in the case $r = s = 0$, and

$$|h_{00}(0)| \leq \alpha_0, \quad |\partial_\mu h_{00}(\mu) - 1| \leq \beta_0 , \tag{24.4}$$

for $\mu \in [0, \infty)$. Here, recall, $\underline{k} = (k_1, \ldots, k_{r+s})$.

Note that, in order to be able to apply our theory to the analysis of resonances of H_ε, the space of operators H should include non-selfadjoint.

Let $D(0, r)$ stand for the disc in $\mathbb{C}$ of the radius r and centered at 0. We denote by $\mathcal{D}_s$ the set of operators of the form (24.1) - (24.4) with $h_{00}(0) = 0$. We define a subset S of the complex plane by

$$S := \{w \in \mathbb{C} | \ \mathrm{Re}w \geq 0, |\mathrm{Im}w| \leq \frac{1}{3}\mathrm{Re}w\}. \tag{24.5}$$

Recall that a complex function f on an open set $\mathcal{A}$ in a complex Banach space $\mathcal{W}$ is said to be *analytic* if $\forall H \in \mathcal{A}$ and $\forall \xi \in \mathcal{W}$, $f(H + \tau\xi)$ is analytic in the complex variable τ for $|\tau|$ sufficiently small (see [40]). We are now prepared to state the main result of this chapter.

Theorem 24.1 Assume that β_0 and γ_0 are sufficiently small. Then there is an analytic map $e : \mathcal{D}_s \to D(0, c\gamma_0^2)$ such that for $H \in \mathcal{D}_s$ the number $e(H)$ is a simple eigenvalue of the operator H and $\sigma(H) \subset e(H) + S$. Moreover, $e(H) \in \mathbb{R}$, for $H = H^*$.

Note that our approach also provides an effective way to compute the eigenvalue $e(H)$ and the corresponding eigenvector.

Operators on the Fock space of the form (24.1) - (24.4) will be said to be in *generalized normal (or Wick) form*. Operators of the form (24.2) will be called (rs)-monomials. Though H_f can be expressed in the standard Wick form, $H_f = \int \omega a^* a$, the corresponding coupling function, $\omega(k_1)\delta(k_1 - k_2)$, is more singular than we allow. But even if this coupling function were smooth, finding the Wick form of operators like $\chi_\rho(H_f)$, or the h_{rs}, is not an easy matter. In what follows we manipulate the operators $a^*(k)$, $a(k)$ and H_f as if they were independent, using only the commutation relations

$$[a(k), H_f] = \omega(k)a(k) ,$$

etc.

24.2 A Banach Space of Operators

In this section we define the Banach space of operators on which the renormalization group acts. We consider formal expressions of the form (24.1) - (24.4) acting on $\mathrm{Ran}E_{[0,1]}(H_f)$.

For $\xi, \nu > 0$, we define the Banach space, $\mathcal{B}_{\xi\nu}$, consisting of formal expressions, (24.1) - (24.2)acting on the space $\mathcal{H}_{\mathrm{red}} := E_{[0,1]}(H_f)\mathcal{F}$ and satisfying $\|H\|_{\xi\nu} < \infty$, where

$$\|H\|_{\xi\nu} := \sum_{r+s\geq 0} \xi^{-(r+s)} \|H_{rs}\|_\nu \tag{24.6}$$

with $\|H_{rs}\|_\nu := \|h_{rs}\|_\nu$ and

$$\|h_{rs}\|_\nu := \sum_{n=0}^{1} \max_j \sup_{\mu,\underline{k}} [\Omega_j^{(r+s)}|\partial_\mu^n h_{rs}|] < \infty , \tag{24.7}$$

where $\Omega_j^{(n)} = 1$ for $n = 0$, and $\Omega_j^{(n)} = \omega(k_j)^{-\nu} \prod_{i=1}^n \sqrt{\omega(k_i)}$ for $r + s \geq 1$.

Similarly to (23.51) we obtain

Proposition 24.2 Let H_{rs} be as above. Then for any $\nu \geq 0$,

$$\|\chi_\rho H_{rs}\chi_\rho\| \leq (r!s!)^{-1/2}(\sqrt{4\pi}\rho)^{r+s}\rho^\nu\|H_{rs}\|_\nu. \tag{24.8}$$

Thus $\sum_{r+s\geq 1} \chi_\rho H_{rs}\chi_\rho$ converges in norm.

Next we state without proof that the map $\{h_{rs}\} \mapsto \sum \int (a^*)h_{rs}a^s$ is one-to-one (see [27, Thm. III.3]). Here the $\{h_{rs}\}$ satisfy (24.7). Hence the normed space $\mathcal{B}_{\xi\nu}$ is indeed a Banach space.

For the future analysis it is important to know how different operators behave under rescaling, more precisely, under the action of the rescaling map $\rho^{-1}S_\rho$, defined in (23.52), as $\rho \to 0$ (or of $(\rho^{-1}S_\rho)^n$ as $n \to \infty$, which in the present case is the same). We have for $r + s \geq 1$

$$\rho^{-1} S_\rho(H_{rs}) = \int (a^*)^r h_{rs}^{(\rho)} a^s, \tag{24.9}$$

where

$$h_{rs}^{(\rho)}(H_f, \underline{k}) = \rho^{3/2(r+s)-1} h_{rs}(\rho H_f, \rho \underline{k}) . \tag{24.10}$$

If $H_{rs} \in \mathcal{B}_{\nu\xi}$, then h_{rs} behaves for small $|k_j|$'s like $\sum_{i=1}^{r+s} \omega(k_i)^\nu \prod_{j=1}^{r+s} \omega(k_j)^{-\frac{1}{2}}$. Since by (23.53), $S_\rho(a^\#(k)) = \rho^{-3/2} a^\#(\rho^{-1} k)$, we have for $r + s \geq 1$

$$\rho^{-1} S_\rho(H_{rs}) \sim \rho^{r+s-1+\nu} H_{rs} . \tag{24.11}$$

In the $r + s = 0$ case, we have to specify the behaviour of the function $h_{00}(\mu)$ at $\mu = 0$. In our case, $h_{00}(\mu) - h_{00}(0) \sim \mu$ and therefore

$$\rho^{-1} S_\rho(H_{00} - \langle H_{00} \rangle_\Omega) \sim H_{00} - \langle H_{00} \rangle_\Omega),$$
$$\rho^{-1} S_\rho(\langle H_{00} \rangle_\Omega) = \rho^{-1} \langle H_{00} \rangle_\Omega. \tag{24.12}$$

where we used the notation $\langle A \rangle_\Omega := \langle \Omega, A\Omega \rangle$. Hence, H_{rs}, $r+s \geq 1$, *contract*, $H_{00} - \langle H_{00} \rangle_\Omega$ are roughly *invariant*, while $\langle H_{00} \rangle_\Omega$ *expand* under our rescaling. This suggests to decompose the Banach space $\mathcal{B}_{\xi\nu}$ into the direct sum

$$\mathcal{B}_{\xi\nu} = \mathbb{C} \cdot \mathbf{1} + \mathcal{T} + \mathcal{W} , \tag{24.13}$$

of the subspaces which expand, are roughly invariant or contract under our rescaling:

$$\mathcal{T} = \{T(H_f) \ \mid \ T : [0, \infty) \to \mathbb{C} \text{ is } C^1 \text{ with } T(0) = 0\} , \tag{24.14}$$

and

$$\mathcal{W} := \{ \sum_{r+s \geq 1} \chi_1 H_{rs} \chi_1 \in \mathcal{B}_{\xi\nu} \}. \tag{24.15}$$

This decomposition will plat an important role below.

Remark 24.3 The subspace $\mathcal{T}$ can be further decomposed into the invariant substace $\mathbb{C} \cdot H_f$ and and contracting one,

$$\mathcal{T}_s = \{T(H_f) \ \mid \ T : [0, \infty) \to \mathbb{C} \text{ is } C^1 \text{ with } T(0) = T'(0) = 0\} . \tag{24.16}$$

The decomposition of the Banach space $\mathcal{B}_{\xi\nu}$ into the subspaces $\mathbb{C} \cdot \mathbf{1}$, $\mathbb{C} \cdot H_f$ and $\mathcal{T}_s + \mathcal{W}$ is related to the spectral decomposition of the map $\rho^{-1} S_\rho$.

24.3 The Decimation Map

In this section we construct a map which projects out the degrees of freedom corresponding to high photon energies. We use the smooth Feshbach-Schur map, which we define here in a greater generality than in Section 23.6, and formulate its important isospectral property. Let $\chi, \overline{\chi}$ be a partition of unity on a separable Hilbert space $\mathcal{H}$, i.e. χ and $\overline{\chi}$ are operators on $\mathcal{H}$ whose norms are bounded by one, $0 \leq \chi, \overline{\chi} \leq 1$, and $\chi^2 + \overline{\chi}^2 = 1$. We assume that χ and $\overline{\chi}$ are nonzero. Let τ be a (linear) projection acting on closed operators on $\mathcal{H}$ with the property that operators in its image commute with χ and $\overline{\chi}$. We also assume that $\tau(1) = 1$. Let $\overline{\tau} := 1 - \tau$ and define

$$H_{\tau, \chi^\#} := \tau(H) + \chi^\# \overline{\tau}(H) \chi^\#, \qquad (24.17)$$

where $\chi^\#$ stands for either χ or $\overline{\chi}$.

Given χ and τ as above, we denote by $D_{\tau, \chi}$ the space of closed operators, H, on $\mathcal{H}$ which belong to the domain of τ and satisfy the following three conditions:

$$D(\tau(H)) = D(H) \text{ and } \chi D(H) \subset D(H), \qquad (24.18)$$

$$H_{\tau, \overline{\chi}} \text{ is (bounded) invertible on Ran} \overline{\chi}, \qquad (24.19)$$

$$\overline{\tau}(H) \chi \text{ and } \chi \overline{\tau}(H) \text{ extend to bounded operators on } \mathcal{H}. \qquad (24.20)$$

(For more general conditions see [27, 139].)

The *smooth Feshbach-Schur map (SFM)* maps operators on $\mathcal{H}$ belonging to $D_{\tau, \chi}$ to operators on $\mathcal{H}$ by $H \mapsto F_{\tau, \chi}(H)$, where

$$F_{\tau, \chi}(H) := H_0 + \chi W \chi - \chi W \overline{\chi} H_{\tau, \overline{\chi}}^{-1} \overline{\chi} W \chi. \qquad (24.21)$$

Here $H_0 := \tau(H)$ and $W := \overline{\tau}(H)$. Note that H_0 and W are closed operators on $\mathcal{H}$ with coinciding domains, $D(H_0) = D(W) = D(H)$, and $H = H_0 + W$. We remark that the domains of $\chi W \chi$, $\overline{\chi} W \overline{\chi}$, $H_{\tau, \chi}$, and $H_{\tau, \overline{\chi}}$ all contain $D(H)$.

Define operator

$$Q_{\tau, \chi}(H) := \chi - \overline{\chi} H_{\tau, \overline{\chi}}^{-1} \overline{\chi} W \chi. \qquad (24.22)$$

The following result ([27]) generalizes Theorem 11.1 of Section 11.4.

Theorem 24.4 (Isospectrality of SFM) Let $0 \leq \chi \leq 1$ and $H \in D_{\tau, \chi}$ be an operator on a separable Hilbert space $\mathcal{H}$. Then we have the following results:

(i) $H\psi = 0 \implies F_{\tau, \chi}(H) \varphi = 0$, $\varphi := \chi \psi \in \text{Ran} \chi$.
(ii) $F_{\tau, \chi}(H) \varphi = 0 \implies H\psi = 0$, $\psi := Q_{\tau, \chi}(H) \varphi \in \mathcal{H}$.
(iii) $\dim \text{Null} H = \dim \text{Null} F_{\tau, \chi}(H)$.
(iv) H is bounded invertible on $\mathcal{H}$ if and only if $F_{\tau, \chi}(H)$ is bounded invertible on $\text{Ran} \chi$. In this case

$$H^{-1} = Q_{\tau, \chi}(H) F_{\tau, \chi}(H)^{-1} Q_{\tau, \chi}(H)^\# + \overline{\chi} H_{\overline{\chi}}^{-1} \overline{\chi}, \qquad (24.23)$$

$$F_{\tau, \chi}(H)^{-1} = \chi H^{-1} \chi + \overline{\chi} \tau(H)^{-1} \overline{\chi}. \qquad (24.24)$$

We also mention the following useful property of $F_{\tau,\chi}$:

$$H \text{ is self-adjoint} \quad \Rightarrow \quad F_{\tau,\chi}(H) \text{ is self-adjoint.} \tag{24.25}$$

The proof of this theorem is similar to the one of Theorem 11.1 of Section 11.4. We demonstrate only the proof of the statement (ii) which we use extensively below and refer for the rest of the proof to [27]. The statement (ii) follows from the relation

$$H Q_{\tau,\chi}(H) = \chi F_{\tau,\chi}(H). \tag{24.26}$$

Now we prove the latter relation, using the shorthand $Q \equiv Q_{\tau,\chi}(H)$, $H_\chi \equiv H_0 + \chi W \chi$, $H_{\overline{\chi}} \equiv H_0 + \overline{\chi} W \overline{\chi}$:

$$\begin{aligned}
H Q = H\chi &- H\overline{\chi} H_{\overline{\chi}}^{-1} \overline{\chi} W \chi = \chi H_\chi \\
&+ \overline{\chi}^2 W \chi - (\overline{\chi} H_{\overline{\chi}} + \chi^2 W \overline{\chi}) H_{\overline{\chi}}^{-1} \overline{\chi} W \chi.
\end{aligned} \tag{24.27}$$

Since $\overline{\chi} H_{\overline{\chi}} H_{\overline{\chi}}^{-1} \overline{\chi} W \chi = (\overline{\chi} H_0 + \overline{\chi}^2 W \overline{\chi}) H_{\overline{\chi}}^{-1} \overline{\chi} W \chi = \overline{\chi}^2 W \chi$, the r.h.s. gives $\chi F_{\tau,\chi}(H)$.

The Feshbach-Schur map is a special case of the smooth Feshbach-Schur map and is obtained from the latter when χ is a projection, $\chi^2 = \chi$, by taking $\tau = 0$. Then $F_{\tau,\chi}$ becomes $F_\chi(H) := \chi(H - H\overline{\chi} H_{\overline{\chi}}^{-1} \overline{\chi} H)\chi$, $H_{\overline{\chi}} := H_{0\overline{\chi}} = \overline{\chi} H \overline{\chi}$, as defined in Section 11.1.

For Hamiltonians of the form $H = \sum_{r+s \geq 0} H_{rs}$ considered in Sections 24.1-24.2, we define the *decimation map* as

$$F_\rho \equiv F_{\tau,\chi}, \tag{24.28}$$

where $F_{\tau,\chi}$ is the smooth Feshbach-Schur map and the operators τ and χ are chosen as

$$\tau(H) = H_{00} := h_{00}(H_f) \text{ and } \chi := \chi_\rho = \chi_\rho(H_f) \equiv \chi_{\rho^{-1} H_f \leq 1}, \tag{24.29}$$

where the cut-off function $\chi_{\lambda \leq 1}$ is defined in Section 23.6. To isolate a set on which the map $\mathcal{R}_\rho$ is defined we write $H = H_{00} + \sum_{r+s \geq 1} \chi_1 H_{rs} \chi_1 \in \mathcal{B}_{\xi\nu}$ as

$$H = E + T + \chi_1 W \chi_1, \tag{24.30}$$

where $E := \langle \Omega, H\Omega \rangle = h_{00}(0)$, $T := H_{00} - E$ and $W := \sum_{r+s \geq 1} H_{rs}$ (cf. (24.13) - (24.15)) and define the following polydisc in $\mathcal{B}_{\xi\nu}$:

$$\mathcal{D}_{\xi\nu}(\alpha, \beta, \gamma) := \Big\{ H \in \mathcal{B}_{\xi\nu} \mid |h_{0,0}[0]| \leq \alpha, \tag{24.31}$$

$$\sup_{r \in [0,\infty)} |h'_{0,0}[r] - 1| \leq \beta, \|W\|_{\xi\nu} \leq \gamma \Big\}.$$

The following lemma shows that the domain of the decimation map F_ρ contains $\mathcal{D}_{\xi\nu}(\alpha, \beta, \gamma)$, for appropriate numbers $\alpha, \beta, \gamma > 0$.

Lemma 24.5 Fix $0 < \rho < 1$, $\nu \leq 1/2$, and $0 < \xi \leq (4\pi)^{-1/2}$. Then it follows that the polydisc $\mathcal{D}_{\xi\nu}(\rho/8, 1/8, \rho/8)$ is in the domain of the decimation map F_ρ.

Proof. Let $H \in \mathcal{D}_{\xi\nu}(\rho/8, 1/8, \rho/8)$. We observe that $\chi_1 W \chi_1 := H - h_{0,0}$ defines a bounded operator on $\mathcal{F}$, and we only need to check the invertibility of $H_{\tau\chi_\rho}$ on $\mathrm{Ran}\,\overline{\chi}_\rho$. Now the operator $h_{0,0}$ is invertible on $\mathrm{Ran}\,\overline{\chi}_\rho$ since for all $r \in [3\rho/4, \infty)$

$$\mathrm{Re}\, h_{0,0}[r] \geq r - |h_{0,0}[r] - r|$$
$$\geq r\left(1 - \sup_r |h'_{0,0}[r] - 1|\right) - |h_{0,0}[0]|$$
$$\geq \frac{3\rho}{4}(1 - 1/8) - \frac{\rho}{8} \geq \frac{\rho}{2}. \tag{24.32}$$

Furthermore, by (24.8), $\|\chi_1 W \chi_1\| \leq \|W\|_{\xi\nu} \leq \gamma = \rho/8$. Hence $\mathrm{Re}(h_{0,0} + \chi_1 W \chi_1) \geq \frac{\rho}{3}$ on $\mathrm{Ran}\,\overline{\chi}_\rho$, i.e. $H_{\tau,\overline{\chi}_\rho}$ is invertible on $\mathrm{Ran}\,\overline{\chi}_\rho$. $\square$

Note that the decimation map, F_ρ maps isospectrally operators which act nontrivially on $\mathrm{Ran}\,\chi_1$ into those which nontrivially on $\mathrm{Ran}\,\chi_\rho$.

24.4 The Renormalization Map

Using that the subspace $\mathrm{Ran} E_{[0,1]}(H_f)$ is invariant under the composition map $S_\rho \circ F_\rho$, we define the renormalization map as a composition of a decimation map and a rescaling map on the domain of the decimation map F_ρ (see (24.28)) as

$$\mathcal{R}_\rho := \rho^{-1} S_\rho \circ F_\rho \,|_{\mathrm{Ran} E_{[0,1]}(H_f)} \tag{24.33}$$

where S_ρ is the rescaling map defined in (23.52). By Lemma 24.5, its domain contains the polydiscs $\mathcal{D}_{\xi\nu}(\alpha, \beta, \gamma)$, with $\alpha, \beta, \gamma \leq \frac{\rho}{8}$. Note that the map $S_\rho \circ F_\rho$ acts on the orthogonal complement, $\mathrm{Ran} E_{[0,1]}(H_f)^\perp$, trivially:

$$\rho^{-1} S_\rho \circ F_\rho = \rho^{-1} S_\rho \circ \tau \quad \text{on} \quad \mathrm{Ran} E_{[0,1]}(H_f)^\perp \tag{24.34}$$

where the map τ is defined in (24.29). Note that while the standard Feshbach-Schur maps have the semigroup property (see Supplement 24.6): $\mathcal{R}_{\rho_2} \circ \mathcal{R}_{\rho_1} = \mathcal{R}_{\rho_2\rho_1}$, the smooth ones have this property only under certain conditions on ρ_1 and ρ_2.

Next, we list some elementary properties of the map $\mathcal{R}_\rho$ which follow readily from the definitions

1. $\mathcal{R}_\rho$ is isospectral in the sense that $\rho \cdot \mathcal{R}_\rho(H)$ and H are isospectral at 0 in the sense of Theorem 24.4,
2. $\mathcal{R}_\rho(w H_f + z\mathbf{1}) = w H_f + \frac{1}{\rho} z\mathbf{1}$ $\forall w, z \in \mathbb{C}$.

In particular, $\mathbb{C}H_f$ is a complex line of fixed points of $\mathcal{R}_\rho$: $\mathcal{R}_\rho(wH_f) = wH_f \; \forall w \in \mathbb{C}$, and $\mathbb{C} \cdot \mathbf{1}$ is (a part of) the unstable manifold. The first property follows from the relations (24.34).

Problem 24.6 Prove the statements above.

Describing the range of the renormalization map $\mathcal{R}_\rho$ on polydiscs $\mathcal{D}_{\xi\nu}(\alpha, \beta, \gamma)$ is considerably harder. The following result, proven in [121], shows that contraction is actually a key property of along 'stable' directions.

Theorem 24.7 Let $\epsilon_0 : H \to \langle \Omega, H\Omega \rangle$, and $\mu := \nu > 0$. Then, for any $\sigma \geq 1$, $0 < \rho < 1/2$, $\alpha, \beta \leq \frac{\rho}{8}$, and $\gamma \ll 1$, we have that

$$\mathcal{R}_\rho - \rho^{-1}\epsilon_0 : \mathcal{D}_{\xi\nu}(\alpha, \beta, \gamma) \to \mathcal{D}_{\xi\nu}(r_\rho(\alpha, \beta, \gamma)). \tag{24.35}$$

continuously, with $\xi := \frac{1}{4}$ and, for an absolute constant c,

$$r_\rho(\alpha, \beta, \gamma) := (c\gamma^2/\rho, \; \beta + c\gamma^2/\rho, \; c\rho^\nu\gamma). \tag{24.36}$$

Moreover, $\mathcal{R}_\rho(H)$ and H are isospectral (modulo rescaling) at 0.

Remark 24.8 Subtracting the term $\rho^{-1}\epsilon_0$ from $\mathcal{R}_\rho$ allows us to control the expanding direction during the iteration of the map $\mathcal{R}_\rho$. In [27], such a control was achieved by changing the spectral parameter λ, which controls $\langle \Omega, H\Omega \rangle$.

The proof of this theorem is similar to the proof of Theorem 23.6 which we outlined above: expand $\mathcal{R}_\rho(H)$ in the perturbation $W := H - H_{00}$, reduce the resulting series to generalized normal form, and estimate the obtained coupling functions. To bring the operators we deal with into generalized normal form, we use the "pull-through" formulae (22.52) and (22.53).

To explain the result above, we, as usual in the study of nonlinear dynamics, consider the linearization (variational derivative) of $\partial\mathcal{R}_\rho(w \cdot H_f)$ in order to understand the dynamics of the map $\mathcal{R}_\rho$ near its near its fixed points $w \cdot H_f$, $w \in \mathbb{C}$. The variational derivative of the map $\mathcal{R}_\rho$ at a point H_0 is defined as

$$\partial\mathcal{R}_\rho(H_0)\xi := \frac{\partial}{\partial s}\mathcal{R}_\rho(H_0 + s\xi)\big|_{s=0}$$

for any $\xi \in \mathcal{B}_{\xi\nu}$. Thus $\partial\mathcal{R}_\rho(H_0)$ is a linear operator on $\mathcal{B}_{\xi\nu}$. By the definition of the map F_ρ we have the relation $F_\rho(w \cdot H_f + s\xi) = w \cdot H_f + s\tau(\xi) + O(s^2)$, which in turn implies (informally) that $\partial\mathcal{D}_\rho(w \cdot H_f)\xi = \tau(\xi) + \chi_\rho\bar{\tau}(\xi)\chi_\rho$. Using this and the relation $\mathcal{S}_\rho(\chi_\rho) = \chi_1$, we obtain

$$\partial\mathcal{R}_\rho(w \cdot H_f)\xi = \rho^{-1}\mathcal{S}_\rho(\xi) . \tag{24.37}$$

Scaling properties of $H_{rs} = \int (a^*)^r h_{rs} a^s$ are given in (24.11) - (24.12). They suggest that for $\nu > 0$, the terms H_{rs}, $r + s \geq 1$ contract, the terms $H_{00} - \langle H_{00} \rangle_\Omega$ are essentially invariant, and the terms $\langle H_{00} \rangle_\Omega$ expand. In the physics terminology and for $\nu > 0$,

$$r + s \geq 1 \quad \leftrightarrow \quad \textit{irrelevant terms,}$$
$$H_{00} - \langle H_{00} \rangle_\Omega \quad \leftrightarrow \quad \textit{marginal terms,}$$
$$\langle H_{00} \rangle_\Omega \quad \leftrightarrow \quad \textit{relevant terms.}$$

where we used the notation $\langle A \rangle_\Omega := \langle \Omega, A\Omega \rangle$.

The following equation follows from Eqs (24.9) - (24.10) :

$$\|\rho^{-1} S_\rho(H_{rs})\|_{\xi\nu} = \rho^{(1+\nu)(r+s)-1}\|H_{rs}\|_{\xi\nu} = \rho^{-1}\|H_{rs}\|_{\rho^{-1-\nu}\xi,\nu} . \qquad (24.38)$$

Applying these equalities to operators of the form $W = \sum_{r+s \geq 1} H_{rs}$, we find

$$\|\rho^{-1}S_\rho(W)\|_{\xi\nu} \leq \|W\|_{\rho^{-\nu}\xi,\nu} \leq \rho^\nu \|W\|_{\xi\nu}. \qquad (24.39)$$

(Recall that $\nu \geq 0$.) Let $\mathcal{R}_\rho^{\mathrm{lin}} := \partial \mathcal{R}_\rho(w \cdot H_f) = \rho^{-1}S_\rho$ and

$$r_\rho^{\mathrm{lin}}(\alpha, \beta, \gamma) := (0,\ \beta,\ \rho^\nu \gamma). \qquad (24.40)$$

The estimate (24.39), together with the relations $\rho^{-1}S_\rho(E1) = \rho^{-1}1$ and $\rho^{-1}S_\rho(wH_f) = wH_f$, implies that

$$\mathcal{R}_\rho^{\mathrm{lin}} - \rho^{-1}\epsilon_0 : \mathcal{D}_{\xi\nu}(\alpha, \beta, \gamma) \to \mathcal{D}_{\xi\nu}(r_\rho^{\mathrm{lin}}(\alpha, \beta, \gamma)). \qquad (24.41)$$

This is the linearization of the estimate (24.35). Theorem 24.7 deals essentially with controlling the nonlinear part of the map $\mathcal{R}_\rho$.

In the next section we address the dynamics of $\mathcal{R}_\rho^n$ as $n \to \infty$ in a vicinity of the fixed point manifold $\mathcal{M}_{fp} \supseteq \mathbb{C}H_f$, and connect this dynamics with spectral properties of operators of interest.

24.5 Dynamics of RG and Spectra of Hamiltonians

To describe the dynamics of $\mathcal{R}_\rho^k$ we need some definitions. Consider an initial set of operators to be $\mathcal{D} := \mathcal{D}_{\xi\nu}(\alpha_0, \beta_0, \gamma_0)$, with $\alpha_0, \beta_0, \gamma_0 \ll 1$. We let $\mathcal{D}_s := \mathcal{D}_{\xi\nu}(0, \beta_0, \gamma_0)$ (the subindex s stands for 'stable'). We fix the scale ρ so that

$$\alpha_0, \beta_0, \gamma_0 \ll \rho \leq \frac{1}{2}. \qquad (24.42)$$

Below, we will use the n-th iteration of the numbers $\alpha_0, \beta_0, \gamma_0$ under the map (24.36): $\forall n \geq 1$

$$(\alpha_n, \beta_n, \gamma_n) := r_\rho^n(\alpha_0, \beta_0, \gamma_0).$$

For $H \in \mathcal{D}$ we denote $H_u := \langle H \rangle_\Omega$ and $H_s := H - \langle H \rangle_\Omega 1$ (the unstable- and stable-central-space components of H, respectively). Note that $H_s \in \mathcal{D}_s$. It is shown in [121] that the following objects are defined inductively in $n \geq 1$ (with $e_0(H_s) = 0 \ \forall H_s \in \mathcal{D}_s$)

$$V_n := \{H \in \mathcal{D} | \; |H_u - e_{n-1}(H_s)| \le \frac{1}{12}\rho^{n+1}\}; \tag{24.43}$$

$$E_n(H) := \left(\mathcal{R}_\rho^n(H)\right)_u, \; H \in V_n; \tag{24.44}$$

$e_n(H_s)$ is the unique zero of the function $\lambda \to E_n(H_s - \lambda 1)$

$$\text{in the disc } D(e_{n-1}(H_s), \frac{1}{12}\rho^{n+1}), \tag{24.45}$$

and have the following properties:

$$V_n \subset V_{n-1} \subset D(\mathcal{R}_\rho^n); \tag{24.46}$$

$$\mathcal{R}_\rho^n(V_n) \subset \mathcal{D}_{\mu,\xi}(\rho/8, \beta_n, \gamma_n); \tag{24.47}$$

$$\partial_\lambda E_n(\lambda) \le -\frac{4}{5}\rho^{-n}; \tag{24.48}$$

$$|e_n(H_s) - e_{n-1}(H_s)| \le 2\alpha_n \rho^n; \tag{24.49}$$

$$e_n(H_s) \in \mathbb{R}, \quad \text{if} \quad H = H^*. \tag{24.50}$$

Proof (Proof of Theorem 24.1). Now we prove the first statement of Theorem 24.1. By (24.49), the limit $e(H_s) := \lim_{j \to \infty} e_j(H_s)$ exists pointwise for $H \in \mathcal{D}$. Iterating Eqn (24.49) and using that $(c\rho^\mu)^2 \rho \le 1/2$, we find the estimate

$$|e_n(H_s) - e(H_s)| \le 2\alpha_{n+1}\rho^{n+1}, \; n \ge 0. \tag{24.51}$$

Since $e_0(H_s) = 0$, for $n = 0$ this estimate gives $|e(H_s)| \le 2\alpha_1\rho = c\gamma_0^2$, which shows that $e : \mathcal{D}_s \to D(0, 2\alpha_1\rho)$. Moreover, (24.50) shows that $e(H_s) \in \mathbb{R}$, if $H = H^*$. We skip the proof of analyticity of $e(H_s)$ in H_s.

Next, we prove that $e(H_s)$ is a simple eigenvalue. We omit the reference to H_s and set $e \equiv e(H_s)$. Let $H := H^{(0)} - e1 \in \bigcap_n V_n$; (24.46) implies that $\bigcap_n V_n \subset D(\mathcal{R}_\rho^k), \forall k$. Hence we can define a sequence of operators $(H^{(n)})_{n=0}^\infty$ in $\mathcal{B}_{\mu,s} \subseteq \mathcal{B}(\mathcal{H}_{\text{red}})$ by $H^{(n)} := \mathcal{R}_\rho^n(H^{(0)})$. Recall the definitions of S_ρ and U_θ in (23.52) and (23.8) and let $\Gamma_\rho := U_\theta$, $\rho = e^{-\theta}$. Then the definition (24.33) of $\mathcal{R}_\rho$ implies that, for all integers $n \ge 0$,

$$H^{(n)} = \frac{1}{\rho} \Gamma_\rho \left(F_\rho(H^{(n-1)}) \right) \Gamma_\rho^*, \tag{24.52}$$

where, recall, F_ρ is defined in (24.28). We will use the operators $Q_{\tau\chi}$ defined in (24.22), and the identity (24.26) ($HQ_{\tau\chi} = \chi F_{\tau\chi}(H)$) which they satisfy. Let

$$Q^{(n)} := Q_{\tau\chi}(H^{(n)}), \tag{24.53}$$

with τ and $\chi = \chi_\rho$ given in (24.29). Then the equation $H^{(n)}Q^{(n)} = \chi_\rho F_\rho(H^{(n)})$ together with (24.52), implies the intertwining property

$$H^{(n-1)} \, Q^{(n-1)} \, \Gamma_\rho^* \;=\; \rho \, \Gamma_\rho^* \, \chi_1 \, H^{(n)} \,. \tag{24.54}$$

Eq. (24.54) is the key identity for the proof of the existence of an eigenvector with the eigenvalue e.

For the construction of this eigenvector, we define, for non-negative integers β, vectors Ψ_k in $\mathcal{H}$ by setting $\Psi_0 := \Omega$ and

$$\Psi_k \;:=\; Q^{(0)} \, \Gamma_\rho^* \, Q^{(1)} \, \Gamma_\rho^* \cdots Q^{(k-1)} \, \Omega \,. \tag{24.55}$$

We first show that this sequence is convergent, as $k \to \infty$. To this end, we observe that $\Omega = \Gamma_\rho^* \chi_\rho \, \Omega$ and hence

$$\Psi_{k+1} - \Psi_k \;=\; Q^{(0)} \, \Gamma_\rho^* \, Q^{(1)} \, \Gamma_\rho^* \cdots Q^{(k-1)} \, \Gamma_\rho^* \, (Q^{(k)} - \chi_\rho) \, \Omega \,. \tag{24.56}$$

Since $\|\chi_\rho\| \leq 1$, this implies that

$$\left\| \Psi_{k+1} - \Psi_k \right\| \;\leq\; \left\| Q^{(k)} - \chi_\rho \right\| \prod_{j=0}^{\beta-1} \left\{ 1 + \left\| Q^{(j)} - \chi_\rho \right\| \right\} \,. \tag{24.57}$$

To estimate the terms on the r.h.s. we consider the j-th step Hamiltonian $H^{(j)}$. By (24.30) and (24.47), we can write $H^{(j)}$ as

$$H^{(j)} \;=\; E_j \cdot \mathbf{1} + T_j + W_j \,, \tag{24.58}$$

with

$$|E_j| \leq \frac{\rho}{8} \text{ and } \|W_j\| \leq \gamma_j \leq \frac{\rho}{16} \,. \tag{24.59}$$

Recalling the definition (24.53) of $Q^{(j)}$, we have

$$\chi_\rho - Q^{(j)} \;=\; \overline{\chi}_\rho \big(E_j + T_j + \overline{\chi}_\rho \, W_j \, \overline{\chi}_\rho \big)^{-1} \overline{\chi}_\rho \, W_j \, \chi_\rho \,. \tag{24.60}$$

By (24.59), for all $j \in \mathbb{N}$, we may estimate

$$\left\| \chi_\rho - Q^{(j)} \right\| \;\leq\; \left(\frac{\rho}{8} - \|W_j\| \right)^{-1} \|W_j\| \;\leq\; \frac{16 \, \gamma_j}{\rho} \,. \tag{24.61}$$

Inserting this estimate into (24.57) and using that $\prod_{j=0}^{\infty}(1 + \lambda_j) \leq e^{\sum_{j=0}^{\infty} \lambda_j}$, for $\lambda_j \geq 0$, we obtain

$$\left\| \Psi_{k+1} - \Psi_k \right\| \leq \frac{16 \, \gamma_k}{\rho} \prod_{j=0}^{k-1} \left\{ 1 + \frac{16 \, \gamma_j}{\rho} \right\}$$

$$\leq \frac{16 \, \gamma_k}{\rho} \exp\left[32 \, \gamma_0 \, \rho^{-1} \right], \tag{24.62}$$

where we have used that $\sum_{j=0}^{\infty} \gamma_j \leq 2\gamma_0$ (recall the definition of γ_j after Eqn (24.42)). Since $\sum_{j=0}^{\infty} \gamma_j < \infty$, we see that the sequence $(\Psi_k)_{k \in \mathbb{N}_0}$ of vectors in $\mathcal{H}$ is convergent, and its limit $\Psi_\infty := \lim_{k \to \infty} \Psi_k$, satisfies the estimate

$$\|\Psi_\infty - \Omega\| = \|\Psi_\infty - \Psi_0\| \leq \frac{32\,\gamma_0}{\rho}\,\exp\left[32\,\gamma_0\,\rho^{-1}\right], \tag{24.63}$$

which guarantee that $\Psi_{(\infty)} \neq 0$.

The vector Ψ_∞ constructed above is an element of the kernel of $H^{(0)}$, as we will now demonstrate. Observe that, thanks to (24.54),

$$\begin{aligned}
H^{(0)}\,\Psi_k &= \left(H^{(0)}\,Q^{(0)}\,\Gamma_\rho^*\right)\left(Q^{(1)}\,\Gamma_\rho^* \cdots Q^{(k-1)}\,\Omega\right) \\
&= \rho\,\Gamma_\rho^*\,\chi_1\left(H^{(1)}\,Q^{(1)}\,\Gamma_\rho^*\right)\left(Q^{(2)}\,\Gamma_\rho^* \cdots Q^{(k-1)}\,\Omega\right) \\
&\;\;\vdots \\
&= \rho^k\left(\Gamma_\rho^*\,\chi_1\right)^k H^{(k)}\,\Omega\,.
\end{aligned} \tag{24.64}$$

Eq (24.58) together with the estimate (24.59) and the relation $T_k\Omega = 0$ implies that

$$\begin{aligned}
\left\|H^{(k)}\,\Omega\right\| &= \left\|\left(W_k + E_k\right)\Omega\right\| \\
&\leq \gamma_k + 8\alpha_k^2 \leq 2\gamma_k\,.
\end{aligned} \tag{24.65}$$

Summarizing (24.64)–(24.65) and using that the operator norm of $\Gamma_\rho^*\,\chi_1$ is bounded by 1, we arrive at

$$\left\|H^{(0)}\,\Psi_k\right\| \leq 2\gamma_k \to 0 \tag{24.66}$$

as $k \to \infty$. Since $H^{(0)} \in \mathcal{B}(\mathcal{H})$ is continuous, (24.66) implies that

$$H^{(0)}\,\Psi_\infty = \lim_{k\to\infty}\left(H^{(0)}\,\Psi_k\right) = 0. \tag{24.67}$$

Thus 0 is an eigenvalue of the operator $H^{(0)} := H - e\mathbf{1}$, i.e. e is an eigenvalue of the operator H, with the eigenfunction Ψ_∞.

Finally, we prove the second statement of Theorem 24.1. To simplify exposition we restrict ourself to self-adjoint operators. We omit the reference to H_s and set $e \equiv e(H_s)$ and $e_n \equiv e_n(H_s)$. Let $H^{(n)}(\lambda) := \mathcal{R}_\rho^n(H_s - \lambda)$ and, recall, $E_n(\lambda) := H^{(n)}(\lambda)_u$. Using the equation $E_n(e_n) = 0$, the mean value theorem and the estimate (24.48), we obtain that $E_n(\lambda) \geq -\frac{4}{5}\rho^{-n}(\lambda - e_n)$, provided $\lambda \leq e_n$. Hence, if $\lambda \leq e_n - \theta_n$, with $\theta_n \gg \gamma_n\rho^n$ and $\theta_n \to 0$ as $n \to \infty$, then $H^{(n)}(\lambda) \geq \frac{4}{5}\rho^{-n}\theta_n - O(\gamma_n) \geq \frac{1}{2}\gamma_n$. This implies $0 \in \rho(H^{(n)}(\lambda))$ and therefore, by Theorem 24.4, $0 \in \rho(H_s - \lambda)$ or $\lambda \in \rho(H_s)$. Since $e_n \to e$ and $\theta_n \to 0$ as $n \to \infty$, this implies that $\sigma(H_s) \subset [e, \infty)$, which is the second statement of the theorem for self-adjoint operators. $\square$

We discuss the geometrical meaning of the results obtained above. Let $H \in V_n \subset D(\mathcal{R}_\rho^{n+1})$. According to (24.30), $H^{(n)} := \mathcal{R}_\rho^n(H)$ can be written as

$$H^{(n)} = E_n \mathbf{1} + T_n + W_n, \tag{24.68}$$

where $T_n \equiv T_n(H_f)$ with $T_n(r) \in C^1$ and $T_n(0) = 0$. By (24.47) we have $|\partial_r T_n(r)-1| \le \beta_n$ and $\|W_n\|_{\mathcal{W}_{op}^s} \le \gamma_n$. Hence the function $\tau_n(r) := T_n(r)/r = \int_0^1 T'(sr)ds$ is continuous and satisfies $|\tau_n(r)-1| \le \beta_n$. One can also show that $\tau_n \to \tau = (\text{constant})$ for $H \in \bigcap_n V_n$. By the definition of V_n, $|E_n - e_{n-1}(H_s)| \le \frac{1}{12}\rho^{n+1}$. Hence

$$\mathcal{R}_\rho^n(H) \to e(H_s) + \tau H_f \text{ in the norm of } \mathcal{B}_{\xi\nu}, \tag{24.69}$$

where $e(H_s) := \lim_{j\to\infty} e_j(H_s)$, as $n \to \infty$. In other words $\mathcal{M}_s := \bigcap_n V_n$ is the (local) stable manifold for the invariant manifold $\mathcal{M}_{fp} := \mathbb{C}H_f$ of fixed points. (Formally, a local stable manifold can be defined as a manifold invariant under $\mathcal{R}_\rho$ and such that $\mathcal{M}_s = \{H \in \mathcal{B}_{\xi\nu} \mid \mathcal{R}_\rho^n(H) \to \mathcal{M}_{fp} \text{ as } n \to \infty\}$.) By the definition of V_n, it is the graph of the map $e : \mathcal{D}_s \to V_u$:

$$\mathcal{M}_s = \{H \in \mathcal{D} \mid \langle H \rangle_\Omega = e(H_s)\}. \tag{24.70}$$

One can show that $\mathcal{M}_s$ is invariant under $\mathcal{R}_\rho$.

Consider the invariant manifolds $\mathcal{M}_u$, $\mathcal{M}_{fp}$ and $\mathcal{M}_s$. Since $\mathcal{M}_s$ is of the codimension one and contains the manifold $\mathbb{C}H_f$ of fixed points, while $\mathcal{M}_u := \mathbb{C}\mathbf{1}$ is invariant and expanding under $\mathcal{R}_\rho$, we see $\mathbb{C}H_f =: \mathcal{M}_{fp}$ and $\mathcal{M}_u := \mathbb{C}\mathbf{1}$ are fixed point and unstable manifolds, respectively.

The subspaces $V_u := \mathbb{C} \cdot \mathbf{1}$, $V_c := \mathbb{C} \cdot H_f$ and $V_c + V_s := T + W$ spanning the Banach space $\mathcal{B}_{\xi\nu}$ (see Section 24.2) are tangent spaces to the manifolds $\mathcal{M}_u$, $\mathcal{M}_{fp}$ and $\mathcal{M}_s$ at $\mathcal{M}_{fp}$.

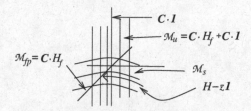

Fig. 24.1. *RG* flow.

24.6 Supplement: Group Property of $\mathcal{R}_\rho$

The Feshbach-Schur maps has the semi-group property:

Proposition 24.9 *(semigroup property)* Assume the projections P_1 and P_2 commute. Then $F_{P_1} \circ F_{P_2} = F_{P_1 P_2}$

Proof. Assume for simplicity that H and $F_{P_2}(H)$ are invertible. Then the statement follows by applying equation (11.36) twice. $\square$

Mathematical Supplement: Elements of Operator Theory

We have seen already in the first chapter that the space of quantum-mechanical states of a system is a vector space with an inner-product (in fact a Hilbert space). We saw also that an operator (a Schrödinger operator) on this space enters the basic equation (the Schrödinger equation) governing the evolution of states. In fact, the theory of operators on a Hilbert space provides the basic mathematical framework of quantum mechanics. This chapter describes some aspects of operator theory and spectral theory that are essential to a study of quantum mechanics. To make this chapter more self-contained, we repeat some of the definitions and statements from the chapters in the main text.

25.1 Spaces

In this section we review briefly some background material related to linear (or vector) spaces. We introduce the simplest and most commonly used spaces, Banach and Hilbert spaces, and describe the most important examples. We begin with the basic definitions.

Vector spaces. A *vector space* V is a collection of elements (here denoted $u, v, w, ...$) for which the operations of addition, $(u, v) \to u + v$ and multiplication by a (real or complex) number, $(\alpha, u) \to \alpha u$, are defined in such a way that

$$u + v = v + u \quad \text{(commutativity)}$$

$$u + (v + w) = (u + v) + w \quad \text{(associativity)},$$

$$u + 0 = 0 + u = u \quad \text{(existence of zero vector)},$$

$$\alpha(\beta u) = (\alpha\beta)u,$$

$$(\alpha + \beta)u = \alpha u + \beta u,$$

$$\alpha(u + v) = \alpha u + \alpha v,$$

© Springer-Verlag GmbH Germany, part of Springer Nature 2020
S. J. Gustafson and I. M. Sigal, *Mathematical Concepts of Quantum Mechanics*, Universitext, https://doi.org/10.1007/978-3-030-59562-3_25

$$0v = 0, \qquad 1v = v.$$

We also denote $-v := (-1)v$. Elements of a vector space are called *vectors*.
Here are some examples of vector spaces:

(a) $\mathbb{R}^n = \{x = (x_1, ..., x_n) | -\infty < x_j < \infty \ \forall j\}$– the Euclidean space of
dimension n;
(b) $C(\Omega)$ – the space of continuous real (or complex) functions on Ω, where
Ω is $\mathbb{R}^n$, or a subset of $\mathbb{R}^n$;
(b) $C^k(\Omega)$ – the space of k times continuously differentiable real (or complex)
functions on Ω, $k = 1, 2, 3, \ldots$.

The addition and multiplication by real/complex numbers in these spaces
is defined in the pointwise way:

$$(x + y)_j = x_j + y_j \quad \text{and} \quad (\alpha x)_j = \alpha x_j \ \forall j$$

and

$$(f + g)(x) := f(x) + g(x) \quad \text{and} \quad (\alpha f)(x) := \alpha f(x) \ \forall x \in \Omega.$$

Problem 25.1 *Show that* $\mathbb{R}^n$, $C(\Omega)$ *and* $C^k(\Omega)$ *are vector spaces.*

Norms. To measure the size of vectors, one uses the notion of norm. A *norm*
on a vector space V is defined to be a map, $V \ni u \mapsto ||u|| \in [0, \infty)$, which has
the following properties:

(a) $||u|| = 0 \iff u = 0$;
(b) $||\alpha u|| = |\alpha| ||u||$;
(c) $||u + v|| \leq ||u|| + ||v||$.

The last inequality is called the *triangle inequality*. We give some examples of
norms:

(a) $||x|| := |x| = (\sum_i x_i^2)^{1/2}$ in $\mathbb{R}^n$;
(b) $||f||_\infty = \sup_{x \in \Omega} |f(x)|$ in $C(\Omega)$ (also denoted $||f||_C$);
(c) $||f||_{C^k} = \max_{0 \leq j \leq k} \sup_{x \in \mathbb{R}} |\frac{d^j}{dx^j} f(x)|$ in $C^k(\mathbb{R})$;
(d) $||f||_p := (\int_\Omega |f(x)|^p dx)^{1/p}$ in $C(\Omega)$

For examples (b) and (d), clearly, $||f||_p = 0 \Leftrightarrow f = 0$, $||\alpha f||_p = |\alpha| ||f||_p$, $\forall \alpha \in$
$\mathbb{C}$, and for $p = 1, \infty$, $||f + g||_p \leq ||f||_p + ||g||_p$. We will prove the triangle
inequality for $1 < p < \infty$ later.

A vector space equipped with a norm is called a *normed vector space*. Here
are some examples of normed vector spaces:

(a) $\mathbb{R}^n$ with the norm $||x|| = |x|$;
(b) The subspace $C_b(\Omega)$ of $C(\Omega)$ consisting of all continuous, bounded func-
tions on $\Omega \subset \mathbb{R}^n$, equipped with the norm $||f||_\infty$;

(c) The space $C^k(\Omega)$, for $\Omega = [0,1]$, equipped with the norm $\|f\|_{C^k}$ or the norm $\|f\|_p$.

Problem 25.2 *Show that $C_b(\Omega)$ is a normed vector space.*

Banach spaces. A normed vector space is said to be *complete*, if Cauchy sequences in it converge; that is, if $u_j \in V$, $j = 1,2,\ldots$ is such that $\lim_{j,k\to\infty} \|u_j - u_k\| = 0$, then there exists $u \in V$ such that $\lim_{j\to\infty} \|u_j - u\| = 0$. A normed vector space which is also complete is called a *Banach space*. Examples of Banach spaces include

(a) $\mathbb{R}^n$ with the norm $\|x\| = |x|$;
(b) $C_b(\Omega)$ with the norm $\|f\|_\infty$;
(c) $C^k(\Omega)$, for $\Omega = [0,1]$ with the norm $\|f\|_{C^k}$;
(d) For $1 \le p < \infty$, the L^p-*space*

$$L^p(\Omega) := \{\, f : \Omega \to \mathbb{C} \mid \int_\Omega |f(x)|^p dx < \infty \,\}$$

with the norm $\|f\|_p$.

Dual spaces. Next we define the important notion of dual space.

1. A *bounded linear functional* on a vector space V, with a norm $\|\cdot\|$, is a map $l : V \to \mathbb{C}$ (or $\to \mathbb{R}$ if V is a real, rather than a complex vector space) such that

$$l(\alpha\xi + \beta\eta) = \alpha l(\xi) + \beta l(\eta)$$

for all $\xi, \eta \in V$, and $\alpha, \beta \in \mathbb{C}$ (or $\mathbb{R}$), and there is $C < \infty$ such that

$$|l(\xi)| \le C\|\xi\|$$

for all $\xi \in V$.
2. The *dual space* of a normed vector space V, is the space V^* of all bounded linear functionals on V.

Note that on a finite dimensional space all linear functionals are bounded. The dual space V^* is also a normed vector space, under the norm

$$\|l\|_{V^*} := \sup_{\xi\in V, \|\xi\|_V=1} |l(\xi)|.$$

If V is a Banach space, then so is V^*, under this norm. One often denotes the action of $l \in V^*$ on $\xi \in V$ by

$$\langle l, \xi \rangle := l(\xi).$$

Inner products and Hilbert spaces. Now let $\mathcal{H}$ be a (complex) vector space. We assume $\mathcal{H}$ is endowed with an *inner product*, $\langle\cdot,\cdot\rangle$. This means the map

$$\langle\, ,\,\rangle : \mathcal{H} \times \mathcal{H} \to \mathbb{C}$$

satisfies the properties

1. linearity (in the second argument):

$$\langle v, \alpha w + \beta z \rangle = \alpha \langle v, w \rangle + \beta \langle v, z \rangle$$

2. conjugate symmetry: $\langle w, v \rangle = \overline{\langle v, w \rangle}$
3. positive definiteness: $\langle v, v \rangle > 0$ for $v \neq 0$

for any $v, w, z \in \mathcal{H}$ and $\alpha, \beta \in \mathbb{C}$. It follows that the map $\| \cdot \| : \mathcal{H} \to [0, \infty)$ given by

$$\|v\| := \langle v, v \rangle^{1/2}$$

is a norm on $\mathcal{H}$. If $\mathcal{H}$ is also complete in this norm – that is, if it is a Banach space – then $\mathcal{H}$ is called a *Hilbert space*.

Our main example of a Hilbert space is the space of square-integrable functions, the L^2-*space* (the *state space* of a a quantum system):

$$L^2(\mathbb{R}^d) := \{\psi : \mathbb{R}^d \to \mathbb{C} \mid \int_{\mathbb{R}^d} |\psi|^2 < \infty\}$$

with the inner-product

$$\langle \psi, \phi \rangle := \int_{\mathbb{R}^d} \bar{\psi}\phi.$$

Here and throughout, we will often use the simplified notations $\int f$ or $\int_{\mathbb{R}^d} f$ for $\int_{\mathbb{R}^d} f(x)dx$.

Another important example of a Hilbert space is the Sobolev space of order n, $n = 1, 2, 3, \ldots$:

$$H^n(\mathbb{R}^d) := \{\psi \in L^2(\mathbb{R}^d) \mid \partial^\alpha \psi \in L^2(\mathbb{R}^d) \, \forall \, \alpha, \, |\alpha| \leq n\}.$$

Here α is a *multi-index*: $\alpha = (\alpha_1, \ldots, \alpha_d)$, α_j non-negative integers, and $|\alpha| := \sum_{j=1}^d \alpha_j$. The expression $\partial^\alpha \psi$ denotes the partial derivative $\partial_{x_1}^{\alpha_1} \cdots \partial_{x_d}^{\alpha_d} \psi$ of order $|\alpha|$. In other words, $H^n(\mathbb{R}^d)$ is the space of functions all of whose derivatives up to order n lie in $L^2(\mathbb{R}^d)$. The inner-product that makes $H^n(\mathbb{R}^d)$ into a Hilbert space is

$$\langle \psi, \phi \rangle_{H^n} := \sum_{0 \leq |\alpha| \leq n} \langle \partial^\alpha \psi, \partial^\alpha \phi \rangle$$

where $\langle \cdot, \cdot \rangle$ is the L^2 inner-product defined above. The Fourier transform – see Section 25.14 – provides a very convenient characterization of Sobolev spaces:

$$\psi \in H^n(\mathbb{R}^d) \iff \int_{\mathbb{R}^d} (1 + |k|^{2n})|\hat{\psi}(k)|^2 dk < \infty. \tag{25.1}$$

We recall here two frequently used facts about Hilbert spaces (see, eg., [106] or [244] for proofs).

Proposition 25.3 (Cauchy-Schwarz inequality) For $v, w \in \mathcal{H}$, a Hilbert space,

$$|\langle v, w \rangle| \leq \|v\|\|w\|.$$

A set $\{v_n\} \subset \mathcal{H}$, $n = 1, 2, \ldots$ is called *orthonormal* if $\|v_n\| = 1$ for all n and $\langle v_n, v_m \rangle = 0$ for $n \neq m$. It is a *complete orthonormal set* (or *basis*) if the collection of finite linear combinations of the v_n's is dense in $\mathcal{H}$. Recall that for a subset $D \subset \mathcal{H}$ to be dense in $\mathcal{H}$ means that given any $v \in \mathcal{H}$ and $\epsilon > 0$, there exists $w \in D$ such that $\|v - w\| < \epsilon$. A Banach space which has a countable dense subset is said to be *separable*. A Hilbert space is separable if and only if it has a a countable orthonormal basis.

Proposition 25.4 (Parseval relation) Suppose $\{v_n\} \subset \mathcal{H}$ is a complete orthonormal set. Then for any $w \in \mathcal{H}$,

$$\|w\|^2 = \sum_n |\langle w, v_n \rangle|^2.$$

If $\mathcal{H}$ is a Hilbert space, then we can identify its dual, $\mathcal{H}^*$, with $\mathcal{H}$ itself, via the map $\mathcal{H} \ni u \to l_u \in \mathcal{H}^*$ with $l_u v := \langle u, v \rangle$ for $v \in \mathcal{H}$ (here the notation $\langle \cdot, \cdot \rangle$ indicates the Hilbert space inner-product). The fact that this map is an isomorphism between $\mathcal{H}$ and $\mathcal{H}^*$ is known as the *Riesz representation theorem*.

25.2 Operators on Hilbert Spaces

In this section we explain the notion of a *linear operator* on a Hilbert space $\mathcal{H}$ (often just called an *operator*), which abstracts some of the key properties of the Schrödinger operator introduced in Chapter 1. Operators are maps, A, from $\mathcal{H}$ to itself, satisfying the linearity property

$$A(\alpha v + \beta w) = \alpha A v + \beta A w$$

for $v, w \in \mathcal{H}$, $\alpha, \beta \in \mathbb{C}$. Actually, we only require an operator A to be defined on a domain $D(A) \subset \mathcal{H}$ which is dense in $\mathcal{H}$:

$$A : D(A) \to \mathcal{H}.$$

An example of a dense subset of $L^2(\mathbb{R}^d)$ is $C_0^\infty(\mathbb{R}^d)$, the infinitely-differentiable functions with compact support. (Recall, the *support* of a function f is the closure of the set where it is non-zero:

$$\mathrm{supp}(f) := \overline{\{x \in \mathbb{R}^d \mid f(x) \neq 0\}}.$$

Thus a function with compact support vanishes outside of some ball in $\mathbb{R}^d$. For $\Omega \subset \mathbb{R}^d$, $C_0^\infty(\Omega)$ denotes the infinitely differentiable functions with support contained in Ω.)

Here are some simple examples of linear operators, A, acting on the Hilbert space $L^2(\mathbb{R}^d)$. In each case, we can simply choose $D(A)$ to be the obvious domain $D(A) := \{\psi \in L^2(\mathbb{R}^d) \mid A\psi \in L^2(\mathbb{R}^d)\}$.

1. The identity map

$$1 : \psi \mapsto \psi$$

2. Multiplication by a coordinate

$$x_j : \psi \mapsto x_j \psi$$

(i.e. $(x_j\psi)(x) = x_j\psi(x)$)

3. Multiplication by a continuous function $V : \mathbb{R}^d \to \mathbb{C}$

$$V : \psi \mapsto V\psi$$

(again meaning $(V\psi)(x) = V(x)\psi(x)$).

4. The *momentum* operators (differentiation)

$$p_j : \psi \mapsto -i\hbar\partial_j\psi$$

5. The Laplacian

$$\Delta : \psi \mapsto \sum_{j=1}^{d} \partial_j^2 \psi$$

6. A Schrödinger operator

$$H : \psi \mapsto -\frac{\hbar^2}{2m}\Delta\psi + V\psi$$

7. An integral operator

$$\mathcal{K} : \psi \mapsto \int K(\cdot, y)\psi(y)dy$$

(i.e. $(\mathcal{K}\psi)(x) = \int K(x, y)\psi(y)dy$). The function $K : \mathbb{R}^d \times \mathbb{R}^d \to \mathbb{C}$ is called the *integral kernel* of the operator $\mathcal{K}$.

The domain of the first example is obviously the whole space $L^2(\mathbb{R}^d)$. The domain of the last example depends on the form of the integral kernel, K. The domains of the other examples are easily seen to be dense, since they contain $C_0^\infty(\mathbb{R}^d)$ (assuming $V(x)$ is a locally bounded function). If the potential function $V(x)$ is bounded, then the largest domain on which the Schrödinger operator, H, is defined, namely $D(H) := \{\psi \in L^2(\mathbb{R}^d) \mid H\psi \in L^2(\mathbb{R}^d)\}$, is the Sobolev space of order two, $H^2(\mathbb{R}^d)$.

Remark 25.5 If the kernel K is allowed to be a *distribution* (a generalized function), then the last example above contains all the previous ones as special cases.

It is useful in operator theory to single out those operators with the property of boundedness (which is equivalent to continuity).

Definition 25.6 An operator A on $\mathcal{H}$ is *bounded* if

$$\|A\| := \sup_{\{\psi \in \mathcal{H} \mid \|\psi\|=1\}} \|A\psi\| < \infty. \tag{25.2}$$

In fact, the expression (25.2) defines a norm which makes the space $B(\mathcal{H})$ of bounded operators on $\mathcal{H}$ into a complete normed vector space (a Banach space). As we will see, bounded operators are, in some respects, much easier to deal with than unbounded operators. However, since some of the most important operators in quantum mechanics are unbounded, we will need to study both.

Problem 25.7 Which of the operators in Examples 1-7 above are bounded? In particular, show that the operators $p_j := -i\hbar\partial_j$ and $H_0 := -\frac{\hbar^2}{2m}\Delta$ are unbounded on $L^2(\mathbb{R}^d)$.

Often we can prove a uniform bound for an operator A on a dense domain, D. The next lemma shows that in this case, A can be extended to a bounded operator.

Lemma 25.8 If an operator A satisfies $\|A\psi\| \leq C\|\psi\|$ (with C independent of ψ) for ψ in a dense domain $D \subset \mathcal{H}$, then it extends to a bounded operator (also denoted A) on all $\mathcal{H}$, satisfying the same bound: $\|A\psi\| \leq C\|\psi\|$ for $\psi \in \mathcal{H}$.

Proof. For any $u \in \mathcal{H}$, there is a sequence $\{u_n\} \subset D$ such that $u_n \to u$ as $n \to \infty$ (by the density of D). Then the relation

$$\|Au_n - Au_m\| \leq C\|u_n - u_m\|$$

shows that $\{Au_n\}$ is a Cauchy sequence, so $Au_n \to v$, for some $v \in \mathcal{H}$ (by completeness of $\mathcal{H}$), and we set $Au := v$. This extends A to a bounded operator on all of $\mathcal{H}$ (with the same bound, C). $\square$

A converse statement – that an operator defined on all of $\mathcal{H}$ must be bounded – holds for certain important classes of operators. One such class is *closed* operators:

Definition 25.9 An operator A on $\mathcal{H}$ is called *closed* if whenever $\{u_j\}_{j=1}^\infty \subset D(A)$ is a sequence with $u_j \to u$ and $Au_j \to v$ as $j \to \infty$, $u \in D(A)$ and $Au = v$. Another way to say this is that the graph of A, $\{(u, A(u)) \mid u \in D(A)\} \subset \mathcal{H} \times \mathcal{H}$, is closed.

A second such class is *symmetric* operators – i.e., those satisfying

$$\langle u,\ Av \rangle = \langle Au,\ v \rangle \tag{25.3}$$

for all $u, v \in D(A)$. As we have seen, the operators of most importance in quantum mechanics are symmetric (indeed, self-adjoint). Symmetric and self-adjoint operators will be discussed in more detail below.

Theorem 25.10 A closed or symmetric operator, defined on the entire Hilbert space, is bounded.

Proof. For a closed operator, this is the "Closed Graph Theorem" of functional analysis. For a symmetric operator, this is the "Hellinger-Toeplitz" theorem – see, eg., [244].

We conclude this section with a useful definition.

Definition 25.11 The *commutator*, $[A, B]$, of two bounded operators A and B is the operator defined by

$$[A, B] := AB - BA.$$

Defining the commutator of two operators when one of them is unbounded requires caution, due to domain considerations. Given this warning, we will often deal with commutators of unbounded operators formally without giving them a second thought.

25.3 Integral Operators

Let $\mathcal{K}$ be an integral operator on $L^2(\mathbb{R}^d)$:

$$(\mathcal{K}\psi)(x) := \int_{\mathbb{R}^d} K(x, y)\psi(y)dy$$

where $K : \mathbb{R}^d \times \mathbb{R}^d \to \mathbb{C}$ is the *integral kernel* of the operator $\mathcal{K}$. Examples include

1. $\mathcal{K} = g(-i\hbar\nabla)$ for which the kernel is

$$K(x, y) = (2\pi\hbar)^{-d/2}\check{g}(x - y) \tag{25.4}$$

(here $\check{g}$ denotes the inverse Fourier Transform of g – see Section 25.14).
2. $K = V$ (multiplication operator) for which the kernel is

$$K(x, y) = V(x)\delta(x - y).$$

The following statement identifies the kernel of the composition of integral operators. The proof is left for the reader.

Proposition 25.12 If $\mathcal{K}_1$ and $\mathcal{K}_2$ are integral operators (with kernels K_1 and K_2), then the integral kernel of $\mathcal{K} := \mathcal{K}_1\mathcal{K}_2$ is

$$K(x, y) = \int_{\mathbb{R}^d} K_1(x, z)K_2(z, y)dz.$$

Problem 25.13 Prove this.

Proposition 25.14 Let $\mathcal{K}$ be an integral operator with kernel, $K(x, y)$, which lies in L^2 of the product space: $K(x, y) \in L^2(\mathbb{R}^d \times \mathbb{R}^d)$. Then $\mathcal{K}$ is a bounded operator on $L^2(\mathbb{R}^d)$, and

$$\|\mathcal{K}\|^2 \le \int_{\mathbb{R}^d \times \mathbb{R}^d} |K(x, y)|^2 dx dy. \tag{25.5}$$

Proof. To show that the operator $\mathcal{K}$ is bounded, we estimate by the Cauchy-Schwarz inequality,

$$\left| \int K(x, y) u(y) dy \right| \le \left(\int |K(x, y)|^2 dy \right)^{1/2} \left(\int |u(y)|^2 dy \right)^{1/2}.$$

This implies

$$\|Ku\|^2 \le \int |K(x, y)|^2 dx dy \int |u(y)|^2 dy$$

which in turn yields (25.5). $\square$

25.4 Inverses and their Estimates

A key notion of theory of operators is that of the inverse operator. Given an operator A on a Hilbert space $\mathcal{H}$, an operator B is called the *inverse* of A if $D(B) = \text{Ran}(A)$, $D(A) = \text{Ran}(B)$, and

$$BA = 1|_{\text{Ran}(B)}, \qquad AB = 1|_{\text{Ran}(A)}.$$

Here $\text{Ran}(A)$ denotes the *range* of A:

$$\text{Ran}(A) := \{ Au \mid u \in D(A) \}.$$

It follows from this definition that there can be at most one inverse of an operator A. The inverse of A is denoted A^{-1}. Put differently, finding the inverse of an operator A is equivalent to solving the equation $Au = f$ for all $f \in \text{Ran}(A)$.

A convenient criterion for an operator A to have an inverse is that A be one-to-one: that is, $Au = 0 \implies u = 0$, or equivalently that it has trivial *kernel* or *nullspace*:

$$\text{Null}(A - z) := \{ u \in D(A) \mid (A - z)u = 0 \} = \{ 0 \}. \tag{25.6}$$

The operator A is said to be *invertible* if A has a *bounded* inverse. Since by definition a bounded operator is defined on all of $\mathcal{H}$, an invertible operator A, in addition to being one-to-one, must also be *onto*: that is,

$$\text{Ran}(A) = \mathcal{H}. \tag{25.7}$$

Remark 25.15 Conditions (25.6) and (25.7) ensure that A^{-1} exists and is defined on all $\mathcal{H}$. In fact, they are enough to ensure that A is actually invertible (i.e. that A^{-1} is also bounded) in some important cases, namely:

- if A is a closed operator, since by Definition 25.9 then so is A^{-1}, and so by Theorem 25.10, A^{-1} is bounded;
- if A is symmetric, since then so is A^{-1}, and so by Theorem 25.10, A^{-1} is bounded.

Problem 25.16 Show that if operators A and C are invertible, and C is bounded, then the operator CA is defined on $D(CA) = D(A)$, and is invertible, with $(CA)^{-1} = A^{-1}C^{-1}$.

The following result provides a widely used criterion for establishing the invertibility of an operator.

Theorem 25.17 Assume the operator A is invertible, and the operator B is bounded and satisfies $\|BA^{-1}\| < 1$. Then the operator $A + B$, defined on $D(A + B) = D(A)$, is invertible.

This theorem follows from the relation $A + B = (1 + BA^{-1})A$, Problem 25.16 above, and Problem 25.18 below.

Problem 25.18 Suppose an operator K is bounded, and satisfies $\|K\| < 1$. Show that the series $\sum_{n=0}^{\infty}(-K)^n$ is absolutely convergent (i.e. $\sum_{n=0}^{\infty}\|(-K)^n\| < \infty$) and provides the inverse of the operator $1 + K$.

In other words, if $\|K\| < 1$, then the operator $1 + K$ has an inverse given by

$$(1 + K)^{-1} = \sum_{n=0}^{\infty}(-K)^n. \tag{25.8}$$

The series (25.8) is called a *Neumann series*, and is used often in this book.

25.5 Self-adjointness

To make this section more self-contained, we repeat the definitions of symmetric and self-adjoint, as well as some basic results for self-adjoint operators from Section 2.2 of the main text. Recall,

1. A linear operator A acting on a Hilbert space $\mathcal{H}$ is *symmetric* if

$$\langle u,\ Av \rangle = \langle Au,\ v \rangle \tag{25.9}$$

   for all $u,\ v \in D(A)$.
2. A linear operator A acting on a Hilbert space $\mathcal{H}$ is *self-adjoint* if it is symmetric, and $\mathrm{Ran}(A \pm i) = \mathcal{H}$.

Note that the condition $\text{Ran}(A \pm i) = \mathcal{H}$ is equivalent to the fact that the equations

$$(A \pm i)\psi = f \tag{25.10}$$

have solutions for all $f \in \mathcal{H}$.

Example 25.19 On $\mathcal{H} = L^2(\mathbb{R}^d)$, the operators x_j, $p_j := -i\hbar\partial_{x_j}$, $H_0 :=$ $-\frac{\hbar^2}{2m}\Delta$ (on their natural domains), $f(x)$ and $f(p)$ for f real and bounded, and integral operators $\mathcal{K}f(x) = \int K(x,y)f(y)\,dy$ with $K(x,y) = \overline{K(y,x)}$ and $K \in L^2(\mathbb{R}^d \times \mathbb{R}^d)$, are all self-adjoint. (See Section 25.14 for the definition of $f(p)$ using the Fourier transform.)

Proof. As an example, we show this for the operator $p = -i\hbar\frac{d}{dx}$ on $L^2(\mathbb{R})$ with domain $D(p) = H^1(\mathbb{R})$. This operator is symmetric, so we compute $\text{Ran}(-i\hbar\partial_x + i)$. For $f \in L^2(\mathbb{R})$, solve

$$(-i\hbar\partial_x + i)\psi = f,$$

which, using the Fourier transform (see Section 25.14), is equivalent to

$$(k + i)\hat{\psi}(k) = \hat{f}(k),$$

and therefore

$$\hat{\psi}(k) = \frac{\hat{f}(k)}{k+i}, \quad \psi(x) = (2\pi\hbar)^{-1/2}\int e^{ikx/\hbar}\frac{\hat{f}(k)}{k+i}\,dk.$$

Notice $(1+|k|^2)|\hat{\psi}|^2 = |\hat{f}|^2$, so since $f \in L^2$ (and hence $\hat{f} \in L^2$), by (25.1), $\psi \in H^1(\mathbb{R}) = D(p)$, and therefore $\text{Ran}(-i\hbar\partial_x + i) = L^2$. Similarly $\text{Ran}(-i\hbar\partial_x - i) = L^2$. $\square$

Problem 25.20 Show that on $L^2(\mathbb{R}^d)$, x_j, $-\frac{\hbar^2}{2m}\Delta$ (on their natural domains), and $f(x)$ and $f(p)$ for $f : \mathbb{R}^d \to \mathbb{R}$ bounded are self-adjoint (the last two are bounded operators, and so have domain all of $L^2(\mathbb{R}^d)$).

The next result provides important information about the invertibility of self-adjoint operators.

Lemma 25.21 Let A be a symmetric operator. If $\text{Ran}(A - z) = \mathcal{H}$ for some z with $\text{Im}\,z > 0$, then it is true for every z with $\text{Im}\,z > 0$. The same holds for $\text{Im}\,z < 0$. Moreover, if A is self-adjoint, then $A - z$ is invertible for every z with $\text{Im}\,z \neq 0$ and satisfies the estimate

$$\|(A - z)^{-1}\| \leq \frac{1}{|\text{Im}\,z|}. \tag{25.11}$$

Proof. Write $z = \lambda + i\mu$ with $\lambda, \mu \in R$. Then, since A is symmetric, we have

$$\|(A-z)u\|^2 = \langle(A-z)u,\,(A-z)u\rangle = \|(A-\lambda)u\|^2 + \|\mu u\|^2 \geq |\mu|^2\|u\|^2. \tag{25.12}$$

Hence, if $\mathrm{Im}\, z > 0$ (or $\mathrm{Im}\, z < 0$), then $\mathrm{Null}(A - z) = \{\,0\,\}$. If also $\mathrm{Ran}(A - z) = \mathcal{H}$, then $(A - z)^{-1}$ is defined on all of $\mathcal{H}$. Further, (25.12) implies that $(A - z)^{-1}$ is bounded, with bound (25.11) (if one defines $v := (A - z)u$), and so in particular $(A - z)$ is invertible. Then, by Theorem 25.17, $A - z' = (A - z) + (z - z')$ is invertible for $|z' - z| < \|(A - z)^{-1}\|^{-1}$. Therefore, $A - z'$ is invertible if $|z' - z| < |\mathrm{Im}\, z|$, so we can extend invertibility of $A - z'$ to all $\mathrm{Im}\, z' > 0$ (or $\mathrm{Im}\, z' < 0$). if A is self-adjoint, then $\mathrm{Ran}(A \pm i) = \mathcal{H}$ and therefore $A - z$ is invertible and satisfies (25.11) for every z in $\mathbb{C}/\mathbb{R}$. $\square$

This lemma shows that if A is self-adjoint, then $\alpha A + \beta$ is self-adjoint for any real $\alpha \neq 0$ and β, and also that

$$A \text{ is self-adjoint } \Rightarrow (A - z)\psi = f \text{ has a unique solution } \forall\, \mathrm{Im}\, z \neq 0. \quad (25.13)$$

The next theorem shows that for bounded operators, self-adjointness is easy to check.

Theorem 25.22 If A is symmetric and bounded, then A is self-adjoint.

Proof. By Lemma 25.21, it suffices to show that $\mathrm{Ran}(A + i\lambda) = \mathcal{H}$ provided $|\lambda|$ is sufficiently large. This is equivalent to solving the equation

$$(A + i\lambda)\psi = f \quad (25.14)$$

for all $f \in \mathcal{H}$ and such a λ. Now, divide this equation by $i\lambda$ to obtain

$$\psi + K(\lambda)\psi = g,$$

where $K(\lambda) = (i\lambda)^{-1}A$ and $g = (i\lambda)^{-1}f$. Let $|\lambda| > \|V\|$. Then $\|K(\lambda)\| = \frac{1}{|\lambda|}\|A\| < 1$ and we conclude that $1 + K(\lambda)$ is invertible, as shown in Problem 25.18 above. $\square$

As an example, we consider an integral operator $\mathcal{K}$ with kernel, $K(x, y)$, which lies in L^2 of the product space: $K(x, y) \in L^2(\mathbb{R}^d \times \mathbb{R}^d)$, and satisfies $K(x, y) = \overline{K(y, x)}$. Then the integral operator $\mathcal{K}$ is symmetric (see Problem 2.3). By Proposition 25.14 it is bounded, and therefore by the theorem above, it is a self-adjoint operator on $L^2(\mathbb{R}^d)$.

The property of self-adjointness can also be described in terms of the general notion of *adjoint* of an operator.

Definition 25.23 The *adjoint* of an operator A on a Hilbert space $\mathcal{H}$, is the operator A^* satisfying

$$\langle A^*\psi, \phi \rangle = \langle \psi, A\phi \rangle \quad (25.15)$$

for all $\phi \in D(A)$, for ψ in the domain

$$D(A^*) := \{\psi \in \mathcal{H} \mid |\langle \psi, A\phi \rangle| \leq C_\psi \|\phi\| \text{ for some constant } C_\psi$$
$$\text{(independent of } \phi\text{)}, \forall \phi \in D(A)\}.$$

It is left as an exercise to show this definition makes sense.

Problem 25.24 Show that equation (25.15) defines a unique linear operator A^* on $D(A^*)$ (hint: use the "Riesz lemma" – see, eg., [244]).

The subtleties surrounding domains in the definition above are absent for bounded operators: if A is bounded, then by Lemma 25.8 we may assume $D(A) = \mathcal{H}$. Since

$$|\langle \psi, A\phi \rangle| \leq \|\psi\|\|A\|\|\phi\|,$$

for $\psi, \phi \in \mathcal{H}$, we have $D(A^*) = \mathcal{H}$.

Not surprisingly, one can show that an operator A is self-adjoint according to our definition above, if and only if $A = A^*$ (that is, A is symmetric, and $D(A^*) = D(A)$).

We conclude this section with a useful definition.

Definition 25.25 A self-adjoint operator A is called *positive* (denoted $A > 0$) if

$$\langle \psi, A\psi \rangle > 0$$

for all $\psi \in D(A)$, $\psi \neq 0$. Similarly, we may define non-negative, negative, and non-positive operators.

Problem 25.26 Show that the operator $-\Delta$ on $L^2(\mathbb{R}^d)$ is positive (take $D(-\Delta) = H^2(\mathbb{R}^d)$). Hint: integrate by parts (equivalently, use the divergence theorem) assuming that $\psi \in D_\beta := \{\psi \in C^2(\mathbb{R}^d) \mid |\partial^\alpha \psi(x)| \leq C_\alpha(1 + |x|)^{-\beta} \ \forall \ \alpha, |\alpha| \leq 2\}$ for some $\beta > d/2$, Then use the fact that D_β is dense in $H^2(\mathbb{R}^d)$ to extend the inequality to all $\psi \in H^2(\mathbb{R}^d)$.

25.6 Exponential of an Operator

In this section we construct the exponential e^{-itA} for a self-adjoint operator A, which allows us to solve the abstract Schrödinger equation

$$i\frac{\partial \psi}{\partial t} = A\psi \tag{25.16}$$

where $\psi : t \to \psi(t)$ is a path in a Hilbert space $\mathcal{H}$ and A is a self-adjoint operator on $\mathcal{H}$. In our applications, $\hbar A$ is a Schrödinger operator. As before, we supplement equation (25.16) with the initial condition

$$\psi|_{t=0} = \psi_0 \tag{25.17}$$

where $\psi_0 \in \mathcal{H}$. Our goal is to prove Theorem 2.16 of Section 2.3 which shows that *self-adjointness of A implies the existence of dynamics*. We restate this theorem here in terms of an operator A:

Theorem 25.27 If A is a self-adjoint operator, then there is a unique family of bounded operators, $U(t) := e^{-itA}$, having the following properties, for $t, s \in \mathbb{R}$,

$$i\frac{\partial}{\partial t}U(t) = AU(t) = U(t)A, \tag{25.18}$$

$$U(0) = 1 \text{ and } U(t)\psi \to \psi, \text{ as } t \to 0, \tag{25.19}$$

$$U(t)U(s) = U(t+s), \tag{25.20}$$

$$\|U(t)\psi\| = \|\psi\|. \tag{25.21}$$

Furthermore, the initial value problem (25.18)-(25.19) has a unique solution.

Proof. We will define the exponential e^{iA} for an unbounded self-adjoint operator A, by approximating A by bounded operators, and then using the power series definition of the exponential for bounded operators:

$$e^A := \sum_{n=0}^{\infty} \frac{A^n}{n!} \tag{25.22}$$

which converges absolutely since

$$\sum_{n=0}^{\infty} \frac{\|A^n\|}{n!} \leq \sum_{n=0}^{\infty} \frac{\|A\|^n}{n!} = e^{\|A\|} < \infty.$$

We have already shown in Section 2.3 that Theorem 25.27 holds for bounded, self-adjoint operators A. (The self-adjointness is only needed for (25.21).) Now, we extend it to unbounded operators. By Lemma 25.21, the operators

$$A_\lambda := \frac{1}{2}\lambda^2[(A+i\lambda)^{-1} + (A-i\lambda)^{-1}]$$

are well-defined and bounded for $\lambda > 0$. The operators A_λ approximate A in the sense that $A_\lambda\psi \to A\psi$ as $\lambda \to \infty$ for all $\psi \in D(A)$. To see this, note first that

$$A_\lambda = B_\lambda A, \qquad B_\lambda := \frac{1}{2}i\lambda[(A+i\lambda)^{-1} - (A-i\lambda)^{-1}] \tag{25.23}$$

and

$$1 - B_\lambda = \frac{1}{2}[(A+i\lambda)^{-1} + (A-i\lambda)^{-1}]A.$$

So using the estimate

$$\|(A \pm i\lambda)^{-1}\| \leq \frac{1}{\lambda},$$

which is established in Lemma 25.21, we find, for any $\phi \in D(A)$,

$$\|(1 - B_\lambda)\phi\| = \|\frac{1}{2}[(A+i\lambda)^{-1} + (A-i\lambda)^{-1}]A\phi\|$$

$$\leq \frac{1}{\lambda}\|A\phi\| \to 0 \quad \text{as} \quad \lambda \to \infty.$$

And since $D(A)$ is dense and $\|B_\lambda\| \le 1$, we have $B_\lambda \phi \to \phi$ as $\lambda \to \infty$ for *any*. $\phi \in \mathcal{H}$. Finally, taking $\phi = A\psi$ for any $\psi \in D(A)$, we conclude by (25.23), that

$$A_\lambda \psi \to A\psi \quad \text{as} \quad \lambda \to \infty \quad \text{for } \psi \in D(A), \tag{25.24}$$

as required.

Since A_λ is bounded, we can define the exponential e^{iA_λ} by power series. We will show now that the family $\{e^{iA_\lambda}, \ \lambda > 0\}$ is a Cauchy family, in the sense that

$$\left\| \left(e^{iA_{\lambda'}} - e^{iA_\lambda} \right) \psi \right\| \to 0 \tag{25.25}$$

as $\lambda, \lambda' \to \infty$ for all $\psi \in D(A)$. To prove this fact, we represent the operator inside the norm as an integral of a derivative:

$$e^{iA_{\lambda'}} - e^{iA_\lambda} = \int_0^1 \frac{\partial}{\partial s} e^{isA_{\lambda'}} e^{i(1-s)A_\lambda} ds. \tag{25.26}$$

Since A_λ is symmetric and bounded, it is self-adjoint (Theorem 25.22). Using (25.18) and (25.21) in (25.26), we find (noting that A_λ and $A_{\lambda'}$ commute)

$$\|(e^{iA_{\lambda'}} - e^{iA_\lambda})\psi\| = \| \int_0^1 e^{isA_{\lambda'}} e^{i(1-s)A_\lambda} i(A_{\lambda'} - A_\lambda)\psi ds \|$$

$$\le \int_0^1 \|e^{isA_{\lambda'}} e^{i(1-s)A_\lambda} i(A_{\lambda'} - A_\lambda)\psi\| ds \tag{25.27}$$

$$= \int_0^1 \|(A_\lambda - A_{\lambda'})\psi\| ds = \|(A_{\lambda'} - A_\lambda)\psi\|.$$

(The inequality used in the first step – the *Minkowski inequality* – can be proved by writing the integral as a limit of Riemann sums and using the triangle inequality – see [106]). Due to (25.24), relation (25.25) follows.

The Cauchy property (25.25) shows that for any $\psi \in D(A)$, the vectors $e^{iA_\lambda}\psi$ converge to some element of the Hilbert space as $\lambda \to \infty$. Thus we can define

$$e^{iA}\psi := \lim_{\lambda \to \infty} e^{iA_\lambda}\psi \tag{25.28}$$

for $\psi \in D(A)$. Since we have already shown that the theorem holds for bounded operators, we have that $\|e^{iA}\psi\| \le \|\psi\|$ for all ψ in $D(A)$, which is dense in $\mathcal{H}$. Thus, as in Lemma 25.8, we can extend this definition of e^{iA} to all $\psi \in \mathcal{H}$. This defines the exponential e^{iA} function for any self-adjoint operator A.

Now we prove (25.18). We use the definition (25.28) and the fact A_λ is a bounded operator and therefore e^{-itA_λ} satisfies (25.18). Formally bringing the differentiation into the limit, we obtain, for $\phi \in D(H)$,

$$i\hbar \frac{\partial}{\partial t} \langle \phi, e^{-itA}\psi_0 \rangle = i \frac{\partial}{\partial t} \lim_{\lambda \to \infty} \langle \phi, e^{-itA}\psi_0 \rangle = i \lim_{\lambda \to \infty} \langle \phi, \frac{\partial}{\partial t} e^{-itA}\psi_0 \rangle$$

$$= \lim_{\lambda \to \infty} \langle \phi, A_\lambda e^{-iA_\lambda t}\psi_0 \rangle = \lim_{\lambda \to \infty} \langle A_\lambda \phi, e^{-iA_\lambda t}\psi_0 \rangle$$

$$= \langle A\phi, e^{-iAt}\psi_0 \rangle.$$

This exchange of limits is readily justified (the reader is invited to supply the details). Furthermore, if $\psi_0 \in D(A)$, one can show that $e^{-iAt}\psi_0 \in D(A)$, and therefore $i\frac{\partial}{\partial t}e^{-itA}\psi_0 = Ae^{-itA}\psi_0$. So $\psi(t) := e^{-itA_\lambda}\psi_0$ satisfies (25.16) and therefore (25.18) holds.

Next, clearly $U(0) = 1$. Moreover, for any $\psi_0 \in D(A)$,

$$U(t)\psi_0 - \psi_0 = i\int_0^t U(s)A\psi_0 ds \to 0 \text{ as } t \to 0.$$

Hence (25.19) holds.

To prove (25.21), we observe that by the self-adjointness of A, the derivative of the square of the l.h.s. of (25.21) w.r.to t is 0. Hence, it is independent of t and therefore (25.21) follows.

This shows that the initial value problem (25.18)-(25.19) has a unique solution. Indeed, if $U_i(t), i = 1, 2$, satisfy (25.18)-(3.11), then $U(t) := U_1(t) - U_2(t)$ satisfies (25.18) with the initial condition $U(t = 0) = 0$, which by (25.21) implies that $U(t) = 0$.

To prove that $U(t)$ has the group property (25.20), we notice that both sides of equation (25.20) satisfy the same differential equation (in t) with the same initial condition. Hence, by the uniqueness, they are equal.

Another way to prove (25.20) is to use representation (25.28). By Problem 2.17, $e^{-iA_\lambda t}$ has the group property (25.20). Given this result, for any $\psi, \phi \in \mathcal{H}$,

$$\langle \psi, U(t)U(s)\phi \rangle = \langle U(t)^*\psi, U(s)\phi \rangle = \lim_{\lambda \to \infty}\langle e^{iA_\lambda t}\psi, e^{-iA_\lambda s/\hbar}\phi \rangle$$

$$= \lim_{\lambda \to \infty}\langle \psi, e^{-iA_\lambda(t+s)}\phi \rangle = \langle \psi, U(t+s)\phi \rangle$$

which proves (25.20). $\square$

The theorem above has the following corollary

Corollary 25.28 *If A is self-adjoint, then the Cauchy problem (25.16)-(25.17) has a unique solution which conserves probability.*

Indeed, the family $\psi(t) := U(t)\psi_0$ is a solution of the Cauchy problem (25.16)-(25.17), which conserves the probability. It is the unique solution of (25.16)-(25.17), since, if there are two solutions, then their difference, $\tilde{\psi}$, solves (25.16) with $\tilde{\psi}|_{t=0} = 0$, and therefore by conservation of probability (a consequence of symmetry of A), $\|\tilde{\psi}(t)\| = \|\tilde{\psi}(0)\| = 0$ for all t and hence $\tilde{\psi} \equiv 0$.

The operator family $U(t) := e^{-itA}$ is called the *propagator* or *evolution operator* generated by A, or for the equation (25.16). The properties recorded in the equations (25.20) and (25.21) are called the *group* and *isometry* properties. The operator $U(t) = e^{-itA}$ is invertible (since $U(t)U(-t) = 1$) and is isometry (i.e. $\|U(t)\psi\| = \|\psi\|$). Such operators are called *unitary*. More precisely,

Definition 25.29 An operator U is called *unitary* if it preserves the inner product: $\langle U\psi, U\phi \rangle = \langle \psi, \phi \rangle$, for all $\psi, \phi \in \mathcal{H}$, or what is the same, $UU^* = U^*U = 1$.

To show that $U(t) = e^{-itA}$ is unitary, we observe that $\|U(t)\psi\| = \|\psi\|$ implies $\langle \psi, \psi \rangle = \langle U(t)\psi, U(t)\psi \rangle = \langle \psi, U^*(t)U(t)\psi \rangle$, from which $U^*(t)U(t) = 1$ follows. Similarly, $U(t)U^*(t) = 1$. So $U^*(t) = U(-t) = (U(t))^{-1}$. This if A is self-adjoint, the operator $U(t) := e^{-iAt}$ exists and is unitary for all $t \in \mathbb{R}$.

The following very simple example illustrates the connection between unitarity and self-adjointness.

Example 25.30 If $\phi : \mathbb{R}^d \to \mathbb{R}$ is continuous, then the bounded operator

$$U : \psi \mapsto e^{i\phi}\psi$$

is easily checked to be unitary on $L^2(\mathbb{R}^d)$ (just note that U^* is multiplication by $e^{-i\phi}$). Now ϕ is bounded as a multiplication operator iff it is a bounded function. Note, however, that U is well-defined (and unitary) even if ϕ is unbounded.

Remark 25.31 If A is a *positive* operator, then we can define the operator e^{-A} in a way similar to our definition of e^{iA} above. We take $e^{-A} := \lim_{\lambda \to \infty} e^{-A_\lambda}$ where $A_\lambda = (A + \lambda)^{-1}\lambda A$ is a family of bounded operators.

To conclude this section we describe an important extension of Theorem 25.27 to the situation when the operator A is time-dependent, $A = A(t)$. Our goal now is to solve the abstract Schrödinger equation

$$i\frac{\partial \psi}{\partial t} = A(t)\psi, \qquad \psi|_{t=0} = \psi_0, \tag{25.29}$$

where $\psi : t \to \psi(t)$ is a path in a Hilbert space $\mathcal{H}$, for every t, $A(t)$ is a self-adjoint operator on $\mathcal{H}$ and $\psi_0 \in \mathcal{H}$. We have the following result

Theorem 25.32 Let $A(t) = A + W(t)$, where A and W are self-adjoint operators and, in addition, W is bounded. Then there is a unique two-parameter family of bounded operators, $U(t, s)$ with the following properties, for $t, s \in \mathbb{R}$,

$$i\frac{\partial}{\partial t}U(t, s) = A(t)U(t, s), \tag{25.30}$$

$$U(s, s) = 1 \text{ and } U(t, s)\psi \to \psi, \text{ as } t \to s, \tag{25.31}$$

$$U(t, s)U(s, r) = U(t, r), \tag{25.32}$$

$$\|U(t, s)\psi\| = \|\psi\|. \tag{25.33}$$

Proof. We construct the two-parameter family of bounded operators, $U(t, s)$ satisfying (25.30) - (25.31). First, we pass to the interaction representation by introducing the two-parameter family $V(t, s) := e^{i(t-s)A}U(t, s)$. This family satisfies the differential equation

$$i\frac{\partial}{\partial t}V(t, s) = W(t, s)V(t, s), \qquad V(s, s) = 1, \tag{25.34}$$

where $W(t,s) := e^{i(t-s)A}W(t)e^{-i(t-s)A}$. Now, we integrate (25.34) in t from s to t and use the fundamental theorem of calculus to obtain

$$V(t,s) = 1 + i\int_s^t W(r,s)V(r,s)dr.$$

We construct a unique solution to this equation. Iterating this equation, we arrive at the series (*Dyson series*)

$$U(t,s) = 1 + \sum_{n=1}^{\infty} U_n(t,s). \qquad (25.35)$$

where $U_n(t,s) := i^n \int_{\Delta(s,t)} d^n r W(r_1,s)\dots W(r_n,s)$, with $\Delta(s,t) := \{(r_1, r_2, \dots r_n$ $s \le r_n \le \dots \le r_2 \le r_1 \le t\}$. Now, using that $W(r,s)$ are bounded (in fact, $\|W(r,s)\| \le \|W(r)\|$), we estimate

$$\|U_n(t,s)\| \le \int_{\Delta(s,t)} d^n r \prod_1^n \|W(r_j)\|.$$

Next, the relation

$$\int_{\Delta(s,t)} d^n r \prod_1^n \|W(r_j)\| = \frac{1}{n!}\int_{[0,t]^n} dr_1 \dots dr_n \prod_1^n \|W(r_j)\|$$

$$= \frac{1}{n!}(\int_0^t dr\|W(r)\|)^n$$

shows that the series above converges in norm and represents a family of bounded operators. Representation (25.35) shows that it satisfies (25.30) - (25.31).

Similarly, one shows that $U(t,s)$ satisfies also the equation

$$i\frac{\partial}{\partial s}U(t,s) = U(t,s)A(s). \qquad (25.36)$$

Using (25.30) and (25.36), one checks easily that the derivative of the l.h.s. of (25.32) w.r.to s is zero. Hence, it is independent of s and therefore (25.32) follows.

Finally, differentiating the square of the l.h.s. of (25.33) w.r.to t, we see that it is independent of t and therefore (25.33) follows.   □

A two-parameter family of bounded operators, $U(t,s)$, satisfying (25.30) - (25.33) is called the *propagator*, or *evolution*, generated by $A(t)$. (Recall that, if A is t-independent then the same term is applied to $U(t) = U(t,0) = e^{-iAt}$.

Problem 25.33 Let $A(t) = A + W(t)$, where A and W are self-adjoint operators on a Hilbert space $\mathcal{H}$ and, in addition, W is bounded. Prove directly (i.e without using Theorem 25.32, but using the ideas of its proof) that the abstract Schrödinger initialvalue problem

$$i\frac{\partial}{\partial t}\psi(t) = A(t)\psi(t), \ \psi(t=0) = \psi_0, \tag{25.37}$$

where $\psi(\cdot) : \mathbb{R} \to \mathcal{H}$ is a differentiable vector function, has a unique solution for every $\psi_0 \in \mathcal{H}$. Show that the solution satisfies $\|\psi(t)\| = \|\psi_0\|$.

25.7 Projections

Let $\mathcal{H}$ be a Hilbert space. A bounded operator P on $\mathcal{H}$ is called a *projection operator* (or simply a *projection*) if it satisfies

$$P^2 = P.$$

This relation implies $\|P\| \leq \|P\|^2$, and so $\|P\| \geq 1$ provided $P \neq 0$. We have

$$v \in \text{Ran}P \iff Pv = v \quad \text{and} \quad v \in (\text{Ran}P)^\perp \iff P^*v = 0. \tag{25.38}$$

Indeed, if $v \in \text{Ran}P$, then there is a $u \in \mathcal{H}$ s.t. $v = Pu$, so $Pv = P^2u = Pu = v$; the second statement is left as an exercise.

Problem 25.34 Prove that (a) $P^*v = 0$ if and only if $v \perp \text{Ran}P$, (b) $\text{Ran}P$ is closed and (c) P^* is also a projection.

Example 25.35 The following are projection operators:

1. let $\mathcal{H} = L^2(\mathbb{R}^d)$ and let E be a subset of $\mathbb{R}^d$. Then

$$\chi_{x\in E} : f(x) \mapsto \chi_E(x)f(x)$$

   where

$$\chi_E(x) := \begin{cases} 1 & x \in E \\ 0 & x \notin E \end{cases}$$

   is a projection.
2. again let $\mathcal{H} = L^2(\mathbb{R}^d)$ and let E be a subset of $\mathbb{R}^d$. Then

$$\chi_{p\in E} = \mathcal{F}^{-1}\chi_{k\in E}\,\mathcal{F} : u(x) \mapsto (\chi_E(k)\hat{u}(k))\check{}(x)$$

   is a projection.
3. let $\mathcal{H}$ be any Hilbert space, and $\varphi, \psi \in \mathcal{H}$ satisfying $\langle \varphi, \psi \rangle = 1$. Then

$$f \mapsto \langle \varphi, f \rangle \psi$$

   is a projection.

4. as in 3), but now let $\{\psi_i\}_{i=1}^{N}$ be an orthonormal set (i.e. $\langle\psi_i,\psi_j\rangle = \delta_{ij}$). Then

$$f \mapsto \sum_{1}^{N} \langle\psi_i, f\rangle\psi_i$$

is a projection.

Definition 25.36 1. A projection P is said to be of *rank* $r < \infty$ if $\dim \mathrm{Ran}P = r$.
2. A projection P is called an *orthogonal projection* if it is self-adjoint, i.e. if $P = P^*$.

If P be an orthogonal projection, then (25.38) implies that

$$v \perp \mathrm{Ran}P \iff Pv = 0, \text{ i.e. Null}P = (\mathrm{Ran}P)^{\perp}. \qquad (25.39)$$

The projections in Examples 1), 2) and 4) above are orthogonal. The projection in Example 3) is orthogonal if and only if $\varphi = \psi$.

Problem 25.37 Let P be an orthogonal projection. Show that

1. $\|P\| \leq 1$, and therefore $\|P\| = 1$ if $P \neq 0$ (Hint: Use (25.38));
2. $1 - P$ is also an orthogonal projection, $\mathrm{Ran}(1 - P)\perp\mathrm{Ran}P$, and $\mathrm{Null}(1 - P) = \mathrm{Ran}P$;
3. $\mathcal{H} = \mathrm{Ran}P \oplus \mathrm{Null}P$.

Remark 25.38 Orthogonal projections on $\mathcal{H}$ are in one-to-one correspondence with closed subspaces of a Hilbert space $\mathcal{H}$. This correspondence is obtained as follows. Let $V = \mathrm{Ran}P$. Then V is a closed subspace of X. To show that V is closed, let $\{v_n\} \subset V$, and $v_n \to v \in X$, and show that $v \in V$. Since P is a projection, we have $v_n = Pv_n$, so $\|v - Pv\| = \|v - v_n - P(v - v_n)\| \leq \|v - v_n\| + \|P\|\,\|v - v_n\| \to 0$, as $n \to \infty$. Therefore $v = Pv$, so $v \in V$, and V is closed. Conversely, given a closed subspace V, define a projection operator P by

$$Pu = v, \quad \text{where} \quad u = v + v^{\perp} \in V \oplus V^{\perp}. \qquad (25.40)$$

Problem 25.39 Show that P defined in (25.40) is an orthogonal projection with $\mathrm{Ran}P = V$. For any given V, show that there is only one orthogonal projection (the one given in (25.40)) such that $\mathrm{Ran}P = V$.

25.8 The Spectrum of an Operator

Again to keep this section self-contained we repeat some definitions and results from Section 6.1 of the main text.

Definition 25.40 The *spectrum* of an operator A on a Hilbert space $\mathcal{H}$ is the subset of $\mathbb{C}$ given by

$$\sigma(A) := \{\lambda \in \mathbb{C} \mid A - \lambda \text{ is not invertible (has no bounded inverse)}\}$$

(here and below, $A - \lambda$ denotes $A - \lambda\mathbf{1}$). The complement of the spectrum of A in $\mathbb{C}$ is called the *resolvent set* of A: $\rho(A) := \mathbb{C}\backslash\sigma(A)$. For $\lambda \in \rho(A)$, the operator $(A - \lambda)^{-1}$, called the *resolvent* of A, is well-defined.

The following exercise asks for the spectrum of our favourite operators.

Problem 25.41 Prove that as operators on $L^2(\mathbb{R}^d)$ (with their natural domains),

1. $\sigma(1) = \{1\}$.
2. $\sigma(p_j) = \mathbb{R}$.
3. $\sigma(x_j) = \mathbb{R}$.
4. $\sigma(V) = \overline{\text{range}(V)}$, where V is the multiplication operator on $L^2(\mathbb{R}^d)$ by a continuous function $V(x) : \mathbb{R}^d \to \mathbb{C}$.
5. $\sigma(-\Delta) = [0, \infty)$.
6. $\sigma(f(p)) = \overline{\text{range}(f)}$, where $f(p) := \mathcal{F}^{-1} f \mathcal{F}$ with $f(k)$, the multiplication operator on $L^2(\mathbb{R}^d)$ by a continuous function $f(k) : \mathbb{R}^d \to \mathbb{C}$.

Theorem 25.42 The spectrum $\sigma(A)$ is a closed set.

Proof. We show that the complement of the spectrum, $\rho(A) := \mathbb{C}/\sigma(A)$, called the resolvent set, is an open set. Indeed, let $z \in \rho(A)$. Then $A - z$ is invertible (has a bounded inverse) and therefore by Theorem 25.17, so is $A - z' = (A - z)[1 + (z - z')(A - z)^{-1}]$, if $|z' - z| < \|(A - z)^{-1}\|^{-1}$. $\square$

We observe that self-adjoint operators have real spectrum.

Theorem 25.43 If $A = A^*$, then $\sigma(A) \subset \mathbb{R}$.

Proof. This follows immediately from Lemma 25.21. $\square$

One familiar reason for $A - \lambda$ not to be invertible is that $(A - \lambda)\psi = 0$ has a non-zero solution $\psi \in D(A) \subset \mathcal{H}$. In this case we say that λ is an *eigenvalue* of A and ψ is called a corresponding *eigenvector*.

Definition 25.44 The *discrete spectrum* of an operator A is

$$\sigma_d(A) = \{\lambda \in \mathbb{C} \mid \lambda \text{ is an isolated eigenvalue of } A \text{ with finite multiplicity}\}$$

(isolated meaning some neighbourhood of λ is disjoint from the rest of $\sigma(A)$).

Here the *multiplicity* of an eigenvalue λ is the dimension of the *eigenspace*

$$\text{Null}(A - \lambda) := \{v \in \mathcal{H} \mid (A - \lambda)v = 0\}.$$

Problem 25.45 1. Show $\text{Null}(A - \lambda)$ is a vector space.
2. Show that if A is self-adjoint, eigenvectors of A corresponding to different eigenvalues are orthogonal.

The rest of the spectrum is called the *essential spectrum* of the operator A:

$$\sigma_{ess}(A) := \sigma(A)\backslash\sigma_d(A).$$

Remark 25.46 Some authors may use the terms "point spectrum" and "continuous spectrum" rather than (respectively) "discrete spectrum" and "essential spectrum'.

Problem 25.47 For the operators x_j and p_j on $L^2(\mathbb{R}^d)$ show that

1. $\sigma_{ess}(p_j) = \sigma(p_j) = \mathbb{R}$;
2. $\sigma_{ess}(x_j) = \sigma(x_j) = \mathbb{R}$;
3. $\sigma_{ess}(-\Delta) = \sigma(-\Delta) = [0, \infty)$.

Hint: Show that these operators do not have discrete spectrum.

Problem 25.48 Show that if $U : \mathcal{H} \to \mathcal{H}$ is unitary, then $\sigma(U^*AU) = \sigma(A)$, $\sigma_d(U^*AU) = \sigma_d(A)$, and $\sigma_{ess}(U^*AU) = \sigma_{ess}(A)$.

Problem 25.49 Let A be an operator on $\mathcal{H}$. If λ is an accumulation point of $\sigma(A)$, then $\lambda \in \sigma_{ess}(A)$. Hint: use the definition of the essential spectrum and Theorem 25.42.

For a self-adjoint operator A the sets {span of eigenfunctions of A} and {span of eigenfunctions of A}$^\perp$, where

$$W^\perp := \{\psi \in \mathcal{H} \mid \langle \psi, w \rangle = 0 \ \forall \, w \in W\},$$

are invariant under A in the sense of the definition

Definition 25.50 A subspace $W \subset \mathcal{H}$ of a Hilbert space $\mathcal{H}$ is *invariant* under an operator A if $Aw \in W$ whenever $w \in W \cap D(A)$.

Problem 25.51 Assume A is a self-adjoint operator. Show that

1. If W is invariant under A, then so is $W^\perp$;
2. The span of the eigenfunctions of A and its orthogonal complement are invariant under A;

25.9 Functions of Operators and the Spectral Mapping Theorem

Our goal in this section is to define functions $f(A)$ of a self-adjoint operator A. We do this in the special case where A is bounded, and f is a function analytic in a neighbourhood of $\sigma(A)$.

Problem 25.52 Let A be a bounded self-adjoint operator. Show $\sigma(A) \subset [-\|A\|, \|A\|]$. Hint: use that if $|z| > \|A\|$, then $\|(A - z)u\| \geq (|z| - \|A\|)\|u\|$.

Suppose $f(\lambda)$ is analytic in a complex disk of radius R, $\{\lambda \in \mathbb{C} \mid |\lambda| < R\}$, with $R > \|A\|$. So f has a power series expansion, $f(\lambda) = \sum_{n=0}^{\infty} a_n \lambda^n$, which converges for $|\lambda| < R$. Define the operator $f(A)$ by the convergent series

$$f(A) := \sum_{n=0}^{\infty} a_n A^n. \tag{25.41}$$

We have already encountered an example of this definition: the exponential e^A discussed in Section 2.3. As another example, consider the function $f(\lambda) = (\lambda - z)^{-1}$, for $|z| > \|A\|$, which is analytic in a disk of radius R, with $\|A\| < R < |z|$, and has power-series expansion

$$f(\lambda) = -\frac{1}{z}\frac{1}{1 - \lambda/z} = -\frac{1}{z}\sum_{j=0}^{\infty}(\lambda/z)^j.$$

The corresponding operator defined by (25.41) is the resolvent

$$(A - z)^{-1} = -\frac{1}{z}\sum_{j=0}^{\infty} z^{-j} A^j. \tag{25.42}$$

(Recall that the series in (25.42) is a Neumann series.) Of course, the resolvent is defined for any z in the resolvent set $\rho(A) = \mathbb{C}\backslash\sigma(A)$. In fact, $(A - z)^{-1}$ is an analytic (operator-valued) function of $z \in \rho(A)$. To see this, we start with the relation

$$(A - z)^{-1} = (A - z_0)^{-1} - (z_0 - z)(A - z_0)^{-1}(A - z)^{-1}$$

for $z, z_0 \in \rho(A)$, which the reader is invited to verify (this relation is called the *first resolvent identity*). Thus

$$(A - z)^{-1} = [1 - (z - z_0)(A - z_0)^{-1}]^{-1}(A - z_0)^{-1}.$$

If $|z - z_0| < (\|(A - z_0)^{-1}\|)^{-1}$, the first inverse on the right hand side can be expanded in a Neumann series, yielding

$$(A - z)^{-1} = \sum_{j=0}^{\infty}(z - z_0)^j (A - z_0)^{-j-1}.$$

Thus $(A - z)^{-1}$ is analytic in a neighbourhood of any $z_0 \in \rho(A)$.

The following useful result relates eigenvalues and eigenfunctions of A to those of $f(A)$.

Theorem 25.53 (Spectral mapping theorem) Let A be a bounded operator, and f a function analytic on a disk of radius $> \|A\|$. If $A\phi = \lambda\phi$, then $f(A)\phi = f(\lambda)\phi$.

Proof. If $\{a_j\}$ are the coefficients of the power series for f, then

$$f(A)\phi = \sum_{j=0}^{\infty} a_j A^j \phi = (\sum_{j=0}^{\infty} a_j \lambda^j)\phi = f(\lambda)\phi.$$

$\square$

We conclude with a brief discussion of alternatives to, and extensions of, the above definition. Suppose f is analytic in a complex neighbourhood of $\sigma(A)$. We can replace definition (25.41) by the contour integral (called **the Riesz integral**)

$$f(A) := \frac{1}{2\pi i} \oint_{\Gamma} f(z)(A - z)^{-1} dz \tag{25.43}$$

where Γ is a contour in $\mathbb{C}$ encircling $\sigma(A)$. The integral here can be understood in the following sense: for any $\psi, \phi \in \mathcal{H}$,

$$\langle \psi, f(A)\phi \rangle := \frac{1}{2\pi i} \oint_{\Gamma} f(z)\langle \psi, (A - z)^{-1}\phi \rangle dz$$

(knowledge of $\langle \psi, f(A)\phi \rangle$ for all ψ, ϕ determines the operator $f(A)$ uniquely).

Problem 25.54 Show that the definition (25.43) agrees with (25.41) when $f(\lambda)$ is analytic on $\{|\lambda| < R\}$ with $R > \|A\|$. Hint: by the Cauchy theorem, and analyticity of the resolvent, the contour Γ can be replaced by $\{|z| = R_0\}$, $\|A\| < R_0 < R$. On this contour, $(A - z)^{-1}$ can be expressed as the Neumann series (25.42).

A similar formula can be used for unbounded operators A to define certain functions $f(A)$ (see [162]).

If A is an unbounded self-adjoint operator, and f is a continuous, bounded function, the bounded operator $f(A)$ can still be defined. One example of this is the definition of e^{iA} in Section 25.6. The definition of $f(i\nabla)$ using the Fourier transform (Section 25.14) provides another example. More generally, the Fourier transform, together with Theorem 25.27, allows one to define functions of self-adjoint operators as follows:

Definition 25.55 Assume A is a self-adjoint operator and $f(\lambda)$ is a function whose inverse Fourier transform, $\check{f}$ is integrable, $\int |\check{f}(t)| dt < \infty$. Then the operator

$$f(A) := (2\pi\hbar)^{-1/2} \int_{-\infty}^{\infty} \check{f}(t) e^{iAt/\hbar} dt \tag{25.44}$$

is well-defined, bounded, and is self-adjoint if f is real.

We present without justification a formula connecting the equations (25.43) and (25.44) (i.e. connecting the propagator and the resolvent):

$$e^{-iAt} f(A) = \frac{1}{\pi} \int_{-\infty}^{\infty} d\lambda f(\lambda) e^{-i\lambda t} \text{Im}(A - \lambda - i0)^{-1}. \tag{25.45}$$

The reader is referred to [244] for the general theory.

25.10 Weyl Sequences and Weyl Spectrum

We now want to address the question of how to characterize the essential spectrum of a self-adjoint operator A. Is there a characterization of $\sigma_{ess}(A)$ similar to that of $\sigma_d(A)$ in terms of some kind of eigenvalue problem? To answer this question, we observe that there is another reason for $A-\lambda$ not to be invertible, besides λ being an eigenvalue of A, namely $(A - \lambda)\psi = 0$ "almost" having a non-zero solution. More precisely, there a sequence $\{\psi_n\} \subset \mathcal{H}$ s.t.

1. $\|\psi_n\| = 1$ for all n
2. $\|(A - \lambda)\psi_n\| \to 0$ as $n \to \infty$
3. $\psi_n \to 0$ *weakly* as $n \to \infty$ (this means $\langle \phi, \psi_n \rangle \to 0$ for all $\phi \in \mathcal{H}$).

We say $\{\psi_n\} \subset \mathcal{H}$ is a *Weyl sequence* for A and λ if these statements hold.

Definition 25.56 The *Weyl spectrum* of an operator A is

$$\sigma_w(A) = \{\lambda \mid \text{there is a Weyl sequence for } A \text{ and } \lambda\}.$$

The following result says that when A is self-adjoint, the sets $\sigma_d(A)$ and $\sigma_w(A)$ are disjoint, and comprise the whole spectrum:

Theorem 25.57 (Weyl) If A is self-adjoint, then $\sigma_{ess}(A) = \sigma_w(A)$, and therefore the spectrum of A is the disjoint union of the discrete spectrum of A and the Weyl spectrum of A:

$$\sigma(A) = \sigma_d(A) \cup \sigma_w(A).$$

Proof. Suppose first that $\lambda \in \sigma_{ess}(A)$. Then $\inf_{\|\psi\|=1, \psi \in D(A)} \|(A - \lambda)\psi\| = 0$, for otherwise $A - \lambda$ would be invertible. Hence there is a sequence $\psi_n \in D(A)$ such that $\|\psi_n\| = 1$ and $\|(A - \lambda)\psi_n\| \to 0$ as $n \to \infty$. By the Banach-Alaoglu theorem (see, eg., [244]), there is a subsequence $\{\psi_{n'}\} \subset \{\psi_n\}$ and an element $\psi_0 \in \mathcal{H}$ such that $\psi_{n'} \to \psi_0$ weakly as $n' \to \infty$ (we drop the prime in n' henceforth). This implies that for all $f \in D(A)$,

$$\langle (A - \lambda)f, \psi_0 \rangle = \lim_{n \to \infty} \langle (A - \lambda)f, \psi_n \rangle = \lim_{n \to \infty} \langle f, (A - \lambda)\psi_n \rangle = 0.$$

Hence $\psi_0 \in D(A)$ (since $D(A) = D(A^*) = \{\psi \in \mathcal{H} \mid |\langle Af, \psi \rangle| \leq C\|f\| \, \forall f \in D(A)\}$ and $|\langle Af, \psi_0 \rangle| = |\lambda \langle f, \psi_0 \rangle| \leq |\lambda| \|\psi_0\| \|f\|$) and $A\psi_0 = \lambda\psi_0$. If $\psi_0 = 0$, then $\{\psi_n\}$ is a Weyl sequence for A and λ, and so $\lambda \in \sigma_w(A)$. If $\psi_0 \neq 0$, this implies that λ is an eigenvalue of A. So λ must therefore have infinite multiplicity or be non-isolated. If λ is an eigenvalue of infinite multiplicity, then an orthonormal basis of Null$(A - \lambda)$ yields a Weyl sequence for A and λ, and therefore $\lambda \in \sigma_w(A)$. If λ is not isolated, then consider a sequence $\lambda_j \in \sigma(A) \setminus \{\lambda\}$ with $\lambda_j \to \lambda$. If there is a subsequence consisting of distinct eigenvalues, a corresponding sequence of normalized eigenvectors is orthonormal, and so converges weakly to 0 – hence it is a Weyl sequence for A and λ.

On the other hand, if the sequence λ_j consists (eventually) of non-eigenvalues, then, arguing as above for each λ_j, one can construct a diagonal sequence which is Weyl for A and λ. So we conclude $\lambda \in \sigma_w(A)$, and we have shown that $\sigma_{ess}(A) \subset \sigma_w(A)$. Now suppose $\lambda \in \sigma_w(A)$, and let ψ_n be a corresponding Weyl sequence. Then certainly $\lambda \in \sigma(A)$, otherwise

$$\|\psi_n\| = \|(A-\lambda)^{-1}(A-\lambda)\psi_n\| \leq \|(A-\lambda)^{-1}\|\|(A-\lambda)\psi_n\| \to 0,$$

a contradiction. Suppose λ is an isolated eigenvalue of finite multiplicity. For simplicity, suppose the multiplicity is one (the argument is straightforward to generalize), and let ψ_0 be a normalized eigenvector. Write $\psi_n = c_n\psi_0 + \tilde{\psi}_n$ with $c_n = \langle \psi_0, \psi_n \rangle$ and $\langle \tilde{\psi}_n, \psi_0 \rangle = 0$. Since $\psi_n \to 0$ weakly, $c_n \to 0$, and so $\|\tilde{\psi}_n\| \to 1$. Also $(A-\lambda_n)\tilde{\psi}_n \to 0$. Because λ is isolated in the spectrum, $(A-\zeta)^{-1}$ is uniformly bounded on $(\psi_0)^\perp$ for ζ near λ, and so

$$\|\tilde{\psi}_n\| = \|(A-\lambda_n)^{-1}(A-\lambda_n)\tilde{\psi}_n\|$$
$$\leq (const)\|(A-\lambda_n)\tilde{\psi}_n\| \to 0,$$

a contradiction. Thus $\lambda \in \sigma_{ess}(A)$, showing $\sigma_w(A) \subset \sigma_{ess}(A)$ and completing the proof of Theorem 25.57.

As an application of the Weyl theorem we consider a Schrödinger operator on a bounded domain, with Dirichlet boundary conditions.

Theorem 25.58 Let Λ be a cube in $\mathbb{R}^d$, and V a continuous function on Λ. Then the Schrödinger operator $H = -\Delta + V$, acting on the space $L^2(\Lambda)$ with Dirichlet boundary conditions, has purely discrete spectrum, accumulating at $+\infty$.

To be precise, the operator "H on $L^2(\Lambda)$ with Dirichlet boundary conditions" should be understood as the unique self-adjoint extension of H from $C_0^\infty(\Lambda)$.

Proof. Suppose $\Lambda = [0, L]^d$. Consider the normalized eigenfunctions of the operator $-\Delta$ on $L^2(\Lambda)$ with Dirichlet boundary conditions:

$$\phi_k(x) = \left(\frac{2}{L}\right)^{d/2} \prod_{j=1}^d \sin(k_j x_j), \qquad k \in \frac{\pi}{L}(\mathbb{Z}^+)^d$$

(see Section 7.1), so that

$$-\Delta\phi_k = |k|^2 \phi_k. \tag{25.46}$$

Now we recall that the eigenfunctions ϕ_k, $k \in \frac{\pi}{L}(\mathbb{Z}^+)^d$, form an orthonormal basis for $L^2(\Lambda)$:

$$\psi = \sum_{k \in \frac{\pi}{L}(\mathbb{Z}^+)^d} \langle \phi_k, \psi \rangle \phi_k$$

for any $\psi \in L^2(\Lambda)$ (this is a special case of a general phenomenon valid for self-adjoint operators).

We show now that the operator H has no essential spectrum. Assume on the contrary that $\lambda \in \sigma_{ess}(H)$, and let u_n be a corresponding Weyl sequence; i.e. $\|u_n\| = 1$, $u_n \to 0$ weakly, and $\|(H - \lambda)u_n\| \to 0$. By the triangle inequality

$$\|(H - \lambda)u_n\| \geq \|(-\Delta - \lambda)u_n\| - \|Vu_n\| \geq \|(-\Delta - \lambda)u_n\| - \max|V| \quad (25.47)$$

since $\|u_n\| = 1$. Writing

$$u_n = \sum_k a_n^k \phi_k$$

where $a_n^k = \langle \phi_k, u_n \rangle$, and using (25.46), we compute

$$(-\Delta - \lambda)u_n = \sum_k (|k|^2 - \lambda)a_n^k \phi_k$$

and so by the Parseval relation (Proposition 25.4)

$$\|(-\Delta - \lambda)u_n\|^2 = \sum_k (|k|^2 - \lambda)^2 |a_n^k|^2. \quad (25.48)$$

The Parseval relation also gives

$$1 = \|u_n\|^2 = \sum_k |a_n^k|^2. \quad (25.49)$$

Now choose K such that $|k|^2 - \lambda \geq \sqrt{2}(\max|V| + 1)$ for $|k| \geq K$. Then by (25.48) and (25.49),

$$\|(-\Delta - \lambda)u_n\|^2 \geq 2(\max|V| + 1)^2 \sum_{|k| \geq K} |a_n^k|^2 = 2(\max|V| + 1)^2 (1 - \sum_{|k| < K} |a_n^k|^2).$$

Since $u_n \to 0$ weakly,

$$a_n^k = \langle \phi_k, u_n \rangle \to 0$$

as $n \to \infty$, for each k. Choose N sufficiently large that $|a_n^k| \leq (2K_\#)^{-1/2}$ for k with $|k| < K$ and $n \geq N$, where

$$K_\# := \#\left\{ k \in \frac{\pi}{L}(\mathbb{Z}^+)^d \mid |k| < K \right\}.$$

Then for $n \geq N$, $\sum_{|k| < K} |a_n^k|^2 \leq 1/2$, and so

$$\|(-\Delta - \lambda)u_n\|^2 \geq (\max|V| + 1)^2.$$

Returning to (25.47), we conclude that for $n \geq N$, $\|(H - \lambda)u_n\| \geq 1$, which contradicts the property $\|(H - \lambda)u_n\| \to 0$. Hence no finite λ can be a point of the essential spectrum of H.

Proceeding as in the proof of Theorem 6.18, one can show that H has an infinite number of eigenvalues which accumulate at ∞. $\square$

Next, we present a result characterizing the essential spectrum of a Schrödinger operator in a manner similar to the characterization of the discrete spectrum as a set of eigenvalues.

Theorem 25.59 (Schnol-Simon) Let H be a Schrödinger operator with a bounded potential. Then

$$\sigma(H) = \text{closure } \{\lambda \mid (H - \lambda)\psi = 0 \text{ for } \psi \text{ polynomially bounded }\}.$$

So we see that the essential spectrum also arises from solutions of the eigenvalue equation, but that these solutions do not live in the space $L^2(\mathbb{R}^3)$.

Proof. We prove only that the right hand side $\subset \sigma(H)$, and refer the reader to [73] for a complete proof. Let ψ be a polynomially bounded solution of $(H - \lambda)\psi = 0$. Let C_r be the box of side-length $2r$ centred at the origin. Let j_r be a smooth function with support contained in C_{r+1}, with $j_r \equiv 1$ on C_r, $0 \leq j_r \leq 1$, and with $\sup_{r,x,|\alpha|\leq 2} |\partial_x^\alpha j_r(x)| < \infty$. Our candidate for a Weyl sequence is

$$w_r := \frac{j_r \psi}{\|j_r \psi\|}.$$

Note that $\|w_r\| = 1$. If $\psi \notin L^2$, we must have $\|j_r\psi\| \to \infty$ as $r \to \infty$. So for any R,

$$\int_{|x|<R} |w_r|^2 \leq \frac{1}{\|j_r\psi\|^2} \int_{|x|<R} |\psi|^2 \to 0$$

as $r \to \infty$. We show that

$$(H - \lambda)w_r \to 0.$$

Let $F(r) = \int_{C_r} |\psi|^2$, which is monotonically increasing in r. We claim there is a subsequence $\{r_n\}$ such that

$$\frac{F(r_n + 2)}{F(r_n - 1)} \to 1.$$

If not, then there is $a > 1$ and $r_0 > 0$ such that

$$F(r + 3) \geq aF(r)$$

for all $r \geq r_0$. Thus $F(r_0 + 3k) \geq a^k F(r_0)$ and so $F(r) \geq (const)b^r$ with $b = a^{1/3} > 1$. But the assumption that ψ is polynomially bounded implies that $F(r) \leq (const)r^N$ for some N, a contradiction. Now,

$$(H - \lambda)j_r\psi = j_r(H - \lambda)\psi + [-\Delta, j_r]\psi.$$

Since $(H - \lambda)\psi = 0$ and $[\Delta, j_r] = (\Delta j_r) + 2\nabla j_r \cdot \nabla$, we have

$$(H - \lambda)j_r\psi = (-\Delta j_r)\psi - 2\nabla j_r \cdot \nabla\psi.$$

Since $|\partial^\alpha j_r|$ is uniformly bounded,

$$\|(H - \lambda)j_r\psi\| \le (const)\int_{C_{r+1}\setminus C_r}(|\psi|^2 + |\nabla\psi|^2) \le (const)\int_{C_{r+1}\ C_r}|\psi|^2.$$

So

$$\|(H - \lambda)w_r\| \le C\frac{F(r + 2) - F(r - 1)}{F(r)} \le C(\frac{F(r + 2)}{F(r - 1)} - 1)$$

and so $\|(H - \lambda)w_{r_n}\| \to 0$. Thus $\{w_{r_n}\}$ is a Weyl sequence for H and λ. $\square$

In the rest of this section, by way of illustration, we construct Weyl sequences for the coordinate and momentum operators. We assume, for simplicity, $d = 1$.

Thus for any $\lambda \in \mathbb{R}$, we will find a Weyl sequence for x and λ. This sequence is such that its square approximates the delta-function $\delta_\lambda(x) = \delta(\lambda - x)$ which formally solves the equation

$$(x - \lambda)\delta_\lambda = 0$$

exactly. Such a sequence is sketched in Fig. 25.1.

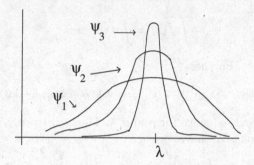

Fig.25.1. Weyl sequence for x, λ.

How do we construct such a sequence ψ_n? Let ϕ be a fixed non-negative function supported on $[-1, 1]$, and such that

$$\int |\phi|^2 = 1.$$

We compress this function, increasing its height, and shift the result to λ:

$$\psi_n(x) := n^{1/2}\phi(n(x - \lambda)).$$

Then

$$\int |\psi_n|^2 = \int |\phi|^2 = 1$$

and

$$\|(x - \lambda)\psi_n\|^2 = \int |x - \lambda|^2 n|\phi(n(x - \lambda))|^2 dx = \frac{1}{n^2}\int |y|^2|\phi(y)|^2 dy \to 0$$

as $n \to \infty$. Thus $\lambda \in \sigma(x)$, at least. Now we show that $\psi_n \to 0$ weakly. Indeed, for any $f \in L^2(\mathbb{R}^d)$,

$$\left| \int \overline{\psi_n} f \right| = \left| \int_{|x-\lambda| \leq 1/n} \overline{\psi_n} f \right| \leq \left(\int |\psi_n|^2 \right)^{1/2} \left(\int_{|x-\lambda| \leq 1/n} |f|^2 \right)^{1/2}$$

which $\to 0$ as $n \to \infty$ by a well-known result of analysis. Thus, $\lambda \in \sigma_{ess}(x)$. It is easy to convince yourself that x has no eigenvalues.

Now we construct a Weyl sequence, $\{\psi_n\}$, for p and λ. Using properties of the Fourier transform, we have

$$\|(p-\lambda)\psi_n\| = \|((p-\lambda)\psi_n)\hat{}\,\| = \|(k-\lambda)\hat{\psi}_n\|.$$

Take for $\hat{\psi}_n$ the Weyl sequence constructed above:

$$\hat{\psi}_n = n^{1/2}\hat{\phi}(n(k-\lambda))$$

for $\hat{\phi}$ supported on $[-1,1]$, and $\int |\hat{\phi}|^2 = 1$. So we have $\|\psi_n\| = \|\hat{\psi}_n\| = 1$ and

$$\int \bar{f}\psi_n = \int \bar{\hat{f}}\hat{\psi}_n \to 0$$

for any $f \in L^2(\mathbb{R}^d)$. Further,

$$\|(k-\lambda)\hat{\psi}_n\| \to 0 \implies \|(p-\lambda)\psi_n\| \to 0$$

and so ψ_n is a Weyl sequence for p and λ. Thus $\sigma(p) = \sigma_{ess}(p) = \mathbb{R}$. Now let us see how ψ_n looks. We have

$$\psi_n(x) = (2\pi\hbar)^{-1/2} \int e^{ik \cdot x/\hbar} n^{1/2}\hat{\phi}(n(k-\lambda))dk = e^{ix \cdot \lambda/\hbar} n^{-1/2}\phi(x/n).$$

Suppose, for example, that $\phi \equiv 1$ for $|x| \leq 1/2$. Then ψ_n looks like a plane wave (with amplitude $n^{-1/2}$ and wave vector λ), cut off near ∞ by $\phi(x/n)$ ($|\psi_n|$ is sketched in Fig. 25.2).

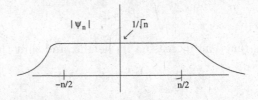

Fig.25.2. Weyl sequence for p, λ.

We remark that the fact $\sigma(p_j) = \sigma_{ess}(p_j) = \mathbb{R}$ also follows directly from the fact $\sigma(x_j) = \sigma_{ess}(x_j) = \mathbb{R}$, together with Problem 25.48 and properties of the Fourier transform.

25.11 The Trace, and Trace Class Operators

This section gives a quick introduction to the notion of the *trace* of an operator, a generalization of the familiar trace of a matrix. More details and proofs can be found in [244], for example.

Let ρ be a bounded operator on a (separable) Hilbert space, $\mathcal{H}$. Since $\rho^*\rho \geq 0$, we can define the positive operator $|\rho| := \sqrt{\rho^*\rho}$ (this operator can be defined by a power series – see [244]; see also Section 25.9). The operator ρ is said to be of *trace class* if

$$\sum_j \langle \psi_j, |\rho|\psi_j \rangle < \infty$$

for some orthonormal basis $\{\psi_j\}$ of $\mathcal{H}$. If ρ is a trace class operator, we define its trace to be

$$\operatorname{Tr}\rho = \sum_j \langle \psi_j, \rho\psi_j \rangle$$

for some orthonormal basis $\{\psi_j\}$ of $\mathcal{H}$. This definition is independent of the choice of basis.

Problem 25.60 Show that the trace is well-defined by showing the that the right-hand side is independent of the choice of basis. Hint: consider another orthonormal basis $\{\phi_j\}$ and let $\psi_i = \sum_j c_{ij}\phi_j$. Show that

$$\sum_i \bar{c}_{ik}c_{il} = \delta_{kl},$$

using the fact that

$$\sum_i \langle \phi_k, \psi_i \rangle \langle \psi_i, \phi_l \rangle = \langle \phi_k, \phi_l \rangle,$$

and then use that to show that

$$\sum_j \langle \psi_j, \rho\psi_j \rangle = \sum_j \langle \phi_j, \rho\phi_j \rangle.$$

Properties of the trace:

1. $\operatorname{Tr}\rho^* = \overline{\operatorname{Tr}\rho}$, and $\operatorname{Tr}\rho \geq 0$ if $\rho \geq 0$.
2. If ρ is trace class and A is bounded, then $A\rho$ and ρA are trace class with $\operatorname{Tr}(A\rho) = \operatorname{Tr}(\rho A)$ (*cyclicity* of the trace).
3. $\operatorname{Tr}(\alpha A + \beta B) = \alpha \operatorname{Tr} A + \beta \operatorname{Tr} B$.
4. If $(Kf)(x) = \int K(x,y)f(y)\,dy$, then $\operatorname{Tr} K = \int K(x,x)\,dx$.
5. If ρ is trace class, then $\sigma_{ess}(\rho) \subset \{0\}$ and $\sum_i |\lambda_i| < \infty$, where λ_i are the eigenvalues of ρ.

The trace class operators form a Banach space under the norm

$$\|\rho\|_1 := \text{Tr}|\rho|,$$

and the trace is a linear functional on this space.

Let us look at a few examples of trace class operators.

The first example is useful in the preceeding sections. Let A be an unbounded, self-adjoint operator, bounded from below, with purely discrete spectrum. Let $E_0 \leq E_1 \leq E_2 \leq \cdots$ be the eigenvalues of A (since there is no essential spectrum, we must have $E_j \to \infty$ if $j \to \infty$). It is a general fact (see [39, 244]) that the set of eigenvectors of A forms a basis in the underlying Hilbert space. Since the eigenvectors of A can be chosen to be mutually orthogonal, there is an orthonormal basis of eigenvectors. Suppose $f : \mathbb{R} \to \mathbb{C}$ is a continuous function.

Proposition 25.61 $f(A)$ is trace class with

$$\text{Tr}(f(A)) = \sum_j f(E_j), \tag{25.50}$$

provided the sum on the r.h.s. converges absolutely.

Proof. Let $\{\psi_j\}$ be an orthonormal basis of eigenvectors corresponding to the eigenvalues $\{E_j\}$ of A. By the spectral mapping theorem (see Section 25.9), $|f(A)|\psi_j = |f(E_j)|\psi_j$, so

$$\sum_j \langle \psi_j, |f(A)|\psi_j \rangle = \sum_j |f(E_j)| < \infty$$

by assumption. Hence $f(A)$ is trace class, and we may compute its trace as

$$\text{Tr}(f(A)) = \sum_j \langle \psi_j, f(A)\psi_j \rangle = \sum_j f(E_j).$$

$\square$

Integral operators provide another useful example.

Proposition 25.62 Let K be a continuous function on $[a,b]^2$. Then the integral operator $\mathcal{K}$ on $L^2([a,b])$ defined by

$$\mathcal{K}f(x) = \int_a^b K(x,y)f(y)dy$$

is trace class, with

$$\text{Tr}\mathcal{K} = \int_a^b K(x,x)dx.$$

A bounded operator $\mathcal{K}$ is called *Hilbert-Schmidt* if $\mathcal{K}^*\mathcal{K}$ is trace class. We have

Proposition 25.63 An integral operator $\mathcal{K}$ on $L^2(\mathbb{R}^d)$ with kernel $K \in L^2(\mathbb{R}^d \times \mathbb{R}^d)$ is Hilbert-Schmidt, and

$$\text{Tr}\mathcal{K}^*\mathcal{K} = \int_{\mathbb{R}^d \times \mathbb{R}^d} |K(x,y)|^2 dxdy.$$

Proof. Let $\{\psi_j\}$ be an orthonormal basis in $L^2(\mathbb{R}^d)$. Then by the definition of the trace,

$$\text{Tr}\mathcal{K}^*\mathcal{K} = \sum_j \langle \psi_j, \mathcal{K}^*\mathcal{K}\psi_j \rangle = \sum_j \langle \mathcal{K}\psi_j, \mathcal{K}\psi_j \rangle$$

$$= \sum_j \|\mathcal{K}\psi_j\|^2 = \sum_j \int_{\mathbb{R}^d} |\int_{\mathbb{R}^d} K(x,y)\psi_j(y)dy|^2 dx$$

$$= \int_{\mathbb{R}^d} \sum_j |\int_{\mathbb{R}^d} K(x,y)\psi_j(y)dy|^2 dx.$$

By the Parseval relation, this is

$$\text{Tr}\mathcal{K}^*\mathcal{K} = \int_{\mathbb{R}^d} |\int_{\mathbb{R}^d} |K(x,y)|^2 dy| dx = \int_{\mathbb{R}^d \times \mathbb{R}^d} |K(x,y)|^2 dxdy$$

as required. $\square$

A final example of a trace-class operator is a finite rank projection.

Problem 25.64 Show that if P is a rank-r projection, then P is trace class. If, in addition, P is an orthogonal projection, then $\text{Tr}P = r$.

We end this section by describing the spectra of trace-class operators.

Theorem 25.65 If ρ is a trace class operator, then its spectrum consists of isolated eigenvalues with finite multiplicity, and possibly the point 0. Thus eigenvalues can accumulate only at 0.

Proof. We prove the theorem for ρ positive. We begin with

Lemma 25.66 If $\rho \geq 0$ is a trace class operator, $\psi_j \to 0$ weakly, and $\|\psi_j\| \leq M$ for all j, then

$$\langle \psi_j, \rho\psi_j \rangle \to 0. \tag{25.51}$$

Proof. Let $\{\phi_n\}$ be an orthonormal basis in our Hilbert space. Writing

$$\psi_j = \sum_n \langle \phi_n, \psi_j \rangle \phi_n,$$

we find

$$\langle \psi_j, \rho\psi_j \rangle = \sum_{n,m} \overline{\langle \phi_n, \psi_j \rangle} \langle \phi_m, \psi_j \rangle \langle \phi_n, \rho\phi_m \rangle.$$

Since $\rho = \rho^{1/2}\rho^{1/2}$, we have

$$|\langle \phi_n, \rho\phi_m \rangle| = |\langle \rho^{1/2}\phi_n, \rho^{1/2}\phi_m \rangle| \leq \|\rho^{1/2}\phi_n\|\|\rho^{1/2}\phi_m\|$$
$$= \langle \phi_n, \rho\phi_n \rangle^{1/2}\langle \phi_m, \rho\phi_m \rangle^{1/2}.$$

The last two relations yield

$$\langle \psi_j, \rho\psi_j \rangle \leq \left(\sum_n |\langle \phi_n, \psi_j \rangle||\langle \phi_n, \rho\phi_n \rangle^{1/2} \right)^2$$
$$\leq 2 \left(\sum_{n \leq N} |\langle \phi_n, \psi_j \rangle||\langle \phi_n, \rho\phi_n \rangle^{1/2} \right)^2$$
$$+ 2 \left(\sum_{n > N} |\langle \phi_n, \psi_j \rangle||\langle \phi_n, \rho\phi_n \rangle^{1/2} \right)^2.$$

Applying the Cauchy-Schwarz inequality, we obtain

$$\langle \psi_j, \rho\psi_j \rangle \leq 2 \left(\sum_{n \leq N} |\langle \phi_n, \psi_j \rangle|^2 \right) \left(\sum_{n \leq N} \langle \phi_n, \rho\phi_n \rangle \right)$$
$$+ 2 \left(\sum_{n > N} |\langle \phi_n, \psi_j \rangle|^2 \right) \left(\sum_{n > N} \langle \phi_n, \rho\phi_n \rangle \right).$$

Since ρ is trace class, given any $\epsilon > 0$, there is $N(\epsilon)$ such that

$$\sum_{n > N(\epsilon)} \langle \phi_n, \rho\phi_n \rangle \leq \epsilon.$$

Since ψ_j converges weakly to zero, for any $\epsilon > 0$ and $N > 0$, there is $J(\epsilon, N)$ such that

$$\sum_{n \leq N} |\langle \phi_n, \psi_j \rangle|^2 \leq \epsilon \quad \text{for all } j \geq J(\epsilon, N).$$

The last three inequalities imply that for all $j \geq N(\epsilon, N(\epsilon))$,

$$\langle \psi_j, \rho\psi_j \rangle \leq 2\epsilon \operatorname{Tr} \rho + 2\|\psi_j\|^2\epsilon, \tag{25.52}$$

where we used

$$\sum_{n \leq N} \langle \phi_n, \rho\phi_n \rangle \leq \sum_n \langle \phi_n, \rho\phi_n \rangle = \operatorname{Tr} \rho$$

and

$$\sum_{n > N} |\langle \phi_n, \psi_j \rangle|^2 \leq \sum_n |\langle \phi_n, \psi_j \rangle|^2 = \|\psi_j\|^2.$$

Since $\|\psi_j\| \leq M$, the inequality (25.52) completes the proof of the lemma. $\square$

Lemma 25.67 If $\rho \geq 0$ is trace class, then so is ρ^2.

Proof. Since

$$\langle \phi, \rho^2 \phi \rangle = \langle \rho\phi, \rho\phi \rangle = \|\rho\phi\|^2$$
$$= \|\rho^{1/2}\rho^{1/2}\phi\|^2 \leq \|\rho^{1/2}\|^2 \|\rho^{1/2}\phi\|^2 = \|\rho^{1/2}\|^2 \langle \phi, \rho\phi \rangle,$$

for any orthonormal basis $\{\phi_j\}$,

$$\sum_j \langle \phi_j, \rho^2 \phi_j \rangle \leq \|\rho^{1/2}\|^2 \sum_j \langle \phi, \rho\phi \rangle < \infty.$$

Hence ρ^2 is trace class.

Now we are ready to prove our spectral statement. Let $\{\psi_j\}$ be a sequence with $\psi_j \to 0$ weakly, and $\|\psi_j\| = 1$. Then by Lemmas 25.66 and 25.67,

$$\|(\rho - \lambda)\psi_j\|^2 = \langle \psi_j, (\rho^2 - 2\lambda\rho + \lambda^2)\psi_j \rangle$$
$$= \langle \psi_j, \rho^2 \psi_j \rangle - 2\lambda\langle \psi_j, \rho\psi_j \rangle + \lambda^2 \to \lambda^2.$$

Hence ρ and $\lambda \neq 0$ have no Weyl sequence. Thus by Theorem 25.57, $\lambda \neq 0$ is not a point of the essential spectrum of ρ. Thus $\sigma_{ess}(\rho) \subset \{0\}$. $\square$

Theorem 25.68 Let ρ be a self-adjoint trace class operator on a Hilbert space $\mathcal{H}$. Then the normalized eigenvectors of ρ form a basis in $\mathcal{H}$.

Proof. The normalized eigenvectors of ρ are independent, since ρ is self-adjoint. Let V denote the span of the normalized eigenvectors of ρ, and define $\rho^\perp$ to be the operator ρ restricted to $V^\perp$. Then $\rho^\perp$ is a self-adjoint trace-class operator on $V^\perp$. It cannot have non-zero eigenvalues, since all such eigenvectors would lie in the space V. Hence $\sigma(\rho^\perp) \subset \{0\}$. Since $\rho^\perp$ is self-adjoint and non-negative, we apply Theorem 8.2 to $\rho^\perp$ and $-\rho^\perp$ to conclude that $\langle \psi, \rho^\perp \psi \rangle = 0$ for all $\psi \in V^\perp$, and so $\rho^\perp = 0$. That means $V^\perp$ must consist of zero-eigenvectors of ρ, and hence must be empty. Thus $V = \mathcal{H}$. $\square$

A self-adjoint trace-class operator ρ can be written in the form

$$\rho = \sum_j p_j P_{\psi_j} \tag{25.53}$$

where $\{\psi_j\}$ are orthonormal eigenfunctions and p_j are the corresponding eigenvalues, $\rho\psi_j = p_j\psi_j$. Indeed, since by Theorem 25.68, $\{\psi_j\}$ forms a basis in $\mathcal{H}$, any $\phi \in \mathcal{H}$ can be written as $\phi = \sum_j \langle \psi_j, \phi \rangle \psi_j$, and so

$$\rho\phi = \sum_j p_j \langle \psi_j, \phi \rangle \psi_j = \sum_j p_j P_{\psi_j} \phi.$$

Since $\mathrm{Tr} P_\psi = 1$ and $\mathrm{Tr}(\sum_j A_j) = \sum_j \mathrm{Tr} A_j$, the last relation implies the relation $\mathrm{Tr}\rho = \sum_j p_j$.

25.12 Operator Determinants

In this section we discuss determinants of differential operators, which appear in the stationary phase expansion of the path integrals developed in Chapter 16.

For a square matrix A, the determinant function has the properties

1. A is invertible iff $\det A \neq 0$
2. $A = A^* \Rightarrow \det A \in \mathbb{R}$
3. $\det(AB) = \det(A)\det(B)$
4. $A > 0 \Rightarrow \det A = e^{\mathrm{Tr}(\ln A)}$
5. $\det A = \prod_{\lambda \text{ ev of } A} \lambda$

We would like to define the determinant of a Schrödinger operator.

Example 25.69 Let $H = -\Delta + V$ on $[0, L]$ with zero boundary conditions (assume V is bounded and continuous). Then using the fact that for $n \in \mathbb{Z}$, $\sin(\pi n x/L)$ is an eigenfunction of $-\Delta$ with eigenvalue $(\pi n/L)^2$, we obtain

$$\|(H-(\pi n/L)^2)\sin(\pi n x/L)\| = \|V(x)\sin(\pi n x/L)\| \leq (\max|V|)\|\sin(\pi n x/L)\|,$$

and spectral theory tells us that H has an eigenvalue in the interval $[(\pi n/L)^2 - \max|V|, (\pi n/L)^2 + \max|V|]$. Since $\{\sqrt{2/L}\sin(\pi n x/L) \mid n = 1, 2, \ldots\}$ is an orthonormal basis in $L^2[0, L]$, we have

$$\sigma(H) = \{(\pi n/L)^2 + O(1) \mid n \in \mathbb{Z}\}.$$

So trying to compute the determinant directly, we get $\prod_{\lambda \text{ ev } H} \lambda = \infty$.

For a positive matrix, A, we can define $\zeta_A(s) := \mathrm{Tr} A^{-s} = \sum_{\lambda \text{ ev of } A} \lambda^{-s}$.

Problem 25.70 Show in this case that $\det(A) = e^{-\zeta_A'(0)}$.

Now for $H = -\Delta + V$ on $[0, L]^d$ with zero boundary conditions,

$$\zeta_H(s) = \mathrm{tr}\, H^{-s} := \sum_{\lambda \text{ ev of } H} \lambda^{-s}$$

exists for $Re(s) > 1/2$ (see Example 25.69 for $d = 1$). If ζ_H has an analytic continuation into a neighbourhood of $s = 0$, then we define

$$\det H := e^{-\zeta_H'(0)}.$$

So defined, $\det H$ enjoys Properties 1-4 above, but not Property 5. It turns out that for $H = -\Delta + V$, ζ_H does have an analytic continuation to a neighbourhood of 0, and this definition applies.

It is difficult, however, to compute a determinant from this definition. In what follows, we describe some useful techniques for computation of determinants.

Using the formula

$$\lambda^{-s} = \frac{1}{\Gamma(s)} \int_0^\infty t^{s-1} e^{-t\lambda} dt$$

for each $\lambda_n \in \sigma(H)$ leads to

$$\zeta_H(s) = \frac{1}{\Gamma(s)} \int_0^\infty t^{s-1} \sum_n e^{-t\lambda_n} dt.$$

Now λ is an eigenvalue of H iff $e^{-t\lambda}$ is an eigenvalue of e^{-tH} (this is an example of the *spectral mapping theorem* – see Section 25.9), and so $e^{-t\lambda_n}$ is the n-th eigenvalue of e^{-tH}. Thus

$$\zeta_H(s) = \frac{1}{\Gamma(s)} \int_0^\infty t^{s-1} \mathrm{Tr} e^{-tH} dt.$$

This formula can be useful, as it may be easier to deal with $\mathrm{Tr}(e^{-tH}) = \int e^{-tH}(x,x) dx$ than $\mathrm{Tr}(H^{-s})$.

Example 25.71 We consider $H = -\Delta$ in a box $B = [-L/2, L/2]^d$ with periodic boundary conditions. In this case

$$e^{-tH}(x,y) \approx (2\pi t)^{-d/2} e^{-|x-y|^2/2t}$$

in $B \times B$, if B is very large, and so

$$\mathrm{Tr} e^{-tH} = \int_{B \times B} e^{-tH}(x,x) dx \approx \int_B (2\pi t)^{-d/2} = (2\pi t)^{-d/2} \mathrm{vol}(B).$$

But calculation of $\det H$ by this method is still a problem.

Remark 25.72 Often (and in all cases we consider here), we have to compute a ratio, $\frac{\det A}{\det B}$, of determinants of two operators A and B, such that $\det A$ and $\det B$ must be defined through a regularization procedure such as the one described above, but the determinant of the ratio AB^{-1} can be defined directly. Since $\frac{\det A}{\det B} = \det(AB^{-1})$, we can make sense of the ratio on the left hand side without going to regularization.

The most useful calculational technique for us is as follows. Let A and B be Schrödinger operators on $L^2([0,T]; \mathbb{R}^d)$ with Dirichlet boundary conditions. Denote by J_A the solution to $AJ_A = 0$ with $J_A(0) = 0$, $\dot{J}_A(0) = \mathbf{1}$ (J a $d \times d$ matrix valued function on $[0,T]$). Then one can show (see, eg., [195, 180])

$$\frac{\det A}{\det B} = \frac{\det J_A(T)}{\det J_B(T)}. \tag{25.54}$$

Remark 25.73 If $A = S''(\bar{\phi})$ for a critical path $\bar{\phi}$, then J_A is the Jacobi matrix along $\bar{\phi}$ (see Section 26.4).

Problem 25.74 Let $A(T)$ be the operator $-\partial_t^2 + q(t)$ defined on $L^2([0,T]; \mathbb{R}^d)$ with Dirichlet boundary conditions, and let J_A be the corresponding Jacobi matrix. Show that the functions $\det A(T)$ and $\det J(T)$ have the same zeros of the same multiplicities (t_0 is a zero of $f(t)$ of multiplicity n if $\partial^k/\partial t^k f(t_0) = 0$ for $k = 0, \ldots, n-1$ and $\partial^n/\partial t^n f(t_0) \neq 0$).

25.13 Tensor Products

We collect here a few facts about tensor products of Hilbert spaces, and tensor products of operators and their spectra (see [244] for details and proofs).

Let $\mathcal{H}_1$ and $\mathcal{H}_2$ be two separable Hilbert spaces. The *tensor product* of $\mathcal{H}_1$ and $\mathcal{H}_2$ is a Hilbert space $\mathcal{H}_1 \otimes \mathcal{H}_2$ constructed as follows. To $\psi_1 \in \mathcal{H}_1$ and $\psi_2 \in \mathcal{H}_2$, we associate a map

$$\psi_1 \otimes \psi_2 : \mathcal{H}_1 \times \mathcal{H}_2 \to \mathbb{C}$$
$$(f_1, f_2) \mapsto \langle f_1, \psi_1 \rangle_{\mathcal{H}_1} \langle f_2, \psi_2 \rangle_{\mathcal{H}_2}$$

which is conjugate linear in each component ($\psi_1 \otimes \psi_2(\alpha f_1, f_2) = \bar{\alpha}\psi_1 \otimes \psi_2(f_1, f_2)$, $\psi_1 \otimes \psi_2(f_1 + g_1, f_2) = \psi_1 \otimes \psi_2(f_1, f_2) + \psi_1 \otimes \psi_2(g_1, f_2)$, and the same for the second component). On the vector space, V, of all finite linear combinations of such conjugate bilinear maps, we define an inner-product by setting

$$\langle \psi_1 \otimes \psi_2, \phi_1 \otimes \phi_2 \rangle := \langle \psi_1, \phi_1 \rangle_{\mathcal{H}_1} \langle \psi_2, \phi_2 \rangle_{\mathcal{H}_2} \qquad (25.55)$$

and extending by linearity (it is straightforward to check that this is well-defined). Then $\mathcal{H}_1 \otimes \mathcal{H}_2$ is defined to be the completion of V in the inner-product determined by (25.55).

A simple example, which appears in Section 13.1, is

$$L^2(\mathbb{R}^m) \otimes L^2(\mathbb{R}^n) \approx L^2(\mathbb{R}^{m+n})$$

for positive integers m, n. This Hilbert space isomorphism is determined by the map

$$f \otimes g \mapsto f(x)g(y)$$

(see, eg., [244] for details).

Given bounded operators A and B acting on $\mathcal{H}_1$ and $\mathcal{H}_2$, the operator $A \otimes B$, which acts on $\mathcal{H}_1 \otimes \mathcal{H}_2$, is defined by setting

$$A \otimes B(\psi_1 \otimes \psi_2) = A\psi_1 \otimes B\psi_2,$$

extending by linearity to all finite linear combinations of elements of this form, and then by density of these finite linear combinations, to $\mathcal{H}_1 \otimes \mathcal{H}_2$. This produces a well-defined operator.

This construction can be extended to unbounded self-adjoint operators A and B, yielding a self-adjoint operator $A \otimes B$ (see, eg., [244]).

Of particular interest for us are operators of the form $A \otimes 1 + 1 \otimes B$, acting on $\mathcal{H} = \mathcal{H}_1 \otimes \mathcal{H}_2$, where A and B are operators acting on $\mathcal{H}_1$ and $\mathcal{H}_2$ respectively. This is an abstract version of the "separation of variables" situation of differential equations. It is intuitively clear that we should be able to reconstruct characteristics of such operators from the corresponding characteristics of the operators A and B. As an example, we have the following important (and simple) description of the spectrum of $A \otimes 1 + 1 \otimes B$ under certain conditions on A and B, and, in particular, for A and B self-adjoint:

$$\sigma(A \otimes 1 + 1 \otimes B) = \sigma(A) + \sigma(B)$$
$$\sigma_d(A \otimes 1 + 1 \otimes B) \subset \sigma_d(A) + \sigma_d(B) \subset \{ \text{ ev's of } A \otimes 1 + 1 \otimes B\}$$
$$\sigma_{ess}(A \otimes 1 + 1 \otimes B) = \sigma_{ess}(A) + \sigma_{ess}(B)$$
$$\cup [\sigma_{ess}(A) + \sigma_d(B)]$$
$$\cup [\sigma_d(A) + \sigma_{ess}(B)].$$

Rather than prove any such statements (an involved task, requiring further assumptions), let us just do a simple, suggestive computation. Suppose $A\psi_1 = \lambda_1 \psi_1$ and $B\psi_2 = \lambda_2 \psi_2$. Then note that

$$(A \otimes 1 + 1 \otimes B)\psi_1 \otimes \psi_2 = A\psi_1 \otimes \psi_2 + \psi_1 \otimes B\psi_2$$
$$= \lambda_1 \psi_1 \otimes \psi_2 + \psi_1 \otimes (\lambda_2 \psi_2)$$
$$= \lambda_1 (\psi_1 \otimes \psi_2) + \lambda_2 (\psi_1 \otimes \psi_2)$$
$$= (\lambda_1 + \lambda_2)\psi_1 \otimes \psi_2,$$

which shows, in particular, that $\sigma_d(A) + \sigma_d(B) \subset \{ \text{ ev's of } A \otimes 1 + 1 \otimes B\}$.

25.14 The Fourier Transform

The Fourier transform is a useful tool in many areas of mathematics and physics. The purpose of the present section is to review the properties of the Fourier transform, and to discuss the important role it plays in quantum mechanics.

The *Fourier transform* is a map, $\mathcal{F}$, which sends a function $\psi : \mathbb{R}^d \to \mathbb{C}$ to another function $\hat{\psi} : \mathbb{R}^d \to \mathbb{C}$ where for $k \in \mathbb{R}^d$,

$$\hat{\psi}(k) := (2\pi\hbar)^{-d/2} \int_{\mathbb{R}^d} e^{-ik \cdot x/\hbar} \psi(x) dx$$

(it is convenient for quantum mechanics to introduce Planck's constant, $\hbar$, into the Fourier transform). We first observe that $\hat{\psi} = \mathcal{F}\psi$ is well-defined if ψ is an integrable function ($\psi \in L^1(\mathbb{R}^d)$, meaning $\int_{\mathbb{R}^d} |\psi(x)| dx < \infty$), and that $\mathcal{F}$ acts as a linear operator on such functions.

In the following exercise, the reader is asked to compute a few Fourier transforms.

Problem 25.75 Show that under $\mathcal{F}$

1. $e^{-\frac{|x|^2}{2a\hbar^2}} \mapsto (\hbar a)^{d/2} e^{-\frac{a|k|^2}{2}}$ $(Re(a) > 0)$. Hint: try $d = 1$ first – complete the square in the exponent and move the contour of integration in the complex plane.

2. $e^{-\frac{1}{2\hbar^2} x \cdot A^{-1} x} \mapsto \hbar^{d/2} (\det A)^{1/2} e^{-\frac{1}{2} k \cdot Ak}$ (A a positive $d \times d$ matrix). Hint: diagonalize and use the previous result.

3. $\sqrt{\frac{\pi}{2\hbar}} \frac{e^{-\sqrt{b/\hbar^2}|x|}}{|x|} \mapsto (|k|^2 + b)^{-1}$ ($b > 0$, $d = 3$). Hint: use spherical coordinates. Alternatively, see Problem 25.77 below.

In the first example, if $Re(a) > 0$ then the function on the left is in $L^1(\mathbb{R}^d)$, and the Fourier transform is well-defined. However, we can extend this result to $Re(a) = 0$, in which case the integral is convergent, but not absolutely convergent.

Properties of the Fourier Transform. The utility of the Fourier transform derives from the following properties.

1. The *Plancherel theorem*: $\mathcal{F}$ is a unitary map from $L^2(\mathbb{R}^d)$ to itself (note that initially the Fourier transform is defined only for integrable ($L^1(\mathbb{R}^d)$) functions – the statement here is that the Fourier transform extends from $L^1(\mathbb{R}^d) \cap L^2(\mathbb{R}^d)$ to a unitary map on $L^2(\mathbb{R}^d)$).

2. The inversion formula: the adjoint $\mathcal{F}^*$ of $\mathcal{F}$ on $L^2(\mathbb{R}^d)$ is given by the map $\psi \mapsto \check{\psi}$ where

$$\check{\psi}(x) := (2\pi\hbar)^{-d/2} \int_{\mathbb{R}^d} e^{ix \cdot k/\hbar} \psi(k) dk$$

(and by the Plancherel theorem, this is also the inverse, $\mathcal{F}^{-1}$).

For the next four statements, suppose $\psi, \phi \in C_0^\infty(\mathbb{R}^d)$.

3. $-i\hbar \widehat{\nabla_x \psi}(k) = k\hat{\psi}(k)$.
4. $\widehat{x\psi}(k) = i\hbar \nabla_k \hat{\psi}(k)$.
5. $\widehat{\phi\psi} = (2\pi\hbar)^{-d/2} \hat{\phi} * \hat{\psi}$.
6. $\widehat{\phi * \psi} = (2\pi\hbar)^{d/2} \hat{\phi}\hat{\psi}$.

Here

$$(f * g)(x) := \int_{\mathbb{R}^d} f(y) g(x - y) dy \qquad (25.56)$$

is the *convolution* of f and g. The last four properties can be loosely summarized by saying that the Fourier transform exchanges differentiation and coordinate multiplication, and products and convolutions.

Proof. The proof of Property 1 is somewhat technical and we just sketch it here (see, eg, [106] for details). In particular, we will show that $\|f\| = \|\hat{f}\|$. Suppose $f \in C_0^\infty$, and let C_ϵ be the cube of side length $2/\epsilon$ centred at the origin. Choose ϵ small enough so that the support of f is contained in C_ϵ. One can show that

$$\{E_k := (\epsilon/2)^{d/2} e^{ik\cdot x/\hbar} \mid k \in \epsilon\hbar\pi\mathbb{Z}^d\}$$

is an orthonormal basis of the Hilbert space of functions in $L^2(C_\epsilon)$ satisfying periodic boundary conditions. Thus by the Parseval equation (Proposition 25.4),

$$\int_{\mathbb{R}^d} |f|^2 = \int_{C_\epsilon} |f|^2 = \sum_{k\in\epsilon\hbar\pi\mathbb{Z}^d} |\langle E_k, f\rangle|^2$$

$$= (\pi\epsilon\hbar)^d \sum_{k\in\epsilon\hbar\pi\mathbb{Z}^d} |\hat{f}(k)|^2 \to \int_{\mathbb{R}^d} |\hat{f}|^2$$

as $\epsilon \to 0$.

Problem 25.76 Show that $\{E_k\}$ is an orthonormal set.

We will prove Property 3, and we leave the proofs of the other properties as exercises. Integrating by parts, we have

$$-i\hbar\widehat{\nabla\psi}(k) = (2\pi\hbar)^{-d/2} \int e^{-ix\cdot k/\hbar} \cdot (-i\hbar\nabla)\psi(x)dx$$

$$= k \cdot (2\pi\hbar)^{-d/2} \int e^{-ix\cdot k/\hbar}\psi(x)dx = k\hat{\psi}(k).$$

$\square$

Problem 25.77 1. Show that for $b > 0$ and $d = 3$, under $\mathcal{F}^{-1}$,

$$(|k|^2 + b)^{-1} \mapsto \sqrt{\frac{\pi}{2\hbar}} \frac{e^{-\sqrt{b/\hbar^2}|x|}}{|x|}$$

(hint: use spherical coordinates, then contour deformation and residue theory).
2. Show that under $\mathcal{F}^{-1}$,

$$\delta(k - a) \mapsto (2\pi\hbar)^{-d/2} e^{ia\cdot x/\hbar}.$$

Here δ is the *Dirac delta function* – not really a function, but a *distribution* – characterized by the property $\int f(x)\delta(x-a)dx = f(a)$. The exponential function on the right hand side is called a *plane wave*.

Functions of the derivative. As an application, we show how the Fourier transform can be used to define functions of the derivative operator. Recall our notation $p := -i\hbar\nabla$. Motivated by Property 3 of the Fourier transform, we define an operator $g(p)$ (for "sufficiently nice" functions g) on $L^2(\mathbb{R}^d)$ as follows.

Definition 25.78 $\widehat{g(p)\psi}(k) := g(k)\hat{\psi}(k)$ or, equivalently,

$$g(p)\psi := (2\pi\hbar)^{-d/2}\check{g} * \psi.$$

Let us look at a few examples.

Example 25.79 1. If $g(k) = k$, then by Property 3 of the Fourier transform, the above definition gives us back $g(p) = p$ (so at least our definition makes some sense).

2. Now suppose $g(k) = |k|^2$. Then $\widehat{g(p)\psi}(k) = |k|^2\hat{\psi}$.

Problem 25.80 Show that $-\hbar^2\widehat{\Delta\psi} = |k|^2\hat{\psi}$.

Thus we have $|p|^2 = -\hbar^2\Delta$. Extending this example, we can define $g(p)$ when g is a polynomial "with our bare hands". It is easy to see that this definition coincides with the one above.

3. Let $g(k) = (\frac{1}{2m}|k|^2 + \lambda)^{-1}$, $\lambda > 0$, and $d = 3$. Then due to Problem 25.77, we have

$$((H_0 + \lambda)^{-1}\psi)(x) = \frac{m}{2\pi\hbar^2} \int_{\mathbb{R}^3} \frac{e^{-\frac{\sqrt{2m\lambda}}{\hbar}|x-y|}}{|x - y|}\psi(y)dy \tag{25.57}$$

where we have denoted $H_0 := \frac{1}{2m}|p|^2 = -\frac{\hbar^2}{2m}\Delta$.

Problem 25.81 Let $y \in \mathbb{R}^d$ be fixed. Find how the operator $e^{iy\cdot p}$ acts on functions (here $y \cdot p = \sum_{j=1}^{d} y_j p_j$, $p_j = -i\hbar\partial_{x_j}$).

Mathematical Supplement: The Calculus of Variations

The calculus of variations, an extensive mathematical theory in its own right, plays a fundamental role throughout physics. This supplement contains an overview of some of the basic aspects of the variational calculus. This material will be used throughout the book, and in particular in Chapters 16 and 17 to obtain useful quantitative results about quantum systems in the regime close to the classical one.

26.1 Functionals

The basic objects of study in the calculus of variations are *functionals*, which are just functions defined on Banach spaces (usually spaces of functions). (Recall that a *Banach space* is a complete normed vector space.) If we specify a space, X, then functionals on X are just maps $S : X \to \mathbb{R}$ (or into $\mathbb{C}$).

In the calculus of variations, one often uses spaces other than $L^2(\mathbb{R}^d)$, and which are not necessarily Hilbert spaces. Among the most frequently encountered spaces are the Sobolev spaces $H^s(\mathbb{R}^d)$, $s = 1, 2, \ldots$, introduced in Section 25.1. Recall that the Sobolev spaces are Hilbert spaces.

Spaces of continuously differentiable functions also arise frequently. For $k \in \{0, 1, 2, \ldots\}$, and an open set $\Omega \subset \mathbb{R}^d$, we define $C^k(\overline{\Omega}; \mathbb{R}^m)$ to be the set of all functions $\phi : \Omega \to \mathbb{R}^m$ such that $\partial^\alpha \phi$ is continuous in Ω for all $|\alpha| \leq k$, and for which the norm

$$\|\phi\|_{C^k(\overline{\Omega};\mathbb{R}^m)} := \sum_{|\alpha| \leq k} \sup_{x \in \Omega} |\partial^\alpha \phi(x)|$$

is finite. Equipped with this norm, $C^k(\overline{\Omega}; \mathbb{R}^m)$ is a Banach space.

Example 26.1 Here are some common examples of functionals, S, and spaces, X, on which they are defined.

© Springer-Verlag GmbH Germany, part of Springer Nature 2020
S. J. Gustafson and I. M. Sigal, *Mathematical Concepts of Quantum Mechanics*, Universitext, https://doi.org/10.1007/978-3-030-59562-3_26

1. $X = L^2([a,b];\mathbb{R})$, $f \in X^* = X$ is a fixed function (X^* denotes the dual space to X, as explained in Section 25.1), and

$$S : \phi \mapsto \int_a^b f(x)\phi(x)dx.$$

Note that $S(\phi)$ is well-defined, by the Cauchy-Schwarz inequality.

2. Evaluation functional: $X = C([a,b])$, $x_0 \in (a,b)$ fixed, and

$$S : \phi \mapsto \phi(x_0)$$

Compare this with the first example by taking $f(x) = \delta(x - x_0)$ ($\in X^*$; in this case $X^* \neq X$).

3. Let $V : \mathbb{R}^m \to \mathbb{R}$ be continuous. Set $X = \{\phi : \mathbb{R}^d \to \mathbb{R}^m \mid V(\phi) \in L^1(\mathbb{R}^d)\}$, and take

$$S : \phi \mapsto \int_{\mathbb{R}^d} V(\phi(x))dx.$$

4. Dirichlet functional: $X = H^1(\mathbb{R}^d)$, and

$$S : \phi \mapsto \frac{1}{2}\int_{\mathbb{R}^d} |\nabla\phi|^2 dx.$$

5. Classical action: fix $x, y \in \mathbb{R}^m$, set $X = \mathcal{P}_{xy} := \{\phi \in C^1([0,T];\mathbb{R}^m) \mid \phi(0) = x, \phi(T) = y\}$, and define

$$S : \phi \mapsto \int_0^T \left\{\frac{1}{2}m|\dot{\phi}|^2 - V(\phi)\right\} dt.$$

6. Classical action: Let $L : \mathbb{R}^m \times \mathbb{R}^m \to \mathbb{R}$ be a twice differentiable function (a *Lagrangian function*), and

$$S : \phi \mapsto \int_0^T L(\phi, \dot{\phi})dt.$$

Here, $X = \mathcal{P}_{xy}$ is as in Example 5.

7. Action of a classical field theory: fix $f, g, \in H^1(\mathbb{R}^d;\mathbb{R}^m)$, set

$$X = \{\phi \in H^1(\mathbb{R}^d \times [0,T];\mathbb{R}^m) \mid \phi(x,0) = f(x), \phi(x,T) = g(x)\}, \quad (26.1)$$

and define

$$S : \phi \mapsto \int_0^T \int_{\mathbb{R}^d} \left\{-\frac{1}{2}|\partial_t\phi|^2 + \frac{1}{2}|\nabla_x\phi|^2 + f(\phi)\right\} dxdt$$

where $f : \mathbb{R}^m \to \mathbb{R}$ is a differentiable function.

8. Lagrangian functional: Suppose $\mathcal{L} : \mathbb{R}^m \times \mathbb{R}^{d+1} \to \mathbb{R}$ (a *Lagrangian density*) is a twice differentiable function, and set

$$S : \phi \mapsto \iint \mathcal{L}(\phi(x,t), \nabla_{x,t}\phi(x,t))dxdt$$

(here X is an appropriate space of vector functions from space-time $\mathbb{R}_x^d \times \mathbb{R}_t$ to $\mathbb{R}^m$, whose specific definition depends on the form of $\mathcal{L}$).

We encountered many of these functionals when we considered the problems of quantization and of (quasi-)classical limit in quantum mechanics and quantum field theory.

26.2 The First Variation and Critical Points

The notion of a *critical point* of a functional is a central one. It is a direct extension of the usual notion of a critical point of a function of finitely many variables (i.e., a place where the derivative vanishes). The solutions of many physical equations are critical points of certain functionals, such as *action* or *energy* functionals.

In what follows, the spaces X on which our functionals are defined will generally be linear (i.e. vector) spaces or affine spaces. By an *affine space*, we mean a space of the form $X = \{\phi_0 + \phi \mid \phi \in X_0\}$, where ϕ_0 is a fixed element of X, and X_0 is a vector space. We will encounter examples of functionals defined on non-linear spaces when we study constrained variational problems in Section 26.5.

Let X be a Banach space, or else an affine space based on a Banach space X_0. (Recall that the notion of Banach space is defined in Section 25.1 of the previous mathematical supplement.)

Definition 26.2 A *path*, ϕ_λ, in X is a differentiable function $I \ni \lambda \mapsto \phi_\lambda$ from an interval $I \subset \mathbb{R}$ containing 0, into X.

Definition 26.3 The tangent space, $T_\phi X$, to X at $\phi \in X$ is the space of all "velocity vectors" at ϕ:

$$T_\phi X := \left\{ \left. \frac{\partial \phi_\lambda}{\partial \lambda} \right|_{\lambda=0} \mid \phi_\lambda \text{ is a path in } X, \phi_0 = \phi \right\}.$$

If X is a Banach space, then $T_\phi X = X$ (and if X is an affine space based on the Banach space X_0, $T_\phi X = X_0$). To see this, just note that for any $\xi \in X$, $\phi_\lambda := \phi + \lambda\xi$ is a path in X, satisfying $\phi_0 = \phi$ and $\partial\phi_\lambda/\partial\lambda = \xi$ (the reader is invited to check the corresponding statement for affine spaces).

So the tangent space to a linear space is not very interesting. The notion of a tangent space is useful when working in non-linear spaces (we will see an example of this shortly).

Definition 26.4 A *variation* of $\phi \in X$ along $\xi \in T_\phi X$ is a path, ϕ_λ in X, such that $\phi_0 = \phi$ and $\partial \phi_\lambda / \partial \lambda |_{\lambda=0} = \xi$.

Example 26.5 One of our main examples is the classical action, Example 6 above (and (4.36)), defined on $\mathcal{P}_{xy} = \{\phi \in C^1([0,T], X) \mid \phi(0) = x, \phi(T) = y\}$ (an affine space). Then $T_\phi \mathcal{P}_{xy} = \mathcal{P}_{00}$ (see Fig. 26.1), and an example of a variation of $\phi \in \mathcal{P}_{xy}$ in the direction $\xi \in \mathcal{P}_{00}$ is $\phi_\lambda = \phi + \lambda \xi \in X_{xy}$.

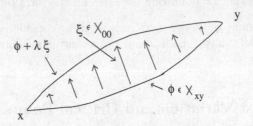

Fig. 26.1. Variations of a path.

We wish to define a notion of differentiation of functionals which is a direct extension of usual differentiation of functions of a finite number of variables. To do so, we use the concepts of the dual space X^* to X and the notation $\langle \cdot, \cdot \rangle$ for the coupling between the X^* to X, described in Section 25.1.

Definition 26.6 Let $S : X \to \mathbb{R}$ be a functional on a real Banach space X, and let $\phi \in X$. We say that S is *differentiable* at ϕ if there is a linear functional, $\partial S(\phi) \in X^*$, such that

$$\frac{d}{d\lambda} S(\phi_\lambda)|_{\lambda=0} = \langle \partial S(\phi), \xi \rangle \qquad (26.2)$$

for any variation ϕ_λ of ϕ along $\xi \in X$. The functional $\partial S(\phi)$ is the *(variational) derivative* of S at ϕ.

Remark 26.7 1. The notion of differentiability introduced here is often called *Gâteaux differentiability*. A stronger notion of differentiability, called *Fréchet differentiability*, demands (for linear spaces) that

$$S(\phi + \xi) = S(\phi) + \langle \partial S(\phi), \xi \rangle + o(\|\xi\|_X)$$

as $\|\xi\|_X \to 0$. The reader can check that if S is continuously (Gâteaux) differentiable at $\phi_0 \in X$ (i.e. $\partial S(\phi)$ exists in a neighbourhood of ϕ_0, and is a continuous map from this neighbourhood into X^*), then S is Fréchet differentiable at ϕ_0.

2. We will sometimes use the notation $S'(\phi)$ for the variational derivative $\partial S(\phi)$.

3. Recall that if X is a Hilbert space, then we can identify its dual, X^*, with X itself, via the map $X \ni \phi \mapsto l_\phi \in X^*$ with $l_\phi \xi := \langle \phi, \xi \rangle$ for $\xi \in X$ (here the notation $\langle \cdot, \cdot \rangle$ indicates the Hilbert space inner-product).

(The fact that this map is an isomorphism between X and X^* is known as the *Riesz representation theorem*.) Thus when X is a Hilbert space, we can identify $\partial S(\phi)$ with an element of X. This element is called the *gradient* of S at ϕ (in the inner product of X), and is sometimes denoted by $grad_X S(\phi)$.

Example 26.8 We compute the derivatives of some of the functionals in Example 26.1. We suppose that whenever X is a space of functions, it is a subspace of an L^2-space. Then the variational derivative can be identified with a function (or distribution), using integration by parts where necessary. This is related to the L^2 gradient as discussed above.

1. For the functional $S(\phi) = \int_a^b f\phi dx$, we compute

$$\langle \partial S(\phi), \xi \rangle = \frac{d}{d\lambda} S(\phi_\lambda)|_{\lambda=0} = \int_a^b f \frac{\partial}{\partial \lambda} \phi_\lambda dx|_{\lambda=0}$$

$$= \int_a^b f\xi dx = \langle f, \xi \rangle.$$

Thus we identify $\partial S(\phi) = f$.

3. For $S(\phi) = \int_{\mathbb{R}^d} V(\phi)$, we compute

$$\langle \partial S(\phi), \xi \rangle = \frac{d}{d\lambda} S(\phi_\lambda)|_{\lambda=0} = \frac{d}{d\lambda} \int_{\mathbb{R}^d} \frac{\partial}{\partial \lambda} V(\phi_\lambda)|_{\lambda=0} dx$$

$$= \int_{\mathbb{R}^d} \nabla V(\phi) \cdot \xi dx = \langle \nabla V(\phi), \xi \rangle$$

and so we identify $\partial S(\phi) = \nabla V(\phi)$.

4. For $S(\phi) = \frac{1}{2} \int_{\mathbb{R}^d} |\nabla \phi|^2$, we compute

$$\langle \partial S(\phi), \xi \rangle = \frac{d}{d\lambda} S(\phi_\lambda)|_{\lambda=0} = \frac{1}{2} \int_{\mathbb{R}^d} \frac{\partial}{\partial \lambda} |\nabla \phi_\lambda|^2|_{\lambda=0} dx$$

$$= \int_{\mathbb{R}^d} \nabla \phi \cdot \nabla \xi dx = \int_{\mathbb{R}^d} (-\Delta \phi)\xi dx = \langle -\Delta \phi, \xi \rangle$$

where we integrated by parts (Gauss theorem), and used the fact that the functions decay at ∞. Thus we identify $\partial S(\phi) = -\Delta \phi$.

We leave the remaining examples as an exercise.

Problem 26.9 Compute the variational derivatives for the remaining functionals in Example 26.1. You should find

5. $\partial S(\phi) = -m\ddot{\phi} - \nabla V(\phi)$
6. $\partial S(\phi) = -\frac{d}{dt}(\partial_{\dot{\phi}} L) + \partial_\phi L$
7. $\partial S(\phi) = \Box \phi + \nabla f(\phi)$ where $\Box := \partial_t^2 - \Delta$ is the *D'Alembertian* operator
8. $\partial S(\phi) = -\nabla_{x,t} \cdot (\partial_{\nabla \phi} L) + \partial_\phi L$

As in the finite-dimensional case, a *critical point* is a place where the derivative vanishes.

Definition 26.10 An element $\phi \in X$ is a *critical point* (CP) of a functional $S : X \to \mathbb{R}$ if $\partial S(\phi) = 0$.

In fact, many physical equations are critical point equations for certain functionals.

Example 26.11 Continuing with the same list of examples of functionals above, we can write down some of the equations describing their critical points:

4. $\Delta\phi = 0$ (Laplace equation, ϕ a harmonic function)
5. $m\ddot{\phi} = -\nabla V(\phi)$ (Newton's equation)
6. $\frac{d}{dt}(\partial_{\dot{\phi}}L) = (\partial_{\phi}L)$ (Euler-Lagrange equation)
7. $\Box\phi + \nabla f(\phi) = 0$ (nonlinear wave/Klein-Gordon equation)
8. $-\nabla_{x,t} \cdot (\partial_{\nabla\phi}L) + \partial_{\phi}L = 0$ (classical field equation)

The following connection between critical points and minima (or maxima) is familiar from multi-variable calculus.

Theorem 26.12 If ϕ locally minimizes or maximizes a differentiable functional $S : X \to \mathbb{R}$, then ϕ is a critical point of S. (We say ϕ is a local minimizer (resp. maximizer) of S if there is some $\delta > 0$ such that $S(\tilde{\phi}) \geq S(\phi)$ (resp. $S(\tilde{\phi}) \leq S(\phi)$) for all $\tilde{\phi}$ with $\|\tilde{\phi} - \phi\|_X < \delta$.)

Problem 26.13 Prove this (hint: it is similar to the finite-dimensional case).

Recall that a function $f(v)$ is called *strictly convex* if

$$f(sv + (1-s)v') < sf(v) + (1-s)f(v'), \ \forall s \in (0,1),$$

$\forall x \in X$, $v, v' \in V$. This condition holds if f is twice differentiable and has positive Hessian, $d_v^2 f(v) > 0$, $\forall v \in V$. A function $f(v)$ is called *strictly concave* iff $-f(v)$ is strictly convex.

Theorem 26.14 Assume $f(v)$ is a differentiable and strictly convex/concave function on a finite-dimensional space V. Then it has a unique critical point, and this critical point minimizes/maximizes f.

We sketch a proof of this theorem. To fix ideas we consider only the convex case. Assume for simplicity that $f(v)$ is twice differentiable. Then as was mentioned above $d_v^2 f(v) > 0$, $\forall v \in V$. Hence every critical point is a (local) minimum. Assume there are two critical points, v_1 and v_2. Then the function $f(sv_1 + (1-s)v_2)$ would have a maximum for some $s \in (0,1)$, a contradiction. Hence $f(v)$ has at most one critical point. One can show furthermore that $f(v) \to \infty$, as $\|v\| \to \infty$ and therefore $f(v)$ has at least one minimizing point. $\Box$

To extend this theorem to the infinite-dimensional case one would have to make some additional assumptions, i.e. that $f(v)$ is weakly lower semicontinuous and V is reflexive (see e.g. [244]).

26.3 The Second Variation

In multi-variable calculus, if one wishes to know if a critical point is actually a minimum (or maximum), one looks at the second derivative. For the same reason, we need to define the second derivative of a functional.

Definition 26.15 Let $\eta, \xi \in T_\phi X$. A *variation of ϕ along η and ξ* is a two-parameter family, $\phi_{\lambda,\mu} \in X$, such that $\phi_{0,0} = \phi$, $\frac{\partial}{\partial\lambda}\phi_{\lambda,\mu}|_{\lambda=\mu=0} = \xi$, and $\frac{\partial}{\partial\mu}\phi_{\lambda,\mu}|_{\lambda=\mu=0} = \eta$.

Definition 26.16 Let $S : X \to \mathbb{R}$ be a functional. We say S is twice differentiable is there is a bounded linear map $\partial^2 S(\phi) : T_\phi X \to (T_\phi X)^*$ (called the *Hessian* or *second variation* of S at ϕ) such that

$$\frac{\partial^2}{\partial\lambda\partial\mu}S(\phi_{\lambda,\mu})|_{\lambda=\mu=0} = \langle \partial^2 S(\phi)\eta, \xi \rangle \tag{26.3}$$

for all $\xi, \eta \in T_\phi X$ and all variations $\phi_{\lambda,\mu}$ of ϕ along ξ and η.

Remark 26.17 1. The Hessian $\partial^2 S(\phi)$ can also be defined as the second derivative of $S(\phi)$, i.e., $\partial^2 S(\phi) = \partial \cdot \partial S(\phi)$. That is, we consider the map $\phi \mapsto \partial S(\phi)$ and define, for $\eta \in T_\phi X$, $\partial^2 S(\phi)\eta := \frac{\partial}{\partial\lambda}\partial S(\phi_\lambda)$, where ϕ_λ is a variation of ϕ along η.
2. We will often use the notation $S''(\phi)$ to denote $\partial^2 S(\phi)$.

Computations of the second derivatives of the functionals in our list of examples are left as an exercise (again, we suppose where appropriate that the action of the dual space is just given by integration).

Problem 26.18 Continuing with our list of examples of functionals above, show that

3. $S''(\phi) = D^2 V(\phi)$ (a matrix multiplication operator).
4. $S''(\phi) = -\Delta$ (the Laplacian).
5. $S''(\phi) = -m\partial_t^2 - D^2V(\phi)$ (a Schrödinger operator) acting on functions satisfying Dirichlet boundary conditions: $\xi(0) = \xi(T) = 0$.
6.

$$S''(\phi) = -\frac{d}{dt}(\partial_\phi^2 L)\frac{d}{dt} - \left(\frac{d}{dt}\partial_{\phi\dot{\phi}}^2 L\right) + \partial_\phi^2 L, \tag{26.4}$$

with Dirichlet boundary conditions. (The first term on the r.h.s. is a product of three operators while the second one is the time-derivative of $\partial_{\phi\dot\phi}^2 L$.

7. $S''(\phi) = \Box + V''(\phi)$, with Dirichlet boundary conditions: $\xi(x,0) = \xi(x,T) = 0$.
8.

$$S''(\phi) = -\nabla_{x,t}\left(\frac{\partial^2 L}{\partial(\nabla\phi)^2}\right)\nabla_{x,t} - \nabla_{x,t}\left(\frac{\partial^2 L}{\partial(\nabla\phi)\partial\phi}\right) + \frac{\partial^2 L}{\partial\phi^2}.$$

The following criterion for a critical point to be a minimizer is similar to the finite-dimensional version, and the proof is left as an exercise.

Theorem 26.19 Let ϕ be a critical point of a twice continuously differentiable functional $S : X \to \mathbb{R}$.

1. If ϕ locally minimizes S, then $S''(\phi) \geq 0$ (meaning $\langle S''(\phi)\xi, \xi \rangle \geq 0$ for all $\xi \in T_\phi X$).
2. If $S''(\phi) > c$, for some constant $c > 0$ (i.e. $\langle S''(\phi)\xi, \xi \rangle \geq c\|\xi\|_X^2$ for all $\xi \in T_\phi X$), then ϕ is a local minimizer of S.

Problem 26.20 Prove this.

Let us now pursue the question of whether or not a critical point of the classical action functional

$$S(\phi) = \int_0^T L(\phi(s), \dot\phi(s))ds$$

(which is a solution of the Euler-Lagrange equation – i.e., a classical path) minimizes the action. As we have seen, the Hessian $S''(\phi)$ is given by (26.4). We call $\partial_\phi^2 L$ the *generalized mass*.

Theorem 26.21 Suppose $\partial_{\dot\phi^2}^2 L > 0$. Suppose further that $\partial_\phi^2 L$ is a bounded function. Then there is a $T_0 > 0$, such that $S''(\phi) > 0$ for $T \leq T_0$.

Proof for $L = \frac{m}{2}\dot\phi^2 - V(\phi)$. In this case $S''(\phi) = -md^2/ds^2 - V''(\phi)$, acting on $L^2([0,T])$ with Dirichlet boundary conditions. Since $\inf \sigma(-d^2/ds^2) = (\pi/T)^2$, we have, by Theorem 8.2, $-d^2/ds^2 \geq (\pi/T)^2$. So $S''(\phi) \geq m(\pi/T)^2 - \max|V''|$, which is positive for T sufficiently small. $\square$

Corollary 26.22 For T sufficiently small, a critical point of S (i.e., a classical path) locally minimizes the action, S.

26.4 Conjugate Points and Jacobi Fields

In this section we study the classical action functional and its critical points (*classical paths*) in some detail. While such a study is of obvious importance in classical mechanics, it is also useful in the quasi-classical analysis of quantum systems that we undertook in Chapters 16 and 17.

Thus we consider the action functional

$$S(\phi) = \int_0^t L(\phi(s), \dot\phi(s))ds.$$

We have shown above that if t is sufficiently small, then $S''(\phi) > 0$, provided $(\partial^2 L/\partial\dot\phi^2) > 0$. So in this case, if $\bar\phi$ is a critical path, it minimizes $S(\phi)$. On the other hand, Theorem 26.19 implies that if $\bar\phi$ is a critical path such that $S''(\bar\phi)$ has negative spectrum, then $\bar\phi$ is not a minimizer. We will show later that eigenvalues of $S''(\bar\phi)$ decrease monotonically as t increases. So the point

t_0 when the smallest eigenvalue of $S''(\bar\phi)$ becomes zero, separates the t's for which $\bar\phi$ is a minimizer, from those for which $\bar\phi$ has lost this property. The points at which one of the eigenvalues of $S''(\bar\phi)$ becomes zero play a special role in the analysis of classical paths. They are considered in this section.

In this discussion we have used implicitly the fact that because $S''(\bar\phi)$ is a Schrödinger operator defined on $L^2([0,t])$ with Dirichlet (zero) boundary conditions, it has a purely discrete spectrum running off to $+\infty$. We denote this spectrum by $\{\lambda_n(t)\}_1^\infty$ with $\lambda_n(t) \to +\infty$ as $n \to \infty$. Note that if $\bar\phi$ is a critical point of S on $[0,t]$, then for $\tau < t$, $\bar\phi_\tau := \bar\phi|_{[0,\tau]}$ is a critical point of S on $[0,\tau]$. Thus for $\tau \le t$, $\{\lambda_n(\tau)\}$ is the spectrum of $S''(\bar\phi_\tau) = S''(\bar\phi)$ on $[0,\tau]$ with zero boundary conditions.

We specialize now to the classical action functional

$$S(\phi) = \int_0^t \{\frac{m}{2}|\dot\phi(s)|^2 - V(\phi(s))\}ds$$

on the space $X = \{\phi \in C^1([0,t];\mathbb{R}^d) \mid \phi(0) = x,\ \phi(t) = y\}$, and continue to denote by $\bar\phi$, a critical point of this functional (classical path).

Theorem 26.23 The eigenvalues $\lambda_n(\tau)$ are monotonically decreasing in τ.

Sketch of proof. Consider $\lambda_1(\tau)$, and let its normalized eigenfunction be ψ_1. Define $\tilde\psi_1$ to be ψ_1 extended to $[0,\tau+\epsilon]$ by 0. So by the spectral variational principle Theorem 8.1,

$$\lambda_1(\tau + \epsilon) \le \langle \tilde\psi_1, S''(\bar\phi)\tilde\psi_1 \rangle = \lambda_1(\tau).$$

Further, equality here is impossible by uniqueness for the Cauchy problem for ordinary differential equations, which states the following: if a solution of a linear, homogeneous, second-order equation is zero at some point, and its derivative is also zero at that point, then the solution is everywhere zero. To extend the proof to higher eigenvalues, one can use the min-max principle. □

Definition 26.24 A point $\bar\phi(\tau_0)$ such that $\lambda_n(\tau_0) = 0$ for some n is called a *conjugate point* to $\bar\phi(0) = x$ along $\bar\phi$.

So if $c = \bar\phi(\tau_0)$ is a conjugate point to x, then $S''(\bar\phi)$ on $[0,\tau_0]$ has a 0 eigenvalue. That is, there is some non-zero $\xi \in L^2([0,\tau_0])$ with $\xi(0) = \xi(\tau_0) = 0$ such that

$$S''(\bar\phi)\xi = 0. \tag{26.5}$$

This is the *Jacobi equation*. A solution of this equation with $\xi(0) = 0$ will be called a *Jacobi vector field*.

Definition 26.25 The *index* of $S''(\bar\phi)$ is the number of negative eigenvalues it has (counting multiplicity) on $L^2([0,t])$ with zero boundary conditions.

We recall that for τ small, $S''(\bar\phi)$ has no zero eigenvalues on $[0,\tau]$ (Theorem 26.21). Combining this fact with Theorem 26.23 gives the following result.

Theorem 26.26 (Morse) The index of $S''(\bar{\phi})$ is equal to the number of points conjugate to $\bar{\phi}(0)$ along $\bar{\phi}$, counting multiplicity (see Fig. 26.2).

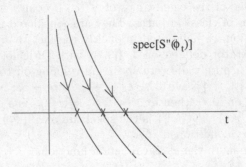

Fig. 26.2. Index = # of conjugate points.

The picture that has emerged is as follows. For sufficiently small times, a classical path $\bar{\phi}(0)$ locally minimizes the action. As time increases, the path might lose this property. This happens if there is a point in the path conjugate to $\bar{\phi}(0)$.

Example 26.27 An example of a conjugate point is a turning point in a one-dimensional potential (see Fig. 26.3, and remember that we are working with the functional of Example 26.1, no. 5).

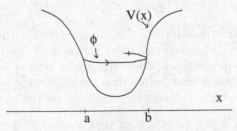

Fig. 26.3. A turning point.

The classical path ϕ starts at a, and turns back after hitting b at time τ. Now

$$S''(\phi) = -m\partial_s^2 - V''(\phi)$$

and it is easy to check that $S''(\phi)\dot{\phi} = 0$ (just differentiate Newton's equation). Since $\dot{\phi}(0) = \dot{\phi}(\tau) = 0$ (the velocity at a turning point is zero), b is conjugate to a.

We return to the Jacobi equation (26.5), and consider its *fundamental solution*, $J(s)$. $J(s)$ is the $d \times d$ matrix satisfying

$$S''(\bar{\phi})J = 0$$

with the initial conditions

$$J(0) = 0 \quad \text{and} \quad \dot{J}(0) = 1.$$

J is called the *Jacobi matrix*.

Proposition 26.28 The Jacobi matrix has the following properties

1. For any $h \in \mathbb{R}^d$, Jh is a Jacobi field. Conversely, any Jacobi field is of the form Jh for some $h \in \mathbb{R}^d$.
2. $\bar{\phi}(\tau_0)$ is a conjugate point to $\bar{\phi}(0)$ iff $J(\tau_0)$ has a zero-eigenvalue, i.e. $\det J(\tau_0) = 0$.

Proof. 1. The first part is obvious. To prove the second part let ξ be a Jacobi field, and let $h = \dot{\xi}(0)$. Then $\tilde{\xi} := Jh$ satisfies the same differential equation as ξ with the same initial conditions. Hence $\xi = \tilde{\xi}$, and therefore $\xi = Jh$.

2. We have shown above that $\bar{\phi}(\tau_0)$ is a conjugate point iff there is a Jacobi field ξ such that $\xi(\tau_0) = 0$. By the previous statement, there is $h \neq 0$ such that $\xi = Jh$, which implies $J(\tau_0)h = 0$. So $J(\tau_0)$ has a zero eigenvalue (with eigenvector h), and $\det J(\tau_0) = 0$.
□

Now we give the defining geometric/dynamic interpretation of J. Consider a family of critical paths $\phi_v(s)$ starting at $\bar{\phi}(0)$ with various initial velocities $v \in \mathbb{R}^d$. Denote $\bar{v} = \dot{\bar{\phi}}(0)$. Then

$$J(s) = \frac{\partial \phi_v(s)}{\partial v}\Big|_{v=\bar{v}}$$

is the Jacobi matrix (along $\bar{\phi}$). Indeed, ϕ_v satisfies the equation $\partial S(\phi_v) = 0$. Differentiating this equation with respect to v, and using that $S''(\phi) = \partial_\phi \partial S(\phi)$, we find

$$0 = \frac{\partial}{\partial v}\partial_\phi S(\phi_v) = S''(\phi_v)\frac{\partial \phi_v}{\partial v}.$$

Thus, $\partial \phi_v/\partial v|_{v=\bar{v}}$ satisfies the Jacobi equation. Next,

$$\frac{\partial}{\partial v}\phi_v(0) = \frac{\partial}{\partial v}\bar{\phi}(0) = 0$$

and

$$\frac{\partial}{\partial v}\dot{\phi}_v(0) = \frac{\partial}{\partial v}v = 1$$

which completes the proof.

26.5 Constrained Variational Problems

Let S and C be continuously differentiable functionals on a real Banach space X. We consider the problem of minimizing the functional $S(\phi)$, subject to the constraint $C(\phi) = 0$. That means we would like minimize $S(\phi)$ for ϕ in the (non-linear) space

$$M := \{\phi \in X \mid C(\phi) = 0\}.$$

We assume that $C'(\phi) \neq 0$ for $\phi \in M$ (here, and below, $C'(\phi)$ and $S'(\phi)$ denote the variational derivatives of the respective functionals considered as functionals on all of X, rather than just M).

If C is a C^2 (twice continuously differentiable) functional, then (see eg. [122])

$$T_\phi M = \{\xi \in X \mid \langle C'(\phi), \xi \rangle = 0\}.$$

To see this, suppose ϕ_λ is a variation of ϕ in M. Differentiating the relation $C(\phi_\lambda) = 0$ with respect to λ at $\lambda = 0$ yields $\langle C'(\phi), \xi \rangle = 0$, where $\xi = \frac{\partial}{\partial \lambda}\phi_\lambda|_{\lambda=0}$. Thus $T_\phi M \subset \{\xi \in X \mid \langle C'(\phi), \xi \rangle = 0\}$. Conversely, given $\xi \in X$ such that $\langle C'(\phi), \xi \rangle = 0$, one can show (using the "implicit function theorem") that there is a path $\phi_\lambda \in M$ satisfying $\phi_0 = \phi$ and $\frac{\partial}{\partial \lambda}\phi_\lambda|_{\lambda=0} = \xi$. So $\xi \in T_\phi M$.

Concerning the constrained variational problem, we have the following result:

Theorem 26.29 (Lagrange multipliers) Let S be a C^1 and C a C^2 functional on a Banach (or affine) space X. Suppose $\bar{\phi}$ locally minimizes $S(\phi)$ subject to the constraint $C(\phi) = 0$ (i.e. $\bar{\phi}$ locally minimizes S on the space M) and $C'(\bar{\phi}) \neq 0$. Then $\bar{\phi}$ is a critical point of the functional $S - \lambda C$ on the space X, for some $\lambda \in \mathbb{R}$ (called a *Lagrange multiplier*). In other words, $\bar{\phi}$ satisfies the equations

$$S'(\bar{\phi}) = \lambda C'(\bar{\phi}) \quad \text{and} \quad C(\bar{\phi}) = 0$$

(the first equation is as linear functionals on X).

Proof. The fact that $\bar{\phi}$ minimizes S over M implies that $\bar{\phi}$ is a critical point of S considered as a functional on M. This means that $S'(\bar{\phi}) = 0$ on $T_{\bar{\phi}}M$. Recall

$$T_{\bar{\phi}}M = \{\xi \in X \mid \langle C'(\bar{\phi}), \xi \rangle = 0\}.$$

Let $\rho \in X$ be such that $\langle C'(\bar{\phi}), \rho \rangle \neq 0$. Then for all $\xi \in X$,

$$\eta := \xi - \frac{\langle C'(\bar{\phi}), \xi \rangle}{\langle C'(\bar{\phi}), \rho \rangle} \rho \in T_{\bar{\phi}}M.$$

Hence

$$0 = \langle S'(\bar{\phi}), \eta \rangle = \langle S'(\bar{\phi}), \xi \rangle - \lambda \langle C'(\bar{\phi}), \xi \rangle, \quad \lambda := \frac{\langle S'(\bar{\phi}), \rho \rangle}{\langle C'(\bar{\phi}), \rho \rangle}.$$

Thus $S'(\bar{\phi})$ is a multiple of $C'(\phi)$. $\square$

Example 26.30 Quadratic form: let B be a a self-adjoint operator on a Hilbert space, $X = D(B)$, and set

$$S : \phi \mapsto \frac{1}{2}\langle \phi, B\phi \rangle \qquad (26.6)$$

and $C(\phi) := \frac{1}{2}(\|\phi\|^2 - 1)$. Easy computations show $\partial S(\phi) = B\phi$ and $\partial C(\phi) = \phi$. Hence by the result above, any critical point of $S(\phi)$ subject to the constraint $C(\phi) = 0$ ($\|\phi\| = 1$) satisfies the equation

$$B\phi = \lambda\phi \qquad (26.7)$$

for some Lagrange multiplier $\lambda \in \mathbb{R}$. This is an eigenvalue equation for B with the eigenvalue being the Lagrange multiplier.

26.6 Legendre Transform and Poisson Bracket

Passing from the lagrangian to hamiltonian constitutes a Legendre transform, defined formally as:

$$g(\pi) := \sup_{u \in X}(\langle \pi, u \rangle - f(u)). \qquad (26.8)$$

Here f is a function (or functional) on a normed vector space, V, while g is defined on the dual space V^*, and $\langle \pi, u \rangle$ is a coupling between V and V^* (See the previous mathematical supplement, Section 25.1 for definitions of the dual space and the coupling $\langle \pi, u \rangle$.) Of course, for the supremum in (26.8) to exist we have to make some assumptions on on the class of functions on which the Legendre transform is defined. In the finite-dimensional case, it suffices to assume that f is differentiable and strictly convex, i.e. $\forall v, v' \in V$,

$$f(sv + (1-s)v') < sf(v) + (1-s)f(v'), \ \forall s \in (0,1).$$

In the infinite-dimensional case, we have to make extra assumptions. In order not to complicate the exposition, we make an assumption which is much stronger than needed, but which suffices for our needs. Namely, we assume that f is of the form $f(u) := \frac{1}{2}\langle u, Lu \rangle$ where the linear operator $L : V \to V^*$ is invertible and satisfies $\langle v, Lw \rangle = \langle Lv, w \rangle$ and $\langle v, Lv \rangle \geq \delta\|v\|^2$, $\delta > 0$. In this case, like in the finite-dimensional one, the function $\langle \pi, u \rangle - f(u)$ has a unique critical point satisfying is $\partial f(u) = \pi$, (in this case, $u = L^{-1}\pi$), and this point is a maximum point. Hence

$$g(\pi) = (\langle \pi, u \rangle - f(u))|_{u:\partial f(u)=\pi}. \qquad (26.9)$$

Problem 26.31 Show that g is differentiable, convex and that

$$(\text{Legendre transform})^2 = 1.$$

Hint: use the fact that $\partial g(\pi) = u(\pi)$ where $u(\pi)$ solves $\partial f(u(\pi)) = \pi$.

Problem 26.32 Show that the Legendre transform maps the functional $f(\psi) = \frac{1}{2}\int|\psi|^2$ on $X = L^2(\mathbb{R}^d)$ into $g(\pi) = \frac{1}{2}\int|\pi|^2$.

The equation (26.9) shows that the classical and Klein-Gordon Hamiltonians, (4.45) and

$$H(\phi,\pi) = \int_{\mathbb{R}^d}\left\{\frac{1}{2}|\pi|^2 + \frac{1}{2}|\nabla\phi|^2 + f(\phi)\right\}dx \qquad (26.10)$$

(see (21.11)), are the Legendre transforms of the classical and Klein-Gordon Lagrange functionals, (4.37) and

$$L(\phi,\chi) = \int\left\{\frac{1}{2}|\chi|^2 - \frac{1}{2}|\nabla\phi|^2 - f(\phi)\right\}dx$$

in the second variables,

$$h(x,k) := \sup_{v\in V}(\langle k,v\rangle - L(x,v)). \qquad (26.11)$$

and

$$H(\phi,\pi) := \sup_{\eta\in V}(\langle\pi,\eta\rangle - L(\phi,\eta)). \qquad (26.12)$$

Now we consider the Poisson brackets, which were defined in Section 4.7. If Z is a real inner-product space, on which there is a linear invertible operator $J : Z^* \to Z$ such that $J^* = -J$ (J is called a *symplectic* operator), then we can define a Poisson bracket of functions (or functionals) F and G as

$$\{F,G\} := \langle\partial F, J\partial G\rangle. \qquad (26.13)$$

For one-particle Classical Mechanics with the phase space $Z = \mathbb{R}^3 \times \mathbb{R}^3$, the symplectic operator is

$$J = \begin{pmatrix} 0 & 1 \\ -1 & 0 \end{pmatrix}, \qquad (26.14)$$

yielding the Poisson bracket (4.47). For the phase space $Z = H^1(\mathbb{R}^d,\mathbb{R}^m) \times H^1(\mathbb{R}^d,\mathbb{R}^m)$, the symplectic operator is (26.14) (but defined on a different space), yielding the Poisson bracket

$$\{F,G\} = \int\{\partial_\pi F \cdot \partial_\phi G - \partial_\phi F \cdot \partial_\pi G\}dx. \qquad (26.15)$$

Suppose that $Z = X \times V^*$ is a space of functions $\Phi(x) = (\phi(x),\pi(x))$ on $\mathbb{R}^d$. Recall that the functional on X which maps $X \ni \phi \mapsto \phi(x)$ is called the *evaluation functional* (at x), which we denote (with some abuse of notation) as $\phi(x)$, and similarly for V. Consider the Hamilton equations

$$\dot{\Phi}_t(x) = \{\Phi(x), H\}(\Phi_t) \qquad (26.16)$$

where $\{F, G\}$ is a Poisson bracket on Z. If this Poisson structure is given by a symplectic operator J as in (26.13) with $\langle \phi, \pi \rangle := \int \phi(x) \cdot \pi(x) dx$, then, by Problem 21.4,

$$\{\Phi(x), H\}(\Phi) = \int (\partial \Phi(x))(y) J \partial H(\Phi)(y) dy$$

$$= \int \delta_x(y) J \partial H(\Phi)(y) = J \partial H(\Phi)(x)$$

which gives

$$\{\Phi, H\}(\Phi) = J \partial H(\Phi),$$

which leads to Hamilton's equation

$$\dot{\Phi} = J \partial H(\Phi). \tag{26.17}$$

Example 26.33 The Klein-Gordon Hamiltonian theory: the phase space is $X = H^1(\mathbb{R}^d, \mathbb{R}^m) \times H^1(\mathbb{R}^d, \mathbb{R}^m)$, the Poisson bracket (26.13), with J given in (26.14), and the Hamiltonian given by (26.10). So for $\Phi(t) = (\phi(t), \pi(t))$ a path in $H^1(\mathbb{R}^n) \times H^1(\mathbb{R}^n)$, Equation (26.17) is

$$\begin{pmatrix} \dot{\phi} \\ \dot{\pi} \end{pmatrix} = J \begin{pmatrix} -\Delta \phi + \nabla f(\phi) \\ \pi \end{pmatrix},$$

and we recover the Klein-Gordon equation $\ddot{\phi} = \Delta \phi + \nabla f(\phi)$.

For a general Hamiltonian system $(Z, \{\cdot, \cdot\}, H)$ Hamilton's equation can be written as

$$\dot{\Phi} = \{\Phi, H\}, \tag{26.18}$$

where we identified the derivation $F \to \{F(\Phi, t), H\}$ with the vector field X_H on Z, determined by this derivation. The map $t \to \Phi_t(\Phi_0)$, where $\Phi_t(\Phi_0)$ is the solution to (26.18) with the initial condition Φ_0 is called the *flow* generated by (26.18). Consider the equation

$$\frac{d}{dt} F(\Phi, t) = \{F(\Phi, t), H\}, \tag{26.19}$$

with an initial condition $F(\Phi, 0) = F_0(\Phi)$, where $F_0(\Phi)$ is a smooth functional on Z. The solution of this equation for various $F_0(\Phi)$ defines the flow $\Phi_t(\Phi_0)$ on Z, by the equation $F_0(\Phi_t(\Phi_0)) = F(\Phi_0, t)$. This is the flow for the vector field X_H on Z.

We conclude this section with brief remarks about an important classical field theory which does not fit the above framework, the Schrödinger CFT. Assume $V = W \oplus W$ and define the Lagrangian on $V \otimes V$ by

$$L(\psi, \chi) = \langle J^{-1} \psi, \chi \rangle - E(\psi),$$

where $J : V^* \to V$ is a symplectic operator as above (i.e. a linear invertible operator satisfying $J^* = -J$, e.g. (26.14) w.r.to the decomposition $V = W \oplus$

W) and $E(\psi)$ is a functional on V. The generalized momentum now is $\pi = J^{-1}\psi$. Then using the equations

$$H(\psi, \pi) = (\langle \pi, \chi \rangle - L(\psi, \chi))|_{\chi : \partial_\chi L(\psi, \chi) = \pi} \qquad (26.20)$$

and $\partial_\chi L(\psi, \chi) = J^{-1}\psi$ we see that $\langle \pi, \chi \rangle$ and $\langle J^{-1}\psi, \chi \rangle$ cancel and we obtain

$$H(\psi) = E(\psi). \qquad (26.21)$$

The phase space here is V and the Poisson bracket is given by (26.13) (where the derivatives are understood to be w.r.to ψ).

26.7 Complex Hamiltonian Systems

In this section we sketch the Lagrangian and Hamiltonian formalism on complex Banach spaces. We begin with a specific complex Banach space and discuss an abstract case briefly at the end of this section. We consider the space

$$X = \{\phi \in H^1(\mathbb{R}^d \times [0, T]; \mathbb{C}) \mid \phi(x, 0) = h(x), \phi(x, T) = g(x)\},$$

where $h, g, \in H^1(\mathbb{R}^d; \mathbb{C})$ are fixed functions. We identify X with the real space (26.1) with $m = 2$ (i.e. with $\operatorname{Re} X \oplus \operatorname{Im} X$), as

$$\phi \Longleftrightarrow \vec{\phi} := (\phi_1, \phi_2), \quad \phi_1 := \operatorname{Re} \phi, \quad \phi_2 := \operatorname{Im} \phi.$$

With this identification, we can define the variational (or Gâteaux or Fréchet) differentiability and derivative, $\partial_{\vec{\phi}} S(\phi)$, for any functional $S(\phi)$ on X, and specifically partial derivatives, $\partial_{\phi_1} S(\phi)$ and $\partial_{\phi_2} S(\phi)$, with respect the real, ϕ_1, and imaginary, ϕ_2, parts of the field ϕ. After that we introduce the derivatives with respect ϕ and $\bar{\phi}$ as follows

$$\partial_\phi S(\phi) := \partial_{\phi_1} S(\phi) - i\partial_{\phi_2} S(\phi) \qquad \partial_{\bar{\phi}} S(\phi) := \partial_{\phi_1} S(\phi) + i\partial_{\phi_2} S(\phi). \quad (26.22)$$

To fix ideas, we consider the complex Klein-Gordon and Schrödinger field theories.

Complex Klein-Gordon CFT. Here the action is given as in the item (7) of Example 26.1, i.e. by

$$S(\phi) := \int_0^T \int_{\mathbb{R}^d} \left\{ |\partial_t \phi|^2 - |\nabla_x \phi|^2 - f(\phi) \right\} dx dt,$$

where $f : \mathbb{C} \to \mathbb{R}$ is a continuous function.

The equation $\partial_{\bar{\phi}} S(\phi) = 0$ for critical points gives the complex non-linear Klein-Gordon equation, as follows from

$$\partial_{\bar{\phi}} S(\phi) = -\partial_t^2 \phi + \Delta_x \phi - \partial_{\bar{\phi}} f(\phi).$$

Here $\partial_{\bar{\phi}} f(\phi)$ is the regular complex derivative of a function. With the Lagrangian $L(\phi, \dot{\phi}) := \int_{\mathbb{R}^d} \left\{ |\partial_t \phi|^2 - |\nabla_x \phi|^2 - f(\phi) \right\}$, the generalized momentum is $\pi = \partial_{\dot{\phi}} L(\phi, \dot{\phi}) = \partial_t \bar{\phi}$. This gives the same Hamiltonian as in (26.10). The Legendre transform is defined as above, (26.8). The Poisson bracket is given now by again by (26.15), but with complex derivatives w.r.to ϕ and π.

Schrödinger CFT. The action for the nonlinear Schrödinger classical field theory, with the self-interaction described by a differentiable function $f : \mathbb{C} \to \mathbb{R}$, is given by

$$S(\psi) := \int_0^T \int_{\mathbb{R}^d} \left\{ -\mathrm{Im}(\bar{\psi} \partial_t \psi) - |\nabla_x \psi|^2 - V|\psi|^2 - f(\psi) \right\} dx dt,$$

on the same space as above. (We think of $\int_0^T \int_{\mathbb{R}^d} \mathrm{Im}(\bar{\psi} \partial_t \psi) dx dt$, modulo an additive constant, as either $\int_0^T \int_{\mathbb{R}^d} \frac{1}{i} \bar{\psi} \partial_t \psi dx dt$ or $\int_0^T \int_{\mathbb{R}^d} \frac{-1}{i} \partial_t \bar{\psi} \psi dx dt$.) The critical point equation $\partial_{\bar{\phi}} S(\psi) = 0$ gives the non-linear Schrödinger equation. Indeed, we compute

$$\partial_{\bar{\psi}} S(\psi) = i \partial_t \psi - (-\Delta_x + V)\psi - \partial_{\bar{\psi}} f(\psi).$$

With the Lagrangian $L(\psi, \dot{\psi}) := \int_{\mathbb{R}^d} \left\{ -\mathrm{Im}(\bar{\psi} \partial_t \psi) - |\nabla_x \psi|^2 - V|\psi|^2 - f(\psi) \right\}$, the generalized momentum is $\pi = \partial_{\dot{\psi}} L(\psi, \dot{\psi}) = \bar{\psi}$. This gives the Hamiltonian

$$H(\psi, \bar{\psi}) := \int_{\mathbb{R}^d} \left\{ |\nabla_x \psi|^2 + V|\psi|^2 + f(\psi) \right\} dx.$$

The Legendre transform is as above and it gives the Hamiltonian above. In this case, the phase space is the complex Hilbert space, $H^1(\mathbb{R}^d, \mathbb{C})$ and the Poisson bracket is given now by

$$\{F, G\} = \int \{\partial_\psi F \partial_{\bar{\psi}} G - \partial_{\bar{\psi}} F \partial_\psi G\} dx. \qquad (26.23)$$

The symplectic operator is $J = $ multiplication by $-i$, so that $\{F, G\} := \mathrm{Re}\langle \partial F, (-i)\partial G \rangle$. Finally the Hamilton equation for this system is $\dot{\psi} = \{\psi, H\}$, or in detail,

$$i \partial_t \psi = (-\Delta_x + V)\psi + \partial_{\bar{\psi}} f(\psi). \qquad (26.24)$$

Now we describe briefly an abstract construction. Let X be a complex Banach space, or else an affine space based on a complex Banach space X_0. Assume there is an antilinear map, C, on X satisfying $C^2 = 1$, so that one can define a real part, $X_1 := \frac{1}{2}(1 + C)X$, and imaginary, $X_2 := \frac{1}{2i}(1 - C)X$, part of X ($\frac{1}{2}(1 + C)$ and $\frac{1}{2}(1 - C)$ are projection operators). Then we can identify X with the real space $X_1 \oplus X_2$, as

$$\phi \iff \vec{\phi} := (\phi_1, \phi_2), \ \phi_1 := \frac{1}{2}(1 + C)\phi, \ \phi_2 := \frac{1}{2i}(1 - C)\phi.$$

With this identification, we can define the variational (or Gâteaux or Fréchet) differentiability and derivatives with respect to ϕ and $\bar{\phi}$ as in (26.22)

26.8 Conservation Laws

As in classical mechanics, an observable $F(\Phi)$ is *conserved* or is a *constant of motion*, i.e. $F(\Phi_t)$, where Φ_t is a solution of the Hamilton equation, is independent of t, if and only if its Poisson bracket with the Hamiltonian, H, vanishes, $\{F, H\} = 0$. This follows from the equation (21.15),

$$\frac{d}{dt}F(\Phi_t) = \{F(\Phi_t), H\}. \tag{26.25}$$

As in classical mechanics, conservation laws often arise from symmetries. Assume a system has a symmetry in the sense that there exists a one-parameter group τ_s, $s \subset \mathbb{R}$, of bounded operators on the space V, s.t. $T_s H = H$, $\forall s \in \mathbb{R}$, where

$$T_s H(\phi, \pi) := H(\tau_s \phi, \tau'_{-s}\pi),$$

with τ'_s being the dual group action of τ_s on V^*, defined by $\langle \tau'_s \pi, \phi \rangle = \langle \pi, \tau_s \phi \rangle$ (recall that $\langle \cdot, \cdot \rangle$ is the coupling between X and X^*). Let A be the generator of the group τ_s, $\partial_s \tau_s = A\tau_s$. Then the classical observable $Q(\phi, \pi) := \langle \pi, A\phi \rangle$ has vanishing Poisson bracket ("commutes") with the Hamiltonian,

$$\{H, Q\} = \partial_s H(\tau_s \phi, \tau'_{-s}\pi)|_{s=0} = 0,$$

and consequently is conserved under evolution: $Q(\phi_t, \pi_t)$ is a constant in t where $\Phi_t = (\phi_t, \pi_t)$ is a solution to (26.18). In analogy with quantum mechanics we formulate this as

$$U_s \text{ is a symmetry of } (26.18) \;\rightarrow\; \langle \pi, A\phi \rangle \text{ is conserved}$$

For the Schrödinger CFT, the one-parameter group τ_s, $s \in \mathbb{R}$, can be chosen to be unitary and the dual group is given by $\tau'_s = C\tau_{-s}C$, where C is the complex conjugation, so that, since $\pi = \bar\psi$, we have $\tau'_{-s}\pi = \overline{\tau_s \psi}$. Hence the operator T_s acts on H as $T_s H(\psi, \bar\psi) := H(\tau_s \psi, \overline{\tau_s \psi})$. The conserved classical observable is defined as $Q(\psi, \bar\psi) := \langle \psi, iA\psi \rangle$ where A is the (anti-self-adjoint) generator of the group τ_s, $\partial_s \tau_s = A\tau_s$, and $\langle \psi, \phi \rangle$ is the standard scalar product on $L^2(\mathbb{R}^d, \mathbb{C})$. As above, it has vanishing Poisson bracket ("commutes") with the Hamiltonian, $\{H, Q\} = \partial_s H(\tau_s \phi, \overline{\tau_s \psi})|_{s=0} = 0$, and consequently is conserved under evolution: $Q(\psi_t, \bar\psi_t)$ is a constant in t where ψ_t is a solution to (26.24).

We list some examples of symmetries and corresponding conservation laws for a CFT with the phase space $Z = H^1(\mathbb{R}^d, \mathbb{C}) \times H^1(\mathbb{R}^d, \mathbb{C})$

- Time translation invariance $((\tau_s \psi)(x, t) := \psi(x, t + s), \; s \in \mathbb{R}) \to$ conservation of energy, $H(\psi, \bar\psi)$;
- Space translation invariance $((\tau_s \psi)(x, t) := \psi(x + s, t), \; s \in \mathbb{R}^d) \to$ conservation of momentum $P(\psi, \bar\psi) := \int \bar\psi(-i\nabla)\psi dx$;

- Space rotation invariance $((\tau_R\psi)(x,t) := \psi(R^{-1}x,t),\ R \in SO(\mathbb{R}^d)) \rightarrow$ conservation of angular momentum $L(\psi,\bar\psi) := \sum \bar\psi(x \wedge -i\nabla)\psi dx$;
- Gauge invariance $((\tau_\gamma\psi)(x,t) := e^{i\gamma}\psi(x,t),\ \gamma \in \mathbb{R}) \rightarrow$ conservation of 'charge/number of particles' $\int |\psi|^2 dx$.

Note that, except for the first case, the families above are multi-parameter groups, but as mentioned in Section 4.7, they can be reduced to one-parameter ones. As an example we also give the momentum field on the phase space $Z = H^1(\mathbb{R}^d, \mathbb{R}) \times H^1(\mathbb{R}^d, \mathbb{R})$:

$$P(\phi,\pi) := \int \pi\nabla\phi dx. \tag{26.26}$$

Sometimes real theories have complex representations and these complex theories have gauge symmetries which lead to the conservation of the number of particles, which is not obvious in the original real representation. We encountered an example of this phenomenon in Section 21.2, where we found that the real Klein-Gordon theory is equivalent to a complex theory on the space $H^1(\mathbb{R}^d, \mathbb{C})$ with the latter theory having gauge symmetry, resulting in the conservation of the number of particles.

27

Comments on Literature, and Further Reading

General references

There is an extensive literature on quantum mechanics. Standard books include [36, 186, 226, 261]. More advanced treatments can be found in [257, 41, 81].

For rigorous treatments of quantum mechanics see [39, 291, 94, 285, 287, 288, 285, 79].

Mathematical background is developed in the following texts: [17, 42, 73, 106, 162, 204, 244, 245].

Further mathematical developments and open problems are reviewed in [95, 169, 201, 208, 273].

Chapter 2

The definition of self-adjointness given in this section and in Section 25.5 of Mathematical Supplement is different from the one commonly used, but is equivalent to it. It allows for a straightforward verification of this property, avoiding an involved argument.

For more on the semi-classical approximation see e.g. [250].

Chapter 4

For a discussion of the relation between quantization and pseudodifferential operators and semi-classical asymptotics, see [39, 49, 50, 51, 78, 95, 158, 164, 250, 263].

Detailed discussions of identical particles can be found in most of the books on quantum mechanics. For a particularly nice discussion of the relation between spin and statistics see [115].

The orthogonal, unitary and symmetric groups and their representations and algebras (in the first two cases) are discussed in most of the books on the group theory, of which there is a considerable number.

© Springer-Verlag GmbH Germany, part of Springer Nature 2020
S. J. Gustafson and I. M. Sigal, *Mathematical Concepts of Quantum Mechanics*, Universitext, https://doi.org/10.1007/978-3-030-59562-3_27

Chapter 5

Extensive development and discussion of coherent states can be found in [70, 150] For more discussion about the relation between the uncertainty principle and the stability of atoms, see [95, 199].

For stability of bulk matter see [95, 199, 201] and the book [208] (the papers [199, 201] can be found in [203]).

Chapter 9

The wave operators for short-range interactions were introduced in [228], and for long-range interactions, in [82, 52].

Scattering theory is presented in the textbooks mentioned above, as well as in [136, 221, 232].

Modern mathematical treatment of scattering theory can be found in [169, 77, 302].

Chapter 10

Certain aspects of the theory of Coulomb systems (atoms, molecules, and aggregates thereof) are reviewed in [95, 199, 202, 203, 264, 291].

The result on existence of bound states of atoms is due to G.M. Zhislin.

Existence of the hydrogen molecule was proven in [249].

Chapter 11

The Feshbach projection method was introduced by H. Feshbach in connection with perturbation problems in Nuclear Physics (see [226]) and by I. Schur in theory of matrices. A similar method, called the method of Lifshitz matrix, was developed independently in systems theory and in linear partial differential equations, where it is called the Grushin problem. See [27, 139] for further extensions and historical remarks

The abstract results presented in this chapter follow the work [29], which introduced the notion of Feshbach-Schur map $F_P : H \rightarrow F_P(H)$ and used it as a basis of a renormalization group (RG) approach to spectral problems (which is presented in Chapter 24).

There is an extensive physics literature on Zeeman effect, and on perturbation of atoms by time-periodic electric fields (see e.g. [186, 226]). The first rigorous results on the Zeeman problem were obtained in [24] and the time-periodic one, in [303, 165]. See [73] and [168] for further discussion and references.

A rigorous approach to the Fermi Golden Rule was proposed by Simon ([270]).

For more on perturbation theory, see [35, 176, 248, 247] and the recent comprehensive review [275].

Chapter 12

The Born-Oppenheimer approximation plays an important role in quantum chemistry, and there is an enormous literature on the subject. We mention here only rigorous works: [69, 148, 149, 151, 179, 218, 237, 280, 289]. This approximation in conjunction with the Feshbach - Schur method is used in [15] to give a rigorous derivation of the van der Waals forces between atoms.

The results (12.40) and (12.41) should be compared with those in [230].

The Born-Oppenheimer approximation for a molecule coupled to the quantized electro-magnetic field (the non-relativistic quantum electrodynamics), including a computation of the probability of spontaneous emission, was developed in [290].

For adiabatic theory and geometrical phases, see excellent review [20] and books [262, 289]. Book [289] covers also application of adiabatic theory to the Born-Oppenheimer approximation and many related topics. For adiabatic theory in relation to thermodynamics, Lindblad dynamics and quantum resonances, see [3, 4], [22] and [5], respectively.

For the relation of the adiabatic theory to the integer quantum hall effect and to charge transport and for the space-time adiabatic theory, see [20, 25] and [237], respectively.

For the adiabatic theory without the spectral gap condition, see [21, 289]

For different aspects of the Born-Oppenheimer approximation, adiabatic theory and geometrical phases, see [104, 256, 84].

Chapter 13

This chapter is somewhat more advanced than the chapters preceding it. Recommended reading for this chapter includes [73, 169, 9].

The key result of quantum many-body theory - the HVZ theorem - is due to W. Hunziker, C. van Winter, and G.M. Zhislin.

Recently, mathematical many-body theory, especially scattering theory for many-particle systems, has undergone rapid and radical development. It is covered in [169, 77, 132, 76].

Chapter 14

For the rigorous derivation of Hartree and Hartree-Fock equations see [159, 279, 133, 34, 33, 86, 87, 90, 91, 89, 117, 123, 124, 1, 253, 238, 239, 182, 14].

The Gross-Pitaevski equation was proposed independently by Gross and Pitaevski in connection with the problem of many bosons (boson gas) which we consider here. It was derived by [90] with the derivation simplified in [238, 239]. (For additional important aspects see [181, 63, 64].) Earlier, [209] have shown that the bosonic many-body ground state looks asymptotically (in the Gross-Petaevskii regime, in which the number of particles $n \to \infty$ and the scattering length $a \to 0$, so that $na =: \lambda/(4\pi)$ is fixed) as the product of the ground states of the Gross-Pitaevski equation. For detailed discussion and references see [210].

For the first result on the Hartree-Fock theory see the paper [211], with further mathematical developments in [214]. (See [203, 191] for more references.)

For asymptotic stability of the Gross-Pitaevski, nonlinear Schrödinger and Hartree equations see [277, 126, 293, 127, 128, 145, 72] and references therein. Other interesting properties (in particular, the blow-up) and generalizations are described in [185, 192, 193].

A small but very effective modification of the Hartree-Fock equations leads to the Kohn-Sham equation, which is at the foundation of the density functional theory, see [55, 65, 85] for reviews and references.

See [286] and [58] for background and results on the nonlinear Schrödinger equations.

A rigorous proof of BEC in the Gross-Pitaevski limit is given in [207] (see also [209]).

Chapter 15

Standard references on path integrals are [101, 180, 259, 254].

Rigorous results can be found in [272].

Important original papers are [188, 68, 195].

Chapter 16

An extensive rigorous treatment of semi-classical asymptotics can be found in [49, 158, 170, 219, 220, 250, 251, 252].

Chapter 17

The mathematical theory of resonances started with the paper [270], whose impetus came from the work [10, 32]. A classic, influential review of the theory of resonances is [271].

See [162] for earlier rigorous references and [48, 71, 225, 276], for some recent ones. Our analysis of the resonance free energy is close to [269], its physical predecessors are [188, 68, 54, 8, 189, 190]. For complex classical trajectories, see, eg., [49, 50, 51, 297, 298]. A more careful treatment of positive temperatures would involve coupling of a quantum system to a thermal reservoir, or, as an intermediate step, replacing the commutator with the Hamiltonian by Lindblad generator, discussed in Chapter 18. Resonances (metastable states) at zero and positive temperatures are a subject of intensive study in condensed matter physics, quantum chemistry, nuclear physics, and cosmology. See [54, 44, 156, 296, 269, 80] for reviews and further references. For formal treatment of tunneling of extended objects see [236].

Moreover, [247] gives an excellent exposition of the perturbation theory of resonances.

Chapter 18

More material on density matrices can be found in [46, 157, 292, 76]. For a rigorous semi-classical analysis in terms of the Wigner functions, see [102, 103, 215, 243, 278]. There is an extensive literature on the Hartree-Fock and Kohn-Sham equations, for relevant discussion and references see [65]. The generalized Hartree-Fock equations were introduced in [65].

Chapter 19

For an abstract definition and discussion of the non-commutative conditional expectation see [143, 175, 76].

For the relation of quantum dynamical maps to completely positive maps see [292], Statement 3.1.4, and [45], Remarks to Section 5.3.1.

The form of generators $\mathcal{L}$ was found by Lindblad and Gorini, Kossakowski and Sudarshan ([213, 137], see [66] for the most general result and [76], for a book discussion).

For the *detailed balance condition* see e.g. [227, 183, 108] and references therein. Inequality (19.36) was shown in [281]. For a discussion of dissipation function (19.41) and dissipation see [76], Lecture 12. For far a reaching generalization, see [174].

The dissipation function $D_{\mathcal{L}}$ was introduced in [213].

Theorem 19.38 was proven in [212].

For important generalizations of the subadditivity of the entropy, (19.46), see [205, 206]. For more on quantum entropy, see book [235] and reviews [56, 255].

For rigorous results on the decoherence - [222, 223, 224] (these papers contain also a very brief review of the relevant physics literature) and on relation between the decoherence and measurement, [107].

The approach to equilibrium is one of the central subjects of theoretical and mathematical physics. In the last 20 or so years its mathematical underpinning underwent a vigorous development on the level of microscopic theory, with robust and elegant theory mainly due to works of V. Jakšić and C.-A. Pillet, see some reviews in [19, 173]

For the relation of the topics covered to quantum information theory, see [116, 157, 177, 233, 242, 295, 301].

Chapter 20

A standard reference for mathematical treatment of the second quantization is [38].

The result on the comparison of the quantum and Hartree (mean-field) dynamics was obtained in [124]. The sketch of its proof follows [16]. The expansion (20.46) - (20.47) was derived in [13] it follows from the symbol composition formula of [38]. It was shown in [124] that the quantum many-body theory with the quantum Hamiltonian (20.25) can be obtained by quantizing the classical field theory described by the classical Hamiltonian (20.38) and Poisson brackets (20.37).

Chapter 22

The renormalization of the electron mass and one-particle states were studied in [109, 110, 240, 241, 152, 161, 28, 59, 60, 61, 62]. See [26] for the review and references.

Chapter 23

The results on the theory of radiation, as well as the renormalization group approach, are taken from [265, 121], which builds upon [29, 30, 27]. (Theorem 23.1 was established for systems with the coupling to photons regularized at spacial infinity in [29] - [30]. It would have been applicable to confined systems, i.e. systems in external potentials, growing at infinity sufficiently fast.) These works assume that the fine-structure constant is sufficiently small. The method in these papers also provides an effective computational technique to any order in the electron charge, something the conventional perturbation theory fails to do. For other approaches and references see the book [283] and a recent reviews [26, 266].

The existence of the ground state for the physical range of the parameters was proved in [140].

The notion of resonance in the non-relativistic QED was introduced in [29, 30].

The generalized Pauli-Fierz transformation was introduced in [265].

Theorem 23.4 was proven in [27].

Chapter 24

The spectral renormalization group method, described in this chapter is due to [29], with further development due to [27, 121, 154, 155]. We follow [121] It shares its philosophy with the standard renormalization group due to Wilson and others which can be found in any standard book on quantum field theory or statistical mechanics (for some papers see e.g. [130, 299]) and which is treated rigorously ina large number of papers, see [37, 100, 47, 97, 96, 31, 100, 129, 258] for a review of rigorous results.

The Feshbach-Schur map could be also used for the problem at hand. In fact, the Feshbach-Schur maps were introduced in [29] for exactly this purpose.

However, for technical reasons, it is convenient to use its present generalization.

Chapters 25 and 26

These chapters cover standard material though many of the proofs are simpler or require less advanced material than those found in standard books.

For a discussion of determinants, see also [53, 45, 180, 195, 254, 259, 260] and references therein.

For elementary facts on the calculus of variations, one can consult [131, 43].

References

1. W.K. Abou Salem, A remark on the mean field dynamics of many body bosonic systems with random interactions in a random potential. Lett Math Phys **84** (2008) 231-243.
2. W. K. Abou Salem, J. Faupin, J. Fröhlich, I. M. Sigal, On the theory of resonances in non-relativistic QED and related models. Advances in Applied Mathematics **43** (2009) 201-230.
3. W. K. Abou Salem and J. Fröhlich, Adiabatic theorems and reversible isothermal processes, Letters in Mathematical Physics **72** (2005) 153-163.
4. W. K. Abou Salem and J. Fröhlich, Status of the fundamental laws of thermodynamics, J. Stat. Phys. **126** (2007) 1045-1068.
5. W. K. Abou Salem and J. Fröhlich, Adiabatic theorems for quantum resonances, Commun. Math. Phys. **273** (2007) 651-675.
6. A. A. Abrikosov, On the magnetic properties of superconductors of the second group. J. Explt. Theoret. Phys. (USSR) **32** (1957) 1147-1182.
7. S. L. Adler, Derivation of the Lindblad generator structure by use of the Itô stochastic calculus, Physics Letters A **265** (2000) 58-61.
8. I. Affleck, Quantum Statistical Metastability, Phys. Rev. Lett. **46** no.6 (1981) 388-391.
9. S. Agmon, *Lectures on Exponential Decay of Solutions of Second Order Elliptic Equations.* Princeton University Press: Princeton, 1982.
10. J. Aguilar and J.M. Combes, A class of analytic perturbations for one-body Schrödinger Hamiltonians. Comm. Math. Phys. **22** (1971) 269-279.
11. A. Ambrosetti, M. Badiale, S. Cingolani, Semiclassical states of nonlinear Schrödinger equations. Arch. Rat. Mech. Anal. **140** (1997) 285-300.
12. A. Ambrosetti, A. Malchiodi, *Nonlinear Analysis and Semilinear Elliptic Problems.* Cambridge Studies in Advanced Mathematics, 104. Cambridge University Press, Cambridge, 2007.
13. Z. Ammari, Nier F., Mean field limit for bosons and infinite dimensional phase space analysis. J. Math. Phys. **50** no. 4 (2009).
14. I. Anapolitanos, Rate of convergence towards the Hartree-von Neumann limit in the mean-field regime. Lett. Math. Phys. **98** (2011) 1-31.
15. I. Anapolitanos and I.M. Sigal, Long range behavior of van der Waals forces. Comm. Pure Appl. Math. **70** no. 9 (2017) 1633-1671.

© Springer-Verlag GmbH Germany, part of Springer Nature 2020
S. J. Gustafson and I. M. Sigal, *Mathematical Concepts of Quantum Mechanics*, Universitext, https://doi.org/10.1007/978-3-030-59562-3

16. I. Anapolitanos and I.M. Sigal, Notes on mean-field limit of quantum many-body systems, unpublished notes, 2008.
17. V.I. Arnold, *Mathematical Methods of Classical Mechanics*. Springer, 1989.
18. S. Attal, Lectures in Quantum Noise Theory, http://math.univ-lyon1.fr/ attal/chapters.html
19. S. Attal, A. Joye, C.-A. Pillet. Open Quantum Systems, I-III, Springer, Lecture Notes in Mathematics 188, 2006.
20. J.E. Avron, Adiabatic quantum transport, *Mesoscopic Quantum Physics* E. Akkermans et al, eds. Les Houches, 1994.
21. J. E. Avron and A. Elgrart, Adiabatic theorem without a gap condition. Comm. Math. Phys. **203** no. 2 (1999) 445-463.
22. J. E. Avron, M. Fraas and G. M. Graf, Adiabatic response for Lindblad dynamics. Journal of Statistical Physics **148** no. 5 (2012) 800-823.
23. J. E. Avron, M. Fraas, G. M. Graf, O. Kenneth, Quantum response of dephasing open systems. New Journal of Physics **13** (2011) 053042.
24. J. E. Avron, I. Herbst, B. Simon, Schrödinger operators with magnetic fields. I. General interactions, Duke Math. J. **45** (1978) 847-883.
25. J. E. Avron, R. Seiler, L.G. Yaffe, Adiabatic theorems and applications to the quantum Hall effect, Comm. Math. Phys. **110** no. 1 (1987) 33-49.
26. V. Bach, Mass renormalization in non-relativistic quantum electrodynamics, in Quantum Theory from Small To Large Scales. Lecture Notes of the Les Houches Summer School, J. Fröhlich et al, eds. **95** 2010.
27. V. Bach, Th. Chen, J. Fröhlich, I. M. Sigal, Smooth Feshbach map and operator-theoretic renormalization group methods. J. Fun. Anal. **203** (2003) 44-92.
28. V. Bach, Th. Chen, J. Fröhlich, and I. M. Sigal. The renormalized electron mass in non-relativistic quantum electrodynamics. *Journal of Functional Analysis* **243** no. 2 (2007) 426-535.
29. V. Bach, J. Fröhlich, I.M. Sigal, Quantum electrodynamics of confined non-relativistic particles. Adv. Math. **137** (1998) 205-298, and **137** (1998) 299-395.
30. V. Bach, J. Fröhlich, I.M. Sigal, Spectral analysis for systems of atoms and molecules coupled to the quantized radiation field. Comm. Math. Phys. **207** no.2 (1999) 249-290.
31. T. Balaban, Constructive gauge theory. II. Constructive quantum field theory, II (Erice, 1988), 5568, NATO Adv. Sci. Inst. Ser. B Phys., **234**, Plenum, New York, 1990
32. E. Balslev and J.M. Combes, Spectral properties of many-body Schrödinger operators with dilation analytic interactions. Comm. Math. Phys. **22** (1971) 280-294.
33. C. Bardos, F. Golse, A. Gottlieb, N. Mauser, Mean field dynamics of fermions and the time-dependent Hartree-Fock equation. J. Math. Pure. Appl. **82** (2003) 665-683.
34. C. Bardos, L. Erdos, F. Golse, N. Mauser, H.-T. Yau, Derivation of the Schrödinger Poisson equation from the quantum N-body problem. C.R. Acad. Sci. Paris, Ser. I **334** (2002) 515-520
35. H. Baumgärtel, *Endlichdimensionale Analytische Störungstheorie*. Akademie-Verlag: Berlin, 1972.
36. G. Baym, *Lectures on Quantum Mechanics*. Addison-Wesley, 1990.

37. G. Benfatto and G. Gallavotti, *Renormalization Group.* Princeton University Press: Princeton, 1995.
38. F. A. Berezin, *The Method of Second Quantization.* Academic Press: New York, 1966.
39. F.A. Berezin and M.A. Shubin, *The Schrödinger Equation.* Kluwer: Dordrecht, 1991.
40. M. Berger, *Nonlinearity and functional analysis. Lectures on nonlinear problems in mathematical analysis.* Pure and Applied Mathematics. Academic Press: New York-London, 1977.
41. H. A. Bethe and R. Jackiw, *Intermediate Quantum Mechanics.* Benjamin/Cummings: Menlo Park, 1986.
42. M. Sh. Birman, and M.Z. Solomyak, *Spectral theory of self-adjoint operators in Hilbert spaces.* Mathematics and its Applications (Soviet series): Reidel, 1987.
43. P. Blanchard, E. Brüning, *Variational methods in mathematical physics: a unified approach.* Springer, 1992.
44. G. Blatter, M.V. Feigel'man, V.B. Geshkenbein, A.I. Larkin, V.M. Vinokur, Vortices in high-temperature superconductors. Rev. Mod. Phys. **66** no.4 (1994) 1125-1389.
45. O. Bratelli and D.W. Robinson, *Operator Algebras and Quantum Statistical Mechanics, 1 and 2.* Springer, 1996.
46. H.-P. Breuer and F. Petruccione, *The Theory of Open Quantum Systems* Oxford Univ. Press, 2003.
47. D. Brydges, J. Dimock, T.R. Hurd, Applications of the renormalization group. CRM Proc. Lecture Notes **7**, J. Feldman, R. Froese, L. Rosen, eds. American Mathematical Society: Providence, 1994, 171-190.
48. N. Burq and M. Zworski, Resonance expansions in semi-classical propagation. Comm. Math. Phys. **223** no.1 (2001) 1-12.
49. V.S. Buslaev, Quasiclassical approximation for equations with periodic coefficients. Russian Math Surveys **42** no. 6 (1987) 97-125.
50. V.S. Buslaev and A.A. Fedotov, The complex WKB method for the Harper equation. St. Petersburg Math. J. **6** no. 3 (1995) 495-517.
51. V.S. Buslaev and A. Grigis, Turning points for adiabatically perturbed periodic equations. J. d'Analyse Mathematique **84** (2001) 67-143.
52. V.S. Buslaev and V.B. Matveev, Wave operators for the Schrödinger equation with a slowly decreasing potential. Theor. Math. Phys. **2** (1970) 266-274.
53. V.S. Buslaev and E.A. Nalimova, Trace formulae in Hamiltonian mechanics. Theoretical and Mathematical Physics **61** (1984) 52-63.
54. A.O. Caldeira and A.J. Leggett, Quantum tunneling in a dissipative system. Ann. Phys. **149** (1983), 374.
55. E. Cancès and N. Mourad, A mathematical perspective on density functional perturbation theory. Nonlinearity **27** no. 9 (2014) 1999-2033.
56. E. Carlen, Trace Inequalities and Quantum Entropy: An Introductory Course. Entropy and the quantum. Contemporary Mathematics **529** (2010) 73-140.
57. L. Cattaneo, G. M. Graf, W. Hunziker, A General resonance theory based on Mourre's inequality. Annales Henri Poincaré **7** no. 3 (2006) 583-601.
58. T. Cazenave, *Semilinear Schrödinger equations.* AMS, 2003.

59. T. Chen, Infrared renormalization in non-relativistic QED and scaling criticality. J. Funct. Anal. **254**, no. 10 (2008) 2555-2647.

60. T. Chen and J. Fröhlich, Coherent infrared representations in non-relativistic QED. Spectral Theory and Mathematical Physics: A Festschrift in Honor of Barry Simon's 60th Birthday, Proc. Symp. Pure Math., AMS, 2007.

61. T. Chen, J. Fröhlich, A. Pizzo, Infraparticle scattering states in non-relativistic QED - I. The Bloch-Nordsieck paradigm. Commun. Math. Phys., **294** (3) (2010) 761-825.

62. T. Chen, J. Fröhlich, A. Pizzo, Infraparticle scattering states in non-relativistic QED - II. Mass shell properties. J. Math. Phys., **50** no.1 (2009) 012103.

63. T. Chen and N. Pavlovic, The quintic NLS as the mean field limit of a boson gas with three-body interactions. J. Funct. Anal. **260** no. 4 (2011) 959-997.

64. T. Chen, N. Pavlovic, N. Tzirakis, Energy conservation and blowup of solutions for focusing Gross-Pitaevskii hierarchies. Ann. Inst. H. Poincar Anal. Non Linaire **27** no. 5 (2010) 1271-1290.

65. I. Chenn and I. M. Sigal, On effective PDEs of quantum physics. in New Tools for Nonlinear PDEs and Application, M. D'Abbicco et al. (eds.), Birkhäuser series "Trends in Mathematics", 2019.

66. E. Christensen and D. E. Evans, Cohomology of operator algebras and quantum dynamical groups. J.Lon.Math. Soc **20** (1979) 358-368.

67. C. Cohen-Tannoudji, J. Dupont-Roc, G. Grynberg, *Photons and Atoms: Introduction to Quantum Electrodynamics* Wiley (New York), 1991.

68. S. Coleman, Fate of the false vacuum: semiclassical theory. Phys. Rev. **D15** (1977), 2929-2936.

69. J.-M. Combes, P. Duclos, R. Seiler. The Born-Oppenheimer approximation. in: Rigorous Atomic and Molecular Physics (eds. G. Velo, A. Wightman), Plenum (New York), 185212, 1981.

70. M. Combescure and D. Robert, A phase-space study of the quantum Loschmidt echo in the semiclassical limit. Ann. Henri Poincar **8** no. 1 (2007) 91108.

71. O. Costin, A. Soffer, Resonance theory for Schrödinger operators. Comm. Math. Phys. **224** no.1 (2001) 133-152.

72. S. Cuccagna, The Hamiltonian structure of the nonlinear Schrödinger equation and the asymptotic stability of its ground states. Comm. Math. Phys. **305** (2011) 279-331.

73. H. Cycon, R. Froese, W. Kirsch, B. Simon, *Schrödinger Operators (with Applications to Quantum Mechanics and Global Geometry.* Springer: Berlin, 1987.

74. F. Dalfovo, S. Giogini, L. Pitaevskii, S. Stingari, Theory of Bose-Einstein condensation in trapped gases. Rev. Mod. Phys. **71** (1999) 463.

75. M. Delbrürk and G. Molèiere, Statistische Quantenmechanik und Thermodynamik, Abhandl. Preuss. Akad. Wissenschaften **1** (1936) 1-42.

76. G. Dell'Antonio, Lectures on the Mathematics of Quantum Mechanics, I and II. Atlantis Press, 2015 and 2016.

77. J. Dereziński and C. Gérard, *Scattering Theory of Classical and Quantum N-Particle Systems.* Springer-Verlag: Berlin, 1997.

78. M. Dimassi and J. Sjöstrand, *Spectral Asymptotics in the Semi-classical Limit*. London Mathematical Society Lecture Note Series, **268**. Cambridge, 1999.
79. J. Dimock, *Quantum Mechanics and Quantum Field Theory*. Cambridge Univ. Press, 2011.
80. T. Dittrich, P. Hänggi, G.-L. Ingold, B. Kramer, G. Schön, W. Zwerger, *Quantum Transport and Dissipation*. Wiley, 1998.
81. W. Dittrich and M. Reuter, *Classical and Quantum Dynamics*. Springer-Verlag: Berlin, 1999.
82. J.D. Dollard, Asymptotic convergence and Coulomb interaction. J. Math. Phys. **5** (1964) 729-738.
83. S. K. Donaldson, *Riemann Surfaces*. Oxford University Press, 2011.
84. W. Dou, G. Miao, J. E. Subotnik, Born-Oppenheimer dynamics, electronic friction, and the inclusion of electron-electron interactions. Phys. Rev. Lett. **119** (2017) 046001.
85. W. E, J. Lu, The Kohn-Sham equation for deformed crystals. Memoirs of the American Mathematical Society, 2013.
86. A. Elgart, L. Erdös, B. Schlein, H.-T. Yau, Nonlinear Hartree equation as the mean field limit of weakly coupled fermions. J. Math. Pure Appl. **83** (2004) 1241.
87. A. Elgart, L. Erdös, B. Schlein, H.-T. Yau, Gross-Pitaevskii equation as the mean filed limit of weakly coupled bosons. Arch. Rat. Mech. Anal. **179** no. 2 (2006) 265283.
88. A. Elgart and B. Schlein, Adiabatic charge transport and the Kubo formula for Landau-type Hamiltonians, Commun. Pure Appl. Math. **28** (2004) 590-615.
89. A. Elgart, B. Schlein, Mean field dynamics of boson stars. Commun. Pure Appl. Math. **60** no. 4 (2007) 500-545.
90. L. Erdös, B. Schlein, H.-T. Yau, Derivation of the cubic non-linear Schrodinger equation from quantum dynamics of many-body systems. Invent. Math. **167** (2007) 515614.
91. L. Erdös, B. Schlein, H.-T. Yau, Derivation of the Gross - Pitaevskii equation for the dynamics of Bose-Einstein condensate. Ann. of Math.(2) **172** no. 1 (2010) 291370.
92. L.C. Evans, *Partial Differential Equations*. AMS, 1998.
93. D. E. Evans, J.T. Lewis, Dilations of Irreversible Evolutions in Algebraic Quantum Theory. Communications of the Dublin Institute for Advanced Studies, Series A: Theoretical Physics **24** Dublin, 1977.
94. L.D. Faddeev and O.A. Yakubovskii, Lectures on Quantum Mechanics for Mathematics Students. With an appendix by Leon Takhtajan. Student Mathematical Library **47**. AMS, Providence RI, 2009.
95. C. Fefferman, Uncertainty Principle. Bull. AMS, **9** no. 2 (1983) 129-206.
96. J. Feldman, in Mathematical Quantum Theory II: Schrödinger Operators. CRM Proc. Lecture Notes **8** J. Feldman, R. Froese, L. M. Rosen, eds. American Mathematical Society: Providence, 1995.
97. J. Feldman, H. Knörrer, E. Trubowitz, *Fermionic Functional Integrals and the Renormalization Group*. AMS-CRM, 2002.
98. E. Fermi, Quantum theory of radiation, Rev. Mod. Phys. **4** (1932) 87-132.
99. H. Feshbach, Unified theory of nuclear reactions, Ann. Phys. **5** (1958) 357-390.

100. R. Fernandez, J. Fröhlich, A. Sokal, *Random Walks, Critical Phenomena, and Triviality in Quantum Field Theory*. Springer-Verlag: Berlin, 1992.

101. R.P Feynman and A.R. Hibbs, *Quantum Mechanics and Path Integrals*. McGraw-Hill: New York, 1965.

102. S. Filippas and G. Makrakis, Semiclassical Wigner function and geometrical optics, Multiscale Model. Simul. **1** (2003) 674-710.

103. S. Filippas and G. Makrakis, On the evolution of the semi-classical Wigner function in higher dimensions. Euro. J. of Applied Mathematics **16** (2005) 1-30.

104. S. Fishman and A. Soffer, Slowly changing potential problems in Quantum Mechanics: Adiabatic theorems, ergodic theorems, and scattering. J. Math. Phys. **57** (2016) 072101.

105. A. Floer and A. Weinstein, Nonspreading wave packets for the cubic Schrödinger equation with a bounded potential. J. Fun. Anal. **69** (1986) 397-408.

106. G. Folland, *Real Analysis*. Wiley, 1984.

107. M. Fraas, G. M. Graf, L. Hänggli, Indirect measurements of a harmonic oscillator. Ann Inst. H. Poincaré **20** (2019) 2937-2970.

108. A. Frigerio, V. Gorini, Markov dilations and quantum detailed balance. Commun. Math. Phys. 93, 517–532 (1984).

109. J. Fröhlich, On the infrared problem in a model of scalar electrons and massless, scalar bosons. Ann. Inst. H. Poincaré, Section Physique Théorique, **19** no. 1 (1973) 1-103.

110. J. Fröhlich, Existence of dressed one electron states in a class of persistent models. Fortschritte der Physik **22** (1974) 159-198.

111. J. Fröhlich, Statistics of fields, the Yang-Baxter equation, and the theory of knots and links. In Non-perturbative Quantum Field Theory (Cargse 1987), G.'t Hooft et al., eds, New York: Plenum Press, 1988.

112. J. Fröhlich, Quantum statistics and locality. In Proceedings of the Gibbs Symposium (New Haven, CT, 1989), 89-142. American Mathematical Society, Providence, 1990.

113. J. Fröhlich, *Nonperturbative quantum field theory. Selecta*. World Scientific: River Edge, 1992.

114. J. Fröhlich, Mathematical aspects of the quantum Hall effect. In Proc. of the first ECM (Paris 1992), Progress in Math., Basel, Boston: Birkhäuser-Verlag, 1994.

115. J. Fröhlich, Spin – or, actually: Spin and Quantum Statistics. Séminar Poincaré XI (2007) 1-50. arXiv:0801.2724.

116. J. Fröhlich, A brief review of the 'ETH - approach to quantum mechanics', 2019, arXiv:1905.06603.

117. J. Fröhlich, S. Graffi, S. Schwartz, Mean Field and classical limit of many body Schrödinger dynamics for Bosons. Commun. Math. Phys. **271** (2007) 681-697.

118. J. Fröhlich, M. Griesemer, B. Schlein, Asymptotic electromagnetic fields in models of quantum-mechanical matter interacting with the quantized radiation field. Advances in Mathematics **164** no. 2 (2001) 349-398.

119. J. Fröhlich, M. Griesemer, B. Schlein, Asymptotic completeness for Rayleigh scattering. Annales Henri Poincar **3** no. 1 (2002) 107-170.

120. J. Fröhlich, M. Griesemer, B. Schlein, Asymptotic completeness for Compton scattering. Comm. Math. Physics **252** no. 1-3 (2004) 415-476.

121. J. Fröhlich, M. Griesemer, I.M. Sigal. Spectral renormalization group analysis. Rev. Math. Phys. **21** no. 4 (2009) 511-548.

122. J. Fröhlich, S. Gustafson, L. Jonsson, I.M. Sigal, Dynamics of solitary waves in external potentials. Comm. Math. Phys. **250** (2004) 613-642.

123. J. Fröhlich, A. Knowles, S. Schwartz, On the mean-field limit of bosons with Coulomb two-body interaction. Comm. Math. Phys. **288** (2009), no. 3, 1023-1059

124. J. Fröhlich, A. Knowles, A. Pizzo, Atomism and quantization. J. Phys. A **40** no. 12 (2007) 3033-3045.

125. J. Fröhlich and P. Pfeifer, Generalized time-energy uncertainty relations and bounds on lifetimes of resonances. Rev. Mod. Phys. **67** no. 4 (1995) 759-779.

126. J. Fröhlich, T.-P. Tsai, H.-T. Yau, On a classical limit of quantum theory and the nonlinear Hartree equation. Geom. Fun. Anal, Special Vol. (2000) 57-78.

127. Zhou Gang and I.M. Sigal, Asymptotic stability of nonlinear Schrödinger equations with potential. Rev. in Math. Phys. **17** no. 20 (2005) 1143-1207.

128. Zhou Gang, I.M. Sigal, Relaxation of solitons in nonlinear Schrödinger equations with potentials, Adv. Math. **216** no. 2 (2007) 443-490.

129. K. Gawedzki and A. Kupiainen, Asymptotic freedom beyond perturbation theory. Phénomènes critiques, systèmes aléatoires, théories de jauge, Part I, II. Les Houches, 1984, 185292, North-Holland, Amsterdam, 1986.

130. S. Glazek and K. Wilson, Renormalization of Hamiltonians. Phys. Rev. **D48** (1993) 5863-5872.

131. I.M. Gelfand and S. Fomin, *Calculus of Variations*. Prentice-Hall, 1963.

132. C. Gérard and I. Laba. *Multiparticle quantum scattering in constant magnetic fields*. AMS, 2002.

133. J. Ginibre and G. Velo, The classical field limit of scattering theory for nonrelativistic many-boson systems I. Comm. Math. Phys. **66** (1979) 37-76.

134. J. Ginibre and G. Velo, The classical field limit for nonrelativistic Bosons II. Asymptotic expansions for general potentials. Annales de l'institut Henri Poincaré (A). Physique Théorique, **33** no. 4 (1980) 363-394.

135. J. Glimm and A. Jaffe, *Quantum Physics: a Functional Integral Point of View*. Springer-Verlag: New York, 1981.

136. M.L. Goldberger and K.M. Watson, *Collision Theory*. Wiley: New York, 1964.

137. V. Gorini, A. Kossakowski, E.C.G. Sudarshan, Completely positive semigroups of N-level systems. J. Math. Phys. **17** no. 5 (1976) 821.

138. S. Graffi and K. Yajima, Exterior complex scaling and the AC-Stark effect in a Coulomb field. Comm. Math. Phys. **89** (1983) 277-301.

139. M. Griesemer and D. Hasler, On the smooth Feshbach-Schur map. J. Funct. Anal. **254** no. 9 (2008) 2329-2335.

140. M. Griesemer, E. Lieb and M. Loss, Ground states in non-relativistic quantum electrodynamics. Invent. Math. **145** no.3 (2001) 557-595.

141. A. Grigis, and J. Sjöstrand, Microlocal analysis for differential operàtors, London Mathematical Society Lecture Note Series, Vol. 196, Cambridge University Press, Cambridge 1994.

142. M. Grillakis, J. Shatah, W. Strauss, Stability theory of solitary waves in the presence of symmetry I. J. Funct. Anal. **74** (1987) 160-197.

143. S. Gudder and J.-P. Marchand, Conditional expectations on von Neumann algebras: a new approach, Rep. on Math. Phys. **12** (1977) 317-329.

144. R. C. Gunning, *Riemann Surfaces and Generalized Theta Functions*, Springer, 1976.
145. S. Gustafson, K. Nakanishi, T.-P. Tsai, Asymptotic completeness in the energy space for nonlinear Schrödinger equations with small solitary waves. Int. Math. Res. Not. **66** (2004) 3559-3584.
146. S. J. Gustafson, I. M. Sigal and T. Tzaneteas, Statics and dynamics of magnetic vortices and of Nielsen-Olesen (Nambu) strings. J. Math. Phys. **51**, 015217 (2010).
147. R. Haag, *Local Quantum Physics, Fields, Particles, Algebras.* Texts and Monographs in Physics. Springer-Verlag: Berlin, 1992.
148. G. A. Hagedorn. High order corrections to the time-dependent Born-Oppenheimer approximation. I. Smooth potentials. Ann. of Math. **124**, no. 3 (1986) 571-590.
149. G. A. Hagedorn. High order corrections to the time-dependent Born-Oppenheimer approximation. II. Coulomb systems. Comm. Math. Phys. **117**, no. 3 (1988) 387-403.
150. G. A. Hagedorn and A. Joye, Exponentially accurate semiclassical dynamics: propagation, localization, Ehrenfest times, scattering, and more general states. Ann. Henri Poincar **1**, no. 5 (2000) 837-883.
151. G. A. Hagedorn and A. Joye, A Time-Dependent Born-Oppenheimer Approximation with Exponentially Small Error Estimates. Commun. Math. Phys. **223** (2001) 583-626.
152. Ch. Hainzl and R. Seiringer, Mass renormalization and energy level shift in non-relativistic QED Adv. Theor. Math. Phys. **6** (5) (2003) 847 -871.
153. B. C. Hall, *Quantum Theory for Mathematicians*, Graduate Texts in Mathematics, 267, Springer, 2013, ISBN 978-1461471158.
154. D. Hasler and I. Herbst, Smoothness and analyticity of perturbation expansions in QED, 2010, ArXiv:1007.0969v1.
155. D. Hasler and I. Herbst, Convergent expansions in non-relativistic QED, 2010, ArXiv:1005.3522v1.
156. P. Hänggi, P. Talkner, M. Borkovec, Reaction-rate theory: fifty years after Kramers. Rev. Mod. Phys. **62** no.2 (1990) 251-338.
157. M. Hayashi, *Quantum Information Theory (Mathematical Foundation)*, Second Edition, 2004.
158. B. Helffer, *Semiclassical Analysis, Witten Laplacians, and Statistical Mechanics.* World Scientific. 2002.
159. K. Hepp, The classical limit for quantum mechanical correlation functions. Comm. Math. Phys. **35** (1974) 265–277.
160. F. Hiroshima, Self-adjointness of the Pauli-Fierz Hamiltonian for arbitrary values of coupling constants. Ann. Henri Poincaré **3**, no. 1 (2002) 171–201.
161. F. Hiroshima and H. Spohn, Mass renormalization in nonrelativistic quantum electrodynamics. J. Math. Phys. **46** (4) (2005).
162. P. Hislop and I.M. Sigal, *Introduction to Spectral Theory. With Applications to Schrödinger Operators.* Springer-Verlag, 1996.
163. P. Hohenberg and W. Kohn, Inhomogeneous electron gas. Phys. Rev. **136** (1964) B864 - B871.
164. L. Hörmander, *The analysis of linear partial differential operators, Vol. I-IV.* Springer, 1983-1985.
165. J. Howland, Scattering theory for Hamiltonians periodic in time. Ind. Univ. Math. J. **28** (1979) 471-494.

166. K. Huang, *Statistical Mechanics*. John Wiley & Sons, 1963.

167. M. Hübner and H. Spohn, Radiative decay: non-perturbative approaches. Rev. Math. Phys. **7** (1995), 289-323.

168. W. Hunziker, Schrödinger operators with electric or magnetic fields. In *Mathematical problems in theoretical physics*, Lec. Not. Phys. **116**. Springer, 1980.

169. W. Hunziker and I.M. Sigal, The quantum N-body problem. J. Math. Phys. **41** no.6 (2000) 3448-3510.

170. V. Ivrii, *Microlocal Analysis and Precise Spectral Asymptotics*. Springer, 1998.

171. R. Jackiw, Quantum meaning of classical field theory. Rev Mod. Pys. **49** (1977), 681-706.

172. A. Jaffe and C. Taubes, *Vortices and Monopoles*. Birkhäuser: Boston, 1980.

173. V. Jakšić and C.-A. Pillet, Mathematical theory of non-equilibrium quantum statistical mechanics. Journal of statistical physics 108 (5-6) (2002) 787–829.

174. V. Jakšić, C.-A. Pillet and M. Westrich, Entropic fluctuations of quantum dynamical semigroups, Journal of Statistical Physics 154 (1-2) (2014) 153–187.

175. R. V. Kadison, Non-commutative conditional expectations and their applications. Contemporary Mathematics Volume 365, 143–179, 2004.

176. T. Kato, *Perturbation Theory for Linear Operators*. Springer-Verlag: Berlin, 1976.

177. M. Keyl, Fundamentals of quantum information theory. Physics Reports **369** (2002) 431-548.

178. A. Kishimoto, Dissipations and Derivations. Commun. Math. Phys. **47**, 25-32 (1976)

179. M. Klein, A. Martinez, R. Seiler and X. P. Wang. On the Born-Oppenheimer expansion for polyatomic molecules. Commun. Math. Phys. **143** (1992) 607-639.

180. H. Kleinert, *Path Integrals in Quantum Mechanics, Statistics, and Polymer Physics*. World Scientific, 1995.

181. S. Klainerman and M. Machedon, On the uniqueness of solutions to the Gross-Pitaevskii hierarchy. Comm. Math. Phys. **279** no. 1 (2008) 169-185.

182. A. Knowles and P. Pickl, Mean-field dynamics: singular potentials and rate of convergence. Comm. Math. Phys. **298** no. 1 (2010) 101138.

183. A. Kossakowski, A. Frigerio, V. Gorini, M. Verri, Quantum detailed balance and KMS condition. Commun. Math. Phys. **57** (1977) 97-110.

184. K. Kraus, *Lecture notes in physics*. NewYork: Springer, 1983.

185. J. Krieger, E. Lenzmann, P. Raphaël, On stability of pseudo-conformal blowup for L^2-critical Hartree NLS. Ann. Henri Poincar **10** no. 6, (2009), 11591205.

186. L.D. Landau and E.M. Lifshitz, *Theoretical Physics 3. Quantum Mechanics: Non-Relativistic Theory*. Pergamon, 1977.

187. O. Lanford and D. Robinson, Mean entropy of states in quantum statistical mechanics, J. Math. Phys. 9, (1968) 1120-1125.

188. J.S. Langer, Theory of the condensation point. Ann. Phys. **41** (1967), 108-157.

189. A.I. Larkin and Yu. N. Ovchinnikov, Damping of a superconducting current in tunnel junctions. Zh. Eksp. Theor. Fiz **85** (1983) 1510.

190. A.I. Larkin and Yu. N. Ovchinnikov, Quantum mechanical tunneling with dissipation. The pre-exponential factor. Zh. Eksp. Theor. Fiz **86** (1984) 719.

191. C. Le Bris and P.-L. Lions, From atoms to crystals: a mathematical journey. Bulletin AMS **42**, N 3, 291-363.

192. E. Lenzmann, Uniqueness of ground states for pseudorelativistic Hartree equations. Anal. PDE **2** no. 1 (2009) 127.

193. E. Lenzmann and M. Lewin, Minimizers for the Hartree-Fock-Bogoliubov theory of neutron stars and white dwarfs. Duke Math. J. **152** no. 2 (2010) 257-315.

194. E Lenzmann and M Lewin, Dynamical ionization bounds for atoms, Analysis & PDE **6** no. 5 (2013) 1183-1211.

195. S. Levit and U. Smilanski, A new approach to Gaussian path integrals and the evaluation of the semiclassical propagator. Ann. Phys. **103**, no. 1 (1977) 198-207.

196. M. Levy, Universal variational functionals of electron densities, first order density matrices, and natural spin-orbitals and solutions of the v-representability problem. Proc. Natl. Acad. Sci. USA (1979) 76: 6062-6065.

197. M. Levy, Electron densities in search of Hamiltonians. Phys. Rev. A **26** (1982) 1200-1208 .

198. M Lewin, Geometric methods for nonlinear many-body quantum systems, Journal of Functional Analysis **260** no. 12 (2011), 3535-3595.

199. E.H. Lieb, Stability of Matter. Rev. Mod. Phys. **48** (1976) 553-569.

200. E. H. Lieb, Density functionals for Coulomb systems. International Journal of Quantum Chemistry, **24** no. 3 (1983) 243-277.

201. E.H. Lieb, Stability of Matter: From Atoms to Stars. Bull. Amer. Math. Soc. **22** (1990) 1-49.

202. E.H. Lieb, *Stability of Matter: From Atoms to Stars. Selecta Elliot H. Lieb*. W. Thirring, ed.5th edition, Springer-Verlag (New York), 2007.

203. E.H. Lieb *Inequalities, Selecta of E.H. Lieb*. M. Loss and M.B. Ruskai, eds.

204. E.H. Lieb and M. Loss, *Analysis*. American Mathematical Society, 2001.

205. E. Lieb and M.B. Ruskai, A Fundamental property of quantum mechanical entropy Phys. Rev. Lett. **30** (1973) 434-436.

206. E. Lieb and M.B. Ruskai, Proof of the strong subadditivity of quantum mechanical entropy? J. Math. Phys. **14**(1973) 1938-1941 .

207. E. Lieb and R. Seiringer, Proof of Bose-Einstein condensation for dilute trapped gases. Phys. Rev. Lett. **88** (2002).

208. E. Lieb and R. Seiringer, Stability of Matter. Cambridge Univ Press, 2010.

209. E. Lieb, R. Seiringer, J. Yngvason, Bosons in a trap: A rigorous derivation of the Gross-Pitaevskii energy functional. Phys. Rev. A **61** (2000).

210. E. Lieb, R. Seiringer, J.P. Solovej, J. Yngvason, *The mathematics of the Bose gas and its condensation*. Oberwolfach Seminars, **34**. Birkhauser Verlag, Basel, 2005.

211. E.H. Lieb and B. Simon. The Hartree-Fock theory for Coulomb systems Comm. Math. Phys. **53** no.3 (1977) 185-194.

212. G. Lindblad, Completely Positive Maps and Entropy Inequalities, Commun. Math. Phys. **40** (1975) 147-151.

213. G. Lindblad, On the generators of quantum dynamical semigroups. Commun. Math. Phys. **48** no. 2 (1976) 119-130.

214. P.L. Lions, Hartree-Fock and related equations. Nonlinear partial differential equations and their applications. Collège de France Seminar, Vol. IX. Pitman Res. Notes Math. Ser. **181** (1988) 304-333.

215. P.L. Lions and T. Paul, Sur les mesures de Wigner. Revista Matematica Iberoamericana, **9** no. 3 (1993).

216. Ch. Lubich, *From Quantum to Classical Molecular Dynamics: Reduced Models and Numerical Analysis.* Zurich Lectures in Advanced Mathematics. European Mathematical Society, 2008.

217. A. Martinez, *An Introduction to Semiclassical and Microlocal Analysis.* Springer, 2002.

218. A. Martinez and V. Sordoni. A general reduction scheme for the time-dependent Born-Oppenheimer approximation, Comptes Rendus Acad. Sci. Paris **334** (2002) 185-188.

219. V.P. Maslov, *Théorie des Perturbations et Méthods Asymtotiques.* Dunod, 1972.

220. V.P. Maslov and M.V. Fedoryuk *Semiclassical approximation in quantum mechanics.* Contemporary mathematics, 5. Reidel: Boston, 1981.

221. R. Melrose, *Geometric Scattering Theory.* Cambridge Univ. Press, 1995.

222. M. Merkli, G. Berman and I.M. Sigal, Dynamics of collective decoherence and thermalization, *Annals of Physics* **323** no. 12 (2008) 3091-3312.

223. M. Merkli, G. Berman and I.M. Sigal, Resonant perturbation theory of decoherence and relaxation of quantum bits, *Adv Math Phys*, Vol. **2010**, Article ID 169710, 2010.

224. M. Merkli, I.M. Sigal and G. Berman, Decoherence and Thermalization, Physical Review Letters **98** no.13, 30 March 2007. (See also the April 2007 issue of Virtual Journal of Quantum Information.) Resonance Theory of Decoherence and Thermalization, *Annals Physics*, **323** no. 2 (2008) 373-412.

225. M. Merkli and I.M. Sigal, A time-dependent theory of quantum resonances. Comm. Math. Phys. **201** (1999) 549-576.

226. A. Messiah, *Quantum Mechanics.* Interscience, 1961.

227. M. Mittnenzweig, A. Mielke, An entropic gradient structure for Lindblad equations and couplings of quantum systems to macroscopic models, Journal of Statistical Physics **167** no. 2 (2017) 205-233.

228. C. Møller, General properties of the characteristic matrix in the theory of elementary particles I. Danske. Vid. Selske. Mat.-Fys. Medd. **23** (1945) 1-48.

229. L. Molnár and P. Szokol, Maps on states preserving the relative entropy II, Linear Algebra and its Applications **432** (2010) 3343-3350.

230. J. Moody, A. Shapere and F. Wilczek, Diatoms and spin precession, Phys. Rev. Lett. **56** (1986) 893-896.

231. P.T. Nam, New bounds on the maximum ionization of atoms, Communications in Mathematical Physics **312** no. 2 (2012) 427-445.

232. R.G. Newton, *Scattering Theory of Waves and Particles.* McGraw-Hill: New York, 1966.

233. M.I.A. Nielsen and I. Chuang, *Quantum Computation and Quantum Information,* Cambridge University Press, 2010.

234. Y.-G. Oh, Existence of semiclassical bound states of nonlinear Schrödinger equations with potentials of the class V. Comm. in PDE **13** (1988) 1499-1519.

235. M. Ohya and D. Petz, *Quantum entropy and its use*. Springer, Berlin (1993).

236. Yu. Ovchinnikov and I.M. Sigal, Decay of vortex states in superconductors, *Phys. Rev. B.* **48** (1993) 1085-1096.

237. G. Panati, H. Spohn and S. Teufel. The time-dependent Born-Oppenheimer approximation, in Mathematical Modelling and Numerical Analysis, Modélisation Mathématique et Analyse Numérique. 2007.

238. P. Pickl, Derivation of the time dependent Gross-Pitaevskii equation without positivity condition on the interaction. J. Stat. Phys. **140** no. 1 (2010) 7689, 82C22

239. P. Pickl, Derivation of the time dependent Gross-Pitaevskii equation with external fields, arXiv:1001.4894, 2010.

240. A. Pizzo, One-particle (improper) States in Nelson's massless model, Annales Henri Poincaré 4, **3** (June, 2003) 439-486.

241. A. Pizzo, Scattering of an infraparticle: The one particle sector in Nelson's massless model, Annales Henri Poincaré 6, **3** (2005) 553-606.

242. J. Preskill. Quantum computation. Lecture Notes, available ` at http://www.theory.caltech.edu/people/preskill/ph229/, California Institute of Technology, Pasadena, CA

243. M. Pulvirenti, Semiclassical expansion of Wigner functions. J. Math. Physics, **47** (2006).

244. M. Reed and B. Simon, *Methods of Modern Mathematical Physics, Vol I. Functional Analysis*. Academic Press, 1972.

245. M. Reed and B. Simon, *Methods of Modern Mathematical Physics, Vol. II. Fourier Analysis and Self-Adjointness*. Academic Press, 1972.

246. M. Reed and B. Simon, *Methods of Modern Mathematical Physics III. Scattering Theory*. Academic Press, 1979.

247. M. Reed and B. Simon, *Methods of Modern Mathematical Physics IV: Analysis of Operators*. Academic Press, 1978.

248. F. Rellich, *Perturbation Theory of Eigenvalue Problems*. Gordon and Breach: New York, 1969.

249. J.-M. Richard, J. Fröhlich, G.M. Graf, Phys Rev Letters, **71** no. 9 (30 August 1993) 1332-1334.

250. D. Robert, *Autour de l'Approximation Semi-Classique*, Progress in Mathematics, Vol. 68, Birkhäuser, 1987.

251. D. Robert, Revivals of wave packets and Bohr-Sommerfeld quantization rules. Adventures in mathematical physics, 219235, Contemp. Math., 447, Amer. Math. Soc., Providence, RI, 2007.

252. D. Robert, On the Herman-Kluk semiclassical approximation. Rev. Math. Phys. **22** no. 10 (2010) 1123-1145.

253. I. Rodnianski and B. Schlein, Quantum fluctuations and rate of convergence towards mean field dynamics. Comm. Math. Phys. **291** no. 1 (2009) 31-61.

254. G. Röpstorff, *Path integral approach to quantum physics: an introduction*. Springer, 1994.

255. M. B.. Ruskai, Inequalities for quantum entropy: A review with conditions for equality. Journal of Mathematical Physics **43** (2002) 4358.

256. Ilya G. Ryabinkin, Loïc Joubert-Doriol, Artur F. Izmaylov, Geometric phase effects in nonadiabatic dynamics near conical intersections. Acc. Chem. Res. **50** (2017) 1785-1793

257. J.J. Sakurai, *Advanced Quantum Mechanics*. Addison-Wesley: Reading, 1987.

258. M. Salmhofer, *Renormalization Group.* Springer-Verlag, 2000.

259. L.S. Schulman, *Techniques and Applications of Path Integration.* John Wiley & Sons, 1981.

260. A. Schwarz, *Quantum Field Theory and Topology.* Springer-Verlag, 1993.

261. L.I. Schiff, *Quantum Mechanics.* McGraw-Hill: New York, 1955.

262. A. Shapere and F. Wilczek, *Geometric Phases in Physics*, World Scientific 1989.

263. M. Shubin, *Pseudodifferential Operators and Spectral Theory.* Springer, 1987.

264. I.M. Sigal, Lectures on Large Coulomb Systems, CRM Proceedings and Lecture Notes **8** (1995) 73-107.

265. I.M. Sigal. Ground state and resonances in the standard model of the non-relativistic quantum electrodynamics. J. Stat. Physics, **134** no. 5-6 (2009) 899-939.

266. I.M. Sigal, Renormalization Group and Problem of Radiation. In *Quantum Theory from Small To Large Scales*, Les Houches Summer School, August, 2010.

267. I. M. Sigal, Magnetic vortices, Abrikosov lattices and automorphic functions. In *Mathematical and Computational Modelling (With Applications in Natural and Social Sciences, Engineering, and the Arts)* A John Wiley & Sons, Inc., 2014.

268. I.M. Sigal, Introduction to Partial Differential Equations in Quantum Physics, www.math.toronto.edu/sigal

269. I.M. Sigal and B. Vasilijevic, Mathematical theory of quantum tunneling at positive temperature. Annales de l'Institut Henri Poincaré **3** (2002) 1-41.

270. B. Simon, The theory of resonances for dilation analytic potentials and the foundations of time-dependent perturbation theory. Ann. Math. **97** (1973) 247-274.

271. B. Simon, Resonances and complex scaling: a rigorous overview. Int. J. Quant. Chem. **14** (1978) 529-542.

272. B. Simon, *Functional Integration and Quantum Physics.* Academic Press, 1979.

273. B. Simon, *Fifteen problems in mathematical physics. Perspectives in Mathematics* 423-454. Birkhäuser, 1984.

274. B. Simon, Holonomy, the quantum adiabatic theorem, and Berry's phase. Physical Review Letters **51**, no. 24 (1983) 2167-2170.

275. B. Simon, Tosio Kato's work on non-relativistic Quantum Mechanics, arXiv:1711.00528

276. A. Soffer and M. Weinstein, Time-dependent resonance theory. Geom. Fun. Anal. **8** (1998) 1086-1128.

277. A. Soffer and M. Weinstein, Selection of the ground state for nonlinear Schrödinger equations. Rev. Math. Phys. (2005).

278. Ch. Sparber, P. Markowich, and N. Mauser, Wigner functions versus WKB-methods in multivalued geometrical optics, Asymptotic Anal. **33** (2003) 153-187.

279. H. Spohn, Kinetic equations from Hamiltonian dynamics. Rev. Mod. Phys **52** no. 3 (1980) 569-615.

280. H. Spohn and S. Teufel, Adiabatic decoupling and time-dependent Born-Oppenheimer theory, Commun. Math. Phys. **224** (2001) 113-132.

281. H. Spohn, An algebraic condition for the approach to equilibrium of an open N-level system, Letters in Mathematical Physics **2** no. 1 (1977) 33-38.

282. H. Spohn. Entropy production for quantum dynamical semigroups. J. Mathematical Phys., 19(5), 1227–1230, 1978.

283. H. Spohn, *Dynamics of Charged Particles*. Cambridge Univ. Press, 2004.

284. N. Steenrod, *The Topology of Fibre Bundles*. Princeton University Press, 2nd printing, 1957.

285. F. Strocchi, *Introduction to Mathematical structure of Quamtum Mechanics. A Short Course for Mathematicians*, World Scientific, 2005.

286. C. Sulem and P.-L. Sulem, *The Nonlinear Schrödinger Equation. Self-Focusing and Wave Collapse*. Springer, New York 1999.

287. L. Takhtajan, *Quantum Mechanics for Mathematicians*. AMS, 2008.

288. G. Teschl, *Mathematical Methods in Quantum Mechanics: With Applications to Schrödinger Operators*, AMS 2009.

289. S. Teufel, Adiabatic perturbation theory in quantum dynamics, Lecture Notes in Mathematics **1821** (2003) Springer.

290. S. Teufel and J. Wachsmuth, Spontaneous decay of resonant energy levels for molecules with moving nuclei, Communications in Mathematical Physics, **315** no. 3 (2012) 699-738.

291. W. Thirring, *A Course in Mathematical Physics, Vol. 3: Quantum Mechanics of Atoms and Molecules*. Springer, 1980.

292. W. Thirring, *A Course in Mathematical Physics, Vol. 4. Quantum Mechanics of Large Systems*. Springer-Verlag, 1983.

293. T.-P. Tsai and H.-T. Yau, Asymptotic dynamics of nonlinear Schrödinger equations: resonance-dominated and dispersion-dominated solutions. Comm. Pure Appl. Math. **55** (2002) 153-216.

294. T. Tzanateas and I.M. Sigal, Abrikosov lattice solutions of the Ginzburg-Landau equations. In *Spectral Theory and Geometric Analysis*, M. Braverman, et al, editors. Contemporary Mathematics, AMS 2011.

295. F. Verstraete and H. Verschelde, On Quantum Channels, https://arxiv.org/pdf/quant-ph/0202124.pdf

296. A. Vilenkin and E.P.S. Shellard, *Cosmic Strings and other Topological Defects*. Cambridge, 1994.

297. A. Voros, Spectre de l'équation de Schrödinger et méthod BKW. Publications Mathématiques d'Orsay **81** (1982).

298. A. Voros, The return of the quartic oscillator: the complex WKB method. Ann. Inst. H. Poincaré Sect. A **39** no. 3 (1983) 211-338.

299. F. Wegner, Ann. Physik (Leipzig) **3** (1994) 555-559.

300. F. Wilczek and A. Zee, a contribution to [262].

301. M.M. Wilde, *Quantum Information Theory* Cambridge University Press, 2013.

302. D. Yafaev, *Scattering Theory: Some Old and New Problems*. Springer Lecture Notes in Mathematics No. 1937, Springer-Verlag, 2000.

303. K. Yajima, Resonances for the AC-Stark effect. Comm. Math. Phys. **87** (1982) 331-352.

304. J. Zinn-Justin, *Quantum Field Theory and Critical Phenomena*. Oxford, 1996.

Index

© Springer-Verlag GmbH Germany, part of Springer Nature 2020
S. J. Gustafson and I. M. Sigal, *Mathematical Concepts of Quantum Mechanics*, Universitext, https://doi.org/10.1007/978-3-030-59562-3

Printed in the United States
By Bookmasters